Studies in Computational Intelligence

Volume 935

Series Editor

Janusz Kacprzyk, Polish Academy of Sciences, Warsaw, Poland

The series “Studies in Computational Intelligence” (SCI) publishes new developments and advances in the various areas of computational intelligence—quickly and with a high quality. The intent is to cover the theory, applications, and design methods of computational intelligence, as embedded in the fields of engineering, computer science, physics and life sciences, as well as the methodologies behind them. The series contains monographs, lecture notes and edited volumes in computational intelligence spanning the areas of neural networks, connectionist systems, genetic algorithms, evolutionary computation, artificial intelligence, cellular automata, self-organizing systems, soft computing, fuzzy systems, and hybrid intelligent systems. Of particular value to both the contributors and the readership are the short publication timeframe and the world-wide distribution, which enable both wide and rapid dissemination of research output.

Indexed by SCOPUS, DBLP, WTI Frankfurt eG, zbMATH, SCImago.

All books published in the series are submitted for consideration in Web of Science.

More information about this series at http://www.springer.com/series/7092

Allam Hamdan · Aboul Ella Hassanien ·
Anjum Razzaque · Bahaaeddin Alareeni
Editors

The Fourth Industrial Revolution: Implementation of Artificial Intelligence for Growing Business Success

Springer

Editors
Allam Hamdan
College of Business and Finance
Ahlia University
Manama, Bahrain

Aboul Ella Hassanien
Information Technology Department
Faculty of Computers and Information
Cairo University
Giza, Egypt

Anjum Razzaque
College of Business and Finance
Ahlia University
Manama, Bahrain

Bahaaeddin Alareeni
Middle East Technical University - Northern Cyprus Campus
Kalkanlı, Güzelyurt, KKTC, Turkey

ISSN 1860-949X ISSN 1860-9503 (electronic)
Studies in Computational Intelligence
ISBN 978-3-030-62795-9 ISBN 978-3-030-62796-6 (eBook)
https://doi.org/10.1007/978-3-030-62796-6

This Springer imprint is published by the registered company Springer Nature Switzerland AG
The registered company address is: Gewerbestrasse 11, 6330 Cham, Switzerland

Foreword

Artificial intelligence is a key driver of the Fourth Industrial Revolution. Although currently witnessing tremendous acceleration in myriad applications in all spheres of human endeavour, artificial intelligence represents an outcome of decades of human intellectual efforts applied in computer science, in mathematics and, with respect to learning theories, in philosophy, psychology and education. These efforts have spurred development of theories in logic, computer science and linguistics.

As a means to palliate the adverse consequences, economic and social, of the current COVID19 pandemic, artificial intelligence holds great promise for humanity, in general, and business, in particular, to adapt to, and to chart a course moving forward in, an environment in which social distancing, working from home and strict limits on group congregation have become the norm.

As a harbinger of strategic opportunities associated with shocks engendered by COVID19, this book explores the immense potential held by the implementation of artificial intelligence, as a generator of profound change, across industry—spanning extractive, manufacturing and services enterprises of all scales of business activity —as well as, on, in relation to the public sector, healthcare, education and commerce.

Prof. Abdulla Yusuf Al Hawaj
Founding President and Chairman
of Board of Trustees
Ahlia University
Manama, Bahrain

Prof. Mansoor Ahmed Alaali
President
Ahlia University
Manama, Bahrain

Preface

Artificial Intelligence (AI), a vitality, fundamental for the Fourth Industrial Revolution. The influence of AI has been observed in our homes, within the processes of our businesses and within the public realm. The quintessence of AI is best reflected through the applications of robotics that drive our cars, stock our warehouses, monitor our mental and physical behaviours, as well as supervising of our health, along with the facilitation of childcare within our homes.

As the revolution of our industries migrated from the third to the fourth, such a revolution launched new avenues giving the rise to plethora opportunities for human development, observed within the incorporation of extraordinary technological advances. The advanced technologies of the Fourth Industrial Revolution occur to amalgamate the business world, enabled through various technologies such as machines learn, big data analytics, Internet of Things and, especially, AI. The Fourth Industrial Revolution has already changed the digital, along with the natural, landscape while introducing wonderful business opportunities for the future entrepreneurs.

Nevertheless, this endeavour comes also with challenges. The Fourth Industrial Revolution is the blending of technologies that end up blurring the thin lines between the physical and the virtual world. This is why scholars who research on IT governance stress on a deeper and a comprehensive investigation on how technologies like AI blend with the Fourth Industrial Revolution to facilitate new business opportunities so as to relish the blurred crossing zone between the physical and the virtual world and enjoy to continue to embrace the governance role of AI within the Fourth Industrial Revolution realm.

This book contains 34 chapters. Each of these chapters was evaluated through an editorial board, and each chapter was passed through a double-blind peer-review process, hence bestowing seven themes:

I. Artificial Intelligence, Entrepreneurship and Business Success.
II. FinTech, RegTech and Artificial Intelligence.
III. Implementation of Artificial Intelligence in Healthcare and Public Sector.

IV. Implementation of Artificial Intelligence in Education and Smart Universities.
V. The Ethical, Professional Practices and Social Implications of Artificial Intelligence.
VI. Artificial Intelligence and Economic Development.
VII. Big Data, Deep Learning and Business Success.

These chapters are reflecting quality research contributing theoretical and practical implications, for those who wisely apply Artificial Intelligence within any business sector, e.g. healthcare, education or even entrepreneurs who wish to blend entrepreneurship with AI, as well as other vital areas, to state the lease. It is our hope that the contribution of this book will be of the academic level which even decision-makers in the various economic and executive level will get to appreciate.

The Editors
Manama, Bahrain — Allam Hamdan
ahamdan@ahlia.edu.bh
Giza, Egypt — Aboul Ella Hassanien
aboitcairo@cu.edu.eg
Manama, Bahrain — Anjum Razzaque
Kalkanlı, Güzelyurt, KKTC, Turkey — Bahaaeddin Alareeni
alareeni@metu.edu.tr

Contents

Artificial Intelligence, Entrepreneurship and Business Success

Artificial Intelligence for Supply Chain Success in the Era of Data Analytics

Usama Awan, Narmeen Kanwal, Suha Alawi, Janne Huiskonen, and Ajantha Dahanayake

Abstract Nowadays, artificial intelligence (AI) is becoming a more effective digital domain promised to facilitate immediate access to information and effective decision making in ever-increasing business environments. While big data analytics for organizational renewal has increasingly received interest from data analytics scholars. Despite the increasing adoption of big data analytics for decision making, relatively little is know about how data management capabilities lead to better data insights for supply chain sustainability and circular economy. The researchers understand the extensive use of big data analytics and artificial intelligence among firms as an essential and necessary tool for shaping the future of the supply chain 4.0 industry. This chapter discusses the role of AI applications for the success of a supply chain in the big data era. From a holistic perspective, today, manufacturers, particularly those with global operations and presence, are under enormous pressure to keep up with the continuous growth of disruptive innovative procurement models. This has open doors for the firms to aggressively seek out big data management capabilities to improve operational efficiencies and to innovate the process. This chapter provides a better understanding related to the application of data analytics in the supply chain context. The research issues are classified into different categories, including big data

U. Awan (✉) · J. Huiskonen · A. Dahanayake
Industrial Engineering and Management, the Lappeenranta-Lahti University of Technology LUT, Lappeenranta 5385, Finland
e-mail: Usama.awan@Lut.fi

J. Huiskonen
e-mail: Janne.Huiskonen@lut.fi

A. Dahanayake
e-mail: Ajantha.Dahanayake@lut.fi

N. Kanwal
University of Trier, Weidengraben 90, A18, 54296 Trier, Germany
e-mail: 4snakanw@uni-trier.de

S. Alawi
Factually of Economics and Administration, King Abdul Aziz University, Jeddah, Saudi Arabia
e-mail: Salawi@kau.edu.sa

A. Hamdan et al. (eds.), *The Fourth Industrial Revolution: Implementation of Artificial Intelligence for Growing Business Success*, Studies in Computational Intelligence 935,
https://doi.org/10.1007/978-3-030-62796-6_1

management and machine learning, a business case for the supply chain and innovation in supply using data. This study also presents machine learning data analysis steps.

Keywords Data analytics · Artificial intelligence · Machine learning · Database management capabilities

1 Introduction to Logic of Artificial Intelligence

The rapid increase of human-computer interaction in recent years has called for further exploration of how humans and machines co-exist in an existing artificial intelligence environment. Artificial intelligence, machine learning, and the internet of things are a major source of generating information among others in a variety of ways, such as the volume of information, diversity of information, and divergence of information are a few data flashpoints for big data that can be used for decision making. For the supply chain problems which are predominant by uncertainty, in that case, Artificial intelligence is more effective than other tools of information technology. The extensive use of artificial intelligence is understood by researchers as important and necessary for shaping the future of the supply chain of industry 4.0. The term" artificial intelligence" (AI) is used to describe the process involved in machines learning to recollect patterns and features directly from the data to take actions using algorithms [48]. The origins of AI as an area of scientific research is not a new concept; its conception goes back to 1965 from "Dartmouth conference" then the term typically referred to as "intelligent machine". However, looking back, the term intelligent machine did not convey the scope of human and machine interface. Therefore, later on, the term artificial intelligence emerged. Figure 1 illustrates the history and developments in AI. Figure 1 is adapted from [58].

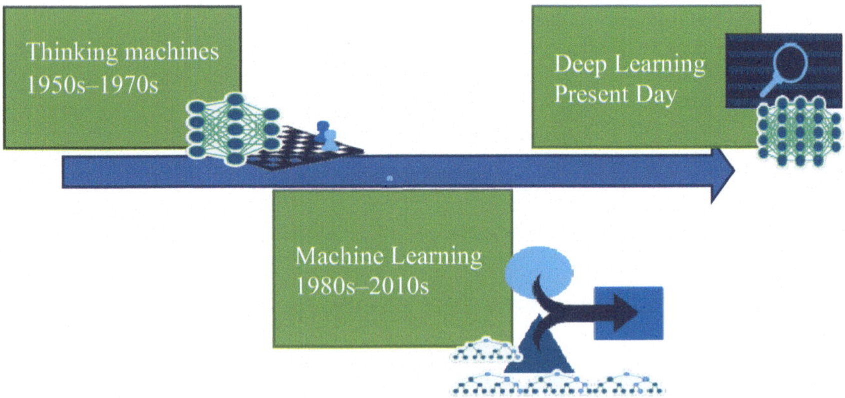

Fig. 1 AI devleopment adapted from (SAS)

More recently, it has been argued that principles of artificial intelligence are suitable for any industrial organization, from software developments to industrial production and supply chain services. There are many layers of supply chain system; every layer provides enormous data, linking data from every layer, and creating connection into useful results that require smart artificial intelligent applications. The role of the AI is a trade-off between the different actors that make up what is commonly called networks for the supply chain. This metaphor is linked with the distribution, logistics management, procurement, monitoring and traceability of material, production inspection and control, fault detection, and predictive maintenance, ideally in a way to optimize the processes, reduce cost and decision making. As Duan et al. [22] states the importance of AI is to replace or support decision-making issues with limited involvement of human interfere. Because of the shift in the use of the internet of things (IoT) in different spheres of the supply chain, such as production and distribution, it is noteworthy that now supply chain networks recognize the value of artificial intelligence with the internet of things. IoT is defined as, a group of infrastructures, interconnecting connected objects and allowing their management, data mining and the access to data they generate" where related objects are "sensor(s) and/or actuator(s) carrying out a specific function that is able to communicate with other equipment" [21]. Thus artificial intelligent supply chain management (AISC) begins with the application of the internet of things concerning the seek optimization and decision making to develop a direct relationship with the individual customers. In AISC, the production process can be tailored to automate to develop customized products for particular customers or in specific situations.

Many tools, methodologies, and terminologies have been developed over the years, including fully digitalized mechanisms and hybrid machine learning techniques [57]. AI is an advanced step in technology digitalization to use computer systems for interpretation, developing and recognizing patterns, conducting or understanding an organization's behaviors from an occurrence. As sentiment analysis, attaining and maintaining acquaintance, and generating enormous sorts of inference to solve the problems in decision-making circumstances where most favorable or accurate solutions are either too luxurious or complicated to construct [45]. In literature, different forms of artificial intelligence have already been discussed for example Roughly, Set theory [51], Machine learning [55], Expert systems [38], Genetic algorithms [46] and Fuzzy logic [63].

The shift from big data to pattern data has undoubtedly been most apparent. The effects of AI and pattern data in manufacturing have produced a striking impact on the development of customized products. AI is a way of representing data and transforming business for further development [14]. An additional advantage of AI is the ability to decoupled data, and to rationally explore using algorithms to improve the organization's ability to reduce the cost of making sustainable decisions [2]. The AI offers an ontological perspective of describing or representation of data as patterns rather than the standard form. As noted by Duan et al. [22], traditional data modeling approaches are abstract, do not encourage specific data patterns. In contrast, data analytics offers a more flexible way out to engage and visualize the influence of different sources of information for decision making and could advance the practical success of artificial intelligence applications in different domain. It's beyond human capacity to analyze huge bulks of data; machine learning can assess huge bulks of data and come up with recommendation. This data includes market trends, trade obligation, advertisement trends, and consumer's sentiment expressions over social media, competitive scenarios and the presence of websites to launch the information about product quality. It does not just supply chain system which needs tracking to reduce the cost and increase the efficiency, but machine learning also supports the norm of labour performance evaluation.

1.1 Application of Data Analytics and Machine Learning in Supply Chain

Big data modeling is an emergent field. Big data refers to as "the information asset characterized by high Volume, Velocity, and Variety to require specific technology and analytical methods for its transformation into Value" [16]. This implies that big data is a plausible set of unstructured large data sets that provide an opportunity for organizations to apply scalable algorithms to capture, store, and analyze the information to gain business intelligence. Big data focuses on the extraction of data from the micro and macro levels of the system and to offer a transparent process to make the process easier. Thus, big data provides a viable picture with details of relevant characteristics to spot trends between different data nodes to find new correlations. Big data and machine learning induce more value-added infrastructure towards a new era of technology. Particularly, when it comes to supply chain performance. The large amounts of collected data sets irrespective of qualifying to be Big data is now an integral part of improvement, competency, and superiority in the supply chain for companies to avail smart benefits [34, 56]. The benefits are manifold as big data deals with massive amorphous data which is required for high-velocity detaining information for business, government agencies, and private entities [12]. This information supports the fundamental models of logistic companies for including estimation of the real-time supply chain, supply chain sanctuary hazards, evaluation, and forecasting of demand-supply determinants, appraising supplier performance, process and cost-effective optimization, resourceful interaction between business to business (B2B) and business to consumer (B2C), and competent strategic decision making. According to [1] large amounts of data sets provide opportunities to manufacturing industries for efficient realization of output by improving process quality in terms of minimising the risk of out of stock. Various companies, software developers and analytics are offering sophisticated and advance tools and platforms to enhance the supply chain performance in stock and operation planning. Like routes and location distance trailing, tracking unforeseen disasters, delivery time cycle, parts assembling operations, storage capacity and approaches, limitation of goods distribution to retailers, customer behavioural patterns, and competitor standing point. For example, 'Blue Yonder' are developers of data-intensive forecasting methods, 'IBM' links production planning and weather forecasts, 'Google trend' provides information on supply disruptions. Similarly, 'Caterpillar' is a massive information provider on an industrial quotation, 'ForkLift 3' is trying to achieve a big data hub for warehouses, 'Logivations' is a developer of cloud-based 3D warehouse layout planning and optimization tool, 'UPS' is a developer of Optimization and Navigation system (Orion), Amazon's Dash service is for consumer's Internet-connected for reordering [5].

Machine learning is a scientific algorithm developed in order to recognize the patterns to forecast the estimates of activity [60]. Machine learning and supply chain management collaborate on the mandatory information to generate high pitch analysis of the system for cost eliminations and for a better forecast of operations [53]. It seems big data, machine learning, and supply chain operations complement each other in a multipath flow where data entry is supplied by humans which in turn generates large volumes of data so that machine learning can be implemented to define the accurate estimations for forecasting in the supply chain management. The relationship between Artificial intelligence, Machine learning, deep learning, and analytics is shown in Fig. 2. Figure 2 is adapted from (Intel). The adoption of big

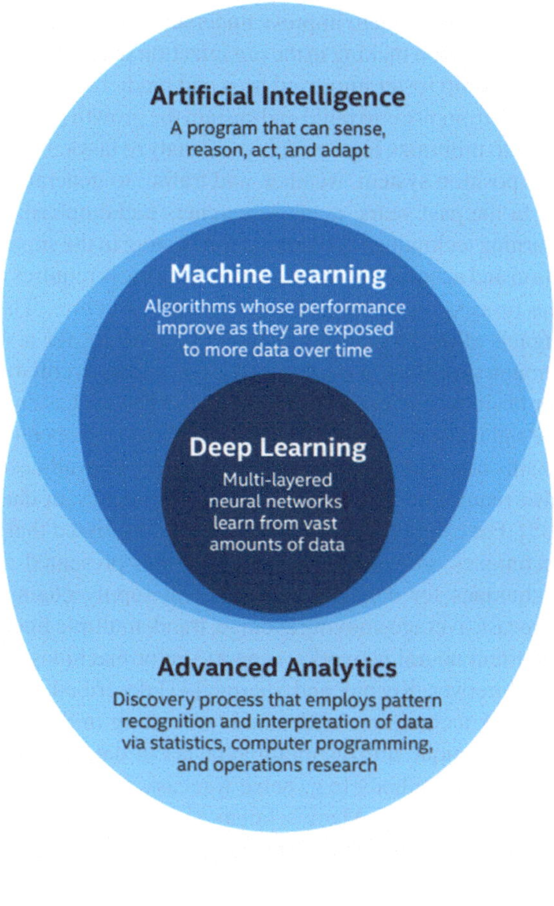

Fig. 2 Advanced analytics relation to artificial intelligence, machine learning and deep learning. Adapted from (Intel)

data applications means that the organizations are able to scrutinize their forecasting, delivery schedule, sourcing and execution, and reverse logistic management.

Business intelligence intends to automate data that can help to support decision making based on various data sources. Business intelligence is defined as a system whose "the major objective is to enable interactive access to data, to enable manipulation of data, and to give business managers and analysts the ability to conduct appropriate analyses" [61, p. 16]. Business intelligence consists of different tools, such as business analytics, data warehouse, business process management, and user interface. The purpose of the BI is to transform large volumes of information into data. Large amounts of data collection and management have brought about a big change in the decision-making process. For example, in the supply chain domain, the use of machine learning is where most of the data pattern is focused while some previous studies have shown that data analytics has significantly contributed to business performance [31, 66].

Large volumes of data are the primary motivation for supply chain organizations to refine its customer services, customization of products with the aim of increasing the customer relationship value [6]. The success of data analytics largely depends on the ability of an organization to manage the supply chain industry's 4.0 tools. Data analytics aim to improve understanding related to business processes to support better decision making in the organization [67]. The data analytics in the supply chain relationship foster improved trust and predict what supply and demand-side partners expect from firms to build and sustain the growth [52]. Data analytics solutions can be used to minimize logistic delays to analyze tasks using environment data (geographical position system, weather, and traffic) to generate value and insights [68].

In the past years, many researchers had emphasized the application of machine learning techniques for demand forecasting in the supply chain procedure. The evaluation and accuracy of demand which business requires to control the capacity enlarge the processing system of supply chain structure. The validation demand through signals allow to generate accurate forecasting for a supply chain to accommodate enough inventory in stock at real-time, and such collaboration has certain advantages for business and organizations [30]. Large data has enabled many industries to apply the right data mining approaches to enhance the operational process [1]. Many supply chains oriented organizations not only try to avail accurate tendency of demand but also require avoidance from demand tendency in data to avoid wrong forecasting [35]. For example Carbonneau et al. [8] discussed that collaborative and networking techniques are worth trying, they further presented many machine learning-based techniques for demand forecasting in supply chain management including naive forecast, average, moving average, trend, multiple linear regression, neural networks, recurrent neural networks, support vector machines.

Moreover, Kochak and Sharma [41] described in their work that artificial neural network techniques for demand forecasting in the supply chain are more effective. This technique utilizes multiple layers of data for assessment and networking than traditional methods (e.g. naive forecast, average, moving average, trend, multiple linear regression). Many scholars are expecting the need for more advanced level artificial machine learning techniques for better demand forecasting in supply chain management [41]. There are many techniques which have been developed and tested by research scholars over time, a few of them are Decision Trees (DT) and Random Forests (RF) [13], HyperBox Classifier [42] Gamma Classifier [65]. Further,Bousqaoui et al. [7] proposed a long-short-term memory prediction model based on machine learning, they argued that data that has been processed with the long lapse of time could generate mistakes on long term dependencies of data in forecasting results. They also referred to the fact that a long-short-term memory prediction model could be a useful tool that will be helpful for further validation as this model emphasizes on the short and long lapse of information overtime periods.

Many manufacturing industries in supply chain 4.0 see data as another source of competitive advantage to proactively comprehend customer requirements and understand market trends and to predict upstream as well as downstream issues regarding scheduling, recovery, Fuel Costs offering Customized Services and compliance with regulations. An improved understanding of data analytics is primarily based on a healthy organization's 'healthy data culture'. Healthy data culture is referred to as, in which organizations are specialized in data deployment for better decision making [19]. Analytics culture is of great concern for organizations aiming to create competitive advantage but also transform data to generate significant insights into decision making [40]. The essential characteristics of data analytics are to provide instant information and creating knowledge in the various domain to take market advantage [68]. The application of data analytics is just starting to get attention, and much of the current research focuses on how to deploy and re-organize data within the organization. Data analytics can visualize hidden patterns of data structures with correct visualization techniques. A schematic flow diagram for machine learning and data analysis steps in a Data-driven decisionmaking methodology is depicted in Fig. 3.

Analytics refers to extracting hidden insights from data [28]. While some researchers advocate the need for healthy data culture or analytics culture [29]. Recent research effort has been directed to predict sales forecasts using analytics in multi-industry data sets [33]. The applications of machine learning in scientific research is used to improve forecasting techniques since early 2000 to understand forecasting issues better. The implementation of different algorithms used in machine learning proved cumbersome and ineffective in representing accurate forecasting issues. When data become large or too complicated, it is challenging to make forecasts using existing data analytics techniques. Machine learning approaches are becoming more effective in this era as they provide different types of machine learning techniques, supervised learning, unsupervised learning, deep learning, and reinforcement learning. The major types of machine learning are summarized in Fig. 4. Decision making via using machine learning is becoming a key performance indicator. Figure 5a–c show different machine learning algorithms applications.

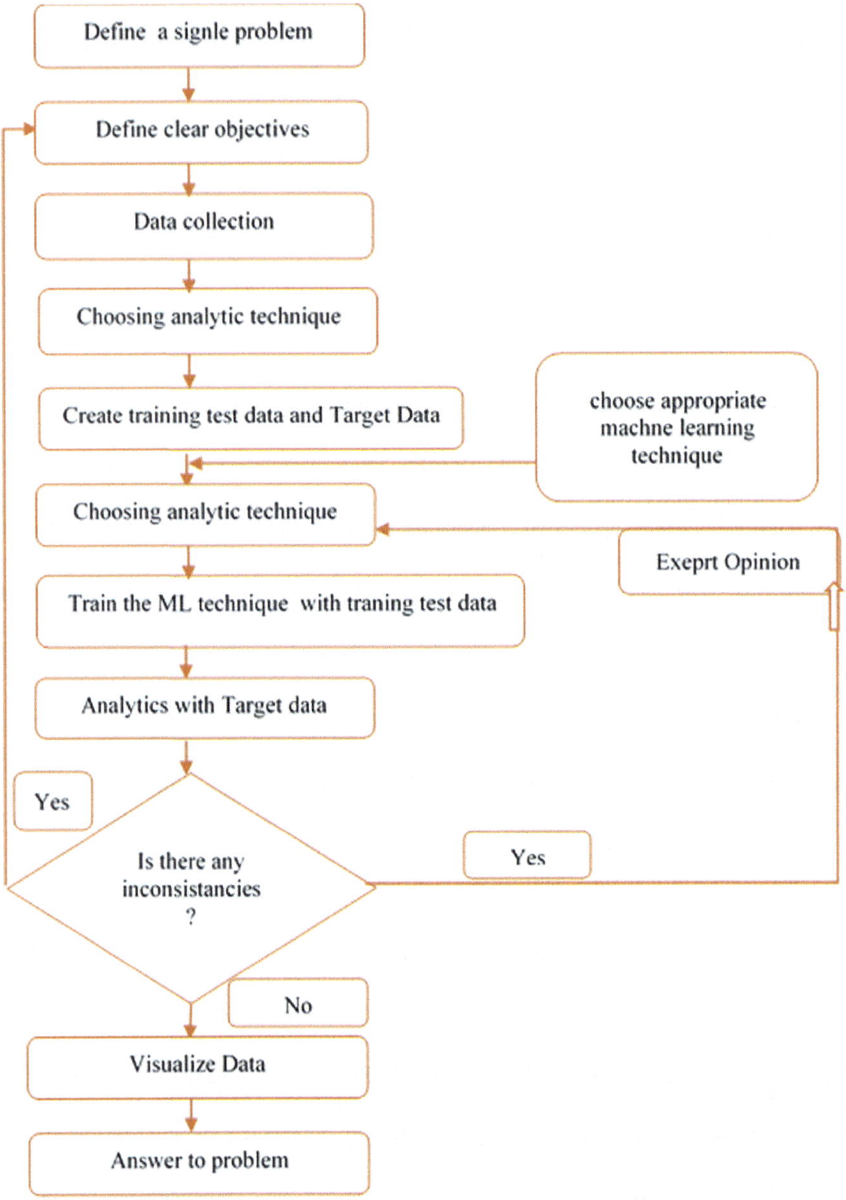

Fig. 3 A schematic flow diagram for machine learning data analysis steps. Figure developed using various literature sources

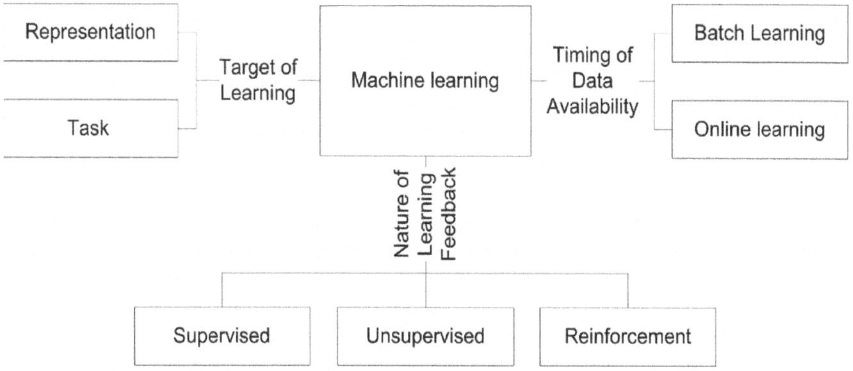

Fig. 4 ML types adapted by [70]

2 Solution Methodology—A Business Case for the Success of Supply Chain

Supply Chain Artificial intelligence (SCAI) is automatic decision-making by making sense of data analytics function for taking action. SCAI enables organizations to develop capacities and processes and to integrate those input data coherently. The input data is used to ensure that the functions and behavior of a complex system are in alignment. In August 2018, JDA acquired Blue yonder [20]. This acquisition helps to automate supply cyber-physical systems to external data to optimized business decisions. AI helps organizations to manage data analytics and the process of gathering data from different sources using different protocols. AI enables organizations to understand the patterns of data distribution, especially those that can get in the way of doing thing correctly. AI helps a firm to collect, analyse and understand the optimal use of resources that is useful to respond quickly in different spheres of operations

2.1 Machine Learning Improve Supply Chain Performance

Machine learning as digital and automatic technology in the future will run algorithms to read the data patterns and operate the functionality on its own to perform supply chain tasks. In future supply chain system will work through machine learning techniques. For example, DHL and Amazon is the giant bidder to facilitate their system with high speed and efficient machine learning system. Russel [54] discussed in the article that machine learning is the future source for unsupervised operational systems for the supply chain, providing high logistic solutions and discoveries for resource and cost-efficiency. Machine learning can support complex supply chain systems with huge data inflow to achieve low error predictability of operations required in future, demand, and supply equilibrium of productivity, cost management, and avoidance from risks in the last-mile delivery system.

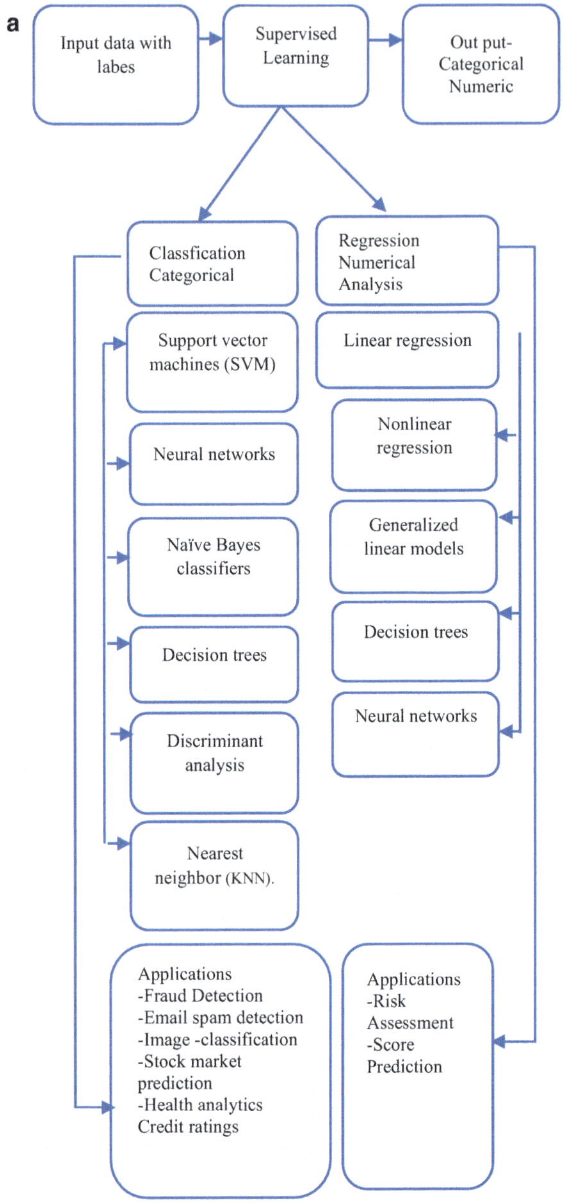

Fig. 5 **a** Supervised learning algorithms, **b** unsupervised learning algorithms, **c** reinforce learning algorithms

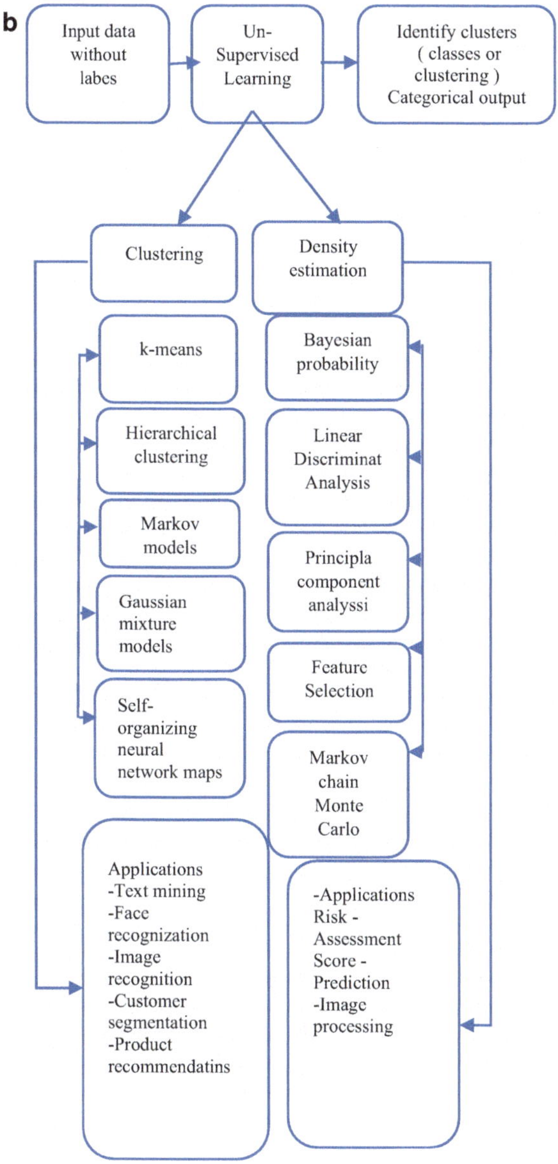

Fig. 5 (continued)

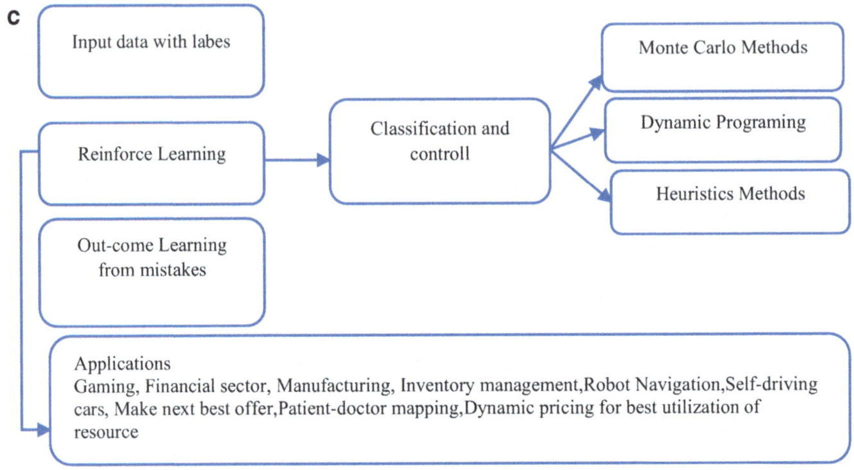

Fig. 5 (continued)

For companies, it is a much better and easy task to evaluate their labor performance as they have been rated by customers, and information is recorded to run machine algorithms. De Treville et al. [17] addressed that delivery leading time management is a chief factor for supply chain management systems if intermediary parties are traceable, then it can reduce risks and possible errors in delivery time. Zhu et al. [71] state that finances in supply chain management for small and medium business is a real point of deep consideration; machine learning is one of the methods which proficiently manages and supervise the credit visibility for the small and medium business in an effective way. Chang et al. [10] states that for supply chain system integration information sharing between stakeholders is a vital source for improvement in supervision, it's imperative to have a clear set of rules between partners, e-procurement can provide this transparency when parties are dealing with intensive contracts. In short, machine learning can be involved in every step of supply chain management system and improve the facts and figures of the business toward more efficency.

3 Discussion

The application of data analytics can be categorized depending upon how it is analyzed. For example, in customization product development, in the production department, products are developed according to the demands of the customer, often associated with a complex set of processes. Customer data could be used for assisting in the customization of designs in the future. The use of customer data could reduce the risk for procurement managers. In supply chain literature form 2000, there is an ever-greater focus on the use of the large volumes of data to exploit an organization's knowledge, to enhance its analytics capability. Data has become one of the primary assets for the organizations to foster strategic alignment of goals [4, 62]. Generation of large volumes of data has been the focuses of supply chain research [27, 37] for some time with great importance. The application of data analytics has become valuable only when an organization realizes a promising value of processing and management of the data. The processing and management of large amounts of data could bring exceptional value and enable organizations to respond to profound opportunities and challenges [62]. There are various characteristics of large data collections well known as big data, and it is widely accepted that big data to have six characteristics: (i) volume; (ii) variety; (iii) velocity; (iv) veracity; (v) variability; and (vi) value [28].

Another source of framing encouraging scenarios for business success is through the idea of 4.0 centered on artificial intelligence and big data. Organizations with better data management and analytics capabilities are able to utilize organization-specific resources, such as database management capabilities. The use of such tangible and intangible organization-specific resources can provide the path to handle performance edge than those who are not focusing on data analytics. The previous research had recognized that the manufacturing industry invests in data analytics and is better able to exploit its resources [23]. Data analytics has been identified as a source of reaching favorable innovation outcomes to predict the future demands of customers [52].

Big data analytics examines large amounts of data to uncover hidden patterns, correlations, and other insights [59]. Big data analytics(BDA) can be defined as a "... a holistic process that involves the collection, analysis, use, and interpretation of data for various functional divisions intending to gain actionable insights, create business value, and establishing competitive advantage ..." [3] (p. 86). Big data analytics has become a major theme in academic research, for example, business process management [18], Performance of coordination in the supply chain [26], impacts on social and environmental sustainability [24], operational performance [25], the effect on firm performance [49], service innovation [43]. The purpose of big data analytics is to re-design the process to deliver and creating avenues for innovation. Previous research shows that the motive behind the use of big data is to reveal the hidden knowledge at the firm level [15]. The literature on bigdata emphasizes how firms explore and reconfigure internally existing understanding in new ways to gain insights from data, which can not be translated through perspective analytics. Big data analytics enables firms to discover related and unrelated patterns of data among different variables to strengthen their supply chain management operations and customer relationship management. Data analytics aims to analyze raw data in order to discover hidden patterns and relationships among different variables. In a word, the main aim of data analysis is to extract useful information from rough data and transfer it to effective knowledge to improve product and process understanding and to support decisions [31, 50].

The benefits that data analytics may bring to firms encompasses to strengthen their supply chain operations [31] and support customization of customer services [6]. The ability to work with agility gives firms an advantage that could have a profound impact on overall business performance. The type of data has a profound effect on business performance. Management must determine what mix of data (textual, multimedia, machine to machine, cyber-physical system) makes sense for each decision making. Different combinations of textual, audio, and video data give rise to significant data management capabilities. A mix of data plan can both motivate data analyst to direct their activities that are consistent with the organizational objective and plan for building long-term data-storage functions. In fact, more and more firms are moving towards big data management capabilities that may drive data analytics to make long term business decisions. Bigdata analyst capabilities(BDAC) are always in the domain and paying the firm to realize their full potential.

Research has demonstrated the potential to use bigdata analyzing technologies, WibiData and Skytree, BigQuery, MapReduce, NoSQL Databases and Hadoop is a means of understanding and improving business strategies [69]. According to [11], data management and processing can facilitate in responding to challenges and opportunities and help bridge structural holes between them. Jukić et al. [39] have demonstrated how the appropriateness of management to analyze the big data could be employed to identify new knowledge, proposing insights to deal with uncertain environmental events. Table 1 helps understand the main steps in data analysis and data management, adapted from [64].

Business analytics is a source of progression in supply chain management and performance efficiency. But what dynamic capabilities should be the sponsorship of business analytics, to avail more cost-efficient and proficient supply chain structure, is still under the focus of many scholars and researchers. As Chae and Olson [9] described in their study that data management capability is the former tool to have competitive advantages over other businesses (e.g. Wal Mart). Furthermore, they discussed that data IT infrastructure and data storage effectiveness is the leading player as data management capability. Supply chain management is crucial when the output of the supply chain procedure is manufactured goods, failure in the production

Table 1 The main steps of data management

Main data analysis steps	Main management steps
Data collection	Establishment and sharing of a cohesive data-oriented culture
Data organization	Selection of an integrated set of analytics in line with strategic goals
Data extraction	Adequate technological infrastructure
Data integration	Computing skills
Data analysis	Analysis and research skills Management's ability to interpret results in line with strategic goals and to catch opportunities
Data sharing	Data report and diffusion
Data storage	Feedback collection
Data reuse	Renewal of the knowledge acquired for continuous improvement

Source adapted from [64]

can be a complex scenario for business to deal with, and it can cost the high prices, compromised quality and delay in delivery of output. Therefore, in the manufacturing industry, industries wanted to make sure that end to end, supply chain management is conscious and properly supervised. Many researchers proposed business and data analytics methodologies to reduce the risk of production, wastage of material, and overlapping in the operational system.

On the other hand,Mikalef et al. [47] extended the definition of BDAC and emphasized the inclusion of organizational resources. They defined it as the ability of a firm to effectively deploy technology and talent to capture, store, and analyze data toward the generation of insight (p. 2). Previous literature has emphasized that when assessing the business performance, it is essential to take a broader view of different capability performance, and particularly to highlight the notion of knowledge management. In the global information age, the business performance hinges on combining data management and analytics capabilities to achieve a long term performance. Research suggests that bigdata analyst capabilities (BDAC) impact upon the understanding operations management and service information [4, 44] and competitive performance [47]. A study by [64] showing the importance of big data and cognitive computing, is particularly relevant for unlocking and accelerating innovation. According to [32] BDA capability encompasses a set of tangible, intangible, and human skills that are useful for the organization to incorporate data-driven culture and organizational learning. In the last decade, big data has emerged as a source of analyzing large volumes of data and has paved the way for exploration of the complex nature of organizational knowledge management practices. Big data, perhaps the most successful application of artificial intelligence, has been employed effectively within the sphere of supply chain management. Many manufacturing industries do seem to be geared efforts within the supply chain to improve efficiency, maintenance, and reduce the risk of theft and accidents. There have been the various deployment of data analytics applications in an organization all over the world. Along with its practical sens, big data analytics and machine learning concept have been found to have profound implications for automatic acquisition of knowledge for decision making. Artificial intelligence fills the gap for the supply chain management system by processing the data in quick results for companies, so advance forecasting of operation can be acquired.

4 Conclusions

This chapter reviews the advance of some artificial intelligence methods in the supply chain. Considering the features of big data and artificial intelligence(AI), this chapter provides a schematic flow of diagram for machine learning data analysis step by step in data analysis. The step by step data methodological filtering approach could provide data analysts with a sense of superiority in handling data. Especially, when carrying out the data analysis in the machine learning environment, the data analyst must understand, why there are inconsistencies in the output. This chapter has provided some remedial measures in case of, if there are data inconsistencies reveals during the data analysis. We also suggest applying in the proposed schematic method in some machine learning applications to improve the data validation and to test its usefulness and applicability. This chapter indicated that the schematic flow diagram could be used to correct the inconsistencies in the model. The chapter highlights that supervised, unsupervised, and reinforce learning algorithms can effectively be used to perform different statistical analyses. Researchers understand the extensive use of artificial intelligence among firms as essential and necessary for shaping the future of the supply chain of industry 4.0. AI strategies should be at the forefront of the management of supply chain practices and must build up through years of data generation and contain a wide variety of machine learning approaches. IA creates a new analytics intelligence to justify the choice of strategic decision making rather than heavily focused on intuition intelligence. AI can also provide an in-depth assessment of the future, at capturing big picture of information and learning within the bounds of paradigm. There is a need for research on the results of which could in selecting only the most appropriate big data analytics and machine learning approach for supply chain management, especially for the forecasting in logistics and production. Academicians and universities should work together and create affordable and reliable artificial intelligence solutions for the supply chain management that have the potential to contribute towards sustainable development goals. Similarly, big data analytics and insights gained from the data plays the role of antecedents of organizational renewal strategies. Decision making, which is gained from the data analytics, increases the likelihood of a firm will engage more in decision making, thus creating new paths ways of achieving performance in the circular economy and sustainability aspects. To be able to grasp new market development opportunities, firms must be able to think beyond the current trends. One of the biggest challenges in the emerging field of a supply chain in industrial revolution 4.0 is to improve firm decision making for a move towards a big data system.

References

1. Addo-Tenkorang, R., Helo, P.T.: Big data applications in operations/supply-chain management: a literature review. Comput. Ind. Eng. **101**, 528–543 (2016). https://doi.org/10.1016/j.cie.2016.09.023
2. Agrawal, A., Gans, J., Goldfarb, A.: Prediction Machines: The Simple Economics of Artificial Intelligence. Harvard Business Press (2018)
3. Akter, S., Wamba, S.F.: Big data and disaster management: a systematic review and agenda for future research. Ann. Oper. Res. **283**, 939–959 (2019)
4. Albergaria, M., Jabbour, C.J.C.: The role of big data analytics capabilities (BDAC) in understanding the challenges of service information and operations management in the sharing economy: Evidence of peer effects in libraries. Int. J. Inf. Manage., 102023 (2019)
5. Alicke, K., Rachor, J., Seyfert, A.: Supply Chain 4.0–the next-generation digital supply chain. McKinsey. available https//www mckinsey com/business-functions/operations/our-insights/supply-chain-40–the-nextgeneration-digital-supply-chain. Accessed 6 Sept 2018

6. Anshari, M., Almunawar, M.N., Lim, S.A., Al-Mudimigh, A.: Customer relationship management and big data enabled: personalization & customization of services. Appl. Comput. Inf. **15**, 94–101 (2018)
7. Bousqaoui, H., Achchab, S., Tikito, K.: Machine learning applications in supply chains: An emphasis on neural network applications. In: 2017 3rd International Conference of Cloud Computing Technologies and Applications (CloudTech), pp 1–7 (2017)
8. Carbonneau, R., Laframboise, K., Vahidov, R.: Application of machine learning techniques for supply chain demand forecasting. Eur. J. Oper. Res. **184**, 1140–1154 (2008)
9. Chae, B., Olson, D.L.: Business analytics for supply chain: a dynamic-capabilities framework. Int. J. Inf. Technol. Decis. Mak. **12**, 9–26 (2013)
10. Chang, H.H., Tsai, Y.-C., Hsu, C.-H.: E-procurement and supply chain performance. Supply Chain Manag. An Int. J. **18**, 34–51 (2013)
11. Chen, J., Chen, Y., Du, X., et al.: Big data challenge: a data management perspective. Front. Comput. Sci. **7**, 157–164 (2013)
12. Chen, M., Mao, S., Liu, Y.: Big data: a survey. Mob Netw. Appl. **19**, 171–209 (2014)
13. Cheng, J.-H., Chen, H.-P., Lin, Y.-M.: A hybrid forecast marketing timing model based on probabilistic neural network, rough set and C4. 5. Expert Syst. Appl. **37**, 1814–1820 (2010)
14. Daugherty, P.R., Wilson, H.J.: Human + machine: reimagining work in the age of AI (2018)
15. Davenport, T.: Big Data at Work: Dispelling the Myths, Uncovering the Opportunities. Harvard Business Review Press (2014)
16. De Mauro, A., Greco, M., Grimaldi, M.: A formal definition of Big Data based on its essential features. Libr. Rev. **65**, 122–135 (2016). https://doi.org/10.1108/LR-06-2015-0061
17. De Treville, S., Shapiro, R.D., Hameri, A.-P.: From supply chain to demand chain: the role of lead time reduction in improving demand chain performance. J. Oper. Manag. **21**, 613–627 (2004)
18. Dezi, L., Santoro, G., Gabteni, H., et al.: The role of big data in shaping ambidextrous business process management: case studies from the service industry. Bus. Process Manag. J. **24**, 1163–1175 (2018)
19. Diaz, A., Rowshankish, K., Saleh, T.: Why data culture matters. McKinsey Q. **3**, 1–17 (2018)
20. Dignan, L.: JDA Software becomes Blue Yonder amid SaaS, autonomous supply chain push. In: ZDNet. https://www.zdnet.com/article/jda-software-becomes-blue-yonder-amid-saas-autonomous-supply-chain-push/. Accessed 4 Mar 2020
21. Dorsemaine, B., Gaulier, J.-P., Wary, J.-P., et al.: Internet of things: a definition & taxonomy. In: 2015 9th International Conference on Next Generation Mobile Applications, Services and Technologies, pp. 72–77 (2015)
22. Duan, Y., Edwards, J.S., Dwivedi, Y.K.: Artificial intelligence for decision making in the era of Big Data—evolution, challenges and research agenda. Int. J. Inf. Manage. **48**, 63–71 (2019). https://doi.org/10.1016/j.ijinfomgt.2019.01.021
23. Dubey, R., Gunasekaran, A., Childe, S.J., et al.: Big data analytics and artificial intelligence pathway to operational performance under the effects of entrepreneurial orientation and environmental dynamism: a study of manufacturing organisations. Int. J. Prod. Econ., 107599 (2019a). https://doi.org/10.1016/j.ijpe.2019.107599
24. Dubey, R., Gunasekaran, A., Childe, S.J., et al.: Can big data and predictive analytics improve social and environmental sustainability? Technol. Forecast Soc. Change (2017)
25. Dubey, R., Gunasekaran, A., Childe, S.J., et al.: Big data and predictive analytics and manufacturing performance: integrating institutional theory, resource-based view and big data culture. Br. J. Manag. **30**, 341–361 (2019)
26. Dubey, R., Luo, Z., Gunasekaran, A., et al.: Big data and predictive analytics in humanitarian supply chains: enabling visibility and coordination in the presence of swift trust. Int. J. Logist. Manag. **29**, 485–512 (2018)
27. Frank, A.G., Dalenogare, L.S., Ayala, N.F.: Industry 4.0 technologies: implementation patterns in manufacturing companies. Int. J. Prod. Econ. **210**, 15–26 (2019). https://doi.org/10.1016/j.ijpe.2019.01.004

28. Gandomi, A., Haider, M.: Beyond the hype: Big data concepts, methods, and analytics. Int. J. Inf. Manage. **35**, 137–144 (2015)
29. Germann, F., Lilien, G.L., Rangaswamy, A.: Performance implications of deploying marketing analytics. Int. J. Res. Mark. **30**, 114–128 (2013)
30. Gunasekaran, A., Ngai, E.W.T.: Information systems in supply chain integration and management. Eur. J. Oper. Res. **159**, 269–295 (2004)
31. Gunasekaran, A., Papadopoulos, T., Dubey, R., et al.: Big data and predictive analytics for supply chain and organizational performance. J. Bus. Res. **70**, 308–317 (2017)
32. Gupta, M., George, J.F.: Toward the development of a big data analytics capability. Inf. Manag. **53**, 1049–1064 (2016). https://doi.org/10.1016/j.im.2016.07.004
33. Hallikainen, H., Savimäki, E., Laukkanen, T.: Fostering B2B sales with customer big data analytics. Ind. Mark. Manag., 1–9 (2019). https://doi.org/10.1016/j.indmarman.2019.12.005
34. Hazen, B.T., Skipper, J.B., Boone, C.A., Hill, R.R.: Back in business: operations research in support of big data analytics for operations and supply chain management. Ann. Oper. Res. **270**, 201–211 (2018)
35. Heikkilä, J.: From supply to demand chain management: efficiency and customer satisfaction. J. Oper. Manag. **20**, 747–767 (2002)
36. Intel Reference Architecture: Machine Learning-Based Advanced Analytics Using Intel Technology. https://www.thailand.intel.com/content/dam/www/public/us/en/documents/guides/machine-learning-based-advanced-analytics-using-technology-ref-arch-fordistribution.pdf%0A. Accessed 3 Mar 2020
37. Ivanov, D., Dolgui, A., Sokolov, B.: The impact of digital technology and Industry 4.0 on the ripple effect and supply chain risk analytics. Int. J. Prod. Res. **57**, 829–846 (2019)
38. Jackson, P.: Introduction to Expert Systems. Harlow, Reading, Menlo Park Addison-Wesley (1999)
39. Jukić, N., Sharma, A., Nestorov, S., Jukić, B.: Augmenting data warehouses with big data. Inf Syst Manag **32**, 200–209 (2015)
40. Kiron, D., Prentice, P.K., Ferguson, R.B.: The analytics mandate. MIT Sloan Manag. Rev. **55**(4), 1–25 (2014)
41. Kochak, A., Sharma, S.: Demand forecasting using neural network for supply chain management. Int. J. Mech. Eng. Robot. Res. **4**, 96–104 (2015)
42. Kone, E.R.S., Karwan, M.H.: Combining a new data classification technique and regression analysis to predict the Cost-To-Serve new customers. Comput. Ind. Eng. **61**, 184–197 (2011)
43. Lehrer, C., Wieneke, A., Vom Brocke, J.A.N., et al.: How big data analytics enables service innovation: materiality, affordance, and the individualization of service. J. Manag. Inf. Syst. **35**, 424–460 (2018)
44. Lozada, N., Arias-Pérez, J., Perdomo-Charry, G.: Big data analytics capability and co-innovation: An empirical study. Heliyon 5 (2019). https://doi.org/10.1016/j.heliyon.2019.e02541
45. Luger, G.F.: Artificial Intelligence: Structures and Strategies for Complex Problem Solving, 4th edn. Harlow, Essex, England (2002)
46. Michalewicz, Z.: Genetic Algorithms + Data Structures = Evolution Programs. Springer Science & Business Media (2013)
47. Mikalef, P., Krogstie, J., Pappas, I.O., Pavlou, P.: Exploring the relationship between big data analytics capability and competitive performance: The mediating roles of dynamic and operational capabilities. Inf. Manag. **57**, 103169 (2020). https://doi.org/10.1016/j.im.2019.05.004
48. Miller, D.D., Brown, E.W.: Artificial intelligence in medical practice: the question to the answer? Am. J. Med. **131**, 129–133 (2018)
49. Müller, O., Fay, M., vom Brocke, J.: The effect of big data and analytics on firm performance: an econometric analysis considering industry characteristics. J. Manag. Inf. Syst. **35**, 488–509 (2018)
50. Nam, D., Lee, J., Lee, H.: Business analytics use in CRM: a nomological net from IT competence to CRM performance. Int. J. Inf. Manag. **45**, 233–245 (2019)

51. Pawlak, Z.: Rough sets. Int. J. Comput. Inf. Sci. **11**, 341–356 (1982)
52. Perera, W.K.R., Dilini, K.A., Kulawansa, T.: A review of big data analytics for customer relationship management. In: 2018 3rd International Conference on Information Technology Research (ICITR), pp. 1–6 (2018)
53. Priore, P., Ponte, B., Rosillo, R., de la Fuente, D.: Applying machine learning to the dynamic selection of replenishment policies in fast-changing supply chain environments. Int. J. Prod. Res. **57**, 3663–3677 (2019)
54. Russell, K.: Machine learning is shaping the future of supply chain and logistics management, improving accuracy, speed, scale, and more. Here's how. https://www.fronetics.com/7-ways-machine-learning-is-improving-supply-chain-management/ (2019)
55. Samuel, A.L.: Some studies in machine learning using the game of checkers. IBM J. Res. Dev. **44**, 206–226 (2000)
56. Sanders, N.R., Ganeshan, R., et al.: Special issue of production and operations management on "big data in supply chain management". Prod. Oper. Manag. **24**, 1835–1836 (2015)
57. Sarhani, M., El Afia, A.: Intelligent system based support vector regression for supply chain demand forecasting. In: 2014 Second World Conference on Complex Systems (WCCS), pp. 79–83 (2014)
58. SAS Artificial Intelligence: What it is and why it matters. https://www.sas.com/en_us/insights/analytics/what-is-artificial-intelligence.html. Accessed 28 Feb 2020
59. SAS Big Data Analytics: What it is and why it matters. https://www.sas.com/en_us/insights/analytics/big-data-analytics.html. Accessed 4 Mar 2020
60. Shahrabi, J., Mousavi, S.S., Heydar, M.: Supply chain demand forecasting: a comparison of machine learning techniques and traditional methods. J. Appl. Sci. **9**, 521–527 (2009)
61. Sharda, R., Delen, D., Turban, E.: Business intelligence, analytics, and data science: a managerial perspective. Pearson (2016)
62. Sivarajah, U., Kamal, M.M., Irani, Z., Weerakkody, V.: Critical analysis of Big Data challenges and analytical methods. J. Bus. Res. **70**, 263–286 (2017)
63. Tanaka, K.: An Introduction to Fuzzy Logic for Practical Applications. Springer, New York City, USA (1997)
64. Troisi, O., Maione, G., Grimaldi, M., Loia, F.: Growth hacking: Insights on data-driven decision-making from three firms. Ind. Mark. Manag. doi:10.1016/j.indmarman.2019.08.005 (2019)
65. Uriarte-Arcia, A.V., López-Yáñez, I., Yáñez-Márquez, C., et al.: Data stream classification based on the gamma classifier. Math. Probl. Eng., ID **939175**, 1–17 (2015). https://doi.org/10.1155/2015/939175
66. Wamba, S.F., Gunasekaran, A., Akter, S., et al.: Big data analytics and firm performance: effects of dynamic capabilities. J. Bus. Res. **70**, 356–365 (2017). https://doi.org/10.1016/j.jbusres.2016.08.009
67. Wang, H., Xu, Z., Fujita, H., Liu, S.: Towards felicitous decision making: An overview on challenges and trends of Big Data. Inf. Sci. (Ny) **367**, 747–765 (2016)
68. Xu, Z., Frankwick, G.L., Ramirez, E.: Effects of big data analytics and traditional marketing analytics on new product success: a knowledge fusion perspective. J. Bus. Res. **69**, 1562–1566 (2016)
69. Yi, X., Liu, F., Liu, J., Jin, H.: Building a network highway for big data: architecture and challenges. IEEE Netw. **28**, 5–13 (2014)
70. Zhou, L., Pan, S., Wang, J., Vasilakos, A.V.: Machine learning on big data: opportunities and challenges. Neurocomputing **237**, 350–361 (2017). https://doi.org/10.1016/j.neucom.2017.01.026
71. Zhu, Y., Xie, C., Wang, G.-J., Yan, X.-G.: Comparison of individual, ensemble and integrated ensemble machine learning methods to predict China's SME credit risk in supply chain finance. Neural Comput. Appl. **28**, 41–50 (2017)

Artificial Intelligence and Firm Performance: A Robotic Taxation Perspective

O. A. Joseph and A. Falana

Abstract Highly robotized environment and breakthroughs in Artificial Intelligence (AI) produce innovative variations of skills, thus, innovations of varying skills influence the performances of business. This study investigated the connection between artificial intelligence (AI) and firm performance. The study employed survey research design using triangulation method for data collection. The choice of the survey research design is justified because it follows a correlational research strategy and help in predicting behavior (Dabor and Adeyemi 2009). The population of this study consisted all companies in Nigeria. The data were analyzed using descriptive and inferential statistics using Pearson correlation matrix test and analysis of Variance (ANOVA). The study showed how firms that adopted AI solutions such as robotic taxation experienced enhanced bottom-line performance. Besides, this study also revealed that robotic involvement led to an increase in employment of skilled workers. The study, therefore, recommends that there is a need to tax robotic involvement that displaces routine workers in order to provide additional income for the government to bring back displaced routine workers into the tax net through training programs. The study suggests that in actively enhancing the performance of businesses in Nigeria, firms should strive towards using artificial intelligence, data analytics and connected technology solutions in driving business processes. Policy and legislations need to enable adoption of technological innovation in businesses are also needed. The conceptualization of Robotic Taxation represents theoretical innovation. This scale can be used in knowledge-intensive firms. This paper concluded that AI has a positive incidence on firm performance.

Keywords Technological innovation · Artificial intelligence · Robotic taxation · Firm performance

O. A. Joseph (✉)
Lead City University, Ibadan, Oyo State, Nigeria
e-mail: joseph.oluseun@lcu.edu.ng

A. Falana
Babcock University, Ilishan, Ogun State, Nigeria

A. Hamdan et al. (eds.), *The Fourth Industrial Revolution: Implementation of Artificial Intelligence for Growing Business Success*, Studies in Computational Intelligence 935,
https://doi.org/10.1007/978-3-030-62796-6_2

1 Introduction

Performance of any company also depends on the increase in financial metrics such as return on equity, return on assets, sales growth, profit margin over several periods. A company can grow rapidly than its peers or beyond geographical location. Firm's growth is also a sustained-growth of total sales, market share, and profit maximization. Firm growth can also be negative [80]. As Cai et al. [17] asserted, firm's growth can negatively be influenced by taxation, in the sense that, increase in taxes lessens profit after tax, thus, resulting in reduced ability to capitalize in innovative technologies or products, and also increase the money that companies devote on evading tax, such as the costs of bribing tax authorities, while, reduction in taxes will reduce organizations' resources devoted on tax evasion. However, reduction in taxes will reduce government resources which in turn, decrease government spending on business expenditure on research and development and decrease government spending on basic amenities.

Globally, firms are faced with global pricing and cost strategies to remain competitive which leads to companies' relocating production, labor, the shipping cost to a cost-effective jurisdictions Reynolds [82], or employment of artificial intelligence (AI) and AI-equipped machines like 60,000 units of imported robots in Australia, 35,000 units of imported robots in Europe and 17,000 units of robot imported to maximize productivity and to be more cost-effective. This has led capital owners to prefer the employment of robotic workforce than human workforce because robots are getting cheaper than human workforce [36, 55, 60, 78, 88]. The use of artificial intelligence or AI fortified machines according to Joshi [47], has greatly contributed to efficiency and convenience in the collection of taxes and reduction of tax defaulters, widening the tax base globally.

In Africa, firms are faced with political and economic instability, in terms of the government's unsupportive role in the development of an equal competitive environment and sudden liberalization of economic policies, [48, 99]. Also, in Nigeria, scholars like Malik et al. [57], Soderbom and Teal [87], and Uma, Chidike et al. [19] opined that firms are faced with corruption, poor infrastructure, insufficient assessment of credit, loose importation of products and lack of investment in human capital which are the key constraints of firm's performance in Nigeria. Besides, the intensity in the uses of AI or robotization has led to the digital divide between the urban and rural communities in sub-Sahara Africa due to lack of effective policies that will foster access to internet technologies, national innovation and firm's participation in the global value chain. Disruptive technologies or industry 4.0 have greatly contributed to the unemployment rate, inequality of income and over-concentration of social amenities in urban areas in Nigeria [74, 92].

AI are computer programs developed to complete a job that would otherwise involve human intelligence [7]. In this study, AI is the brain of artificially intelligent machines or robots. Historically, Perez et al. [79] traced the birth of AI from the automata theory in computing, common-sense machine use for codifying, stowing and decoding German trigrams of Enigma machines during World War 2 by Ramon

Cajal to 1958 when Marvin Minsky, John McCarthy, Claude Shannon and Nathan Rochester first coined the term AI at the Dartmouth conference. A robot is a sense—think-act machine, according to Bekey [13] or a sophisticated machine that materializes full or semi-autonomous operations of observing multifaceted tasks and makes appropriate decisions in the production of goods and services, delivery of goods and services [16].

Highly robotized environment and breakthroughs in AI erupt innovative varieties of skills from human workers, thus, this kind of innovativeness of skills influences the performances of business [34]. There is also a high growth of health care startup as AI disrupts operational, financial healthcare systems and also changes the way illnesses are diagnosed and are prevented, enhancing value and revolutionize the way health care workers and top management collect health data, interprets and even relate with patients [37]. Concerning enhancing value and AI changing the early detection and diagnoses of illness, Vaishya et al. [95] stress the fact that AI plays a germane role not only in the detection of the cluster of cases of illness but also in the prediction of where the deadly virus of COVID-19 will affect in future by gathering and analyzing of data through the support of Big data, internet of things and machine learning. Furthermore, AI through the use of the internet, big data and machine learning are employed to flatten the curve of the spread of COVID-19 and suggest the creation of a vaccine for the deadly virus [95].

Apart from AI been used to suggest the creation of vaccines, AI-equipped machines or robots are also used in Rwanda to fight the novel coronavirus by reducing doctor and nurse contact with the deadly virus [65], while, 113 Nigeria's healthcare givers are infected with the virus [2]. The novel coronavirus (COVID 19) in Nigeria according to Nigeria Centre for Disease Control NCDC [66], has 6,677 cases with 200 deaths.

Conversely from the health sector in which artificially intelligent machines are used to curb the virus from spreading and infecting health caregivers, Prior studies like Sylvester and Okorie [88] and Nnamani and Nwoha [69] have asserted that artificially intelligent machines or robots had a great effect on the firm's performance in Nigeria, however, the World development report [97], reinforces technological unemployment of 65% of jobs have been lost in the country. For example, a milk company in Lagos Nigeria according to Alli [8], robotized its production process which resulted in factory robot expansion across 30 African countries and technological unemployment of over 230 routine staff.

Scholars like Zhou [99], have also opposed robotic intervention. He claimed it has harmed business growth in emerging economies like Nigeria, especially, when the emerging economies opened up the size of the domestic market to developed countries. Countries like Nigeria heavily imports manufactured goods from the developed countries who employ robotic workforce to produce more and to be cost-effective than the manufacturing companies in emerging economies, thus disrupting the manufacturing sector growth in developing countries like Nigeria.

Apart from heavy importation of manufactured goods, Chiedozie [20] stated that 95% of technologies used in the country are imported, loose importation of products is one of the key constraints of firm's growth in Nigeria. Also, Nawakitphaitoon and

Ormiston [67] posited that these workers are vulnerable to declining demand, and also huge earnings losses between their pre- and post-up-skilling or getting a new routine job. Thus, this has made scholars like Csillag and Scharle [26] and Jaimovich and Siu [46] to posit financial incentive for displaced workers most especially in developing countries, by subjecting company's income or the income generated by the robot that distorts personal income tax enjoyed by the government into tax in form of a robot tax. Robot taxation is an increase in income taxes on hyper-scale companies or the provision of a tax credit for hyper-scale companies who employs more workers [1, 59].

Robot taxation is also expanding tax burden on capital income and minimizing the tax burden on labor and tax gaming chances according to Mazur [60], which is in contrast to taxing the company' s income tax but on policies that seek to develop human capital in which South Korea has implemented a charge of 2–7% on a capital allowance.

2 Statement of the Problem

The major problem of this study is addressing Nigeria's low tax to GDP of about 3.4% compared to other developing economies like Egypt (15%), Ghana (18%), Kenya (18%), Morocco (27.6%), Tunisia and South Africa (29%) and also an average of 34% in developed economies (OECD) countries [71, 84]. These low tax to GDP is said to be caused by technological unemployment and employment of aggressive tax planning strategies erupting into socio-economic inequality. Davies [29], and United Nations [93], stated that about 2,153 billionaires are getting richer, reaping from the benefit of "industry 4.0" than 4.6 billion people, in which women and girls are overburdened with the disparity of economic opportunities of robotization. Oxfam [75], also stated the wealth of 22 richest men in the world is more than the wealth of all the women in Africa.

Secondly, despite, the strategic implementation plan (SIP), economic recovery and growth plan (ERGP), erratic power supply resulted in businesses spending $14 billion on heavy use of generator and fuel which puts the country's firm's performance and climate in jeopardy. Also, the medium-term economic plan of (2017–2020) has also contributed to the unemployment growth in the country by harnessing on technological innovations in which 95% of the technologies or robotic technologies used in all sectors of the economy are imported [20] to alleviate the unemployment rates of the workforce in a country faced with a rising population growth rate projected by UN to be 398 million people in 2050 and also a rising percentage of unemployed youth from 23.63% in 2014 to 38% in the third quarter of 2018 according to Trading Economics, 2019. Government and business support for research and development (R&D) have been tracked by Postel-Vinay [81] to contributing to unemployment. On top of this concern, tax incentive and other R&D investment on robotization aggravate economic, social and ethnic imbalances and reducing R&D incentives enjoyed by hyper-scale companies [1]. Hoeschl et al. [45], asserted that robots are getting cheaper and better, while employment of human workforce is becoming less attractive to capital owners, also, the scholar stated that labor income is overburden with tax deductibility. Thus, this study will examine the level of robotic intervention among companies in Nigeria.

One of the major problems on robotic intervention on the human workforce is that robotic adoption has been traced to lead to boredom Hewitt [44], high rate of suicide in Japan [49, 27]. Conversely, like Augustine et al. [11] and Gordon and Slemrod [39] emphasized, robotic adoption boost productivity and labor competitiveness and increase in manufacturing productivity led to the creation of new jobs for human workers while the study of Uwe [94] substituted robots with routine work employing intensive margin labor supply endogenous wages and occupational choice model to show that it is optimal to tax robots substituting routine workers in an organization. Thus, this will determine the effect of a robotic intervention on the level of employment among companies in Nigeria.

3 The Objective of the Study

The main objective of this study is to investigate the effect of robot taxation on the performance of companies in Nigeria. The specific objectives are to

(i) examine the level of robotic involvement (Number of robots, Cost of robots and Output or productivity of robot) on the performance of companies in Nigeria.
(ii) Investigate the effect of pre and post robotic involvement (Number of robots, Cost of robots and Output or productivity of robot) on the performance of companies in Nigeria.
(iii) determine the effect of a robotic involvement (Number of robots, Cost of robots and Output or productivity of robot) on the level of employment among companies in Nigeria.
(iv) Investigate the effect of pre- and post-reduction in capital allowance on the performance of companies in Nigeria.

4 Research Hypotheses

The following hypothesis will be tested

H_{01}: There is no robotic intervention (Number of robots, Cost of robots and Output or productivity of robot) in companies in Nigeria.

$H_0 2$: There is no significant relationship between robotic involvement (Number of robots, Cost of robots and Output or productivity of robot) and performance of companies in Nigeria.

$H_0 3$: There is no significant relationship between robotic involvement (Number of robots, Cost of robots and Output or productivity of robot) and the level of employment in companies in Nigeria.
$H_0 4$: There is no significant relationship between reduction in capital allowance and the performance of companies in Nigeria.

5 Operationalization of Variables

To accomplish the objectives of this research, two main variables will pilot this study which is robotic intervention and performance of companies, with robotic intervention serving as the explanatory variable, while the performance of companies as the explained variables.

This is operationalized below:

Y = f(X)

$Y = y_1, y_2,$ (Explained Variables surrogate)

$X = x_1, x_2, x_3, x_4$ (Explanatory Variable indicators)

where:

Y = vectors of performance companies (PC)

y_1 = profit before tax (PBT)

y_2 = level of Employment of human labor (E)

X = robotic taxation (RT)

x_1 = number & cost of robots (NCR)

x_2 = level of Employment (EL)

x_3 = Output of robot (OR) or productivity of robots (PR)

x_4 = Reduction in capital allowance (RCA)

Functional Relationship

Objective one

$$PBT_{it} = f(NCR_{it},) \tag{1}$$

Objective two

$$\begin{aligned} PBT_{it} - RT &= f(NCR_{it} - RT, CR_{it} - RT, OR_{it} - RT) \quad \text{pre} - \text{RI functional from} \\ PBT_{it} + RT &= f(NCR_{it} + RT, CR_{it} + RT, OR_{it} + RT) \quad \text{post RI functional form} \end{aligned} \tag{2}$$

Objective Three

$$\text{EL}_{\text{it}} = f(\text{NCR}_{\text{it}}) \tag{3}$$

Objective Four

$$\begin{aligned} \text{PBT}_{\text{it}} - \text{RT} &= f(\text{RCA}_{\text{it}} - \text{RT}) \quad \text{pre} - \text{RT functional from} \\ \text{PBT}_{\text{it}} + \text{RT} &= f(\text{RCA}_{\text{it}} + \text{RT}) \quad \text{post RT functional form} \end{aligned} \tag{4}$$

6 Conceptual Review

6.1 Artificial Intelligence

Artificial intelligence is defined by Wang [96] as an aspect of developing sophisticated machines that can adapt to the environment and also changing the environment with an inadequate data moreover, Dobrev [30] defined AI as a world of an algorithm that controls the hardware. However, Grewal [40] defined AI as a computational replica of human thought, behavior. AI is further defined as a software that collects data about the location or situation and takes actions founded on that data gathered and processed. The data gathered and processed consist of 74% of the environment (data of dusky energy, 22%, data of dark matter and 4% data of perceptible matter through the use of sensors [40]. Furthermore, AI is computer programs developed to complete a job that would otherwise involve human intelligence [7].

The accounting profession has completely disrupted by artificial intelligence in the aspect of strategic and analytical roles like management accounting, financial accounting and taxation through the use of various accounting software [32, 86]. Accounting software has been historically excavated by Eamonn [31], from 1974 when accounting was performed manually on paper-based trial balances and rapidly transformed into spreadsheets on Lotus 123 as the original double entry system during the early era of PC. During this era, Turbo Cash was launched around 1987 with an automated system for trial balance, balance sheet, it took the 15-day process of consolidating ledgers. It was the first software for consolidating trial balance and this software was limited with some important features like calculating tax or invoicing for the organization.

6.2 Robot or Artificial Fortified Machine

Oberson [72], defined robots as the physical emanation of software that have the cognitive and reasoning capacities, having the capacity to process information, sense the environment and act on the environment intelligently. Nevejans [68], expanded the definition of robots by legally defining robots as a physical machine alimented by energy, having the capacity to intelligently learn, render decisions, analyze the environment, make decisions to act in the human world. In addition to the legal definition of a robot, Oberson [73], opined that robots should be granted legal status to pay personal income (PIT) in which Bryson et al. [16] opposed granting robot personhood, asserting that it will create legal loopholes that robot and robotic manufacturers will capitalize to avoid accountability and direct liability,

Beside the definition, Isaac Asimov who introduced the three laws of the robot which are robot must not cause harm to humans being and must obey the orders from humans excepts where such order flaunts the first order of causing harm to human beings and also a robot must protect its existence [10]. The three laws of robots have been critiqued by prior studies like Mark [58], who stated that Asimov's ethical guidelines listed have been outdated through constant technological innovations of simple robots like a vacuum cleaner and complex robots like the military robots for spying and killing their enemies, bomb detectors, carrying wounded soldiers, in which was supported by Murphy and Woods [64], stating that the present close interaction robots have with humans, firstly when robots violates the first law, robots won't be held accountable or liable but the manufacturers and so, therefore, manufacturers make sure roots undergo series of test to minimize error, secondly robots should have the capability of responsiveness to different societal roles and cues interacting with the environment.

In addition to the legal definition of a robot, Oberson [73], opined that robots should be granted legal status to pay personal income (PIT) in which Bryson et al. [16] opposed granting robot personhood, asserting that it will create legal loopholes that robot and robotic manufacturers will capitalize to avoid accountability and direct liability, Pagallo [76], also, opposed by stating that it is morally needless granting robot personhood and legally difficult defining robots and opined that regulatory policymakers should grant robot legal agent hood which hoods robots and robotics personally responsible for accountability and robotic makers liable for its harmful behaviour. Conversely, Chopra and White [21], and Wurah [98], supported Oberson [73], by stating that robot should be granted personhood in order not to be treated like slaves in the industrial sector.

6.3 Robot Taxation

A robot tax is an increase in companies' income tax rates on companies that fully automate their manufacturing processes [1, 59]. Apart from increasing CIT of companies that employ a robotic workforce, Florian and Victoria [35], stated that robot tax can come in form of reducing the degree of expensing that the company can do for, particularly sophisticated machines. [62]. A robot can be tax through an increase in VAT on robots, impose a charge on expenditure on robots [38].

6.4 *Firm's Performance*

Lebans and Euske [53] and Alareeni [6] defined firm performance as a dynamic combination of financial and non-financial pointers that provides knowledge on the level of achievement of the set objectives and outcome while Ben [14] defines firm's performance as a bag-word due to the reason that it can be replaced with these words such as the growth of the organization, profitability of the organization, firm's productivity, competitiveness and competence. Similarly, Taouab and Issor [89] defines a firm's performance as bringing products of higher superiority to present and potential customers and also have large and long-term cost-effectiveness.

7 Relevant Theory

7.1 *Social Contract Theory*

Laskar [52], opined that social contract theory was associated to Thomas Hobbes in 1651 during the civil war in Britain, the theorist believed that the state of nature man lived in was based on selfishness, fear and ruthless until followed human beings voluntarily surrendered their right to some authority to gain the protection of property and lives. The theory was expanded by John Locke in 1776 which viewed man's state of nature as a peaceful and selfless. The social contract theory was further expanded by Jean-Jacques in 1762 which the theorist explained that individual has agreed either clearly or silently to surrender some of their freedoms and submit to the authority [52]. Social contract theory according to Harrison [43], is a relation among the legal and natural right, a scrutiny of the human condition excluding any political order. Conversely to this definition, El-Haddad [33], stated that social contract theory is an agreement between the citizens and the government by that band's individuals live calmly with each other in the society based on the agreed law that establishes moral and political rules of behavior or by the individual own choice because a divine being requires such co-existence. The scholar also contributed to the expansion of the theory by opening the populist-authoritarian social contract for development of the state.

7.2 *Robotic Involvement Among Companies in Nigeria*

Empirically, the study of Nomaler and Verspagen [70] studied the effect of robotization, machine learning and artificial intelligence in substituting human workforce through a Solow growth model to capture the decreasing share of wage income and also predicts the rising wages in a consistent growth path. The study used cost of the robot, the output of the robot, segment of human work that can potentially be robotized and the investment in robots in the model to determine the rate at which robots substitutes human workforce. The study discovered that perpetual increase in robots contributes to the socio-economic disparity between the capital owners and the human workers in which perpetual growth in robotization causes redundancy of human workforce through substitution through more efficient output, thus, making capital owners richer. The study opined that instead of taxing robots and transferring the proceeds to people depending only on salaries which hurt investment in robotization, the study recommended social protection policy because the income of the human workers remains small to even generate much savings.

Similarly, to the study of Nomaler et al. [70] who opposed generating income from robots, the study of Uwe [94], studied the optimal taxation of robots and labour income, the study employed quantitative analysis of data from US Current Population Survey (CPS), Merged Outgoing gropes (MORG) made by National Bureau of Economic Research (NBER) from 1979 to 2016 but the study concentrates on 1993 of individuals from 16 to 64 and also excluded labour hours of 35 h and estimated by Maximum probability. The study substituted robots with routine work the scholar used intensive margin labour supply endogenous wages and occupational choice model to show that it is optimal to disrupt robots substituting routine workers in an organization. The scholar recommended taxes on robots which compresses wages, raises welfare and relaxes incentive constraints because the use of robot polarizes wage distribution.

On the other opinion [3, 4], and also Author and Dorn [12], stated that the higher the robot tax the more there will be a wide wage gap between the routine and non-routine cognitive workers and if the government decides to collects lower or no tax on robots the more the routine or manual workers will lose their jobs to automation but this was empirically opposed by Uwe [94], that taxing robots is inelastic, looking at the US robot tax rate of 4% in the short run, the country having fixed jobs but now the tax rate on a robot is 0.4% and at the medium run, the rate of fixed jobs approaches zero because the robots are getting cheaper. Recently, the study of Guerreiro et al. [41], which estimated the decline in the price of robots o be 0.0083 in the US, empirically opposed the study of Uwe [94] and discovered that the drastic reduction in the price of robots will result in huge economic disparity even though routine workers may not lose their jobs, but there will be a drastic fall in the wages they collect to compete with a robot, so that the employee will see routine workers be cheaper than the cost of purchasing, installing or employing robots or robotic technologies in the manufacturing processes. The study employed recommended a Mirrlessian optimal income tax approach where the transfer of lump sum independent of income but a form of universal basic income.

8 Robotic Involvement on Productivity of Companies in Nigeria

The research work of Acemoglu and Restrepo [4] investigated the effect of the perpetual growth in industrial robots' employment in the USA on the country's human workforce market from 1990 to 2007. The study employed the Autarky equilibrium model to capture the elasticity of substitution of labor by robot productivity. The study also used a total number of robots, total number of labors, price of the robot and wage rate in the commuting zone as indicators for the supply of robot and labor. The study discovered that changes in the number of robots and also the price of robots lowers the cost of production and increases output. The study also discovered that changes in the number of robots and their price (robots getting cheaper) strongly displaces human workforce and decline wages, conversely the price of robots and the perpetual growth in robot employment could have a positive impact on wages and employment of non-routine laborer because of the enhanced productivity effect.

Prior research works have investigated the effect of robotization on the economic growth of OECD countries, studies like Adachi et al. [5], investigated the impact of employment of industrial robot, employing shift-share instrumental variable approach and regional labor market approach to explore the variation in the influence of robotic infiltration by regions due to the previous heterogeneity in manufacturing composition to discover that robot infiltration or penetration increases employment in the service industries but reduces employment in the manufacturing industry. The scholar also argued that even though robotization reduces local labor demand, it increases productivity and because of the effect of an increase in productivity results into an upsurge in the employment of the human workforce. Compagnucci, Gentili et al. [25], supported Adachi et al. [5], by examining the impact of robotization on 16 OECD countries from 2011–2016. The study collected sectorial human labor engagement and wages data from IFR and STAN database and employed panel VAR analysis to discover that 1% growth in robotization in the industrial companies of the selected OECD countries reduces the human workforce periods by 0.16 in addition to the selling prices and the real values of the reimbursements of the human workforce. The scholars argued that manufacturers indulge heavily in robotization as a result of expanding human workforce remunerations and relative selling prices and concluded that robotization disrupts not only the human workforce but compensation growth of human workers.

In the US, the study of Leigh et al. [54], the study empirically opposed the studies of Adachi et al. [5], and Compagnucci et al. [25], by investigating on the impact of robotization on skill demand in manufacturing region. Data on robot stock were extracted from IFR from 2010 to 2016 post-recession period. The study employed the Graetz and Michaels approach to measuring robot density which uses millions of hours worked as the human labor input while the A&R analytical approach to measuring robot exposure by creating a ratio of the robot to the human workforce in each manufacturing sector, and also employed the Initial condition (IC) framework and the different model to estimate the influence of skill demand of non-routine workers on regional human workforce market. The study discovered that robotization have created more jobs in the United States' urban level.

Prior studies of Csillag et al. [26], employed a case study of countries with a relatively long tradition of preventing unemployment and working with employed job seekers and also employed interview to collect data from PES experts to investigate on the vulnerability of robotization among the European countries and how technological unemployment have been curtained through public employment services (PES) measures of employing technological advancement of channeling skill needs by effectively matching both the demand and supply of jobs, quick registration of unemployment for analytical and preventive purposes, counselling and training purposes. The study discovered that jobs that are paid higher remunerations are at a lesser risk of automation because of the increase in robotic skill demand. The study also discovered an increase in non-routine workers after the introduction of robotization and that liberalized labor market makes it easier to sack human labor.

The study of Choudhury et al. [22], investigated the potential influence of disruptive technologies on human capital productivity by employing US patent and investigating the trademark office processes and discovered that one of the results showed speed and exactness in patent judgment irrespective of the skill of the non-routine workers needed for recognizing prior art that overturns the patent being inspected and output in the robotic technologies needs engineering and computer science. Pauline et al. [78], investigated changes in job security and their cause in France from 1982 to 2002 and discovered a rise in the loss of jobs of industrial companies, significantly have a large share of R&D worker' technological innovations. Also, Sabine [83] researched on why assembly work is more than routine work, the study employed 62 qualitative interviews in the assembly plants and opined that the more companies automate their business processes, the more the increase in highly-skilled, well-trained workers.

However, the study of Cirillo et al. [23], opposed the study of Paul et al. [77], by investigating the impact of innovation and wage inequalities within both small and large firms in European countries, comparing the effect technological advancement across gamut between equal and more inegalitarian companies through quantile regression which was complimented with standard OLS estimate to records for the relationship of innovation, size of the company and other influences with internal companies' wage distribution to examine whether technological creative and not technological creative industries shows variations in wage distribution and also whether the size is a moderative construct between technologically advanced companies and not technological creative companies. Data were obtained from the Structure of Earnings Survey (SES) to measure wage distribution. The data was obtained based on the criteria that the business unit that has 10 full-time employees from each country. The scholars, therefore, discovered that the influence of technological advancement does diverge to the diverse measures of wage disparity and also across the size of the company.

9 Robotic Intervention in the Level of Employment Among Companies in Nigeria

The study of Brussels [15] studied the effect of manufacturing robots on EU employment and wages employing the local labour market equilibrium approach to determine which of the domineering effect on the labour market. The scholar discovered that the quality of robot increases while the price of robots drastically reduces which results in robots displaces human workforce especially routine and middle educated workers. Moreover, the study discovered a negative influence of robot exposure on wages, implying that the increase in robot density because robots are becoming more attractive to employers than human workforce, negatively leads to the fall wages of laborers. On the contrary to the study of Brussels [15], the study of Dauth et al. [28] investigated the impact of robotic exposure of 25 industrial robots on the labor, the market of 50 countries throughout 1994 to 2014. The data set used for the study was derived from the International Federation of Robotics and discovered firstly that automobile industries in Germany were exposed to additional 60 to 100 robots while robot exposure is lower in East Germany and West Germany from 0 to 78.1 extra robots exposure per thousand of human laborers in the manufacturing industry.

Concerning the effect of robotic invention on laborer's, the scholars discovered positive effects of robotic exposure in the local employment growth in the US because it is an automobile industry bit, on the contrary, the study also discovered a strong negative influence on employment growth in Germany's manufacturing sectors, disrupting the composition of employment in Germany of about 23% decline in manufacturing jobs. However, the scholars opined that robotic exposure in Germany manufacturing sector are not the major destroyers so far when it comes to the overall number of jobs in the country.

Concerning the aspect of the impact of robotic exposure on wages, Dauth et al. [28] discovered that robotic intervention contributes to the wage gains of highly educated human workforce especially highly skilled workers from the management and scientific positions because their skill complements to the technology and the task they perform which makes them not easily discarded. While robotic exposure has a negative influence on the wages of medium-skilled workers in Germany. However, the study of Michaels and Graetz [61] studied the effect of robotic exposure on productivity and employment of 17 EU countries spanning 14 industries majorly manufacturing sector. The dataset used for the study was obtained from economic performance indicators from EUKLEMS from 1993 to 2007. The study employed Ordinary Least Square to discovered that industrial robot fosters the productivity of the human workforce and also earnings of workers, the study discovered even though, robotic intervention increases total factor productivity and wages of human workforce, however, robotic intervention disrupts employment of middle educated workers and routine workers. Similarly, the study of Cockshott and Renaud [24] discovered that robotic technology does not only displaces human workforce but uses Shannon's model to also discover that robotic intervention crushes hopes and aspiration of humans.

10 Reduction of Capital Allowance and Performance of Companies in Nigeria

The study of Tennant and Tracey [90], employed 1105 large companies in which their gross sales are above J$ 500 m during 2010 to 2011 to examine the impact of a positive shock to LTU enforcement and also employing regression discontinuity (RD) design to test whether results respond to such a discontinuous jump at a cutoff of LTU. The stud recommended that stricter policy to increase the effective tax rate of larger firms and further suggested that should that LTU strength in combating tax avoidance should be tested in another tax environment. Reduction of effective tax rates of large companies have been traced to government support in business expenditure on research and development (BERD) and also provide capital allowance to improve in the total sales of highly demanded goods and services with the lowest cost conceivable but support in BERD and granting capital preferences through various incentives have also been traced in contributing to economic disparity increases unemployment Chen et al. [18], Postel-Vinay [81], Munongo et al. [63], and reducing R&D incentives which increases large companies' ETR is root tax [1].

Aside from Tennant et al. [90], the study Kou and Sun [50], studied how organizational leadership change or prefecture leaders' turnovers use that opportunity to evade or avoid taxes and discovered that during the year of company's prefecture leader's turnover, large companies evade more tax payments and also private companies. The study was carried out in China from the period of 2005 to 2012 on 300 prefectures in the country which is collected by the country's Ministry of Finance and State administration in charge of taxation which contains companies-level tax and financial data. Furthermore, the study also employed the endogeneity test by using a Heckman two-step model to solve for endogeneity bias and recommended fostering of policies that will generate more tax revenue and collection. On the other hand, the study of Liu and Mao [56], the study investigated the effect of tax incentives or capital preferences given to capital on productivity and investment instead of taxing capital. The study's scope of the study span from 2005 to 2012 used firm-level dataset evidence from China. Also, the study employed a quasi-experimental design to examine the effect of the reform on companies' investment and productivity and therefore, asserted that tax incentives and capital preferences are given to businesses promote productivity and investment by 8.9 and 38.4%. Base on the second results of the research, the study also asserted a positive significant impact tend to be reinforced for companies with financial difficulties than taxing capital of such companies.

Research and development are vital towards the growth of total sales of businesses, in which study of Alvarez and Argothy [9], investigated the impact of investing in R&D and the sales growth of 86 state-owned enterprises in Ecuador from the period of two years (2012 to 2014) and data was derived from the survey of activities of science and technology (ACTI), employing the endogenous growth model and OLS to empirically discovered that there is a positive impact of investing in R&D and the total sales growth in state-owned enterprises in Ecuador Similar to the study of Kovermann et al. [51], the study of Thomsen and Watrin [91] examined whether changes can be observed over time between the tax avoidance behaviours of 34,209 U.S. companies and those of 25,717 companies from 12 European countries from 2005 to 2016. The study employed panel analysis and correlation matrix to discovered a reduction of ETR in nearly all the OECD countries despite their widely differing STR, this implies that companies in nearly all the OECD countries enjoy capital preferences which boost total sales of those companies, thus contributing to tax avoidance of companies.

10.1 Gaps in Literature

Nomaler et al. [70] employed trend analysis to discover that the robot contributes to unemployment. Based on the trend of tax revenue derived from labour and business income, the study justified that it is optimal to tax robot and robotic technologies employed in manufacturing sectors of selected countries like Span, France, Singapore, Germany, Japan, Hong Kong and South Korea This study is different in which we want to examine robot taxation and productivity of listed industrial companies in Nigeria.

The study of Acemoglu and Restrepo [4] investigated the effect of the perpetual growth in industrial robots' employment in the USA on the country's human workforce market from 1990 to 2007. The study employed the Autarky equilibrium model to capture the elasticity of substitution of labour by robot productivity. The study also used the total number of robots, total number of labours, price of the robot and wage rate in the commuting zone as indicators for the supply of robot and labour in which this study adapted several robots as one of the indicators for measuring robotic intervention on the productivity of listed industrial companies in Nigeria.

11 Methodology

This chapter explains the methodology employed in this study. The chapter documents all its activities involving in the collection and analyzing of data needed to achieve the anticipated research objectives. This section will present the following sequence namely research design, description of the population of the study, the sampling frame, sample size and sampling technique, sample size, source of data, data analysis techniques, model specification, A priori expectation.

11.1 Research Design

This study will employ *ex-post facto* design. This is because according to Sharma [85] *ex-post facto* research design is employed to refer to studies which investigate possible cause and effect relationship by observing an existing condition or state of affairs, observing back in time for plausible causal factors.

11.2 Population

The population of this study will consist of all companies in Nigeria. The President of National Union of Food, Beverages and Tobacco Employees (NUFBTE), Lateef Oyelekan said that robot and robotic technologies are increasingly disrupting jobs and aggravating unemployment rate, raising the alarm that the country is not mature for robot or an automated machine to be employed in local industries instead of the human workforce.

11.3 Sources of Data

Triangulation method will be employed for this study which comprises Primary data through questionnaire and secondary data will be employed from the financial statements of companies in Nigeria for the period of 2008 to 2018. Data from secondary sources will be appropriate for this study because it captured past development which provide the needed background to establish the degree of validity and reliability hence, they are prepared by professionals and validated by relevant regulatory agencies decisions.

11.4 Reliability of the Research Instrument

The reliability index for this study was 0.957 for the overall scale, this shows the consistency of measure of concepts for this study.

11.5 Econometric Model

$$PBTit = \alpha 0 + \alpha 1 NCRit + \varepsilon it \tag{5}$$

$$PBTit = \alpha 0 + \alpha 1 NCRit - RTit, CRit - RTit, ORit - RTit + \varepsilon it \tag{6}$$

$$PBTit = \alpha 0 + \alpha 1 NCRit + RTit, CRit + RTit, ORit + RTit + \varepsilon it \tag{7}$$

$$ELit = \alpha 0 + \alpha 1 NCRit + \varepsilon it \tag{8}$$

$$PBTit = \alpha 0 + \alpha 1 RCAit - RTit, CRit - RTit, ORit - RTit + \varepsilon it \quad (9)$$

$$PBTit = \alpha 0 + \alpha 1 RCAit - RTit, CRit - RTit, ORit - RTit + \varepsilon it \quad (10)$$

11.6 Apriori Expectation

This study expects a positive effect of robotic involvement and performance of companies in Nigeria. This study expects the positive influence of robotic involvement on the level of employment of skilled or non-routine workers. Lastly, the study also expects a positive effect of reduction in capital allowance of large companies. However, companies may experience a reduction in profitability due to the pandemic outbreak of coronavirus and curfew to flatten the spread of the virus.

12 Data Analysis and Discussion of Findings

In this section, the data of the study was presented, analyzed and results were interpreted and discussed. To begin with, the descriptive statistics of variables in the model was presented and discussed. Data Analysis and interpretation of results will follow.

13 Response Rate

From the data derived, out of 200 questionnaires were sent through email messages, just 159 responses were received. This showed 75% response rate. This rate of response is considered very good to make deduction. The response rate indexes are shown in the Table 1.

Table 1 Questionnaire response rate

Categories	Frequency	Percentage (%)
Copies of questionnaire administered	209	100
Copies of questionnaire filled and received via email	150	75
Copies of questionnaire that were not filled	50	25

13.1 Descriptive Statistics of Variables

See Tables 2 and 3.

Table 2 Summary of variables

Variable	Obs	Mean	Std. Dev.	Min	Max
gender	156	1.371795	0.4848405	1	2
age	156	3.903846	1.29405	1	5
EDULEVEL	156	4.288462	1.135978	1	6
mgtlevel	156	2.211538	0.6721138	1	3
COYSECTOR	155	3.851613	1.598675	1	6
HYPOla	156	3.102564	1.250442	1	5
HYPO1b	156	2.705128	1.281219	1	5
HYPO1c	156	2.660256	1.204812	1	5
HYPO1d	156	4.301282	0.9327024	1	5
HYPO1e	156	4.108974	0.8767515	1	5
HYPO1f	156	4.044872	0.9527072	1	5
HYPO2a	156	3.730769	0.999007	1	5
HYPO2b	156	3.589744	1.20146	1	5
HYPO2c	156	3.237179	1.260375	1	5
HYPO2d	156	3.884615	1.002973	1	5
HYPO3a	156	3.378205	1.235721	1	5
HYPO3b	156	3.923077	1.092724	1	5
HYPO3c	156	2.076923	1.177961	1	5
HYPO3d	156	3.961538	1.168868	1	5
HYPO3e	156	3.955128	0.9859853	1	5
HYPO3f	156	3.923077	1.050582	1	5
HYPO4a	156	3.012821	1.348773	1	5
HYPO4b	156	2.923077	1.379754	1	5
HYPO4c	156	2.320513	1.519708	1	11
HYPO4d	156	3.096154	1.376198	1	5
HYPO4e	156	3.012821	1.405001	1	5
GEN1	154	1.876623	0.6698739	1	3
GEN2	156	1.660256	0.9401221	1	4
FP	155	2.464516	1.482703	1	5
weight srs	240	500	0	500	500
fpc	240	0.002	0	0.002	0.002

Table 3 Correlation matrix of variables

	HYPO1a	HYPO1b	HYPO1c	HYPO1d	HYPO1e	HYPO1f	HYPO2a	HYPO2b	HYPO2c	HYPO2d	HYPO3a	HYPO3b
HYPO1a	1.0000											
HYPO1b	0.1005	1.0000										
HYPO1c	−0.0434	0.4471	1.0000									
HYPO1d	0.0756	−0.0047	0.0365	1.0000								
HYPO1e	0.0929	−0.0603	0.0123	0.3384	1.0000							
HYPO1f	0.0352	−0.0434	0.0452	0.3885	0.5182	1.0000						
HYPO2a	0.0584	−0.0544	0.0782	−0.0259	0.1583	0.2252	1.0000					
HYPO2b	−0.0459	−0.1688	−0.1057	0.1881	0.3474	0.2501	0.3821	1.0000				
HYPO2c	0.1070	−0.0225	0.0830	0.1318	0.1402	0.3024	0.1998	0.2476	1.0000			
HYPO2d	−0.0326	−0.1752	0.0485	0.3230	0.1247	0.3526	0.2184	0.2738	0.2433	1.0000		
HYPO3a	−0.1549	−0.1474	−0.0364	−0.0217	−0.1056	−0.1229	0.1069	0.0777	0.0233	0.0031	1.0000	
HYPO3b	−0.0857	−0.0574	−0.0181	0.3836	0.2996	0.3268	0.0845	0.1358	0.0589	0.2240	0.1400	1.0000
HYPO3c	0.0303	0.0547	0.0275	−0.0627	−0.1338	−0.1302	−0.0435	−0.0638	−0.0777	−0.1082	0.1103	0.0245
HYPO3d	0.1341	0.0714	0.1603	0.1699	0.0019	−0.0470	0.1278	0.0989	0.0211	0.0492	0.1851	0.0976
HYPO3e	0.1103	−0.0384	−0.0583	0.0946	0.2095	0.2304	0.1926	0.1358	0.0671	0.0004	−0.1128	0.0146
HYPO3f	0.0204	−0.0067	0.0532	−0.1111	−0.1346	−0.0363	0.1136	0.0166	0.2003	0.0106	0.0708	−0.0159
HYPO4a	−0.0526	0.1798	0.0910	0.1479	0.0527	0.0193	0.1194	0.0497	0.0150	0.0610	0.0439	0.2009
HYPO4b	−0.0993	0.0959	0.0144	0.0203	0.1083	0.0467	−0.0561	0.0463	−0.0294	0.0039	−0.1590	0.1355
HYPO4c	0.1111	0.0650	0.0558	−0.1060	−0.1422	−0.0945	−0.1071	−0.1269	0.0852	−0.0435	0.1106	−0.0134
HYPO4d	0.2072	0.0342	−0.0599	0.0292	−0.0344	−0.1065	−0.0346	0.0322	−0.0868	0.0831	0.0722	0.0517
HYPO4e	0.1271	0.0350	−0.0129	−0.0530	−0.0331	−0.0935	0.1444	0.0938	0.0728	−0.1297	−0.0331	−0.1616

(continued)

Table 3 (continued)

	HYPOla	HYPOlb	HYPOlc	HYPOld	HYPOle	HYPOlf	HYPO2a	HYPO2b	HYPO2c	HYPO2d	HYPO3a	HYPO3b
GENl	0.0139	0.0189	0.0243	−0.1633	0.0757	−0.1878	0.0017	−0.0648	−0.0468	−0.1381	−0.0906	−0.2530
GEN2	−0.0784	−0.1061	−0.0097	0.0071	−0.0146	−0.0660	−0.0729	−0.1554	−0.0574	−0.0299	0.0595	0.0592
FP	0.0812	0.0828	0.0802	0.0770	−0.0607	−0.0407	−0.0299	−0.0017	−0.0841	−0.0243	−0.0120	0.0690
	HYPO3c	HYPO3d	HYPO3e	HYPO3f	HYPO4a	HYPO4b	HYPO4c	HYPO4d	HYPO4e	GENl	GEN2	FP
HYPO3c	1.0000											
HYPO3d	0.1316	1.0000										
HYPO3e	−0.1197	−0.1105	1.0000									
HYPO3f	0.0098	0.2436	0.0216	1.0000								
HYPO4a	0.1343	0.1573	0.0701	0.0943	1.0000							
HYPO4b	0.0676	0.0033	0.0936	0.0783	0.5241	1.0000						
HYPO4c	0.1208	0.0940	−0.1979	0.0189	0.1839	0.0322	1.0000					
HYPO4d	0.0917	0.0217	0.0561	0.1217	0.1073	0.0650	0.0325	1.0000				
HYPO4e	−0.0562	0.0655	0.0668	0.1882	0.2017	0.1707	0.2142	0.1484	1.0000			
GENl	−0.1462	0.0055	−0.1483	−0.0028	−0.0832	−0.0512	0.0251	−0.0947	0.0790	1.0000		
GEN2	0.1268	−0.0434	0.0517	−0.1471	−0.0645	−0.1361	0.0015	−0.0519	−0.0877	−0.1512	1.0000	
FP	−0.0174	0.1055	−0.0323	−0.0203	0.0120	0.0215	0.1032	0.0587	0.0255	−0.0101	0.0713	1.0000

13.1.1 Discussions

Table 1 above showed mean, standard deviation, maximum and minimum statistics for the variables, the number of observations is 156. Table 2 showed the correlation matrix between the variables of interest to test for bivariate analysis. In this type of analysis, the main interest was to observe the direction in which the variables move with one another. From this analysis, two main separate ends were expected: to test the presence of multicollinearity among the variables and to predict the possible effect of the variables on one another, using the direction and coefficient of association. The presence of multicollinearity involved the existence of a perfect or exact linear relationship among some or all the explanatory variables of a regression model [42]. From Table 2, there were cases of multicollinearity, as the extent of association between some variables varied above 0.5 (+ or −). The graphical illustration depicted the level of the responses per variable.

14 Inferential Analysis

14.1 Hypothesis One

There is no robotic intervention (Number of robots, Cost of robots and Output or productivity of robot) in companies in Nigeria (Tables 4 and 5).

<u>Interpretation</u>

The overall model was significant, showing a significant relationship with firm performance. Also, R square gives 0.734, which meant that 73% of the change in the firm value could be attributed to artificial intelligence solutions.

Table 4 Analysis of Variance (ANOVA)

	Number of obs = 155 Root MSE = 1.44771			R-squared = 0.7855 Adj R-squared = 0.7462	
Source	Partial SS	df	MS	F	Prob > F
Model	1005.44132	24	41.8933885	19.99	0.0000
HYPO1a	2.76381762	4	0.690954405	0.33	0.8576
HYPO1b	37.0796813	4	9.26992031	4.42	0.0022
HYPO1c	16.8637668	4	4.21594171	2.01	0.0965
HYPO1d	31.0560627	4	7.76401568	3.70	0.0068
HYPO1e	6.6246473	4	1.65616183	0.79	0.5335
HYPO1f	8.54464702	4	2.13616176	1.02	0.3999
Residual	274.558676	131	2.09586776		
Total	1280	155	8.25806452		

Table 5 Regression analysis

Equation	Obs	Parms	RMSE	"R-sq"	F	P
FP	155	6	1.512485	0.7337	68.42267	0.0000
FP	Coef.	Std. Err.	t	P > |t|	[95% Conf. Interval]	
HYPO1a	0.1727426	0.0948001	1.82	0.070	−0.0145836	0.3600688
HYPO1b	0.0953501	0.1055479	0.90	0.368	−0.1132139	0.3039141
HYPO1c	0.1285096	0.1118493	1.15	0.252	−0.0925062	0.3495253
HYPO1d	0.3089149	0.1330081	2.32	0.022	0.046089	0.5717407
HYPO1e	−0.0067833	0.1559191	−0.04	0.965	−0.3148815	0.3013149
HYPO1f	−0.003918	0.1522072	−0.03	0.979	−0.3046814	0.2968455

14.2 Hypothesis Two

There is no significant relationship between robotic involvement (Number of robots, Cost of robots and Output or productivity of robot) and performance of companies in Nigeria (Tables 6 and 7).

Table 6 Analysis of Variance (ANOVA)

	Number of obs = 155 Root MSE = 1.5494			R-squared = 0.7393 Adj R-squared = 0.7093	
Source	Partial SS	df	MS	F	Prob > F
Model	946.311766	16	59.1444854	24.64	0.0000
HYPO2a	24.1370686	4	6.03426714	2.51	0.0444
HYPO2b	10.0340758	4	2.50851894	1.04	0.3864
HYPO2c	8.26815819	4	2.06703955	0.86	0.4891
HYPO2d	27.2749167	4	6.81872919	2.84	0.0266
Residual	333.688234	139	2.40063477		
Total	1280	155	8.25806452		

Table 7 Regression Analysis

Equation	Obs	Parms	RMSE	"R-sq"	F	P
FP	155	4	1.588969	0.7021	88.99158	0.0000
FP	Coef.	Std. Err.	t	P > |t|	[95% Conf. Interval]	
HYPO2a	0.2422784	0.1267125	1.91	0.058	−0.0080799	0.4926367
HYPO2b	0.1127848	0.1183091	0.95	0.342	−0.1209702	0.3465399
HYPO2c	−0.0067968	0.1048984	−0.06	0.948	−0.2140549	0.2004614
HYPO2d	0.2767254	0.1164578	2.38	0.019	0.0466282	0.5068225

Interpretation

The model was significant. The coefficient of the independent variables was mostly positive showing a positive relationship with firm performance. Also, R square gives 0.70, which means that 70% of the change in the firm value can be attributed to the productivity of artificial intelligence solutions.

14.3 Hypothesis Three

There is no significant relationship between robotic involvement (Number of robots, Cost of robots and Output or productivity of robot) and the level of employment in companies in Nigeria (Tables 8 and 9).

Table 8 Analysis of Variance (ANOVA)

	Number of obs = 155 Root MSE = 1.50185			R-squared = 0.7269 Adj R-squared = 0.7269	
Source	Partial SS	df	MS	F	Prob > F
Model	984.523095	24	41.0217956	18.19	0.0000
HYPO3a	5.58203341	4	1.39550835	0.62	0.6500
HYPO3b	18.6758683	4	4.66896707	2.07	0.0884
HYPO3c	4.0097464	4	1.0024366	0.44	0.7763
HYPO3d	12.7146971	4	3.17867427	1.41	0.2344
HYPO3e	4.72635779	4	1.18158945	0.52	0.7184
HYPO3f	18.8221505	4	4.70553763	2.09	0.0862
Residual	295.476905	131	2.25554889		
Total	1280	155	8.25806452		

Table 9 Regression Analysis

Equation	Obs	Parms	RMSE	"R-sq"	F	P
FP	155	6	1.51594	0.7325	67.99802	0.0000
FP	Coef.	Std. Err.	t	P > |t|	[95% Conf. Interval]	
HYPO3a	0.0137046	0.0985921	0.14	0.890	−0.1811148	0.2085239
HYPO3b	0.1882051	0.104284	1.80	0.073	−0.0178615	0.3942717
HYPO3c	0.011972	0.1031631	0.12	0.908	−0.1918796	0.2158236
HYPO3d	0.238829	0.1070592	2.23	0.027	0.0272785	0.4503794
HYPO3e	0.130695	0.1013659	1.29	0.199	−0.0696054	0.3309954
HYPO3f	0.0396298	0.1112443	0.36	0.722	−0.1801904	0.2594499

Interpretation

The model was significant. The coefficient of the independent variables is all positive showing a positive relationship with employment. Also, R square gives 0.732, which means that 73% of the change in the level of employment can be attributed to artificial intelligence solutions involvement.

14.4 Hypothesis Four

There is no significant relationship between reduction in capital allowance and the performance of companies in Nigeria (Tables 10 and 11).

Interpretation

The model is significant. The coefficient of the independent variables is all positive showing a positive relationship with firm performance. Also, R square gives

Table 10 Analysis of Variance (ANOVA)

	Number of obs = 155 Root MSE = 1.66358			R-squared = 0.7103 Adj R-squared = 0.6649	
Source	Partial SS	df	MS	F	Prob > F
Model	909.154667	21	43.2930794	15.64	0.0000
HYPO4a	8.68313684	4	2.17078421	0.78	0.5372
HYPO4b	8.74424102	4	2.18606026	0.79	0.5337
HYPO4c	22.3379531	5	4.46759062	1.61	0.1604
HYPO4d	40.7684121	4	10.192103	3.68	0.0070
HYPO4e	24.3843018	4	6.09607544	2.20	0.0720
Residual	370.845333	134	2.76750248		
Total	1280	155	8.25806452		

Table 11 Regression Analysis

Equation	Obs	Parms	RMSE	"R-sq"	F	P
FP	155	5	1.590179	0.7037	71.23907	0.0000
FP	Coef.	Std. Err.	t	P > |t|	[95% Conf. Interval]	
HYPO4a	0.0759115	0.1123788	0.68	0.500	−0.1461384	0.2979614
HYPO4b	0.1694085	0.1058004	1.60	0.111	−0.039643	0.3784601
HYPO4c	0.1995868	0.084992	2.35	0.020	0.0316505	0.367523
HYPO4d	0.2509932	0.0824792	3.04	0.003	0.088022	0.4139644
HYPO4e	0.1188737	0.0903537	1.32	0.190	−0.0596567	0.297404

Table 12 Analysis of Variance (ANOVA)

	Number of obs = 153 Root MSE = 1.97676			R-squared = 0.5354 Adj R-squared = 0.5292	
Source	Partial SS	df	MS	F	Prob > F
Model	679.955514	2	339.977757	87.00	0.0000
GEN1	679.955514	2	339.977757	87.00	0.0000
Residual	590.044486	151	3.90757938		
Total	1270	153	8.30065359		

Table 13 Regression Analysis

Equation	Obs	Parms	RMSE	"R-sq"	F	P
FP	153	1	1.708351	0.6507	283.1604	0.0000
FP	Coef.	Std. Err.	t	P > |t|	[95% Conf. Interval]	
GEN1	1.163934	0.0691691	16.83	0.000	1.027277	1.300591

0.70, which means that 70% of the performance of companies can be attributed to a reduction in capital allowance.

14.5 General Question 1

In your opinion, which of the following enhances the chances of implementing robot taxation in Nigeria (Tables 12 and 13)?

Interpretation

The model is significant. The coefficient of the independent variables is all positive showing a positive relationship with firm performance. Also, R square gives 0.65, which means that 65% of the performance of companies can be attributed to implementing artificial intelligence solutions.

14.6 General Question 2

Given the present argument for and against taxing of robots, what is your attitude towards robot tax (Tables 14 and 15)?

Interpretation

The model is significant. The coefficient of the independent variables is all positive showing a positive relationship with firm performance. Also, R square gives

Table 14 Analysis of Variance (ANOVA)

	Number of obs = 155 Root MSE = = 2.32574			R-squared = 0.3577 Adj R-squared = 0.3450	
Source	Partial SS	df	MS	F	Prob > F
Model	457.824638	3	152.608213	28.21	0.0000
GEN2	457.824638	3	152.608213	28.21	0.0000
Residual	822.175362	152	5.40904844		
Total	1280	155	8.25806452		

Table 15 Regression Analysis

Equation	Obs	Parms	RMSE	"R-sq"	F	P
FP	155	1	1.873271	0.5778	210.7614	0.0000
FP	Coef.	Std. Err.	t	P > |t|	[95% Conf. Interval]	
GEN2	1.14311	0.0787394	14.52	0.000	0.9875607	1.298658

0.578, which means that 58% of the performance of companies can be attributed to implementing artificial intelligence solutions.

15 Discussion of Findings

From Table 4, the result shows the size of the coefficients that implies that a unit increase in hypo1e, hypo1f. hypo1a, hypo1b, hypo1c and hpo1d, will cause a 0.0067833 decrease, 0.003918 decreases, 0.1727426 increase, 0.0953501 increase, 0.1285096 increase, 0.3089149 increase companies performance, The overall model is significant, showing P-value of the F-statistics of 0.0000. the findings likewise show that an increase in robotic involvement may lead to an increase in companies' performance. This finding supports the study of Acemoglu and Restrepo [4] investigated the effect of the perpetual growth in industrial robots' employment in the USA on the country's human workforce market from 1990 to 2007. The study discovered that changes in the number of robots and also the price of robots lowers the cost of production and increases output. The results of this study were consistent with apriori expectations.

From Table 6, The model is a significant P-value of the F-statistics of 0.000. this finding shows that pre-post robotic involvement has a significant effect on the performances of companies in Nigeria. This finding supports the study of Leigh et al. [54], who investigated the impact of robotization on skill demand in the manufacturing region. Data on robot stock were extracted from IFR from 2010 to 2016 post-recession period. The study employed the Graetz and Michaels approach to measuring robot density which uses millions of hours worked as the human labor input while the A&R analytical approach to measuring robot exposure by creating a ratio of the robot to the human workforce in each manufacturing sector, and also employed the Initial condition (IC) framework and the different model to estimate the influence of skill demand of non-routine workers on regional human workforce market. The study discovered that robotization foster's companies' profitability thus, creating more jobs in the United States' urban level through companies' expansion. The size of the coefficients implies that a unit increase in hypo2c. hypo2a, hypo2b and hypo2d may lead to 0.0067968 decrease, 0.2422784 increase, 0.1127848 increase, 0.2767254 increase. The findings from this study were consistent with apriori expectations.

From Table 8, The model has a significant P-value of the F-statistics of 0.000. the findings show that an increase in robotic involvement may lead to an increase in the employment level of the skilled human workforce in Nigeria. These findings are consistent to Acemoglu and Restrepo [4] who discovered that changes in the number of robots and their price (robots getting cheaper) strongly displaces human workforce and decline wages, conversely, the price of robots and the perpetual growth in robot employment could have a positive impact on wages and employment of non-routine laborer because of the enhanced productivity effect. Further, the study of Brussels [15] studied the effect of manufacturing robots on EU employment and wages employing the local labor market equilibrium approach to determine which of the domineering effect on labor market supports these findings. The scholar discovered that the quality of robot increases while the price of robots drastically reduces which results in robots displacing human workforce especially routine and middle educated workers. Moreover, the study discovered a negative influence of robot exposure on wages, implying that the increase in robot density because robots are becoming more attractive to employers than human workforce, negatively leads to the fall wages of laborers.

Concerning the aspect of the impact of robotic exposure on wages, Dauth et al. [28] discovered that robotic intervention contributes to the wage gains of highly educated human workforce especially highly skilled workers from the management and scientific positions because their skill complements to the technology and the task they perform which makes them not easily discarded. While robotic exposure has a negative influence on the wages of medium-skilled workers in Germany. However, the study of Michaels and Graetz [61] studied the effect of robotic exposure on productivity and employment of 17 EU countries spanning 14 industries majorly manufacturing sector. The dataset used for the study was obtained from economic performance indicators from EUKLEMS from 1993 to 2007. The study employed Ordinary Least Square to discovered that industrial robot fosters the productivity of the human workforce and also earnings of workers, the study also discovered even though, robotic intervention increases total factor productivity and wages of human workforce, however, robotic intervention disrupts employment of middle educated workers and routine workers. Similarly, the study of Cockshott and Renaud [24] discovered that robotic technology does not only displace human workforce but uses Shannon's model to also discover that robotic intervention crushes hopes and aspiration of humans.

Furthermore, this study also supports Adachi, et al. [5] who investigated the impact of employment of industrial robot, employing shift-share instrumental variable approach and regional labor market approach to explore the variation in the influence of robotic infiltration by regions due to the previous heterogeneity in manufacturing composition to discover that robot infiltration or penetration increases employment in the service industries but reduces employment in the manufacturing industry. The scholar also argued that even though robotization reduces local labor demand, it increases productivity and because of the effect of an increase in productivity results into an upsurge in the employment of the human workforce. The size of the coefficients implies that a unit increase in hypo3a. hypo3b, hypo3c, hypo3d, hypo3e, hypo3f may lead to 0.0137046 increase, 0.1882051 increase, 0.011972 increase, 0.238829 increase, 0.130695, 0.0396290 increase. This finding was consistent with the apriori expectation.

From Table 10, The model is a significant P-value of the F-statistics of 0.000, the findings show that a reduction in capital allowance leads to the performance of companies in Nigeria. This finding is in support of Kou and Sun [50], who studied how organizational leadership change or prefecture leaders' turnovers use that opportunity to evade or avoid taxes and discovered that during the year of company's prefecture leader's turnover, large companies evade more tax payments and also private companies. The study was carried out in China from the period of 2005 to 2012 on 300 prefectures in the country which is collected by the country's Ministry of Finance and State administration in charge of taxation which contains companies-level tax and financial data. Furthermore, the study also employed the endogeneity test by using a Heckman two-step model to solve for endogeneity bias and recommended fostering of policies that will generate more tax revenue and collection. On the other hand, the study of Liu et al. [56], the study investigated the effect of tax incentives or capital preferences given to capital on productivity and investment instead of taxing capital. The study's scope of the study span from 2005 to 2012 used firm-level dataset evidence from China. Also, the study employed a quasi-experimental design to examine the effect of the reform on companies' investment and productivity and therefore, asserted that tax incentives and capital preferences are given to businesses promote productivity and investment by 8.9 and 38.4%. Base on the second results of the research, the study also asserted a positive significant impact tend to be reinforced for companies with financial difficulties than taxing capital of such companies.

This study agrees with the study of Nomaler and Verspagen [70] who studied the effect of robotization, machine learning and artificial intelligence in substituting human workforce through a Solow growth model to capture the decreasing share of wage income and also predicts the rising wages in a consistent growth path. The study discovered that perpetual increase in robots contributes to the socio-economic disparity between the capital owners and the human workers in which perpetual growth in robotization causes redundancy of human workforce through substitution through more efficient output, thus, making capital owners richer. The study opined that instead of taxing robots and transferring the proceeds to people depending only on salaries which hurt investment in robotization, the study recommended social protection policy because the income of the human workers remains small to even generate much savings. The size of the coefficients implies that a unit increase in hypo3a. hypo4a, hypo4b, hypo4c, hypo4d and hypo4e may lead to 0.0759115 increase, 0.1694085 increase, 0.1995868 increase, 0.2509932 increase, 0.1188737 increase. This study is in line with our apriori expectation.

16 Recommendations

1. It is recommended that robot should be tax to provide additional revenue for the government in program that brings back displaced routine workers in Nigeria into the tax net.
2. There is a need for organizations to strive towards artificial intelligence, data analytics and connected technology to foster performance. Findings from the Digital Transformation 2020 survey show that a company's level of digital maturity and by extension technological innovation is one of the most significant indicators of growth and financial success according to a new study from Deloitte. Companies identified by their executives as well along the road to digital transformation were three times more likely to achieve growth in the last year. This implies that connected technologies are one of the critical enablers of business success.
3. Creating awareness for the benefits of technological innovation is recommended. Technological innovation helps to improve business processes while adding value to existing products or services. Success comes from identifying those business processes to focus on and applying resources to exploit them.
4. It is recommended that policy reforms encouraging the implementation of technological innovation should be enacted and implemented. Additionally, including technological innovation as part of the parameters of assessing ease of doing business index which ranks countries according to how they effectively and efficiently create enabling environment for business to thrive is highly encouraged by the authorities involved.
5. Besides, to implement policy reforms, policies on the regulation of technological innovation solutions are critical. Regulation is considered necessary to both encourage Artificial Intelligence (AI) and other associated technologies and manage associated risks but challenging. Regulation of AI through mechanisms such as review boards can be seen as a social means to approach the AI control problem.

17 Contribution to Knowledge

This study was able to make conceptual, theoretical and empirical contributions to the field of performance management, within the Nigerian context. To be explicit, the following are some of the specific areas where this study makes its contributions.

17.1 Theoretical Contribution

This study theoretically contributed to the body of literature in the area of performance management within the Nigerian context. This was based on observing the dynamic relationships between the role of Artificial Intelligence (AI) and its resultant effect on companies' performance. This study helped to explain AI and firm performance, beaming the searchlight on the role of AI associated technologies such as robotic taxation. It supported social contract theory that opined that individual voluntary agreed to silently surrender some of their freedom to an authority to enjoy socio-economic inclusiveness. Policies should be enacted that will promote socio-economic inclusiveness like reducing redundancy of human workforce through robot taxation.

17.2 Empirical Contribution

This study has empirically revealed the importance of Artificial Intelligence (AI) to business performance. This study expanded the understanding of how AI affects the bottom-line of businesses, employment and capital allowance. This is an important dimension that had not been sufficiently explored.

17.3 Conceptual Contribution

This study contributed to the conceptual theory of the AI and firm performance within the Nigerian context. This study was able to expand on the conceptual literature surrounding the determination of a company's performance.

References

1. Abbott, R., Bogenschneider, B.: Should robots pay taxes? Tax policy in the age of automation (2017)
2. Abu-bashal, A.: Nigeria: 113 healthcare workers infected with COVID-19 (2020). https://www.aa.com.tr/en/africa/nigeria-113-healthcare-workers-infected-with-covid-19/1825398. Accesseed 21 May 2020
3. Acemoglu, D., Author, D.: Chapter 12 – skills, tasks and technologies: implications for employment and earning. In: Ashenfelter, O., Card, D. (eds.) Handbook of Labor Economics. $ (B), pp. 1043–1171. Elsevier (2017)
4. Acemoglu, D., Restrepo, P.: Robots and Jobs: evidence from US labor markets. NBER Working paper 23285 (2017)
5. Adachi, D., Kawaguchi, D., Saito, Y.U.: Robot, employment, and population: evidence from articulated robot in Japan's local labor markets. Asian Aust. Soc. Labour Econ. **21**, 1–43 (2019)

6. Alareeni, B.: The impact of firm-specific characteristics on earnings management: evidence from GCC countries. Int. J. Manag. Finan. Acc. **10**(2), 85–104 (2018)
7. Alex, O.H.: What is the difference between robotics and artificial intelligence? (2017). https://blog.robotiq.com/whats-te-difference-between-robotics-and-artificial-intelligence. Accessed 14 Oct 2019
8. Alli, F.: Robots taking over jobs from factory workers. Vanguard (2017, January 12). http://www.vanguardngr.com/2017/robots-taking-jobs-factory-workers/amp/. Accessed 9 Oct 2018
9. Alvarez, N.G., Argothy, A.: Research, development and growth in state-owned enterprise: empirical evidence from Ecuador. Ind. Innov. **26**(2), 158–175 (2018)
10. Asimov, I.: Robots and empire. Harper Collins, London (1996)
11. Augustine, A.C., Maryrose, O.C., Felix, U.C.: Evaluation of the extent of integration of humanoids in tackling social-economic challenges confronting the Africa continent. Int. Robot. Autom. J. **4**(2), 110 (2018)
12. Author, D.H., Dorn, D.: The growth of low skill service jobs and the polarization of the US labor market. Am. Econ. Rev. **103**(5), 1553–1333 (2013)
13. Bekey, G.A.: Autonomous robots: from biological inspiration to implementation and control. Cambridge University Press, Cambridge, MA, USA, London, UK (2005)
14. Ben, S.H.: Determinants of firm performance A comparison of European countries. Int. J. Econ. Fin., 243–249 (2014)
15. Brussels, B.: The impact of industrial robots on EU employment and wages: a local labor market approach. Bruegel Working Pap. **2**, 1–33 (2018)
16. Bryson, J.J., Diamonds, M.E., Grant, T.D.: Of, for and by the people: the legal lacuna of synthetic person. Arttf. Intel. Law **23**, 273–291 (2017)
17. Cai, J., Chen, Y., Wang, X.: The impact of corporate taxes on firm innovation: Evidence from the corporate tax collection reform in China (2018). https://voxeu.org/article/impact-corporate-taxes-firm-innovation. Accessed 7 March 2020
18. Chen, P., Chu, A.C., Chu, H., Lai, C.: Short run and long run effect of capital taxation on innovation and economic growth. J. Macroecon. (2017). https://doi.org/10.1016/j.jmacro.2017.07.002
19. Chidike, O., Ikechukwu, U., Abayomi, O.: Assessment of the information and communication technology (ICT) literacy of librarians in national root crop research institute, library. Int. J. Res. Rev. **7**(7), 64–74 (2020)
20. Chiedozie, I. (2019). 95% of technology used in Nigeria imported–NOTAP. Punch. https://punchng.com/95%-of-technoogy-used-in-nigeria-imported-notap/
21. Chopra, S., White, L.F.: A Legal Theory for Autonomous Artificial Agents. The University of Michigan Press, Ann Arbor, MI, USA (2017)
22. Choudhury, P., Starr, E., Agawal, R.: Machine learning and human capital experimental evidence on productivity complementarities. Harvard business School working paper, pp. 18–065 (2018)
23. Cirillo, V., Sostero, M., Tamagni, F.: Innovation and within-firm wage inequalities: empirical evidence from major European countries. Lab. Econ. Manag. ISSN 2284-0400
24. Cockshott, P., Renaud, K.: Humans, robots and values. Technol. Soc. **45**, 19–28 (2016)
25. Compagnucci, F., Gentili, A., Valentini, E., Gallegati, M.: Robotization and labour dislocation in the manufacturing sectors of OECD countries: a panel VAR approach. Appl. Econ. (2019). ISSN 0003-6846
26. Csillag, M., Scharle, A.: How does PES act to prevent unemployment in the changing world of work? An analytical paper of European commission (2019). ISBN 978-92-79-98484-6
27. Danaher, J.: Robotic rape and robotic child sexual abuse: should they be criminalised? Crim. Law Philos. **11**(1):71–95 (2017)
28. Dauth, W., Fideisen, S., Sudekum, J., Woessner, N.: The rise of robots in the German labor market (2017). https://voxeu.org/article/rise-robots-german-labour-market. Accessed 19 March 2020
29. Davies, M.D.: The world's 2153 billionaires are richer than 4.6 billion people (2020). https://www.bloomberg.com/news/articles/2020-01-20/2-153-billionaires-are-richer-than60-of-the-world-says-oxfam. Accessed 1 Jan 2020

30. Dobrev, D.: A definition of artificial intelligence. Institute of Mathematics and Informatics, Bulgarian Academy of Science (2004)
31. Eamonn, R.: The Evolution of Accounting Software: Past, Present and Future. GAA Accounting (2012). www.gaaaccounting.com/the-evolution-of-accounting-software-past-present-and-future. Accessed 25 Dec 2018
32. Eastwood, G.: How accountants can survive industry 4.0 (2017). Accounting web. https://www.accountingweb.com/community/blogs/garyastwood/howaccountants-can-urvive-indusry-40. Accessed 26 Sept 2019
33. El-Haddad, A.: Redefining the social contract in the wake of the Arab Spring: the experiences of Egypt, Morocco and Tunisia. World Dev **127** (2020)
34. Elvira, N., Cristina, M., Cristian, S.I.: A labor less society? How highly automated environments and breakthroughs in artificial intelligence bring about innovative kinds of skills and employment disruptions, altering the nature of business process and affecting the path of economic growth. J. Self-Govern. Manag. Econ. **6**(4), 25–30 (2018)
35. Florian, S., Victoria, W.: In Theory, There Is an OPTIMAL Taxation System. University of Zurich, Zurich (2018). https://doi.org/10.5167/uzh-152997
36. Francesco, C., Ekkehard, E., Enzo, W.: Robots worldwide: The impact of automation on employment and trade. Research Department working paper 36 (2018)
37. Garbuio, M., Lin, N.: Artificial intelligence as a growth engine for health care startups: Emerging business models. California Manag. Rev. (2018). https://doi.org/10.1177/0008125618811931
38. Germana, B.: (2017). A tax on robot? Some food for thought (2017). http://www.finanze.it/opencms/it/il-dipartimento/documentazione/collana-di-lavori-e-di-ricerca/. Accessed 4 Feb 2020
39. Gordon, R.H., Slemrod, J.: Do we collect any revenue from taxing capital income? University of Michigam and NBER (2019)
40. Grewal, D.S.: A critical conceptual analysis of definitions of artificial intelligence as applicable to computer engineering. IOSR J. Comput. Eng. **2**(1), 09–13 (2014)
41. Guerreiro, J., Rebelo, S., Teles, P.: Should a robot be taxed? Working paper 23806, National Bureau of Economic Research (2017)
42. Gujarati, D.N., Porter, D.C.: Basic Econometrics, 5th edn. McGraw Hill Inc, New York (2009)
43. Harrison, R.: (2003) Hobbes Seems to Have Invented this Useful Term, p. 70. Cambridge University Press
44. Hewitt, J.: This is the hidden risk of automation that no one is talking about (2017). https://www.weforum.org/agenda/2017/11/automation-automated-job-risk-robot-bored-boredom-ffort-fourth-industrial-revolution. Accessed 2 Feb 2020
45. Hoeschl, M.B., Tania, C.D.B., Hoeschl H.C.: Fourth industrial revolution and the future of Engineering: Could robots replace human jobs? How ethical recommendations can help engineers' rule on artificial intelligence. In: 7th World Engineering Education Forum (WEEF) (2017)
46. Jaimovich, N., Siu, H.E.: Disappearing routine jobs: who, how and why? J. Monetary Econ. **91**, 69–87 (2017)
47. Joshi, N.: How AI and robotics can change taxation (2020). https://www.forbes.com/sites/cognitiveworld/2020/01/09/how-ai-and-robotics-can-change-taxation/. Accessed 10 Feb 2020
48. Karabag, S.F.: Factors impacting firm failure and technological development: a study of three emerging-economy firms. J. Bus. Res. (2018) https://doi.org/10.1016/j.jbusres.2018.03.008
49. Koo, J., Cox, W.M.: An economic interpretation of suicide cycles in Japan. Contemp. Econ. Policy **26**(1), 34 (2007)
50. Kou, E., Sun, Y.: Policy uncertainty and tax non-compliance: evidence from China (2019). https://pdfs.semanticscholar.org/727c/8f990e87a21508c86f808a87cda1d5b00148.pdf. Accessed 28 June 2020
51. Kovermann, J., Velte, P.: The impact of corporate governance on corporate tax avoidance—A literature review. J. Int. Acc. Auditing Taxation **36**, 100270 (2019)
52. Laskar, M.E.: Summary of social contract theory by Hobbes, Locke and Rousseau. Research Gate. https://www.researchgate.net/publication/261181816. Accessed 9 Feb 2019

53. Lebans, M., Euske, K.: A Conceptual and Operational Delineation of Performance. Business Performance Measurement Cambridge University Press (2006)
54. Leigh, N.G., Kraft, B., Lee, H.: Robots, skill demand and manufacturing in USregional labour markets. Cambridge J. Reg. Econ. Soc. (2019). https://doi.org/10.1093/cjres/rez019
55. Lin, B., Jia, Z.: Tax rate, government revenue and economic performance: a perspective of Laffer curve. China Econ. Rev. **56** (2019)
56. Liu, Y., Mao, J.: How do tax incentives affect investment and productivity? Firm-level evidence from China. Am Econ. J. **11**(3), 261–91 (2019)
57. Malik, A., Teal, F., Baptist, S.: The performance of Nigerian manufacturing firms: report on the Nigerian manufacturing enterprise survey 2004. University of Oxford, UNIO and CSAE Department of Economics (2006)
58. Mark, R.A.: After 75 years, Isaac Asimov's three laws of robotics need updating. The Conversation (2017). https://www.google.com.ng/amp/s/theconversation.com/amp/after-75-years-isaac-asimovs-three-laws-of- robotics-need-updating-74501. Accessed 19 Sept 2018
59. Marwala, T.: On robot revolution and taxation. Allen Institute for Artificial Intelligence (2018). https://www.arxiv.org/pdf/1808.0166. Accessed 26 Sept 2018
60. Mazur, O.: Taxing robots. Pepperdine Law Rev. **26**, 277 (2019)
61. Michaels, G., Graetz, G.: Estimating the impact of robots on productivity and employment (2015). https://voxeu.org/article/robots-productivity-and-jobs. Accessed 19 March 2020
62. Mitha, S.: Robots, technological change and taxation. Tax J (2017). https://www.taxjournal.com/articles/robotys-technological-change-and-taxation-14092017. Accessed 24 Jan 2019
63. Munonongo, S., Akabi, O.A., Robinson, Z.: Do tax incentive matter for investment? A literature review. Bus. Econ. Horizons **13**(2), 152–168 (2018)
64. Murphy, R.R., Woods, D.D.: Beyond Asimov: the three laws of responsible robotics. IEEE Computer Society, 14–20 (2009)
65. Mwangi, N. (2020). Rwanda to use high-tech robots in fight against COVID-19. https://Africa.cgtn.com/2020/05/19/Rwanda-to-use-high-tech-robots-in- fight-against-covid-19/
66. NCDC: Emergency: Coronavirus disease (COVID-19) Pandemic (2020). https://ncdc.gov.ng/. Accessed 21 May 2020
67. Nawakitphaitoon, K., Ormiston, R.: Occupational human capital and earnings losses of displaced workers: does the degree of similarity between pre-and post- displacement occupations impact. J. Labor Mark. Res. **48**(1), 57–73 (2015)
68. Nevejans, N.: Les robots tentative de definition. In: Bensamoun,A. (ed) Les robots.Objets scientifiques, Objets de droits. Collection des Presses Universitaires de Scceaux, 100 (2016)
69. Nnamani, E., Nwoha, C.: Trend in technological changes and its effect on organizational performance of manufacturing firms in Enugu State, Nigeria. Int. Digital Organ. Sci. Res. IDOSR J. Appl. Sci. **4**(1), 44–55 (2019)
70. Nomaler, O., Verspagen, B.: Perpetual growth, distribution, and robots. Maastricht Graduate School of Governance (MGSoG) (2018). http://www.maastrichtuniversity.nl/governance. Accessed 20 March 2020
71. OECD: Revenue Statistics in Africa 2019. OECD (2019). www.oecd.org/tax/tax-policy/oecd-classification-taxes-interpretative-guide.pdf
72. Oberson, X.: Taxing Robots: Helping the Economy to Adapt to the Use of Artificial Intelligence. Edward Elgar Publishing Inc (2019). https://doi.org/10.4337/9781788976527
73. Oberson, X.: Taxng robots? From the emergence of an electronic ability to pay to a tax on robots or use of robots. University of Geneva. World Tax J., 247–260 (2017)
74. Oledele, P.O., Akeke, I., Oladunjoye, O.: Entrepreneurship development: a panacea for unemployment reduction in Nigeria. J. Emerg. Trends Econ. Manag. Sci. (JETEMS) **2**(4), 251–256 (2011)
75. Oxfam: Time to care: Unpaid and underpaid care work and the global inequality crisis (2019). https://indepth.oxfam.org.uk/time-to-care/. Accessed 28 Jan 2020
76. Pagallo, U.: Vital, Sophia, and Co.—The quest for the legal personhood of robots. Information **9**(9). (2013). https://doi.org/10.3390/info9090230

77. Paul, A., Anantharaman, R.: Impact of people management practices on organizational performance: analysis of a causal model. Int. J. Human Res. Manage. **14**(7), 1246–1266 (2003)
78. Pauline, G., Eric, M.: Changes in job security and their causes: an empirical analysis for France, 1982–2002. Eur. Econ. Rev. **48**(3), 595–615 (2003)
79. Perez, J.A., Deligianni, F., Ravi, D., Yang, G.: Artificial intelligence and Robotics. UK- RAS White Papers (2017). ISSN 2398–4422
80. Pervan, M., Pervan, I., Curak, M.: The influence of age on firm performance: evidence from the Croatian food industry. J. Eastern Eur. Res. Bus. Econ. **2017**(2017), 2–10 (2017)
81. Postel-Vinay, F.: The dynamics of technological unemployment. Int. Econ. Rev. **43**, 737–760 (2002)
82. Reynolds, K.: 11 biggest challenges of international business in 2017. Hult International Business School (2017). https://www.hult.eds/blog/international-busines-challenges/
83. Sabine, P.: Robots, industry 4.0 and humans or why assembly work is more than routine work. Mdpi.com. Societies **6**(2), 16 (2018). https://www.mdpi.com/2075-4698/6/2/16/htm. Accessed 2 July 2019
84. Schraer, R., Goodman, J., Jakhar, P., Giles, C., Morris, C., Mwai, P.: Nigeria: Why is t struggling to meet its tax targets? BBC News (2019). https://www.bbc.com/news/world-africa-49566927
85. Sharma, S.: Experimental and ex-post Facto Design. Being a Ph.D. thesis of Horizons University, Paris (2019)
86. Simon, M.: A future in accounting without human intervention. Master's Dissertation. University of Ghent (2017)
87. Soderbom, M., Teal, F.: The performance of Nigerian manufacturing firms: report on the Nigerian Manufacturing enterprise survey. United Nations Industrial Development Organization (UNIDO) and Centre for the Study of African Economies (CSAE) Department of Economics, University of Oxford (2002)
88. Sylvester, O., Okorie, J.: Comparative analysis on the trend in global technology speed and its effects on organizational performance of manufacturing firms in South East Nigeria (JUHEL Nigeria-Enugu- South East Nigeria Camila vegetable Oil LTD – Owerri Imo – State Rice Cluster Company – Abakiliki Ebonyi State). Int. J. Manag. Soc. Sci. **7**(5), 2321–1784 (2019)
89. Taouab, O., Issor, Z.: Firm performance: definition and measurement models. Eur. Sci. J. **15**(1), 1857–7881 (2019)
90. Tennant, S.N., Tracey, M.R.: Corporate profitability and effective tax rate: the enforcement effect of large taxpayer units. Acc. Bus. Res., 1–20 (2019). https://doi.org/10.1080/00014788.2018.1512398
91. Thomsen M., Watrin C.: Tax avoidance over time: a comparison of European and U.S. firms. J. Int. Acc. Auditing Taxation **33**, 40–63. (2018). https://www.journals.elsevier.com/journal-of-international-accounting-auditing-and-taxation. Accessed 06 Jun 2019
92. Uddin, P.S.O.: Causes, effects and solutions to youth unemployment problems in Nigeria. J. Emerg. Trends Econ. Manag. Sci. (JETEMS) **4**(4), 397–402 (2013)
93. United Nations: Rising inequality affecting more than two-thirds of globe, but it's not inevitable: new UN report (2020). https://news.un.org/en/story/2020/01/1055681. Accessed 25 Jan 2020
94. Uwe, T.: Optimal taxation of robots (2018).http://uwethuemmel.com/wpcontent/uploads/2017/02/Thuemmel2018_OptimalTaxationofRobots_June.pdf
95. Vaishya, R., Javaid, M., Khan, I.H., Haleem, A.: Artificial intelligence (AI) applications for COVID-19 pandemic. Diab. Metab. Syndr. **14**(4), 337–339 (2020)
96. Wang, P.: On defining artificial intelligence. J. Artif. Gen. Intell. **10**(2), 1–37 (2019)
97. World Development Report: Digital dividends. World Bank (2016). ISSN: 0163-5085
98. Wurah, A.: We hold these truths to be self-evident, that all robots are created equal. J. Fut. Stud. **22**(2), 61–74 (2017)
99. Zhou, H.: Unemployment and economic integration for developing countries. Munch Personal RePEc Archive (MPRA) Paper 83313 (2017). https://mpra.ub-muenchen.de/83313/

The Impact of Innovative Technology on the Aviation Industry and on Customers Preference

Ahmed Fuad, Mukhtar Al-Hashimi, and Allam Hamdan

Abstract This research has investigated the impact of innovative technology on customer preference. Innovative technologies used by other carriers in the world were explored and populations from the Kingdom of Bahrain were asked if these technologies have any impact on their selection. The study showed that there's a significant relationship between customer preference and innovative technology. Our chapter also has shown that the adoption of these technologies will increase the number of customers, which subsequently will have an impact on the revenue of the company. Other technologies also have a slight impact on the customer's decision such as Near Field communication. The researchers were able to provide recommendations to the carrier in order to focus on several areas in order to attract more customers and gain a competitive advantage over carriers in the region.

Keywords Innovative technology · Airline industry · Airline passengers · Bahrain

1 Introduction

Nowadays, the business world is a very competitive environment. While facing many difficulties and struggling to survive the economic crisis, Businesses and industries also struggle in generating income and profit while providing the customers with the optimum services. Many industries in order to survive have provided lower standards in their services; many industries have eliminated the special services they are well known for in order to minimize the cost and reduce it to the maximum [1]. These factors have a high impact on all industries and many countries. The change in the conditions of the market includes technologies, sanctioned customers, and distribution channels, which are new. The Factors forces an airline to constantly adopt

A. Fuad
Customer and Priority Care Officer, Gulf Air, Washington, DC, Bahrain

M. Al-Hashimi · A. Hamdan (✉)
College of Business and Finance, Ahlia University, Manama, Bahrain
e-mail: allamh3@hotmail.com

A. Hamdan et al. (eds.), *The Fourth Industrial Revolution: Implementation of Artificial Intelligence for Growing Business Success*, Studies in Computational Intelligence 935,
https://doi.org/10.1007/978-3-030-62796-6_3

and operation improvement of their models of business [2]. In the aviation industry, more other factors impact the airline itself such as the rising of operating costs due to several reasons like fuel price, cost of aircraft maintenance and its renovation, the replacement of parts, landing or parking fees, etc. Airlines nowadays have to work harder to maximize profit, while on the same hand they need to keep into consideration avoiding any unwanted additional costs and minimizing the costs. Like many industrial sectors, airlines have been relying heavily on innovative technology to minimize unnecessary costs that could burden the yearly revenue of the industry. For example, all airlines now unable the customer to make a booking via the internet, customers no longer have to visit local office or contact a call center to make the booking, this move made many airlines minimize the costs by terminating local offices which had many expenses such as rent, salary of employees and due bills. Few years ago, online check in was introduced and now has been adopted by almost every most of the carriers in the world regardless being international or domestic. The technology now plays a huge role in how companies compete and build strategies or even gain new customers; companies who fail to keep up with the technology eventually don't survive in the market. Technologies plays a role for many individuals in their decision whether to select the company that offers the service or select another company that the individual feels like it's easier to complete his tasks and it's easier to get assistance. As an aviation industry where many struggle to maintain its market share and to minimize the impact of the economic crisis, the rise of thc fuel and many additional expenses. The airline faces a vicious competition with other similar companies in the region [3]. In a competitive aviation market, each airline competes to maintain its loyal frequent customers and gain the trust of customers. Keeping into consideration the expenses it comes along gaining and maintaining these customers.

Many aviation companies use traditional ways in keeping its customer, such as loyalty award system, special offers, and season tickets. However, these ways are available in almost every airline, which no longer helps in gaining a competitive advantage. This study will mainly focus on innovation and how explores how innovative technologies that are cost efficient impacts on the selection of the consumer towards a specific airline. Keeping into consideration that these technologies will not be expensive to adopt and also easy to implement. Giving an airline a competitive advantage over other airlines in the region to maintain and gain new customers, which help the company increase its revenue and minimize the impact caused by the rising of many other expenses. This study will also identify several innovative technologies that are being used by leading international carriers in Europe and the United States of America including how they are implemented and the impact they play on the traveler's trip. The innovative technologies that are going to be identified will not be currently used in the national carrier of Bahrain. As this study will also evaluate the impact they have on consumers in the country if these technologies were adapted and used by the national carrier of the kingdom of Bahrain [4].

The remainder of the chapter is organized as follows. The next section briefly discusses the research methods are discussed and results presented. The chapter ends with a conclusion and limitations of the study.

2 Methodology

In the proposed research, data has been collected by conducting a survey that has been consisted of several research questions. Closed ended questions and open-ended questions with text response have been used to collect responses from the respondents. The questionnaire has been handed to the respondents who were randomly picked from the population of Bahrain that will be discussed further in the next part. The questionnaire has also been distributed on social media communication application as a virtual link.

2.1 Sampling

In selection of the sample from the population, a probability sampling technique has been used which is Simple Random Sampling which will assure that each element of the population will have an equal chance of being included in the sample. Sampling is an elementary idea which involves selecting of few elements in a which will reflect the population, the sampling method may provide the researcher with a general conclusion of how the entire populations perception. There are several reasons in which the researcher may need the sampling methodology such as a great speed of collecting data, lower effort and cost and finally of population selection [5]. The participants of the questionnaire were randomly selected in Bahrain by distributing the questionnaire manually or sharing a virtual link. The sample size was selected through probability sampling, more precisely random selection in order to provide an unbiased selection of businesses from the Bahraini market. The sample size was selected based the size of the population but also taking into consideration the time and resource limitations of the research project.

In order to get a more accurate result, the questionnaire has been distributed to different age groups and genders amongst people in Bahrain, including the citizens as well as the residences. The total received responses through the randomly sampling method were 384 out of 384 all incomplete and invalid data were eliminated from the research.

2.2 Research Hypothesis

In order to examine the impact of innovation technology on aviation industry the following hypothesis were tested in the research, namely:

- H1: There is a positive relationship between innovative technology and customer's satisfaction.
- H2: There is a significant difference in Customer Preference between male and female respondents.

- H3: There is a significant difference in Innovative Technology between male and female respondents.

3 Results, Interpretation of Data and Analysis

3.1 Demographic Characteristics

Refer to Table 1. There are 382 valid respondents in the study data out of which 201 are male (about 52.6%) and the rest of the 181 respondents (about 47.4%) are female. Out of the 382 respondents who participated in the survey, we categorized the age of the respondents into four categories—19 and below, 20–29, 30–39 and 40–49 years.

A total of 138 (36.1%) of the respondents are aged 19 and below, 154 (40.3%) were aged between 20 to 29 years, 11 (2.9%) of the respondents were aged between 30 to 39 years and rest of the 79 (20.7%) were aged between 40 and 49 years. So, peoples aged between 20 years to 39 years are participated more in the study while people aged between 30 to 39 had minimum representation in the study. The data

Table 1 Sample characteristics

Demographic Characteristics		Frequency	Percent
Gender	Male	201	52.6
	Female	181	47.4
	Total	**382**	**100**
Age	19 and below	138	36.1
	20–29	154	40.3
	30–39	11	2.9
	40–49	79	20.7
	Total	**382**	**100**
Nationality	Bahraini	251	65.7
	Non-Bahraini	131	34.3
	Total	**382**	**100**
Income	Unemployed	28	7.3
	Less than BHD 500	51	13.4
	BHD 501 - BHD1,000	165	43.2
	BHD 1001 and above	138	36.1
	Total	**382**	**100**
Travel	Satisfaction	73	19.1
	2–3 Times a year	160	41.9
	4–5 Times a year	74	19.4
	6 or more times a year	75	19.6
	Total	**382**	**100**

was collected from Bahraini and Non-Bahraini residents. The results revealed that a total of 251 (65.7%) respondents in the study were Bahraini while 131 (35.7%) were non- Bahraini. Out of the 382 respondents who participated in the survey, Monthly income was categorized into four different categories.—Unemployed, Less than BHD 500, BHD 501—BHD 1000, BHD 1001 and Above. A total of 28 (7.3%) of the respondents were unemployed, 51 respondents (13.4%) has monthly income less than BHD 500, 165 (43.2%) of the respondents had income in the range of BHD 501—BHD 1000 and rest of the 138 (36.1%) had income over BHD 1001. People with monthly income between BHD 501—BHD 1000 participated more in the study while people with no income had minimum representation in the study. Out of the 382 respondents who participated in the survey, Average Travel by place per year was categorized into four different categories.—Once a Year, 2–3 Times a Year, 4–5 Times a Year, and 6 or More times a Year. A total of 73 (19.1%) of the respondents travel once a year, 160 respondents (41.9%) travel 2–3 times a year, 74 (19.4%) of the respondents travel 4–5 times a year and rest of the 75 (19.6%) travel 6 or times a year. People who travel between 2 and 3 times a year participated more in the study while people with who travel once a year had minimum representation in the study.

3.2 *Other Characteristics*

Refer to Table 2. Out of the 382 respondents who participated in the survey, place of purchase for airline tickets was categorized into three different categories.—Directly from Airline, Through Travel Agent, and Some Other Way. A total of 136 (35.6%) of the respondents purchase tickets directly from airline, 183 respondents (47.9%)

Table 2 Other characteristics

Characteristics		Frequency	Percent
Purchase of airline tickets	Directly from the airline	136	35.6
	Through a Travel Agent	183	47.9
	Some other way	63	16.5
	Total	**382**	**100**
Member of rewards program	Yes	136	35.6
	No	218	57.1
	Missing	28	7.3
	Total	**382**	**100**
Booking class	First Class	6	1.6
	Business Class	46	12
	Economy Class	330	86.4
	Total	**382**	**100**

purchased tickets from travel agents and rest of the 63 (16.5%) purchased through some other way. People who purchased tickets through travel agents participated more in the study. The results revealed that 136 respondents (38.4%) were member of airline rewards program while 218 (61.6) were not part of any airline benefits program. 28 respondents didn't answer the question. Out of the 382 respondents who participated in the survey, booking class was categorized into three different categories.—First Class, Business Class, and Economy Class. A total of 330 (86.4%) respondents purchase Economy Class tickets, 46 respondents (12%) purchased business class tickets and only 6 (1.6%) purchased First class tickets. People who purchased economy class tickets participated more in the study.

3.3 Customer Preference

Refer to Table 3. The respondents were asked for their preference with respect to Self Service Check In. 208 (54.6%) of the respondents reported that it is important in Medium Trip, while 149 (39%) respondents pointed out that it is important in Short Trip. Only 8 (2.1%) deemed it important in long flights while only 17 (4.5%) called it important in all flights. This shows that self-service check-in according to the respondents is important in Medium Trips. The respondents were asked when they find Seat Comfort and Leg Space more important. 126 (33%) of the respondents reported that it is important in Medium Trip, while 101 (26.4%) respondents pointed out that it is important in Short Trip. 60 (15.7%) deemed it important in long flights while 95 (24.9%) called it important in all flights. This shows that Seat Comfort and Leg Space to a certain extent is important in all types of flights.

The respondents were asked when internet connectivity is onboard flight is important. 158 (41.4%) of the respondents reported that it is important in Medium Trip, while 79 (20.7%) respondents pointed out that it is important in Short Trip. 75 (19.6%) deemed it important in long flights while 70 (18.3%) called it important in all flights. This shows that Internet Connectivity to a certain extent is more important in medium trips.

The respondents were asked when in-flight entertainment is important. 108 (28.3%) of the respondents reported that it is important in all Trips, while 99 (25.9%) respondents pointed out that it is important in medium trips. 98 (25.7%) deemed it important in long flights while 77 (20.2%) called it important in short flights. This shows that Internet Connectivity to a certain extent is more important in all trips. The respondents were asked when prompt flight notifications via mobile are important. 156 (40.8%) of the respondents reported that it is important in all Trips, while 124 (32.5%) respondents pointed out that it is important in long trips. 68 (17.8%) deemed it important in medium flights while 34 (8.9%) called it important in short flights. This shows that flight notifications is important in all trips. The respondents were asked when online check-in is important. 155 (40.6%) of the respondents reported that it is important in medium Trips, while 143 (37.4%) respondents pointed out that it is important in short trips. 44 (11.5%) deemed it important in all flights while

Table 3 Customer preference

Variables	Percent%			
	Short trip	Medium trip	Long trip	All flights
Airport Self Service Check in is important in	39.0	54.5	2.1	4.5
Seat Comfort and Leg Space are important in	26.4	33.0	15.7	24.9
Internet connectivity onboard flights is important in	20.7	41.4	19.6	18.3
Inflight entertainment system is important in	20.2	25.9	25.7	28.3
Prompt flight notifications via mobile (Check in, Last call, flight status) is important in	8.9	17.8	32.5	40.8
Online Check in is important in	37.4	40.6	10.5	11.5
Private Media can be brought along and viewed in the inflight monitors is important in	27.7	60.7	2.4	9.2
Quality of meals and variety of meal choices important in	23.0	39.8	14.1	23.0
In flight Duty Free is important in	26.7	50.5	13.1	9.7
Safety and reputation of an airline is important in	23.0	34.3	22.0	20.7
Nationality of an airline is important in	36.4	37.4	19.6	6.5
Convenient departure and arrival time is important in	27.7	28.3	19.6	24.3
Real time baggage tracing mobile application is important in	58.1	33.0	5.2	3.7
Ticket Price is important in	3.9	6.0	15.7	74.3
Ticket Restriction is important in	7.9	13.4	14.4	58.9
Inflight Customer Service is important in	22.0	46.9	16.5	14.7

***Short Trip**: Dubai, Kuwait, Cairo, Muscat. etc. **Medium Trip**: Istanbul, Pakistan, India, etc. **Long Trip**: London, Paris, Bangkok etc

40 (10.5%) called it important in long flights. The respondents were asked whether private media can be viewed on in-flight monitors. 232 (60.7%) of the respondents reported that it is important in medium Trips, while 106 (27.7%) respondents pointed out that it is important in short trips. 35 (9.2%) deemed it important in all flights while only 9 (2.4%) called it important in long flights. The respondents were asked about importance of quality of meals in different flights. 152 (39.8%) of the respondents reported that it is important in medium Trips, while 88 (23.0%) respondents pointed out that it is important in short trips and all flights. 54 (14.1%) deemed it important in long flights. The respondents were asked about importance of in-flight duty free in different flights. 193 (50.5%) of the respondents reported that it is important in medium Trips, while 102 (26.7%) respondents pointed out that it is important in short trips. 50 (13.1%) called for its importance in Long flights while 37 (9.7) deemed it important in all types of flights. The respondents were asked about importance of safety and reputation of an airline in different flights. 131 (34.3%) of the respondents

reported that it is important in medium Trips, while 88 (23%) respondents pointed out that it is important in short trips. 84 (22%) called for its importance in Long flights while 79 (20.7%) deemed it important in all types of flights. The respondents were asked about importance of nationality of airline in different flights. 143 (37.4%) of the respondents reported that it is important in medium Trips, while 139 (36.4%) respondents pointed out that it is important in short trips. 75 (19.6%) called for its importance in Long flights while 25 (6.5%) deemed it important in all types of flights.

The respondents were asked about convenient departure in different flights. 108 (28.3%) of the respondents reported that it is important in medium Trips, while 106 (27.7%) respondents pointed out that it is important in short trips. 75 (19.6%) called for its importance in Long flights while 93 (24.3%) deemed it important in all types of flights. The respondents were asked about importance of real time baggage tracking mobile application in different flights. 222 (58.1%) of the respondents reported that it is important in short trips, while 126 (33%) respondents pointed out that it is important in medium trips. 20 (5.2%) called for its importance in Long flights while 14 (3.7%) deemed it important in all types of flights. The respondents were asked about importance of ticket price in different flights. 225 (62.3.1%) of the respondents reported that it is important in all trips, 55 (15.2%) respondents pointed out that it is important in long trips, 51 (14.1%) of the respondents reported that it is important in medium trips, and 30 (8.3%) of the respondents reported that it is important in short trips. The respondents were asked about importance of ticket restrictions in different flights. 225 (62.3%) of the respondents reported that it is important in all trips, while 55 (15.24%) respondents pointed out that it is important in long trips, followed by 51% (14.1%) of the respondents reported that it is important in medium trips, and the last answer was 30 (8.31%) of the respondents reported that it is important in short trips. The respondents were asked about importance of in-flight customer service in different flights. 179 (46.9%) of the respondents reported that it is important in medium trips, while 84 (22%) respondents pointed out that it is important in short trips. 63 (16.5%) called for its importance in Long flights while 56 (14.7%) deemed it important in all types of flights.

3.4 Innovative Technology

The respondents were asked whether RFID impacts their decision in selecting an airline for travel. 180 (47.1%) of the respondents Strongly Agreed that it does impact air travel decision. 90 (23.6%) agreed that it impacts their decision while 40 (10.5%) were neutral in their opinion. A total of 72 (18.9%) respondents either disagreed or strongly disagreed to the statement that it impacts their decision of selecting an airline. The respondents were asked whether Mobile Wallet impacts their decision in selecting an airline for travel. 115 (30.1%) of the respondents Strongly Agreed that it does impact their travel decision. 117 (30.6%) agreed that it impacts their decision while 26 (6.8%) were neutral in their opinion. A total of 124 (32.5%) respondents either disagreed or strongly disagreed to the statement that it impacts their decision of selecting an airline. The respondents were asked whether KIOST impacts their decision in selecting an airline for travel. 209 (54.7%) of the respondents Strongly Agreed that it does impact their travel decision. 71 (18.6%) agreed that it impacts their decision while 34 (8.9%) were neutral in their opinion. A total of 68 (17.8%) respondents either disagreed or strongly disagreed to the statement that it impacts their decision of selecting an airline (Table 4).

The respondents were asked whether private media player impacts their decision in selecting an airline for travel. 83 (21.7%) of the respondents Strongly Agreed that it does impact their travel decision. 177 (46.3%) agreed that it impacts their decision while 18 (8.9%) were neutral in their opinion. A total of 104 (27.2%) respondents either disagreed or strongly disagreed to the statement that it impacts their decision of selecting an airline. The respondents were asked whether Wi-Fi on-board impacts their decision in selecting an airline for travel. 117 (30.6%) of the

Table 4 Innovative technology

Variables	Percent%				
	Strongly agree	Agree	Neutral	Disagree	Strongly disagree
RFID impacts your decision of selecting an airline	47.1	23.6	10.5	16.0	2.9
Mobile wallet impacts your decision of selecting an airline	30.1	30.6	6.8	19.4	13.1
Kiosk impacts your decision of selecting an airline	54.7	18.6	8.9	8.6	9.2
Private Media player impacts your decision of selecting an airline	21.7	46.3	4.7	17.8	9.4
WiFi onboard impacts your decision of selecting an airline	30.6	29.1	7.1	18.6	14.7
NFC impacts your decision of selecting an airline	32.5	34.8	7.3	15.7	9.7
iPad enabled crew impact your decision of selecting an airline	31.2	21.2	13.1	22.5	12.0
Baggage tracing mobile application impacts your decision of selecting an airline	29.8	27.2	8.6	18.3	16.0
Mobile Pay impacts your decision of selecting an airline	36.4	26.7	8.9	15.4	12.6

respondents Strongly Agreed that it does impact their travel decision. 111 (29.1%) agreed that it impacts their decision while 27 (7.1%) were neutral in their opinion. A total of 127 (33.3%) respondents either disagreed or strongly disagreed to the statement that it impacts their decision of selecting an airline. The respondents were asked whether NFC on-board impacts their decision in selecting an airline for travel. 124 (32.5%) of the respondents Strongly Agreed that it does impact their travel decision. 133 (34.8%) agreed that it impacts their decision while 28 (7.3%) were neutral in their opinion. A total of 97 (25.4%) respondents either disagreed or strongly disagreed to the statement that it impacts their decision of selecting an airline. The respondents were asked whether iPAD enabled crew on-board impacts their decision in selecting an airline for travel. 119 (31.2%) of the respondents Strongly Agreed that it does impact their travel decision. 81 (21.2%) agreed that it impacts their decision while 50 (13.1%) were neutral in their opinion. A total of 13 (34.5%) respondents either disagreed or strongly disagreed to the statement that it impacts their decision of selecting an airline. The respondents were asked whether baggage tracing mobile application impacts their decision in selecting an airline for travel. 114 (29.8%) of the respondents Strongly Agreed that it does impact their travel decision. 104 (27.2%) agreed that it impacts their decision while 33 (8.6%) were neutral in their opinion. A total of 131 (34.3%) respondents either disagreed or strongly disagreed to the statement that it impacts their decision of selecting an airline. The respondents were asked whether mobile pay impacts their decision in selecting an airline for travel. 139 (36.4%) of the respondents Strongly Agreed that it does impact their travel decision. 102 (26.7%) agreed that it impacts their decision while 34 (8.9%) were neutral in their opinion. A total of 107 (28%) respondents either disagreed or strongly disagreed to the statement that it impacts their decision of selecting an airline.

3.5 Hypotheses Testing

The first hypothesis under study for the research is there exists a positive relationship between customer preference and innovative technology. Person Correlation Coefficient is utilized to ascertain the significant and strength of relationship between customer preference and innovative technology (refer to Table 5).

The results of Pearson correlation analysis reveal a negative but however, due to the questionnaire strongly agree being the first option the results are considered as Positive strong relationship which is significant between customer preference and innovative technology. Pearson correlation coefficient, $r = -0.111$. The p-value for the test is 0.030 which is less than 0.05. As the p-value less than the desired level of 0.05 (at 5% level of significance), so, we can reject the null hypothesis.

Two differential hypotheses are proposed. H2 seeks to ascertain whether there exists significant difference in customer preference between male and female respondents in the study while H3 seeks to ascertain whether there exists significant difference in innovative technology between male and female respondents in the study.

Table 5 Testing of hypotheses

Variables	Male	Female	Mean Difference	t-Test	Sig.	Correlation between customer's preference and innovative technology
Customer Preference						
No.	201	181	0.011	0.425	0.671	
Mean	2.137	2.125				
Std. Deviation	0.252	0.261				
Innovative Technology						
No.	201	181	-0.058	−1.228	0.220	
Mean	2.378	2.436				
Std. Deviation	0.483	0.431				
Pearson Correlation						−0.111
Sig.						0.03

Independent Sample T-Test is utilized to discover differences between the male and female groups.

The results indicate that for Customer Preference, there exists no significant difference (t (380) = 0.425, p > 0.05) between male and female respondents. Hence H2 is not proved. For Innovative Technology, results of independent sample t-test reveal that there exist no significant differences (t (380) = 0.220, p > 0.05) between male and female respondents. Hence, H3 is not proved.

This Study has investigated the impact of innovative technology on the customer's preference in the Kingdom of Bahrain. A thorough literature review was conducted in order to find out which innovative technologies were used by other airlines and not the airline that this research is based on. Several technologies were found keeping into consideration the price and if its possible to adopt. The main measurement tool was questionnaire which was based on the literate contained the technologies and the respondents were to answer if these technologies have any impact on their selection of airline.

A quantitative approach and a probability sampling technique has been used which is Simple Random Sampling The questionnaire was distributed on the population of the Kingdom of Bahrain and the sample size was 382 the total number of the sample. The results of the study revealed that 52.6% of the respondents were male. The questionnaire was distributed randomly and to no specific age in order to get a more thorough result of all desired preference 36.1% of the respondents were Below 19 year old. The majority was between the age of 20 to 29 years old 40.3%, the rest were 20.7% 40 to 49 years old and the rest are above 50 years old which were the least category. This this research population is the entire population of Bahrain the 251 of the respondents were Bahraini the rest 131 were not. Only 28 respondent 7.3% were unemployed where the majority of 43.2% have a monthly income between BHD 501- BHD 1000. 41.9% of all the respondents travel two to three times a year.

More than half of the respondents prefer making the booking through travel agent, as only 35.7% book there tickets directly through the airline. 36.8% of the responds are members of a loyal program with the airline where the majority is not. A significant 86.4% of the total number of respondents travels on economy class. 12% travel on Business class and only 1% of the entire respondents prefer traveling in first class.

The results of pearson correlation analysis reveal a positive negative but significant relationship between customer preference and innovative technology. Pearson correlation coefficient, $r = -0.111$. The p-value for the test is 0.030 which is less than 0.05. As the p-value less than the desired level of 0.05 (at 5% level of significance), so, we can reject the null hypothesis of no relationship. However, the study proposed a positive relationship, which has been proved by the results. Hence, hypothesis 1 is approved. Difference in customer preference between male and female respondents in the study while H3 seeks to ascertain whether there exists significant difference in innovative technology between male and female respondents in the study. Independent Sample T-Test is utilized to discover differences between the male and female groups.

4 Conclusion, Recommendations and Future Research

The evolution of technology has enabled the airline to provide an efficient service, the portable innovative technologies have become very important to gain the customers loyalty and maximize the satisfaction, which enable users to choose and allocate their services in terms of seat allocating or even in tracing personal baggage. One of this studies main objective is to understand the behavior of consumer when it comes to selection of an airline whether it's a low cost or a high cost airline. This study also focused on answering the problem in whether or not the innovative technologies have an impact on the customer choice behavior [6–8]. The conclusion of the carried out distributed surveys on the population of Bahrain has outlined that innovation and innovative technology are important with regards economic growth and increase in revenue. As the analysis has shown that it is in fact a reason that determines whether the most will select an airline or not.

Its important to understand how passengers make their decisions as this has also been outlined in this research and this data were obtained from the survey. As airlines should be nowadays more alerted and proactively responsive to the constant changes. Its vital to understand the customer decision behavior and analyzing the decision made by the customers in order further understanding the innovative technology impact on the decision. The outcome of the data analysis of the distributed surveys has proved that the majority of the respondents would make heir decision based on the innovative technology available in the chosen carrier. As the test conducted proves that there is a relationship between customer preference and innovative technology.

However, the result has also showed respondents also highly take into consideration the ticket prices and fare regulations.

The airline should consider adopting the technologies and testing them the outcome it has and the impact on customer preference and selection.

The airline management should prepare the staff and train them to handle new technologies that allow them to learn more about the environment where they are developing new areas of work. The airline should also consider the possibility of having the services of professionals specialized in new technologies that will ensure the technologies are used in the right manner and for a constantly proactive update in the innovative technology. Good management of technological tools also enables organizations to work proactively in retaining customer loyalty. This research has confirmed that innovative technology does in fact have an impact on customer preference in terms of selecting an airline.

As future researched may study the impact innovative technology has on the carrier once the innovative technology has been adopted, future researcher may include additional technologies, which were not included in the study. On the other hand, researcher may find new methods of increasing customer satisfaction and loyalty other than the traditional methods conducted such as the special offers or new destinations. The researcher may also conduct interviews with management of the decision to view their perspective in order to adopt these technologies.

References

1. Gaskill, L., Auken, H.E., Manning, R.A.: A factor analytic study of the perceived causes of small business failure. J. Small Bus. Manage. **31**(4), 18–31 (1993)
2. Kossmann, M.: Delivering Excellent Service Quality in Aviation: A Practial Guide for Internal and External Service Providers. Ashgate, Hampshire (2006)
3. Al-Naser, M.: Public governance and economic growth: conceptual framework. Int. J. Bus. Ethics Gov., **2**(2), 1–15 (2019). https://doi.org/10.51325/ijbeg.v2i2.21
4. Salman, M., Battour, M.: Career excellence between leadership roles and achievement motivation for employees in the ministry of education in the united arab emirates. Int. J. Bus. Ethics Gov., **3**(1), 46–96 (2020). https://doi.org/10.51325/ijbeg.v3i1.33
5. Cooper D.R., Schindler, P.S.: Business Research Methods; Eighth edition; ISBN: 0- 07- 249870-6; McGraw-Hill higher Education (2003)
6. Alzaneen, R., Mahmoud, A.: The role of management information systems in strengthening the adminstrative governance in minstry of education and higher education in gaza. Int. J. Bus. Ethics Gov., **2**(3), 1–43 (2019). https://doi.org/10.51325/ijbeg.v2i3.44
7. Al-Afifi, A.A.M.: Factors affecting decision makers preference of msmes in financing sources choice. Int. J. Bus. Ethics Gov., **2**(2), 16–29 (2019)
8. Anjum, R., Mukhtar, Al-Hashimi, Allam, M., Abdulmuttaleb, M., Ahlam, H., Esra, S., Bahaa, Awwad, Sameh, Reyad: Learning readiness when sharing knowledge while e-learning. J. Option, Bolivarian, Acceptable search for publication (2019)

The Relationship Between Intellectual Capital in the Fourth Industrial Revolution and Firm Performance in Jordan

Kamelia Moh'd Khier Al Momani, Abdul-Naser Ibrahim Nour, Nurasyikin Jamaludin, and Wan Zanani Wan Abdullah

Abstract This paper attempts to determine if intellectual capital (IC) has an impact on firm performance in the industrial companies in Jordan. A quantitative approach was used, with value-added intellectual coefficient ($VAIC^{TM}$) methodology of IC measurement. Secondary data were used in measuring the study variables. The relationship between $VAIC^{TM}$ in fourth industrial revolution (IR 4.0) and the ratios of earnings per share (EPS) and return on assets (ROA) were examined. The study sample comprised 50 industrial companies listed on the Amman Stock Exchange (ASE) covering the 2008–2017 period. Multiple Regression analysis was used in data analysis, involving panel data models. The obtained results prove a positive significant influence of $VAIC^{TM}$ on EPS and ROA, and a positive significant relationship between human capital efficiency (HCE) and structural capital efficiency (SCE), and EPS and ROA, while no relationship was found between capital employed efficiency (CEE) and EPS and ROA. The results suggest the need for industrial companies in Jordan and in the Middle East to focus on intellectual capital (IC) elements, particularly human capital (HC), which is core to $VAIC^{TM}$. Moreover, this study is valuable to policymakers and managers in determining the industry development for better corporate returns.

Keywords Intellectual capital · Earning per share · Return on assets · Human capital · Structural capital · Capital employed

---

K. M. K. Al Momani (✉) · N. Jamaludin · W. Z. W. Abdullah
University Malaysia Terengganu, Kuala Terengganu, Malaysia
e-mail: Kamelia.momani@gmail.com

N. Jamaludin
e-mail: asyikin@umt.edu.my

W. Z. W. Abdullah
e-mail: zanani@umt.edu.my

A.-N. I. Nour
Al-Najah National University, Al-Najah, Palestine
e-mail: a.nour@najah.edu

A. Hamdan et al. (eds.), *The Fourth Industrial Revolution: Implementation of Artificial Intelligence for Growing Business Success*, Studies in Computational Intelligence 935,
https://doi.org/10.1007/978-3-030-62796-6_4

1 Introduction

The developed countries have made great progress due to their dependence on the experienced and scientists, as the strength of any country under the Fourth Industrial Revolution (IR 4.0) became dependent on the extent of its possession of science and knowledge and its use. In the era of IR 4.0, relying on digital progress and huge artificial intelligence, many organizations have come to realize an important fact that their true value depends on something more important than physical and financial capital, known as intellectual capital, which relies mainly on employees knowledge. And their creativity, skills, experience, patents, and organizational functions in the company, in addition to the company's relationship with customers [1].

In these days, companies need to take advantage of all financial and intellectual resources to achieve their goals and continue in the world of competition, which we live because of technological progress in the information and communications, therefore has become creativity and innovation are a characteristic of success and development in organizations, as the attention of organizations has shifted from physical and financial resources to IC, and thus human resources became the most important assets of companies, as the basis for creativity in it [2].

Thus, the companies' interest in IC has become necessary to keep pace with the scientific and technical developments witnessed in the world in various fields. Therefore, it is expected that the efficiency of IC will have a direct impact on the companies' performance [3].

Pulic at 1998 argues that the best way to measure the IC efficiency should depend on the efficiency of the employees, where IC defined as employees' knowledge and expertise that can be used to raise or add value to the company stored by employees [4, 5]. Thus, the value-added intellectual coefficient model (VAIC^{TM}) was developed by Anti Pulic in 1998, this model came to reflect the extent of the contribution of experienced and efficient employees in creating new value in the company [6].

The VAIC^{TM} model suggests that the value of the company is formed by two main resources namely financial and physical capital (CE), and Intellectual capital (IC), which comprises HC, and SC, both resources assure company efficiency as they link between the recourses and their created value [4, 7].

According to VAIC^{TM} model, employees are the primary components of IC owing to their knowledge and experience in the utilization of the company infrastructure or SC in the generation of new profitability products through the utilization of CE [5, 8, 9]. Moreover, HC cannot complete their task without SC, so employees are key in value creation [5, 8, 9].

Pulic employed the Skandia Navigator Model in presenting VAIC^{TM}, but with a different meaning, where the components of IC in the Skandia Navigator comprise HC which includes all HC characteristics, for instance, ideas, experience, ability, skills, and so forth, also, SC that comprises the characteristics of other intangible assets, which consist of two categories of RC and organization capital. But in VAIC^{TM} model, HC comprises the investment in company employees who can transform knowledge into value through enhancing the current and expected products [10, 11].

Oppositely, the $VAIC^{TM}$ model depends on the company's VA, and it is an aggregate of capital employed efficiency (CEE), and intellectual capital efficiency (ICE) that comprises HCE and SCE [12, 13].

Within the context of Jordan, the industrial sector has been the leader in supplying products to the neighboring Arab countries, and globally as well [14]. Meanwhile, among industrial companies in Jordan, IC management has been lacking and managers were found to lack knowledge on the concept of $VAIC^{TM}$ [6, 7, 15, 16]. In addition, the impact of $VAIC^{TM}$ on firm performance within the Jordanian context has not been examined, particularly among the ASE listed industrial sector companies.

According to the previous problem discussed above, it is crucial to study the relationship between $VAIC^{TM}$ and firm performance. Therefore, this study aims to answer the following questions:

1. What is the relationship between $VAIC^{TM}$ and firm performance of the industrial sector in ASE?
2. What is the relationship between $VAIC^{TM}$ components and firm performance of the industrial sector in ASE?

Therefore, this paper is laid out as follows: Sect. 2 sets the paper objectives, while Sect. 3 highlights the literature review of the paper. Section 4 describes the theoretical background of the study and the conceptual framework in Sect. 5. While Sects. 6 and 7 are display the hypotheses and methodology of the study respectively. The study results coms in Sect. 8. Finally, Sects. 9 and 10 highlights the study's conclusions and limitations respectively.

2 The Paper Objectives

The main objective of this paper is to determine the influence of $VAIC^{TM}$ and firm performance in Jordan, the firm performance used is EPS and ROA of the industrial sector companies that listed in ASE during the period of 2008–2017.

So, this study aims to achieve the following objective to:

1. To understand the role of IC on the Jordanian industrial companies firm performance.
2. To examine the relationship between $VAIC^{TM}$ and firm performance of the industrial sector in ASE.
3. To examine the relationship between $VAIC^{TM}$ components and firm performance of the industrial sector in ASE.

3 Literature Review

The new concepts in accounting tend to be more concerned with intangible assets than physical assets because maximizing the value of physical assets comes because of intellectual assets. Moreover, the researchers found that the difference between the book values and the market value of companies, due to the intangible assets that are known as IC [17].

Many researchers have studied the issue of intellectual capital, but they have not agreed on a common definition for it. It is defined as the capacity of the employees that allow them to make intellectual contributions that increase the productivity of the company [18]. Moreover, Stewart defined IC as the knowledge that converts physical assets and makes it more valuable to the companies [19]. Also, IC was defined as a comprehensive and important strategy for organizations represented by the knowledge and skills that employees possess [20].

Most economists have classified IC in three main sections, which are: HC, that includes the company's employees knowledge, their skills, and experience, and its importance lies in what the company achieves from competitive advantages because of it, and SC, where it represents all values that remain in the company after employees leave it, such as machines, databases, company software, and many others, finally CC, which is the value the company obtains from loyal customers and suppliers, and any external sources that provide added value to the company [6]. Various studies have suggested more than 30 ratings and estimates of IC, one of which is the VAIC^{TM} model [5].

The VAIC^{TM} model created by Pulic (1998) measures the power of company in value creation [5]. Specifically, it measures the efficiency of value creation through the utilization of accounting numbers obtained from annual reports [8]. In the measurement of company performance, this method looks into financial and physical capital efficiency and Intellectual Capital efficiency (ICE) [4, 8, 21]. In other words, the value of IC itself is not measured by this method, rather, IC efficiency and financial and physical capital impact performance for the company are the measured constructs [22].

The VAIC^{TM} model partially follows Skandia Navigator, it is based on the following: IC is market value minus capital used for the company while IC is equivalent to HC plus SC [23]. Accordingly, (Fig. 1) illustrates the part of Skandia Navigator that underpins the VAIC^{TM} model. The equation below illustrates the concept of IC by way of Skandia Navigator [24].

$$Market\ Value = Capital\ Employed + IC$$
$$Market\ Value = Capital\ Employed + (HC + SC)$$

In the model, Pulic described IC as the knowledge of employees, and the knowledge can be converted into products that generate value for the company. As can thus be stated, employees are the actual investment of the company, and they are the

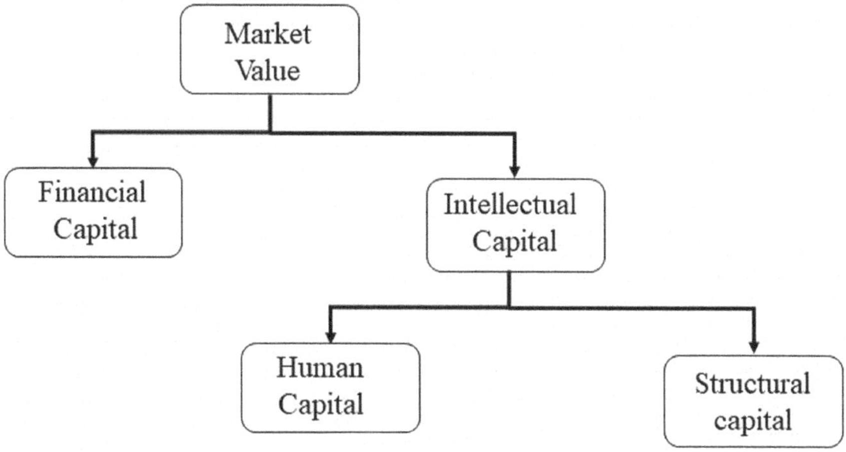

Fig. 1 Skandia navigator

major source of IC [8]. Within the knowledge economy, [11] explained that companies integrate two resources for value creation, and these resources are the physical and financial resources and IC. In regards to Skandia Navigator, it differs in terms of IC definition, whereby IC is described as a set of characterizes for HC, SC and CC [11].

VA as an appropriate tool for measuring value creation in the economy knowledge for the companies, considering that VA measures the productivity for all work types at the company level [4, 8, 9].

Then, VA is an indicator for the measurement of the knowledge performances of employees. In other words, VA is an indicator of knowledge investment. It is computed by dividing VA over the cost of labor. According to Pulic, VA refers to the value generated by intellectual knowledge investments per time unit [11], and therefore, it reflects the newly created value of certain period [8].

VA is computable as the difference between output, and the output entails the revenue or the total sales, while input entails all cost or expenses with the exception of employee cost [9]. In the process of value creation, employees play an integral role, and therefore, labor expenses cannot be classed as costs but as an investment [8].

Pulic methodology classes HC as employee expenses. In VA, employee expenditure is the new element, and is classed as HC. Hence, expenditure is not an expense but an investment of employees. The other component, which is SC, measures as the difference between VA and HC [25].

VAIC^{TM} model proposes that after VA determination, the efficiency of the used resources must be computed. This includes ICE which contains two components, and HCE that measures the company's ability in value creation through one monetary unit investment in its employees. The calculation is as follows: VA divided by HC.

Considering that SC refers to the situations that allow the employee or HC to create VA, SCE is measured by dividing SC by VA [8].

As the third component of VAIC^{TM}, CEE illustrates the created value per one monetary unit of shareholders capital. As demonstrated in many past works e.g., [4, 9, 11, 23], CEE is computed as VA divided by the book value for the company, and is referred as CE.

VAIC^{TM} has been regarded as a unique method as it allows the data to be attained from audited annual reports. IC is also included in financial statements at the company level. Such inclusion can provide affirmation of the degree of efficiency of value creation of a given company. Furthermore, the method allows comparison among companies within the same industry. For this reason, IC has been examined using VAIC^{TM} [4].

Somehow, owing to the misinterpretation towards Skandia Navigator concept and Pulic model, VAIC^{TM} has been criticized. In particular, the aforementioned models employ similar items but with different meaning of IC particularly concerning the items of HC and SC. In Skandia Navigator, HC and SC encompass all the characteristics associated with HC and SC, while in VAIC^{TM}, it is measure the value created from HC and SC. Additionally, in Skandia Navigator, SC is HC subtracted from IC [26], while in VAIC^{TM} model, SC refers to VA minus HC [8].

As stressed in [27], VAIC^{TM} parameters and IC are not related. In fact, the primary concept of VAIC^{TM} encompasses the measurement of value generated by knowledge possessed by employees as these employees have the capability in converting such knowledge into product. It creates value not to measure IC or it is components, but to measure the efficiency of IC and its components [11].

3.1 The Advantages of VAIC^{TM} Model

The advantages of using VAIC^{TM} model has been described in several studies e.g., [28–30], and some of these advantages are as discussed below:

1. The results generated by VAIC^{TM} model are in a form of numerical indicator which will be of value to stakeholders including customers, creditors, investors, and shareholders, as it offers them basis for making comparison between IC components.
2. VAIC^{TM} model entails a quantitative method that utilizes statistical analysis and calculations for significant number of companies, and it allows the use of historical and real data items obtained from financial statements of the companies over time.
3. The measurement of VAIC^{TM} model is quantitative in nature. Hence, comparison can be made between the model's results and the traditional financial ratios presented in company's financial statements.
4. The VAIC^{TM} model is simple enough that management and interested stakeholders can easily understand.

5. The VAIC^{TM} model is applicable at all levels (i.e., departmental, industrial, country) and the results can be compared as well. Hence, the model presents a standard for effective IC management.
6. The usage of accessible information provided by audited financial statement increases the reliability and efficiency of the results.
7. The VAIC^{TM} model is grounded upon VA that follows Resource-Based View theory (RBV) which describes the significance of IC for the companies.

4 Theoretical Background

This study employs three theories as follows: resource-based view (RBV), knowledge-based view (KBV), and Skandia Navigator. In specific, this study employs resource-based view and knowledge-based view to describe the significance of the shift in company assets and resources, as evidenced by the addition of intangible assets in financial statements of companies. Meanwhile, Skandia Navigator as utilized in Pulic is applied in this study as a basis for the VAIC^{TM} model.

4.1 Resource-Based View (RBV)

RBV introduced by Penrose (1959) has been the most popularly referred theory in the works related to strategic management concepts, and this theory has been a popular reference among scholars ever since the mid-1980s [31]. RBV is based on the premise that the competitiveness of the company is dictated by the resources and capabilities that the company possesses, and these resources can be in tangible and intangible forms [32]. Examples of intangible resources are client trust, relationships, and capabilities [33].

In the RBV theory the forthcoming competition for successful companies relies on the development of unique capabilities, and such capabilities are commonly implicit or intangible in nature. Among the examples of intangible assets are reputation, brand, and intellectual property [34].

In the attainment of sustainable competitive advantage, and the resources that a company possesses must have the following criteria: (a) the resources must be of value to the company or carry a positive value to the company, (b) the resources must be unique when compared with those of prospective or present rivals, (c) the resources must be hard to replicate, and (d) the resources must be irreplaceable by those of rivals [35].

RBV essentially revolves around the connection between IC and resources, capabilities, companies' profitability and competitive advantage, and also the investigation of the impact of such connections in the formation of sustainable competitive advantage and in the achievement of superior firm performances [34, 36]. The focal point of RBV is the resources within the company as follows: material resources, human resources, and structural resources. As mentioned earlier, the resources can be of tangible or intangible type. Besides, the need to clarify the profit and the value of the company [34].

4.2 Knowledge-Based View (KBV)

As suggested by RBV, sustainable competitive advantage of a company is determined by the company's assets, both tangible and intangible. Accordingly, the company knowledge is a primary component of IC, and for this reason, the knowledge can be used in establishing and preserving the core competencies, to assure the survival of the company in the competitive world [34].

A company comprises a group of union between entities that form knowledge through the utilization of material resources subordinated to the provided services [37]. Appositely, knowledge-based view (KBV) is an expansion of RBV, while the KBV concerns the knowledge that cannot be easily imitated which can be used in generating a competitive advantage [36]. The KBV of company distinguishes four characteristics of skills: knowledge and skills of employees, technical systems, management systems and database, knowledge construction and control processes [34, 38].

As such, sustainable competitive advantage highly depends on company knowledge, and company knowledge is a key component to IC. Hence, IC is arguably a key component in value creation of a company. Company knowledge is also among the key elements for competitive advantage. Moreover, the KBV illustrates the primary source for leveraging knowledge resources and performances considering that IC is a novel method for value creation. KBV can facilitate success [34, 39].

4.3 Skandia Navigator

Skandia navigator can be linked to a financial service company based in Sweden called Skandia, this company was the first to have measured knowledge assets. In 1985s, the director of IC department of Skandia, Mr. Edvinsson, created a new report on IC called Skandia Navigator. The report provided by Skandia included both financial and non-financial assets in attachments form to the traditional financial reporting of the company [40].

The non-financial assets (IC) are classed into two categories namely HC and SC. HC refers to the knowledge, experiences, and information possessed by the employees that are convertible to wealth, while SC is associated with information system, software, policies and patents [41].

SC includes customer capital or RC, and RC is associated with the link between a company and its customers, partners, owners and suppliers. Further, in the creation of competitive advantages, companies regard RC as key, and RC encompasses an

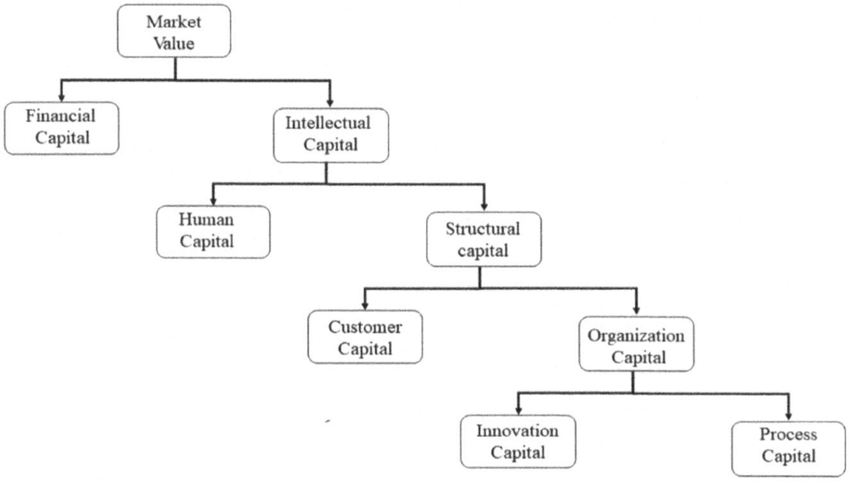

Fig. 2 Skandia navigator

organizational capital that assists employees in achieving their mission. As described in [42], organizational capital can be classed into two types namely innovation capital and process capital.

Skandia Navigator facilitates value creation comprehension especially for companies. Skandia Navigator proposes that for most companies, a gap exists between market value and book value, and this gap is particularly prevalent among companies that are undergoing mergers and acquisitions. The aforementioned gap is describable as IC which encompasses a sum of HC and SC. Meanwhile, market value relates to the aggregate of book value, IC and book value, and the concept follows Pulic but the meaning differs [43]. Notably, VAIC^{TM} model is grounded upon the key IC definition following Skandia Navigator that is similar in terms of HC and SC (see Fig. 2.), but with meaning that differs [8].

5 Conceptual Framework

The conceptual framework of this study is illustrated in (Fig. 3). Accordingly, to address the study questions, the figure illustrates the relationship between the variables. As can be observed in the proposed relationship, VAIC^{TM} comprises three components as independent variables and they include HCE, SCE, and CEE. The control variables comprise company size and company age, while the dependent variable to the relationship is firm performance represented by EPS and ROA.

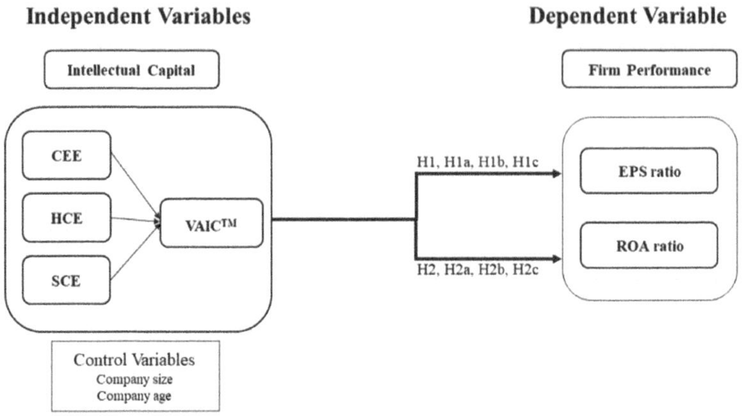

Fig. 3 Conceptual framework

6 Hypotheses Development

The relationship between VAIC™ and firm performance is the focal point of this study. Accordingly, the value of VAIC™ and its components to firm performance have been examined by other scholars (e.g., [44, 45]). In examining the proposed relationship, a negative link between VAIC™ and its components and firm performance has been reported in [46]. Meanwhile, the conceptual framework shows firm performance being represented by EPS and ROA. Relevantly, the hypotheses that this study is proposing are as follows:

Relationship between VAIC™ and EPS. EPS ratio refers to an index for the achieved income amount on a common share stock via an accounting period for the companies. Accordingly perceived EPS as a very important variable of interest to the investor as EPS assists investor in determining the amount that the market is able to pay for every dollar of profits [47].

Most past works found a positive relationship between VAIC™ and EPS ratio, which means that the increasing gap observed between market and book value relation is describable via the IC value measured by VAIC™.

In [48] reported a positive significant relationship between VAIC™ and ROE and EPS. In another study in Pakistan, found that VAIC™ impacted ROE, ROA and EPS when they examined 78 listed financial industries in the country in the period between 2008 and 2013. The authors reported the significant correlation between HCE, CEE and VAIC™ and all constructs of performance namely ROE, ROA and EPS [49].

Meanwhile, the relationship between VAIC™ and EPS ratio among industrial companies in Jordan has not been explored. For this reason, the present study expects to find a significant relationship between the constructs of VAIC™ and firm performance represented by EPS. The hypothesis is as follows:

H1: VAIC^{TM} has a significant relationship with EPS ratio in the industrial sector in Jordan.

This hypothesis (H2) is further divided into three sub hypotheses as follows:

H1a: HCE has a significant relationship with EPS ratio in the industrial sector in Jordan.

H1b: SCE has a significant relationship with EPS ratio in the industrial sector in Jordan.

H1c: CEE has a significant relationship with EPS ratio in the industrial sector in Jordan.

Relationship between VAIC^{TM} and ROA. ROA ratio functions as indicator of the company's profitability over the company's total assets. As described in [50] ROA can be used in the evaluation of company's assets' efficiency in the generation of earning. In theory, the knowledge industry including food and beverage, pharmaceutical, or chemical, efficient utilization of resources can increase production quantity, reduce cost, while also increasing quality. In other words, the provision of high-quality products can increase company revenue, and in turn, the company can attain more net income [51].

ROA ratio, therefore, provides an indication to investors and stockholders on the effectiveness of the company in transforming assets investment into earning. Additionally the higher ROA ratio is better because it means that the company has higher net income generated from assets investment [47]. Interestingly, examining 21 listed banks in Italy, found no link between IC, ROI, and return on asset ROA for the period between 2005 and 2007 [52].

In their study involving 20 industrial companies in Jordanian, [53] attempted to ascertain the impact of VAIC^{TM} on financial performance represented by ROA and ROE. From the results, the authors concluded a positive relationship between VAIC^{TM}, HCE, and SCE on ROA. On the other hand, there was no link between SCE and financial performance. While, [8] highlighted the need to understand how VAIC^{TM} and its different elements. For the purpose, the author presented a conceptual model that utilizes ROA ratio as a financial performance measure.

As such, the components of VAIC^{TM} are expected to be significantly linked to ROA ratio. Hence, among managements, investors, and stockholders, the utilization of VAIC^{TM} will give them idea on the effectiveness of the company in transforming the investment in assets into earning. Considering the past findings, the study presents the following hypothesis:

H2: VAIC^{TM} has a significant relationship with ROA ratio in the industrial sector in Jordan.

It is of particular interest of this study to gain understanding on how the diverse elements of VAIC^{TM} based on the conceptual model developed by [8] would relate to financial performance ROA ratio. Accordingly, H2 is further broken down into three sub hypotheses as follows:

H2a: HCE has a significant relationship with ROA ratio in the industrial sector in Jordan.

H2b: SCE has a significant relationship with ROA ratio in the industrial sector in Jordan.

H2c: CEE has a significant relationship with ROA ratio in the industrial sector in Jordan.

7 Research Methodology

The research methodology that the present study has chosen to employ is detailed in this section. Accordingly, this is further broken down into four subsections. Specifically, Sect. 7.1 discusses the study population, sampling, and data collections; Sect. 7.2 provides the details on the measurement of study variables which are broken down into independent, dependent, and control variables; and Sect. 7.3 provides the description on the multiple regression models.

7.1 Population and Sample of Research

The quantitative research approach is utilized in this study, and the data used are secondary data obtained from ASE database covering the period between 2008 and 2017. Similar to [54], this study will employ panel data in observing the variables in the companies, over different times. In order to achieve the study purpose, ASE listed companies in industrial sector in Jordan are the unit of analysis in this study.

Study population described as the entire group that comprises people, events, or things of interest of research [55]. Accordingly, all ASE listed companies in industrial sector in Jordan make up the study population. The sector comprises 11 sub-sectors, and there were 77 companies altogether. The period of study is between 2008 and 2017. This study also attempts to make a comparison between industrial companies, prior to and following the episode of Arab Spring. Table 1 presents the total number of ASE listed industrial companies in Jordan.

The data were obtained from annual reports of the sample companies in the period of 2008–2017. During the study period, there were 50 industrial company operating in Jordan. In choosing the companies as sample, this study followed the criteria below:

1. The company must be listed and active during the study period, that is, their shares must be published and traded in ASE during that period.
2. The financial year of the company must end on December 31st every year.

The sample used in this study comprises fifty companies with the following breakdown: five companies representing the Pharmaceutical and Medical Industries, eight companies representing the Chemical Industries, one company representing the Paper and Cardboard Industries, one company representing the Printing and Packaging industries, eight companies from Food and Beverage, two companies associated with Tobacco and Cigarettes, 11 companies representing the Mining and Extraction Industries, six companies that are involved in Engineering and Construction, three

Table 1 Population of the study

Sub-sectors	Population	N.	Percent (%)
1	Pharmaceutical and medical industries	6	7.79
2	Chemical industries	11	14.28
3	Paper and cardboard industries	3	3.89
4	Printing and packaging	2	2.59
5	Food and beverages	13	16.88
6	Tobacco and cigarettes	2	2.59
7	Mining and extraction industries	16	20.77
8	Engineering and construction	10	12.98
9	Electrical industries	5	6.49
10	Textiles, leathers and clothing	7	9.09
11	Glass and ceramic industries	2	2.59
Total		77	100

Table 2 Sample of the Study

Number	Sample	N.	Percent (%)
1	Pharmaceutical and medical industries	5	10
2	Chemical industries	8	16
3	Paper and cardboard industries	1	2
4	Printing and packaging	1	2
5	Food and beverages	8	16
6	Tobacco and cigarettes	2	4
7	Mining and extraction industries	11	22
8	Engineering and construction	6	12
9	Electrical industries	3	6
10	Textiles, leathers and clothing	5	10
Total		50	100

companies representing the Electrical Industries, and five companies involved in textiles, Leathers and Clothing (see Table 2).

7.2 *Measurement of Research Variables*

The variables were measured in order to determine the impact of VAIC^{TM} on firm performance of ASE listed industrial sector in Jordan during the period of 2008–2017.

The independent variable in this study is VAIC^{TM} based on [5]. The model functions as a performance measure for companies, and it is appropriate for measuring the efficiency of company's key resources in the new economy. VAIC^{TM} model comprises three basic resources namely HCE, SCE, and CEE [56]. Accordingly, there are five steps to follow in its usage as follows:

First step: Calculation of Value added (VA) in order to determine the most fitting index to achieve success. The measurement of VA is as follows:

$$\text{VA it} = \text{OUT it} - \text{IN it} \tag{1}$$

where,

VA it = Value Added of the company i in year t.

OUT it = Output of the company i in year t.

IN it = Input of the company i in year t.

In VAIC^{TM} model, employee expenses are classed as output. This is because employees play an active role in value creation process through the utilization of company infrastructure. Also, these employees utilize their intellectual abilities to create more value to the company [4, 57].

As explained in [4, 21], VA demonstrates the capability of the company in creating value from the investments of resources including dividends, interests and salaries, taxes. The following equation expresses the formula of VA [58]:

$$\text{VA it} = \text{OP it} + \text{EC it} + \text{D it} + \text{I it} + \text{Div it} + \text{T it} \tag{2}$$

where,

OP it = operating profit of company i in year t.

EC it = Employee cost (employee expenses) of company i in year t.

D it = Depreciation of company i in year t.

I it = Interest of company i in year t.

Div it = Dividend of company i in year t.

T it = Tax of company i in year t.

Second step: HCE estimation.

HCE functions as an indicator of the value amount in every unit of monetary formed in HC. HC is regarded as an indicator for the creation of value added because it provides estimation of the average profit per employee or the contribution made by the employee in the company's created value [9]. HCE is expressed as follows:

$$\text{HCE it} = \text{VA it} / \text{HC it} \tag{3}$$

where, HCE it = Human Capital Efficiency of company i in year t.

HC it = Human Capital determine by total salaries and wages of company i in year t.

Third step: SCE estimation

SC encompasses what stays within the company after the employees have left, for instance, database, infrastructure programs and software [59]. First, SC is estimated via the computation of the difference between VA and HC. The following equation is referred:

$$SC\ it = VA\ it - HC\ it \tag{4}$$

where SC it = Structural Capital of the company i in year t.

Next, SCE is estimated, through measuring the amount of required SC to generate a monetary unit of VA [60]. The estimation of SCE is based on the following equation:

$$SCE\ it = SC\ it\ /\ VA\ it \tag{5}$$

where SCE it = Structural capital Efficiency of company i in year t.

Fourth step: CEE estimation

The capital employed (CE) encompasses the book value of the company's net assets [4]. Meanwhile, CEE denotes the amount of generated VA via one monetary unit of the company's physical and financial capital. Then, CEE is estimated using the equation below:

$$CEE\ it = VA\ it\ /\ CE\ it \tag{6}$$

where, CEE it = Capital Employed Efficiency of company i in year t.

CE it = the book value of company i in year t.

Fifth step: VAIC^{TM} estimation

VAIC^{TM} denotes the amount of generated VA which per one monetary unit invested in each resource in the company [3, 11], and estimated as the following equation:

$$VAIC^{TM}it = HCE\ it + SCE\ it + CEE\ it \tag{7}$$

Dependent Variables: The effect of VAIC^{TM} on firm performance has been examined in many studies, by using traditional accounting ratios have been used. Among these ratios include Earnings Per Share (EPS), Return on Assets (ROA), and Market to Book Ratio (MB), [61, 62]. Accordingly, two ratios are used in this study, and they are EPS and ROA as described below:

1. Earnings per Share (EPS): This ratio refers to the amount earned income on the common stock within an accounting period [47]. The EPS is generally utilized by financial market analysts as a tool to evaluate companies [48]. EPS is based on the following equation [47]:

$$EPS = (\text{Net Income} - \text{preferred dividends})/\text{Weighted Average Number of share} \tag{8}$$

2. Return on Assets (ROA): This ratio relates to the competency of a company in their usage of total assets, while also showing the company's profitability, ROA is computable as net income divided by average total assets [63]. The estimation of this construct is based on the following equation [47]:

$$\text{ROA} = \text{Net Income / Average Total Assets} \tag{9}$$

Control variables: Control variables are variables that minimize the external influences which may impact the relationship between intellectual capital and corporate performance [64]. Accordingly, two control variables are included in this study, as they have been reported to affect the link between VAIC^{TM} and firm performance in past studies [65]. These variables are company size and company age, as described below:

1. Company size, which can be measured using diverse methods. In this study, this variable is measured using natural logarithm for book value of the company's total assets [66].
2. Company age, which can be measured using the formula below [49]:

$$\text{Company Age} = \text{current Year} - \text{Formation Year of the Company} \tag{10}$$

This variable will be measured using natural logarithm [67].

7.3 Multiple Regression Analysis Models

Several hypotheses were formulated in this study and then tested. For the purpose, multiple regression analyses were carried out, generating outcomes that determine the relationship between IC and firm performance in ASE listed industrial companies in Jordan. The models are as illustrated as follows:

First models: To examine the influence of VAIC^{TM} on EPS and ROA of the company i in year t.

Model 1: EPS it = $\alpha 0 + \beta 1$ VAIC^{TM} it + $\beta 2$ logsize it + $\beta 3$ logage it + εit.

Model 2: ROA it = $\alpha 0 + \beta 1$ VAIC^{TM} it + $\beta 2$ logsize it + $\beta 3$ logage it + εit.

Second models: To examine the influence of VAIC^{TM} components on EPS and ROA of the company i in year t.

Model 3: EPS it = $\alpha 0 + \beta 1$ HCE it +$\beta 2$ SCE it + $\beta 3$ CEE it + $\beta 4$ logsize it + $\beta 5$ logage it + εit.

Model 4: ROA it = $\alpha 0 + \beta 1$ HCE it +$\beta 2$ SCE it + $\beta 3$ CEE it + $\beta 4$ logsize it + $\beta 5$ logage it + εit.

EPS it = Earning Pare Share of company i in year t.

ROA it = Asset Turn Over Ratio of company i in year t.

VAIC^{TM} it = Value-Added Intellectual Coefficient of company i in year t.

HCE it = Human Capital Efficiency of company i in year t.

SCE it = Structural Capital Efficiency of company i in year t.
CEE it = Capital Employed Efficiency of company i in year t.
Logsize = Natural logarithm for book value of the total assets in the company.
Logage = Natural logarithm for company age.

8 Results and Discussion

The results of the study are discussed in this section, in three subsections as follows: Section 5.1 highlights the descriptive analysis, Sect. 5.2 discusses the correlations matrix between variables, and Sect. 5.3 highlights the results of regression.

8.1 *Descriptive Statistics*

Table 3 displays the descriptive statistics after the outliers have been deleted.

As demonstrated by the results, the main component of VAIC™ in the industrial sector in Jordan is HCE at 77%, and this is followed by CEE at 22.98%, and SCE is at only 0.02%. As can be observed in Table 3, the mean value for VAIC™ is 1.091 with a standard deviation of 0.447, while the minimum value is 0.477 and the maximum value is 1.7917. The mean value of HCE is 1.026, with a standard deviation of 0.634. Meanwhile, the minimum value of HCE is 0.111 while its maximum value is 2.507.

Table 3 Descriptive statistics

Variable	Obs	Mean	Std. Dev.	Min	Max
Independent variables					
VAIC™	464	1.090851	0.447161	0.47748	1.7917
HCE	464	1.02616	0.63374	0.11093	2.50787
SCE	464	−0.00001	0.09461	−0.224748	0.92219
CEE	464	0.306294	0.225237	0.021497	0.873476
Dependent variables					
EPS	464	0.037437	0.147857	−0.27	0.33
ROA	464	0.185332	0.86606	−2.866	4.3940
Control variables					
Size	464	16.50316	1.358569	11.94746	20.63084
Age	464	3.007863	0.7384435	0	4.189655

VAIC™ = Value added intellectual coefficient. **HCE** = Human capital efficiency. **SCE** = Structural capital efficiency. **CEE** = Capital employed efficiency. **EPS** = Earnings per share. **ROA** = Return on assets. **Size** = Natural logarithm company size. **Age** = Natural logarithm company age

This is followed by SCE with mean value of −0.00001, and standard deviation of 0.095, while the minimum value is −0.2247 and the maximum value is 0.92219. Further, the mean value of CEE is 0.3063, with standard deviation of 0.225, and the minimum value is 0.0215 and the maximum value is 0.873. For EPS, it scores a mean value of 0.0374, with a standard deviation of 0.1478, while the minimum value is 0.27 and the maximum value is 0.33. ROA has a mean value of 0.1853, with a standard deviation of 0.8661, while the minimum value is −2.866 and the maximum value reaches 4.39. Finally, the mean size is 16.5031 with a standard deviation of 1.3586 while the minimum value is 11.9475 and the maximum value is 20.6308. The mean age is 3.007863, with a standard deviation of 0.7384435, while the minimum value is 0 and the maximum value is 66.

8.2 Correlation Analysis

Table 4 displays the Pearson correlation matrix among variables with 464 observations covering 50 industrial companies for the period of 10 years. Gujarati (2001) accordingly stated that Pearson's correlation that is more than 0.8 denotes the presence of Multicollinearity problem.

Table 4 shows that in the correlation matrix, all correlation coefficients among the study's independent variables are lower than 0.80, with the exception of the correlation between VAIC^{TM} and HCE (0.955), and between VAIC^{TM} and SCE (0.810). Based on the models utilized in this study, this is not considered as a problem considering that each variable is formed within an isolated regression model. In addition, this study finds that the correlation between EPS and ROA is 0.855, and this is not considered a problem considering that each variable is of discrete dependent variables, as each is formed in a distinct regression model. Contrariwise, Table 4 shows that the correlation between the variables is positive and significant, especially between VAIC^{TM} and HCE, SCE, CEE, EPS, ROA, and company size at p <

Table 4 Pearson correlation matrix among variables (N = 464)

Variables	1	2	3	4	5	6	7	8
VAIC^{TM}	1							
HCE	0.96*	1						
SCE	0.81*	0.76*	1					
CEE	0.43*	0.39*	0.43*	1				
EPS	0.54*	0.54*	0.48*	0.31*	1			
ROA	0.53*	0.54*	0.52*	0.37*	0.86*	1		
SIZE	0.44*	0.45*	0.39*	0.21*	0.19*	0.23*	1	
Age	−0.12**	−0.09***	−0.07	−0.10*	0.06	0.05	0.05	1

Notes ***, **, * significant at the levels 1%, 5%, and 10% respectively

Table 5 Multicollinearity test

Variables	VIF	1/VIF
VAIC™	1.27	0.786648
HCE	2.60	0.384991
SCE	2.53	0.395448
CEE	1.26	0.794496
Size	1.29	0.775990
Age	1.04	0.957726

0.01 level. However, with company age, the correlation is negative and significant at $p < 0.05$ level.

Multicollinearity problem occurs if one of the independent variables is linked to one or more another independent variable [68]. Hence, in models of regression, Multicollinearity is not considered as an issue, as shown through Pearson matrix. And it is clearly by variance inflation factors (VIF) test for independent variables, and the value of VIF must be less than 10. Table 5 can be referred.

8.3 *Regression Results*

Panel data which comprise data series are used in this study. The use of panel data allows the detection of attitude of entities (individuals, companies, countries, etc.) across time [69]. The entities examined in this study comprise companies that were in operation during the study period.

The analytical model for panel data comes in three models, namely pooled model, random effect model, and fixed effect model. Each of them is as described below:

1. Pooled regression model (OLS): This model is called the Ordinary Least Square (OLS). In deciding between pooled model and random effect model, Breusch and Pagan Lagrangian multiplier test must be carried out to determine the random effects. As mentioned in several studies (e.g., [54, 70, 71], if the probability value for ch2 is significant, then random effect model should be chosen over pooled model.
2. Random effect model and Fixed effect model: Both types of model differ in a sense that the time independent in dependent variables in random effect model can be examined in regression model [41, 63].

A Hausman test is carried out to decide the appropriate model to use. Based on the result, if the probability value for ch^2 does not show significance, then, a random effect model should be used. However, if the probability value for ch^2 shows significance, then, a fixed effect model should be used [54, 72]. Results from Breusch and Pagan Lagrangian multiplier test and Hausman test can be viewed in Table 6.

Table 6 Breusch and Pagan and Hausman Test

Models	Breush & Pagan test		Results	Hausman test		Results
	Chi^2	Prob > chi^2		Chi^2	Prob > chi^2	
1	231.66	0.0000	Hausman	0.6800	0.8779	Random. M
2	202.59	0.0000	Hausman	2.8400	0.4162	Random. M
3	237.27	0.0000	Hausman	3.9700	0.5537	Random. M
4	174.23	0.0000	Hausman	4.1600	0.5264	Random. M

Table 7 Regression Results between VAIC™ and Firm Performance

Variables	Model 1 (EPS)		Model 2 (ROA)	
	Coeff	Z-stat	Coeff	Z-stat
VAIC™	0.1988612	13.36***	1.119056	12.36***
Size	−0.0044499	−0.53	0.0021736	0.04
Age	0.0237977	−1.85*	0.0837324	1.08
Constant	−0.1781613	−1.31	−1.323553	−1.06
Time effect	No		No	
R^2	38.34%		37.56%	
Chi^2 value	184.95		167.44	
Prob Chi^2	0.0000		0.0000	

Notes ***, **, * significant at the levels 1%, 5%, and 10%

The results of regression models are displayed in Table 7 for the hypotheses on the relationship between VAIC™ and EPS, and ROA as displayed in Models 1 and 2, for the period between 2008 and 2017 for ASE listed industrial companies in Jordan.

As can be viewed in Table 7, model 1 shows a good fit as shown by R^2 of 38.34% with Chi^2 probability 0.0000, company size, and age included in the model, while the 61.66% of the changes is explained by other reasons. Further, the coefficient of VAIC™ is 0.1988612 with Z-statistics of 13.36 at 1% significance level. This means that when VAIC™ increases by one JD, EPS of Jordanian industrial companies will increase by 19.89% JD, and it is statistically significant at 1% significance level. Company size shows insignificant impact on EPS but the company age has a negative and significant effect on EPS. H1 is therefore is supported.

The aforementioned outcome is expected as many past studies reported findings that are in line with this hypothesis, by concluding a significant relationship between VAIC™ and EPS ratio. Among these studies include [49, 61, 73]. Conversely, [10] reported no link between VAIC™ and EPS.

Moreover, the percentage of independent VAIC™ has a positive and significant relationship with ROA, and a significant level 1%. As can therefore be concluded, a positive relationship exists between VAIC™ and ROA. The obtained result show that the coefficient of VAIC™ is 1.119056, and this means when VAIC™ increases by

Table 8 Regression results between VAIC™ components and firm performance

Variables	Model 3 (EPS)		Model 4 (ROA)	
HCE	0.0912264	4.21***	0.4986238	3.26***
SCE	0.3001	2.57***	2.070613	1.86*
CEE	0.1071256	1.37	0.6300608	1.39
Size	−0.0083428	−1.03	−0.0299447	−0.75
Age	0.0152744	0.86	0.0068099	1.33
Constant	0.0022406	0.02	−0.1572038	−0.17
R^2	38.09%		45.67%	
Chi^2 value	88.32©		93.89©	
Prob Chi^2	0.0000		0.0000	
N- obs	464		464	

Notes ***, **, * significant at the levels 1%, 5%, and 10%

one JD, the ROA ratio of Jordanian industrial companies will increase by 1.119056 JD and it is statistically significant at 1% significance level. Hence, H2 is supported. Similar findings were also reported in past studies e.g., [74, 75].

H1 and H2 were tested to affirm if a relationship exists between HCE, SCE, and CEE and firm performance of ASE listed industrial companies during the period of 2008–2017.

Table 6 displays the results of Breusch and Pagan Lagrangian multiplier test and Hausman test. As shown by the table, the sub hypotheses H1a, H1b, and H1c (model 3) and the sub hypotheses H2a, H2b, and H2c (model 4) need to be tested using the random effect model. The results of multiple models of regression for the direct sub-hypotheses are displayed in Table 8.

As displayed in Table 8, R^2 is 38.09%, implying that 38.09% of variations in the EPS ratio of industrial companies in ASE were explained by the components of VAIC™ namely HCE, SCE, and CEE along with the controlling variables, while 61.91% of variations were explained by other factors. Also, the table shows that the probability of Ch^2 is 0.0000, which means that the overall model shows significance at 1%. As such, the model is affirmed to be fit.

The results of regression displayed in Table 8 show that H1a and H1b are supported, as a significant effect is observed. This means that there exists a relationship between HCE and SCE, and the EPS ratio, and the coefficient for HCE, SCE, is 0.0912264, 0.3001, with Z-statistic (6.41 and 3.27) respectively. However, H2c is not supported because $Z = 1.37$. As such, H1c is rejected. Comparatively, several past works reported a significant relationship between HCE, and CEE, and EPS as shown in [76]. Also, the results obtained in this study are different from those by [49] which concluded HCE as the only component with a significant relationship with EPS, while other components did not have a significant relationship. On the other hand, [73] reported SCE as the most effective factor, as opposed to HCE and CEE, before and after the subprime mortgage crisis, particularly with EPS.

The results for H2a, H2b, and H2c are presented in Table 8. As shown, R^2 is 45.67%, and this means that 45.67% of the variations in the ROA ratio of ASE listed industrial companies were explained by the components of VAIC^{TM} namely HCE, SCE, and CEE along with the controlling variables. Also, the probability of Ch^2 is 0.0000, and this means that the overall model demonstrates significance at 1%. As such, the model is fit.

The results of regression displayed in Table 8 denote support for H2a and H2b, as there exists a positive significant effect which denotes a relationship between HCE and SCE and the ROA ratio, and the coefficient for HCE and SCE is 0.4986238 and 2.070613, with Z-statistic (3.26 and 1.86) respectively. However, H3c is not supported as $Z = 1.39$. So, H2c is rejected.

Comparable results were reported in past studies, as a significant relationship between HCE, SCE, and CEE with ROA was found, as can be observed in [60, 77, 78]. Contrariwise, [73] reported SCE as an effective factor when compared to HCE and CEE, in relation to ROA. Meanwhile, [79] reported a significant relationship between HCE and CEE, and ROA, but the relationship was not significant between SCE and ROA. In several studies e.g., [49, 52] a significant link was found between CEE and HCE and ROA, while SCE did not show such link. On the other hand [80] reported that all components of VAIC^{TM} were not significant with ROA among SSE (Saudi Stock Exchange) listed companies. Meanwhile, company size and age as control variables do not show significance.

9 Conclusion and Recommendation

This study, discussed the relationship between IC and firm performance among ASE listed industrial companies was examined, involving the use of VAIC^{TM} in IR 4.0 for IC measurement. Data were obtained from both the ASE database and annual reports of the examined companies during the period of 2008–2017.

The obtained results generally show the significance of VAIC^{TM} in determining the firm performance of industrial companies operating in Jordan. Essentially, the role of VAIC^{TM} and the firm performance are shown to increase profit, and these can be made a positive indicator that supports the firm performance of ASE listed industrial companies. Additionally, the present study found a positive and significant relationship between HCE and SCE and both EPS and ROA increase the profit as a positive indicator that supports the firm performance of the industrial companies throughout the study period.

Within the context of IR 4.0, the present study greatly contributes to the knowledge of IC in the context of firm performance. Firstly, the direct impact of IC on firm performance of industrial companies operating in Jordan was found in this study. Also, the present study found an increasing interest towards the technological developments caused by IR 4.0 in Jordan. Such interest was manifested following the economic crisis during the first quarter of 2020 following the outbreak of Coronavirus19 pandemic. Specifically, the pandemic has disrupted all economic and social activities globally, irrespective of country. Nonetheless, Jordan has infrastructure that can somewhat adapt to the new condition. For instance, the infrastructure that Jordan has allows education to continue, through distant education, while many companies can still operate as meetings and certain works can still be carried out online.

The results of this study have several implications, it might be helpful for policymakers in their decisions regarding the development of the industrial sector in Jordan. Moreover, the results of the current study are important to understand the current situation of VAIC^{TM} components' in Jordanian industrial companies in ASE and the significance of VAIC^{TM} components on the firm performance. So, the study suggests that the managers and business leaders in the industrial companies in Jordan can develop strategies to confirm the importance of IC to the enhancement of the most significant IC components to get better performance in the companies.

For future researches, it is suggested that all ASE listed companies from various sectors (industry, education, healthcare, etc.) are included in future studies, and the results of these companies are compared, particularly on the impact of intellectual capital on the firm performance. Also, policymakers in Jordan and Middle East could benefit from utilizing the results of this study especially in their decisions related to the development of the industrial sector in the countries. Additionally, this study found that awareness of managers of industrial companies in Jordan will ease them in enriching their knowledge, and this in turn will improve and increase corporate returns. As such, methods of management will increase the implementation success of new ideas and new knowledge.

10 Limitation of Study

1. Considering that the used sample was limited to the ASE listed industrial companies, the obtained results cannot be generalized to other non-listed companies operating the Middle East region.
2. The data utilized in the present study were quantitative data of VAIC^{TM}, and ratio of firm performance furnished by the ASE database and published annual reports. On the other hand, the qualitative portions of VAIC^{TM} were not included in this study.
3. The period of the study was when two major crises occurred. The first crisis was the mortgage crisis impacting the economy of Jordan from 2008 to 2011. The second crisis was the Arab Spring that began since late 2011 until the present time.

References

1. Hussinki, H., Ritala, P., Vanhala, M., Kianto, A.: Intellectual capital, knowledge management practices and firm performance. J. Intellect. Cap. **18**(4), 904–922 (2017)
2. Guoping, L., Yun, H., Aizhi, W.: Fourth industrial revolution: technological drivers, impacts and coping methods. Chin. Geogr. Sci. **27**(4), 626–637 (2017)
3. Clarke, M., Seng, D., Whiting, R.H.: Intellectual capital and firm performance in Australia. J. Intellect. Cap. **12**(4), 505–530 (2011)
4. Pulic, A.: The principles of intellectual capital efficiency—A brief description. Croat. Intellect. Cap. Center, Zagreb, 2–25 (2008)
5. Pulic, A.: Measuring the performance of intellectual potential in the knowledge economy. In: The 2nd World Congress on the Management of Intellectual Capital, pp. 1–20 (1998)
6. Momani, K.M.K., Nour, A.N.I.: The influence of intellectual capital on the return of equity among banks listed in Amman Stock Exchange. Int. J. Electron. Bank. **1**(3), 220 (2019)
7. Momani, K.M.K., Jamaludin, N., W.Z. @ W. Zanani Wan Abdullah, W., Nour, A.-N.I.: The effects of intellectual capital on firm performance of industrial sector In Jordan. Hum. Soc. Sci. Rev. **8**(2), 184–192 (2020)
8. Pulic, A.: VAIC ^{TM} – an accounting tool for IC management Ante Pulic. Int. J. Technol. Manag. **20**, 702–714 (2000)
9. Pulic, A.: Intellectual capital—Does it create or destroy value? Meas. Bus. Excell. **8**(1), 62–68 (2004)
10. Bakhsha, A., Afrazeh, A., Esfahanipour, A.: A criticism on Value Added Intellectual Coefficient (VAIC) model a criticism on Value Added Intellectual Coefficient (VAIC) model. IJCSNS Int. J. Comput. Sci. Netw. Secur. **17**(6), 59–71 (2017)
11. Iazzolino, G., Laise, D.: Value added intellectual coefficient (VAIC): A methodological and critical review. J. Intellect. Cap. **14**(4), 547–563 (2013)
12. Mondal, A.: Application of modified VAICTM model for measuring intellectual capital performance. Int. J. Res. Financ. Mark. **6**(11), 19–30 (2016)
13. Sharabati, A.-A.A., Nour, A.-N.I., Jabali, S.M.: Relational capital and Jordanian pharmaceutical manufacturing organisations' business performance. Eur. J. Bus. Innov. Res. **1**(2), 85–100 (2013)
14. Al-mahrouq, M.: Industrial Sector in Jordan and Export Potential To The EU Market (2017)
15. Al-shubiri, F.N.: Testing the relationship between the efficiency of value added intellectual coeffi cient and corporate performance at commercial banks in Amman Stock Exchange (ASE). Zagreb Int. Rev. Econ. Bus. **14**(1), 1–22 (2011)
16. Sharabati, A.-A.A., Jawad, S.N., Bontis, N.: Intellectual capital and business performance in the pharmaceutical sector of Jordan. Manag. Decis. **48**(1), 105–131 (2010)
17. Smriti, N., Das, N.: The impact of intellectual capital on firm performance: a study of Indian firms listed in COSPI. J. Intellect. Cap. **19**(5), 935–964 (2018)
18. Bramhandkar, A., Erickson, S., Applebee, I.: Intellectual capital and organizational performance: an empirical study of the pharmaceutical industry. J. Knowl. Manag. **5**(4), 357–362 (1990)
19. Stewart, T., Ruckdeschel, C.: Intellectual capital: The new wealth of organizations. Perform. Improv. **37**(7), 56–59 (1998)
20. Bontis, N.: Intellectual capital: an exploratory study that develops measures and models. Manag. Decis. **36**(2), 63–76 (1998)
21. Pulic, A.: Do we know if we create or destroy value? Int. J. Entrep. Innov. Manag. **4**(4), 349–359 (2004)
22. Pratiwi, E., Kadry, R.: Islamic Banking - Value Added Intellectual Coeficient (IB-VAIC) as an intellectual capital proxy indonesian islamic banking. Glob. Rev. Islam. Econ. Bus. **2**(2), 127–135 (2014)
23. Nazari, J.A., Herremans, I.M.: Extended VAIC model: Measuring intellectual capital components. J. Intellect. Cap. **8**(4), 595–609 (2007)

24. Anghel, I.: Intellectual Capital and Intangible Assets Analysis and Valuation (2014)
25. Nadeem, M., Gan, C., Nguyen, C.: The importance of intellectual capital for firm performance: evidence from Australia. Aust. Account. Rev. **28**(3), 334–344 (2018)
26. Edvinsson, L.: Developing intellectual capital at Skandia. Long Range Plann. **30**(3), 366–373 (1997)
27. Ståhle, P., Ståhle, S., Aho, S.: Value added intellectual coefficient (VAIC): A critical analysis. J. Intellect. Cap. **12**(4), 531–551 (2011)
28. Firer, S., Stainbank, L.: Testing the relationship between intellectual capital and a company' s performance: Evidence from South Africa University of Natal Durban. Meditari Account. Res. **11**, 25–44 (2003)
29. Utama, A., Mirhard, R.: The influence of sustainability report disclosure as moderating variable towards the impact of intellectual capital on company's performance. Int. J. Econ. Finan. Issues **6**(3), 1262–1269 (2016)
30. Joshi, M., Kansal, M., Sharma, S.: Awareness of intellectual capital among bank executives in India: a survey. Int. J. Account. Inf. Manag. **26**(2), 291–310 (2018)
31. Boxall, P.F., Purcell, J.: Strategic HRM and the resource based view of the firm. Strateg. Hum. Resour. Manag. **1999**(10), 71–88 (2003)
32. Barney, J.: Firm resources and sustained competitive advantage. J. Manage. **17**(1), 99–120 (1991)
33. Clulow, V., Barry, C., Gerstman, J.: The resource-based view and value: The customer-based view of the firm. J. Eur. Ind. Train. **31**(1), 19–35 (2007)
34. Radjenovic, T., Krstic, B.: Intellectual capital in the theory of the firm. Ekonomika **63**(4), 13–27 (2017)
35. Madhani, P.M.: Resource based view (RBV) of competitive advantage: an overview. Open J. Syst., 3–22 (2014)
36. Marr, B., Gray, D., Neely, A.: Why do firms measure their intellectual capital? J. Intell. Capital **4**(4) (2003)
37. Spender, J.: Making knowledge the basis of a dynamic theory of the firm. Strateg. Manag. J. **17**(S2), 45–62 (1996)
38. Curado, C.: The knowledge based-view of the firm: from theoretical origins to future implications. Chiropr. Osteopat. **14**(1), 3 (2006)
39. Evaggelia, F.: Intellectual capital & organizational advantage : an economic approach to its valuation and measurement. Int. J. Inf. Bus. Manag. **7**(1), 36–57 (2015)
40. Chen, J., Zhu, Z., Xie, H.Y.: Measuring intellectual capital: a new model and empirical study. J. Intellect. Cap. **5**(1), 195–212 (2004)
41. Muhammad, N.M.N., Ismail, M.K.A.: Intellectual capital efficiency and firm's performance: Study on Malaysian financial sectors. Int. Rev. Econ. Financ. **1**(2), 206–212 (2009)
42. Tripathy, T., Gil-Alana, L.A., Sahoo, D.: The effect of intellectual capital on firms' financial performance: An empirical investigation in India. Int. J. Learn. Intellect. Cap. **12**(4), 342–371 (2015)
43. Iazzolino, G., Laise, D., Migliano, G.: Measures of Value Creation: a comparison between VAIC and EVA. In: Proceedings of the IFKAD-ISSN, vol. 2280, 787X, pp. 501–514 (2013)
44. Khanqah, V.T., Khosroshahi, M.A., Ghanavati, E.: An empirical investigation of the impact of intellectual capital on firms' market value and financial performance: evidence from Iranian companies. Int. J. Manag. Bus. Res. **2**(1), 1–12 (2012)
45. Nimtrakoon, S.: The relationship between intellectual capital, firms' market value and financial performance: Empirical evidence from the ASEAN. J. Intellect. Cap. **16**(3), 587–618 (2015)
46. Mehralian, G., Reza, H.: The impact of intellectual capital efficiency on market value: an empirical study from iranian pharmaceutical companies. Iran. J. Pharm. Res. **11**(1), 195–207 (2012)
47. Gibson, C.H.: Financial Reporting and Analysis, 11th edn. South-Western Cengage Learning (2011)
48. Tan, H.P., Plowman, D., Hancock, P.: Intellectual capital and financial returns of companies. J. Intellect. Cap. **8**(1), 76–95 (2007)

49. Ahmad, M., Ahmed, N.: Testing the relationship between intellectual capital and a firm's performance: an empirical investigation regarding financial industries of Pakistan. Int. J. Learn. Intellect. Cap. **13**(2/3), 250 (2016)
50. Heikal, M., Khaddafi, M., Ummah, A.: Influence analysis of Return on Assets (ROA), Return on Equity (ROE), Net Profit Margin (NPM), Debt To Equity Ratio (DER), and current ratio (CR), against corporate profit growth in automotive In Indonesia Stock Exchange. Int. J. Acad. Res. Bus. Soc. Sci. **4**(12), 101–114 (2014)
51. Mehralian, G., Rasekh, H.R., Akhavan, P., Ghatari, A.R.: Prioritization of intellectual capital indicators in knowledge-based industries: evidence from pharmaceutical industry. Int. J. Inf. Manage. **33**(1), 209–216 (2013)
52. Puntillo, P.: Intellectual capital and business performance. Evidence from Italian banking industry. Electron. J. Corp. Financ. **4**(12), 97–115 (2009)
53. Haan, A.-A., AL-Sakin, S.A.-K., AL-sufy, F.J.H.: The effect of value added intellectual coefficient on firms' performance : evidence from Jordanian Industrial Sector. Res. J. Finan. Acc. **7**(14), 163–174 (2016)
54. Baltagi, B.H.: Econometric Analysis of Panel Data, 3rd edn. Wiley (2005)
55. Sekaran, U.: Research Methods for Business, 4th edn., vol. 65, no. 3. United States of America (2013)
56. Fijałkowska, J.: Value Added Intellectual Coefficient (VAIC^{TM}) as a tool of performance measurement. Przedsiebiorczosc i Zarz. **15**(1), 129–140 (2014)
57. Makki, A.M., Lodhi, S.A., Rohra, C.L.: Impact of intellectual capital on shareholders earning. Aust. J. Basic Appl. Sci. **3**(4), 3386–3398 (2015)
58. Belkaoui, A.R.: Intellectual capital and firm performance of US multinational firms. J. Intellect. Cap. **4**(2), 1–26 (2003)
59. Abdulsalam, F., Al-Qaheri, H., Al-Khayyat, R.: The intellectual capital performance of KuwaitiBanks: an Application of vaicTM1 Model. iBusiness **3**(1), 88–96 (2011)
60. Sherif, M., Elsayed, M.: The impact of intellectual capital on corporate performance: evidence from the Egyptian insurance market. Int. J. Innov. Manag. **20**(03), 1650034 (2016)
61. Amin, S., Aslam, S.: Relationship between intellectual capital and financial performance: the relationship between intellectual capital and financial performance: the moderating role of knowledge assets. Pak. J. Commer. Soc. Sci. **12**(August), 521–547 (2018)
62. Kamath, G.B.: Impact of intellectual capital on financial performance and market valuation of firms in India. Int. Lett. Soc. Hum. Sci. **48**(1997), 107–122 (2015)
63. Steger, R., Mankin, J., Jewell, J.: How to organize a real life problem-based learning project in a business class using strength assessment to determine team assignment. J. High. Educ. Theory Pract. **11**(1), 45–55 (2011)
64. Hill, R., Murphy, G., Trailer, J.: Measuring performance in entrepreneurship research. J. Bus. Res. **36**(1), 15–23 (1996)
65. Dženopoljac, V., Muhammed, S., Janošević, S.: Intangibles and performance in oil and gas industry. Manag. Decis. **57**(5), 1267–1285 (2019)
66. Dang, C., Li, Z. (Frank), Yang, C.: Measuring firm size in empirical corporate finance. J. Bank. Finan. **86**, 159–176 (2018)
67. Yasuda, T.: Firm growth, size, age and behavior in Japanese manufacturing. Small Bus. Econ. **24**(1), 1–15 (2005)
68. Gujarati, D.N.: Basic Econometrics, 4th edn. Gary Burke, Americas, New York (2001)
69. Nwakuya, M.T., Ijomah, M.A.: Fixed effect versus random effects modeling in a panel data analysis; a consideration of economic and political indicators in six african countries. Int. J. Stat. Appl. **7**(6), 275–279 (2017)
70. Baltagi, B.H., Li, Q.: On instrumental variable estimation of semiparametric dynamic panel data models. Econ. Lett. **76**(1), 1–9 (2002)
71. StataCorp: Data Analysis and Statistical Software (2008)
72. Hsiao, C.: Analysis of Panel Data, 2nd edn. Cambridge University Press (2003)
73. Nassar, S.: The impact of intellectual capital on firm performance of the Turkish real estate companies before and after the crisis. Eur. Sci. J. **14**(1), 29–45 (2018)

74. Alipour, M.: The effect of intellectual capital on firm performance: An investigation of Iran insurance companies. Meas. Bus. Excell. **16**(1), 53–66 (2012)
75. Kehelwalatenna, S.: Intellectual capital performance during financial crises. Meas. Bus. Excell. **20**(3), 55–78 (2016)
76. Rehman, W., Abdul Rehman, C., Rehman, H., Zahid, A.: Intellectual capital performance and its impact on corporate performance: An empirical evidence from Modaraba sector of Pakistan. Afr. J. Bus. Manag. **5**(20), 8041–8049 (2013)
77. Chan, K.H.: Impact of intellectual capital on organisational performance: An empirical study of companies in the Hang Seng Index (Part 1). Learn. Organ. **16**(1), 4–21 (2009)
78. Sany, Hatane, S.E.: The impact of value added of intellectual capital to firm S. In: 4th International Conference on Management, pp. 342–355 (2014)
79. Konde, Y.T., Oktavianti, B., Hijrah, L.: The analysis of the effect of NPL on ROA in the banking companies listed on The Indonesia stock exchange. J. Appl. Philos. Manag. Innov. **1**(1), 43 (2018)
80. Buallay, A.: The relationship between intellectual capital and innovation: a review. Corp. Gov. Organ. Behav. Rev. **1**(June), 32–41 (2018)

Potentials of Artificial Intelligence for Business Performance

Arezou Harraf and Hasan Ghura

Abstract The Fourth Industrial Revolution and its impact on business, has been a subject of much consideration. As with many other innovations and earlier industrial revolutions, there is much uncertainty in regards to the benefits and adverse impact of the new technologies. In this chapter, we provide an overview of what is Artificial Intelligence and its potentials for business. The theoretical background in support of its utilization for businesses, as well as prominent frameworks which guide the efficient integration of AI into organizations. In addition, we discuss the empirical evidence based on to-date research that supports the potential of AI in improving firm performance, only if the organization has an adequate human capital capability in integrating these new technologies across its processes.

Keywords Artificial intelligence · Big data · Performance

1 Introduction

The term the 'Fourth Industrial Revolution' and its impact on businesses and human capital has been the subject of considerable discussion among researchers and practitioner alike. A leading research firm IDC predicts Artificial Intelligence (AI) will account for $52 Billion in global economy by 2021. In the same vein, Forrester-a market research company predicted that companies will expand their focus on adopting AI's practical solutions for immediate business benefits across the board [1]. Moreover, in an interview with Carlos Melendez the co-founder and COO of Wovenware an artificial development company, he predicted that the relevance and use of Artificial Intelligence (AI) will skyrocket within multiple industries as they will increasingly rely on AI and big data to make more informed business decisions. He predicted AI would be in great need in such areas as image, object, and facial recognition to more clearly identify and analyze movement for such purposes as

A. Harraf (✉) · H. Ghura
Department of Business Studies, Box Hill College Kuwait, Abu Halifa, Kuwait
e-mail: a.harraf@bhck.edu.kw

A. Hamdan et al. (eds.), *The Fourth Industrial Revolution: Implementation of Artificial Intelligence for Growing Business Success*, Studies in Computational Intelligence 935,
https://doi.org/10.1007/978-3-030-62796-6_5

fraud prevention and security and pattern assessment, to healthcare and law enforcement applications. Moreover, Carlos contended, given the hyper competition in the market, firms can no longer just look at static intelligence that focuses on past performance for the purposes of business intelligence, and would rather be able predict as accurately as possible what is expected to happen in the future to better anticipate and plan changes through use of AI algorithms [1]. As companies grow their reliance on data, their focus would shift to quantifiable and data-driven performance matrices, heightening the need for constant data gathering and processing and more scientists to tackle the amount of data needed to run various organizations. Consequently, it is needless to say, that AI is changing the market and has great impact on businesses and the above-mentioned points are just starting points of AI potential for business [1]. This said, however, there are numerous challenges as how this technology would impact the human capital and how organizations will balance between using technologies such as AI for growth with the needs and welfare of their workforce. This as [23] contented that nearly 47% of U.S employment is at risk from some form of Automation. In this vein, citing growth theory, using the singularity hypotheses, as AI replaces human labor firms would experience great growth which propels organizations to adopt the technology for superior growth [21]. On the other hand, based on empirical data [42] contends that this replacement is far from imminent in the near future, and in fact automation is a quite complimentary addition to a solution-oriented, creative and skilled workforce highlighting the importance of workforce education parallel with the technological advancement and adoption. Moreover, based on an empirical study of 3000 organizations in Japan, [40] found that organizations with more skilled and creative workforce are more open about absorption of Big Data Analytics and Artificial Intelligence as they see it as a great addition to the current skillset of the workforce, once again, highlighting the importance of workforce development in adaptability and capability of working with these technologies. Furthermore, lessons from previous industrial revolutions have proved that advent of new technologies bring about new products and services and consequently many new occupations as a result [39].

This said, however, before delving into an in-depth analysis of how the Fourth Industrial Revolution in general and Bid Data Analytics and Artificial Intelligence (AI) in particular– impact financial and non-financial matrices of performance within organizations; it is pertinent to give a brief overview of theories and frameworks that advocate for the use of such technologies and recommended methodologies for incorporating (AI) in organizations and determining what are BDA and AI potentials, before proceeding to analyze its impact on business performances. This chapter therefore, aims to shed light on the current state of Big Data and AI in firms and its potential uses and benefits in the future uses and various frameworks for successful application of these technologies within firms.

2 Intelligent Machines—How Are They Different?

The fourth industrial revolution is focused on data and its interactions with machines. While there are plenty of machines in use today, the primary difference between an intelligent machine is its ability to learn from experiences as opposed to only performing a certain function, and applying the new level of knowledge into action [14]. While this characteristic sets intelligent machines such as AI apart, it allows them much potential to impact all aspects of society in general and the businesses in particular.

Most of the research to date has focused on the potential for increased efficiency and growth from incorporation of Big Data Management (BDM) and Artificial Intelligence (AI), while little attention has been made in reference to the interplay between human intelligence and AI throughout this process [3]. This said, the interplay between Human and Artificial Intelligence is not a one size fit all affairs, the extent of this interaction depending on the required task and type of industry takes different shapes.

In this vein, [37] has classified the extent of this interaction in four categories as shown in Fig. 1 and as evidenced in

- Standard: refers to instances when AI takes over with little to no human interaction,

  For example, in automated manufacturing machines.
- Substitute: This type of intelligence tends to substitute human intelligence in performing low-skilled tasks such as those used in chatbots.
- Superiority: In this type of interaction, human intelligence is far superior to AI, particularly in tasks requiring, creativity, wisdom and empathy.

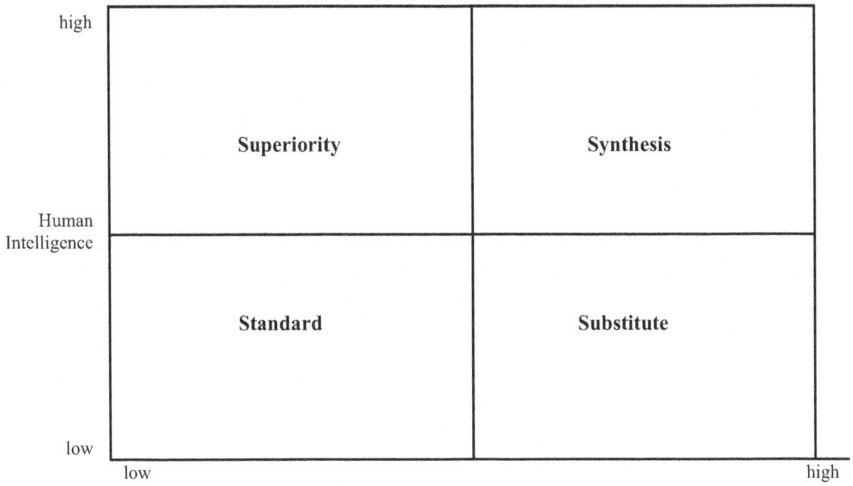

Fig. 1 Human and Artificial Intelligence Interaction

- Synthesis: It occurs when AI augments the human intelligence through helping with decision-making precision as a result of data-mining algorithms.

In this chapter, therefore, we are going to discuss various aspects of AI as it relates to performance, and some models of absorption and integration of this new technology into firms for maximum potential. As well as, best ways to determine risks from its integration and possible mitigation from such endeavors.

3 Theoretical Background

Various theoretical frameworks have been suggested and utilized in determining the benefits of incorporating artificial intelligence in improving organizational outcomes such as sustainability and performance. While number of theoretical perspectives such as dynamic capabilities theory, cognitive theory as well as market orientation perspective [24, 30]. However, two theoretical approaches have received more momentum in linking benefits of AI to firm performance. Resource-based view theory (RBV) [54] as well as Information System Success Model (ISSM) [19] are these theories that are used in advocating for the use of Big Data Analytics inclusive of AI in achieving improved organizational performance and sustainable competitive advantage [8]. Use of BDA and AI as a source of competitive advantage in accordance with RBV is further supported by [4, 55], as they posited such capabilities cover many organizational facets such as its management, human talent and technology which have a positive influence on organizational performance. In the same vein empirical research by [27] suggested that utilization of BDA led to improved market and operational performance in the studied firms. Additional studies have adopted RBV to better understand organizational capabilities that lead to improved performance in such industries as health care [9], manufacturing [44], and retail [55]. While other studies have opted to use the ISSM model in understanding the role of data, information, and system quality on organizational performance, thus looking at the role of Big Data and AI through the lens of information and system quality in determining its benefits [32].

Despite the underpinning theory used to determine BDA capabilities that impact organizational performance [6] found three key themes based on their extensive literature review in describing BDA capabilities. The first emergent theme is possession of adequate personnel and technology in using the information from BDA and transforming it into a competitive force [4, 55]. The second common theme is the ability to efficiently acquire, store and analyze large amounts of data in discovering business values and insightful information. The third theme is capability to integrate and organize big-data based resources such as management, organizational, talent pool and technology capabilities in big firms [27]. Athique [6] therefore, recommended developing a framework regarding the factors and elements that affect Big Data Implementation Assessment Model (BDIAM) for organizations drawing on RBV and ISSM application in the literature. This model takes into consideration the common themes and success factors of: organization, people and technology and their effective interaction on decision-making in firms. These themes are further divided into three factors of: 1—Perform Data Strategy, 2—Collaborate Knowledge Worker, and 3- Execute Data Analytics.

Perform Data Strategy factor eludes to alignment of the BDA goals with the goals of the business [27, 44, 55]. While the Collaborative Knowledge Worker refers to the securing the support of the senior management and organization leaders in providing support and adequate resources for implementing the BDA system [4, 31, 34]. Additionally, the Execute Data Analytics factor refers to adequate management of resources which is inclusive of people, technology and competency needed for BDA implementation [4, 36, 55].

4 Integration of Artificial Intelligence with Business Processes

Before organizations can benefit from AI in an impactful way, organizations must set in place an integrated intelligence model that combines and augments various intelligences with AI for maximum impact [56]. As contended by the authors, Integrated Intelligence Model encompasses various forms of intelligence such as Artificial Intelligence, Business Intelligence, Competitive Intelligence, Decision Intelligence and lastly, Emotional Intelligence [56]. Accordingly, leveraging each type of intelligence is imperative in addressing specific organizational issues and their integration makes the process of strategic decision-making more effective across organizations [56]. In making their case for the need to combine the above forms of intelligence for successful strategy setting across organizations the authors relied on the wisdom pyramid also known as Data, Information, Knowledge and Wisdom (DIKW) which is commonly used in information technology research [2, 17, 46, 51, 56, 57]. This framework focuses on different spectrum of human experience which uses different elements of understanding in the process of sound decision-making.

As illustrated by Fig. 2, the data analytics is a three phase process. The first phase is knowledge discovery phase, which the phase where knowledge is transformed into information. Phase 2, entails knowledge creation which is the transformation of information to knowledge. Phase 3, alludes to knowledge creation where knowledge is transformed to wisdom that helps with the process of effect strategic decision-making based on the given data. This process is completed as aforementioned by integrating the five type of intelligence.

Artificial Intelligence: In AI technology, Machine Learning pivots as a bridge in bringing closer information and knowledge. Machine Learning, in turn is comprised of three types of: supervised learning, unsupervised learning, and reinforcement learning [14]. These three types of learning within the realm of AI gives humans the ability to fill in the gap where they have shortcomings depending on the tasks they are undertaking. For example, supervised learning is suitable when performing

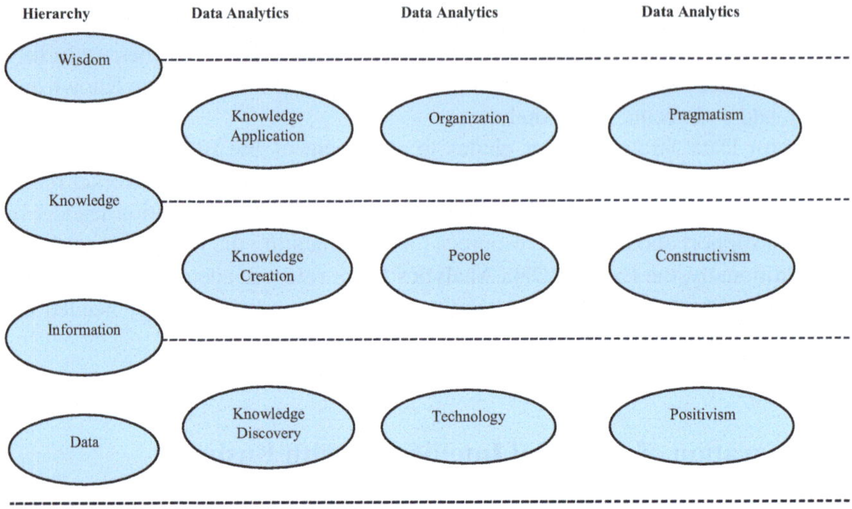

Fig. 2 Three phases of Data Analyses [56]

statistical tasks that can aide with ascertaining correlations between different variables. The unsupervised learning however, could be used in instances of clustering large amounts of data and determining certain patterns that can help with business decision-making that is data driven. While, enforcement learning focuses on learning through maximizing the reward function of an action, and is simply done through a trial and error strategy [14]. This said, however, for any type of AI learning to be successful, it would need to interact with human intelligence for it to bear fruit.

Business Intelligence: Business intelligence refers to mining the ever increasing data and its analytics for commercial intelligence purposes. This type of intelligence collects and analyzes data in determining patterns based on internally gathered data from organizations operations and determines course of actions accordingly [16, 53]. This type of intelligence however, is increasingly reliant on AI in determining data and patterns and knowledge discovery from various sources, through different sets of statistical and AI tools in making more informed decisions for the organization [5, 35].

Competitive Intelligence: Unlike business intelligence, completive intelligence seeks to gather, analyze and understand data from external sources to determine the organizations capability in comparison to other firms in regards to market analysis, technology analysis and capability analysis [49]. This form of intelligence is integral for firms' competitive advantage as they highlight the strengths and shortcomings of current capabilities in comparison with competitors and market trends [38]. While, human intelligence, to-date is integral to gathering this type of intelligence, more and more organizations really on AI technology in perfecting this type of data gathering in making informed data-driven decisions [33].

Decision Intelligence: The best way to define decision intelligence is the application knowledge management in compiling the data and information from business intelligence which is the current data based on internal operations and competitive intelligence which is data and information based on the external knowledge information in forecasting future decisions that are data-driven and forward looking [41]. This type of intelligence is highly relied upon by organizations such as Google [41]. and U.S. Army [20] not only to understand past activities but rather to disseminate new knowledge that fosters informed decision-making.

Emotional Intelligence: Emotional intelligence is a soft skill which is unfortunately, overlooked at times in the recruitment process that enables the managers in charge in making informed decisions [48, 56] define emotional intelligence as the ability to understand one's own emotions and emotions of others. Goleman et al. [26] posited that emotional intelligence describes the capabilities of individuals in understanding and managing feelings regarding to self and others and managing of those information as demonstrated in Table 1.

Wang et al. [56] therefore posit that in a dynamic and ever changing world, adaptability is key to organizational success. This adaptability however, would not be possible with adaptation of technological tools alone, and would require organizational leaders to be adaptable in using the various forms of intelligence in fostering capable and high performing organizations.

Table 1 Goleman's Framework of Emotional Competencies [26]

	Self (Personal competence)	Other (Social competence)
Recognition	Self-Awareness • Emotional self-awareness • Accurate self-awareness • Self-confidence	Social Awareness • Empathy • Service orientation • Organizational awareness
Regulation	Self-Management • Self control • Trust worthiness • Conscientiousness • Adaptability • Achievement drive • Initiative	Relationship Management • Developing others • Influence • Communication • Conflict Managemnet • Leadership • Change catalyst • Building bonds • Teamwork and collaboration

5 Artificial Intelligence and Firm Performance Empirical Evidence

Empirical research has demonstrated that involving external resources and innovative capabilities is a good choice for improving firms' economic performance [13, 18, 43]. However, the external innovations cannot replace the internal organizational capabilities and those internal capabilities must be honed and trained to increase their absorptive capacity to use the knowledge from external sources for effective performance to take place [52], as the ability of the firm to innovate is highly dependent upon its internal ability to exploit the new knowledge and capability [12, 50].

This is also the case when it comes to utilizing Big Data and AI as it denotes to internal capabilities to use these technologies for new and efficient decision-making [44]. Azma and Mostafapour [7] posited that there are two ways in which organizations can utilize and benefit from Big Data and AI. First, organizations can use data for the discovery and dissemination of new knowledge for increased innovation. Second, the smart analyses of data for increased efficiency in decision making and adoption of best data-based approaches for the organization.

To better understand this process, we need to look at it from different perspectives. Process perspective indicates the adoption of Big Data technology aims at making the decision making process more efficient [45]. The product prospective on the other hand looks at the adoption of Big Data technologies is an IT component used to generate data to aid and expedite decision-making my managers [28]. In contrast, the organizational perspective looks at the adoption of Bid Data technology as holistic organizational approach that combines the technology with human capabilities and capacity to make decisions that are conducive to the firms' goals [11]. This is in addition to other studies that have found both artificial intelligence (AIC) and big data analysis capabilities (BDAC) to have a positive impact on organizational outcomes and performance [4, 22, 55].

In this vein, [10] conducted an empirical study to assess the interplay between organizations human capability and AI to assess the implications for the organization. Their findings suggested that companies that desire to benefit from digital transformation, must do so, first by investing and developing their human capabilities, to create an adequate condition for effective utilization of these technologies. Moreover, [58] conducted an empirical study to determine the big data analyses capability (BDAC) and artificial intelligence capability (AIC)'s ability to lead to improved sustainability and performance. Their findings suggested that both BDAC and AIC lead to sustainable growth and improved performance should the organizational capabilities and the design of each technology be adequate.

6 Conclusion

As evidence in this chapter artificial intelligence and big data have many potentials for creating improvement across organizations [25]. However, since we are many years away from full automation, it is pertinent for organizations to realize the importance of their internal capabilities and human development, to truly realize the performance benefits of AI and Big Data in their organization, in a sustainable manner. This said however, once AI use becomes more prevalent, many areas of business and such firms' performance would not only use AI in its activities, it would rely on it as a form of competitive advantage. In this realm, AI could impact such areas as marketing and sales, and accounting and finance within organizations to name a few. In terms of marketing and sales the future role of AI would be to pinpoint the perfect customer for a certain product and service and determine the best ways to reach this prospective customer through intelligent use of algorithms [15, 47]. AI is also anticipated to play a major role in the areas of accounting and finance as humans are prone to errors, more and more reliance would be given to AI to assist with financial modeling as well as rudimentary tasks within the organizations [25]. It is noteworthy, however, that for all these changes to happen successfully for businesses, societal adjustments would need to occur in terms of updating the skills of workforce to work in conjunction with AI, as well as new skills to manage and lead organizations that are derived by data and AI.

References

1. 4-ways-artificial-intelligence-will-impact-your-business (2018). Inc. Magazine. Accessed 26 Dec 2018
2. Ackoff, R.L.: From data to wisdom. J. Appl. Syst. Anal. **16**(1), 3–9 (1989)
3. Agrawal, A., Gans, J., Goldfarb, A.: What to expect from artificial intelligence. MIT Sloan Manag. Rev. (2017)
4. Akter, S., Wamba, S.F., Gunasekaran, A., Dubey, R., Childe, S.J.: How to improve firm performance using big data analytics capability and business strategy alignment. Int. J. Prod. Econ. **182**, 113–131 (2016)
5. Alnoukari, M., Hanano, A.: Integration of business intelligence with corporate strategic management. J. Intell. Stud. Bus. **7**(2) (2017)
6. Athique, A.: The dynamics and potentials of big data for audience research. Media Cult. Soc. **40**(1), 59–74 (2018)
7. Azma, F., Mostafapour, M.A.: Business intelligence as a key strategy for development organizations. Proc. Technol. **1**, 102–106 (2012)
8. Barney, J.: Firm resources and sustained competitive advantage. J. Manag. **17**(1), 99–120 (1991)
9. Cao, G., Duan, Y.: Gaining competitive advantage from analytics through the mediation of decision-making effectiveness: an empirical study of UK manufacturing companies. In: PACIS, p. 377 (2014)
10. Caputo, F., Cillo, V., Candelo, E., Liu, Y.: Innovating through digital revolution. The role of Soft Skills and Big Data in increasing firm performance (2018)
11. Chen, H., Chiang, R.H., Storey, V.C.: Business intelligence and analytics: from big data to big impact. MIS Q., 1165–1188 (2012)

12. Chesbrough, H.W.: The era of open innovation. Manag. Innov. Change **127**(3), 34–41 (2006)
13. Chesbrough, H., Vanhaverbeke, W., West, J.: (Eds.): Open Innovation: Researching a New Paradigm. Oxford University Press on Demand (2006)
14. Cognilytica: Artificial intelligence and machine learning project management training and certification (In-class training material) (2018). https://www.cognilytica.com
15. Conick, H.: The past, present and future of AI in marketing. Mark. News **51**(1), 26–35 (2017)
16. Da Mota Pinto, F.J., Mato Méndez, F.J.: Pervasive business intelligence as a competitive advantage (2016)
17. Davenport, T.H., Prusak, L.: Working knowledge: managing what your organization knows, p. 210. Harvard Business School Press, Boston, MA (1998)
18. Del Giudice, M., Della Peruta, M.R.: The impact of IT-based knowledge management systems on internal venturing and innovation: a structural equation modeling approach to corporate performance. J. Knowl. Manag. (2016)
19. Delone, W.H., McLean, E.R.: The DeLone and McLean model of information systems success: a ten-year update. J. Manag. Inf. Syst. **19**(4), 9–30 (2003)
20. Farrell, M.: Leadership reflections: leadership skills for knowledge management. J. Lib. Adm. **57**(6), 674–682 (2017)
21. Fernald, J.G., Jones, C.I.: The future of US economic growth. Am. Econ. Rev. **104**(5), 44–49 (2014)
22. Ferraris, A., Mazzoleni, A., Devalle, A., Couturier, J.: Big data analytics capabilities and knowledge management: impact on firm performance. Manag. Decis. (2019)
23. Frey, B.S., Stutzer, A.: Happiness and Economics: How the Economy and Institutions Affect (2010)
24. Garud, R., Tuertscher, P., Van de Ven, A.H.: Perspectives on innovation processes. Acad. Manag. Ann. **7**(1), 775–819 (2013)
25. Geisel, A.: The current and future impact of artificial intelligence on business. Int. J. Sci. Technol. Res. **7**(5), 116–122 (2018)
26. Goleman, D., Boyatzis, R., McKee, A.: Primal Leadership. Harvard Business School Press, Boston, MA (2002)
27. Gupta, H., Barua, M.K.: Identifying enablers of technological innovation for Indian MSMEs using best–worst multi criteria decision making method. Technol. Forecast. Soc. Chang. **107**, 69–79 (2016)
28. Hazen, B.T., Boone, C.A., Ezell, J.D., Jones-Farmer, L.A.: Data quality for data science, predictive analytics, and big data in supply chain management: An introduction to the problem and suggestions for research and applications. Int. J. Prod. Econ. **154**, 72–80 (2014)
29. https://www.inc.com/entrepreneurs-organization/4-ways-artificial-intelligence-will-impact-your-business-in-2019.html
30. Huizingh, E.K.: Open innovation: state of the art and future perspectives. Technovation **31**(1), 2–9 (2011)
31. Janssen, M., Konopnicki, D., Snowdon, J.L., Ojo, A.: Driving public sector innovation using big and open linked data (BOLD). Inf. Syst. Front. **19**(2), 189–195 (2017)
32. Ji-fan Ren, S., Wamba, S.F., Akter, S., Dubey, R., Childe, S.J.: Modelling quality dynamics, business value and firm performance in a big data analytics environment (2016)
33. Keiser, B.E.: Competition among competitive intelligence platforms. Online Searcher **37**(1), 16–21 (2013)
34. Koronios, A., Gao, J., Selle, S.: Big data project success-A meta analysis. In: Pacis, p. 376 (2014)
35. Kuruşcu, M., Yenilmez, Ö.: Duygusal Zeka ile Örgütsel Vatandaşlık Davranışı İlişkisi. MANAS Sosyal Araştırmalar Dergisi **8**(1), 284–307 (2018)
36. Kwon, O., Lee, N., Shin, B.: Data quality management, data usage experience and acquisition intention of big data analytics. Int. J. Inf. Manage. **34**(3), 387–394 (2014)
37. Lichtenthaler, U.: Substitute or synthesis: the interplay between human and artificial intelligence. Res.-Technol. Manag. **61**(5), 12–14 (2018)

38. Liebowitz, J.: Strategic Intelligence: Business Intelligence, Competitive Intelligence, and Knowledge Management. CRC Press (2006)
39. Mokyr, J., Vickers, C., Ziebarth, N.L.: The history of technological anxiety and the future of economic growth: Is this time different? J. Econ. Perspect. **29**(3), 31–50 (2015)
40. Morikawa, M.: Firms'expectations about the impact Of AI and robotics: evidence from a survey. Econ. Inq. **55**(2), 1054–1063 (2017)
41. Moser, R.: Decision intelligence (2019). https://www.alexandria.unisg.ch/256178/1/HSG-Decision%20Intelligence-Concept.pdf
42. Nordhaus, W.D.: Are we approaching an economic singularity? Information technology and the future of economic growth (No. w21547). National Bureau of Economic Research (2015)
43. Perkmann, M., Walsh, K.: University–industry relationships and open innovation: Towards a research agenda. Int. J. Manag. Rev. **9**(4), 259–280 (2007)
44. Popovič, A., Hackney, R., Tassabehji, R., Castelli, M.: The impact of big data analytics on firms' high value business performance. Inf. Syst. Front. **20**(2), 209–222 (2018)
45. Provost, F., Fawcett, T.: Data science and its relationship to big data and data-driven decision making. Big Data **1**(1), 51–59 (2013)
46. Rowley, J.: The wisdom hierarchy: representations of the DIKW hierarchy. J. Inf. Sci **33**(2), 163–180 (2007)
47. Ruth, J.P.: Examples of AI in Business Intelligence Applications. Tech Emergence, August 24, 2017
48. Salovey, P., Mayer, J.D.: Emotional intelligence. Imag. Cognit. Pers. **9**(3), 185–211 (1990)
49. Sassi, D.B., Frini, A., Abdessalem, W.B., Kraiem, N.: Competitive intelligence: History, importance, objectives, process and issues. In 2015 IEEE 9th International Conference on Research Challenges in Information Science (RCIS), pp. 486–491. IEEE (2015)
50. Scuotto, V., Del Giudice, M., Della Peruta, M.R., Tarba, S.: The performance implications of leveraging internal innovation through social media networks: An empirical verification of the smart fashion industry. Technol. Forecast. Soc. Chang. **120**, 184–194 (2017)
51. Skovira, R.J.: Ontological grounding of a knowledge mapping methodology: defining data, information, and knowledge. Issues Inf. Sci. VII **2**, 258–264 (2007)
52. Spithoven, A., Clarysse, B., Knockaert, M.: Building absorptive capacity to organise inbound open innovation in traditional industries. Technovation **30**(2), 130–141 (2010)
53. Visinescu, L.L., Jones, M.C., Sidorova, A.: Improving decision quality: the role of business intelligence. J. Comput. Inf. Syst. **57**(1), 58–66 (2017)
54. Wade, M., Hulland, J.: The resource-based view and information systems research: Review, extension, and suggestions for future research. MIS Q. **28**(1), 107–142 (2004)
55. Wamba, S.F., Gunasekaran, A., Akter, S., Ren, S.J.F., Dubey, R., Childe, S.J.: Big data analytics and firm performance: Effects of dynamic capabilities. J. Bus. Res. **70**, 356–365 (2017)
56. Wang, C., Hoffman, F., Lim, A., Kwon, J.: From artificial to emotional intelligence: integrating five types of intelligence to achieve organizational excellence. Manag. Knowl. Learn., 125 (2019)
57. Zeleny, M.: Knowledge-information autopoietic cycle: towards the wisdom systems. Int. J. Manag. Decis. Making **7**(1), 3–18 (2006)
58. Zhang, H., Song, M., He, H.: Achieving the success of sustainability development projects through big data analytics and artificial intelligence capability. Sustainability **12**(3), 949 (2020)

Collaborative and Social Media SaaS (Software as a Service) Cloud Computing Services' Adoption and Acceptance Model on the Millennials: Conceptual Model

Ghilan Al-Madhagy Taufiq-Hail, Shafiz Mohd Yusof, Ramadhan Abdo Musleh Alsaidi, Saleh R. Alanazi, and Adel M. Sarea

Abstract Collaborative and social media SaaS cloud services emerged as one promising technology to meet the demand of high connectivity, collaboration, and reliability, while achieving cost-effective solutions. However, scarcity of empirical works has been devoted to the higher education sector represented by the millennials at university campus. This work tries to fill this gap and proposes an integrated model of Theory of Planned Behaviour (TPB) and Technology Acceptance Model (TAM) to understand the dominant factors of the adoption of the collaborative and social media that are cloud-based services and applications. The purpose of the study is to formulate the conceptual model: A review of the literature.

Keywords Collaborative and social media services · Integrated models · SaaS cloud computing · TAM and TPB · Technology acceptance and adoption

1 Introduction

With the rapid pace of technology advancements and the daily usage of applications and services that are cloud-based in different businesses, cloud computing gained the spotlight in the past decade and came into the frontier of technology importance that infiltrated the individuals', organizations', and Governments' daily usage of

G. A.-M. Taufiq-Hail (✉)
University Utara Malaysia (UUM), Changlun, Malaysia
e-mail: towfeeq2k5@yahoo.com

S. Mohd Yusof
University of Wollongong Dubai, Wollongong, United Arab Emirates

R. A. M. Alsaidi · S. R. Alanazi
Technical and Vocational Training Corporation, Tabuk College of Technology, Tabuk, Saudi Arabia

A. M. Sarea
Ahlia University, Manama, Bahrain

A. Hamdan et al. (eds.), *The Fourth Industrial Revolution: Implementation of Artificial Intelligence for Growing Business Success*, Studies in Computational Intelligence 935,
https://doi.org/10.1007/978-3-030-62796-6_6

technology. In other words, cloud services and applications have revolutionized the use of ICT on the organizational and individual levels [51]. In the past two decades, the intense need for technology that is accessible from anywhere, available at any time or place, reliable, and cost-effective, led researchers, practitioners, and professionals to think of technology that meets this demand. With the advent of Virtualization, this demand became feasible especially with the growing need for high computation speeds and smart networks that have the burden to transport the immense data. As an outcome, cloud computing surfaces as a solution that utilizes virtualization in its core [69].

Furthermore, cloud computing garnered the spotlight in the past decade and cloud-based applications and services gained the popularity in every aspect in our life on the personal side of this technology. Take for example, the smart phone, tablets, and smart devices and appliances that utilize various applications and services are widely spread as an outcome of the usage of cloud-based SaaS services. Despite the cloud computing immense benefits, it has gained less scholarly empirical investigation [69]. Therefore, researchers regard this an issue that warrants further investigation empirically.

The main purpose of this article is to explore the salient factors affecting the adoption and acceptance of SaaS cloud-based collaborative and social media applications and services among university students and to suggest a conceptual model in this article.

Moreover, the selection falls on university communities in Malaysia represented by students because they offer respondents from different areas in Malaysia with different beliefs, perceptions, languages, and cultures. These culture differences are more likely to influence the perceptions of the individuals of the same social environment [29]. In this way, the differences in culture creates differences in the way that an individual thinks and acts and, accordingly, affects the behaviour that is socially acceptable or not [60]. Consequently, the respondents from universities offer a mixture that would provide a clear clue of the phenomena investigated. That is, having different cultures provides different perceptions and attitudes towards a behaviour, in the case of the current study the acceptance and adoption process of collaborative and/or social media SaaS cloud services. Based on the above-mentioned justifications, the researchers argue the inclusion of the social influence, i.e. the social norms influence, in the current model and exploration of its role in the acceptance process of SaaS could computing is one of the objectives of the study.

Besides, the individuals at the universities are perceived to have the adequate skills to deal with innovative technology of SaaS applications and services. These capabilities they possess are apparent in the collaborative services of cloud computing by using online applications in their respective university's tasks and assignments (e.g., Google Forms or other Google Apps, Microsoft Office 365 applications, university portals utilities or Apps, etc.) and social media SaaS applications (e.g., Facebook, twitter, WhatsApp, telegram, Instagram, Skype, etc.). These capabilities and skills are arguably reflecting the perceived behavioural control that the participants have, thereby, this construct is included in the model proposed and probing its role is perceived to add more insights of the user's perceptions. Therefore, this is another objective of the study.

Understanding the perceptions of users, would lead to better decisions at an organization adopting or planning to adopt SaaS services as well as the providers themselves. Hence, one objective is to probe the salient factors in the Malaysian contexts from the students' standpoint as they are heavily using collaborative and/or social media SaaS applications in their academic tasks or private use. Also, studying their perceptions gives a concrete and wide view of the acceptance factors of this innovative technology.

When talking about SaaS cloud computing, the education sector is not studied broadly in terms of the individual perceptions especially students. This view is indicated in previous works that the higher education sector received a little attention to investigate the acceptance of SaaS cloud computing empirically [52, 74]. Likewise, Sabi et al. [59] asserts on the dearth of empirical investigation of the adoption and diffusion of cloud computing at higher education sector. Hence, the focus of this sector by selecting the bachelor students at the university is warranted to be investigated.

More importantly, the mediating effect of behaviour intention, i.e. precisely studying it as a mediator, is hardly found in literature and overlooked in many research areas [47]. Hence, the last objective is to explore the role of behaviour intention as a mediator between the social norms influence and perceived behaviour control, from one side, and the actual behaviour of accepting or adopting collaborative and/or social SaaS services.

Based on the above, the study aims to fill these gaps and issues presented above to explore the paucity of empirical studies utilizing individuals at Malaysian higher education sector by investigating the students' attitudes, social norms, their perceived skills and capabilities (i.e. perceived behaviour control), and intentions regarding the acceptance and adoption of collaborative and social media SaaS cloud computing services. The scope, thereby, is focused on personal level of SaaS application used on daily bases such as the collaborative and social media SaaS services. Also, the individuals with bachelor's degree from different colleges are the unit of sampling at a university campus in Northern Malaysia. Additionally, the study utilizes the Theory of Planned Behaviour (Ajzen [5, 6] augmented by constructs from Technology Acceptance Model to shed more light on the salient factors affecting the adoption and acceptance process as perceived by individuals with presumed skills and variety of cultures (i.e., the different races of Malaysia) that exist inside the university campuses' community.

It is presumed that the findings would give clues to the beneficiaries of SaaS services to adopt or develop applications that lie within their business plans as well as the providers to develop or update their services that align with business trends of different sectors, private and public, to boost their view and long-term investment with different entities. More importantly, the higher education sector represented by the universities' administration is specifically important beneficiaries of this study.

2 Literature Review

2.1 Theoretical Background and Research Model

By reviewing the literature, although number of theories and models used to investigate the adoption of technology or information system innovations such as Theory of Reasoned Actions (TRA) (Fishbein and Ajzen [24], TAM [23], and TPB [6] that are apparently similar, they seem to have distinctive variables to explain and predict the behaviour under consideration and interest [55]. TRA (Fishbein and Ajzen [24] was the pioneer theory that other models and theories are based on with extensions. For example, TAM was primarily aiming to predict the adoption of Information Systems and TPB was developed to give a special focus on the intention and subsequent behaviours in using Information Systems (IS) [25]. Besides, the central focus of TRA is the individual's intention which is translated as how hard they are willing to try, in which the case applies for TAM [35]. Also, Leung and Chen [42] emphasized the importance of the concept that the prudent considerations of available information and motivations from the perspective of the adopter, are elements that drive the human behaviour in which TRA and TPB are derived from. Moreover, TPB is considered as an augmentation of TRA [24] and developed to overcome its limitations as emphasized by Alzahrani et al. [10]. Therefore, TPB extended TRA with perceived behavioural control that gave it strength over TRA.

On the other hand, it was well documented that TAM and TPB are integrated in many studies as they complement each other and, therefore, give wider view of the IS or innovative technology acceptance, usages, or adoption under consideration. The integration gives more exploratory power and predictability over the individual use of any of the theories [25] and further identify cross-correlation of the predictive variables of the behaviour investigated [55]. Therefore, the suggested integrative model of the current study with a single model framework gives the estimated strength of both theories, enhance our understanding of the interested behaviour, and take the advantage of both theories.

Based on the above, the theoretical framework proposed departs from TPB and augmented by including constructs from TAM, i.e. the perceived ease of use and perceived usefulness, and their relationships to bring insightful understanding on the innovative services of SaaS cloud computing (i.e. collaborative and/or social media services as an example) and better understand the perception of students regarding the adoption and acceptance of this technology. Besides, there are little works have been conducted to investigate these perceptions based on the technology combined by the need to shed more light on the lower adoption rates in Malaysian context in the business sector. That is, when the motivations and perceptions of the real users or adopters of the SaaS cloud computing are known, this would give broader view for business sectors to develop and market their services based on the real users' needs and perceptions and not the managerial level's perception. Figure 1 depicts the conceptual model of the current study.

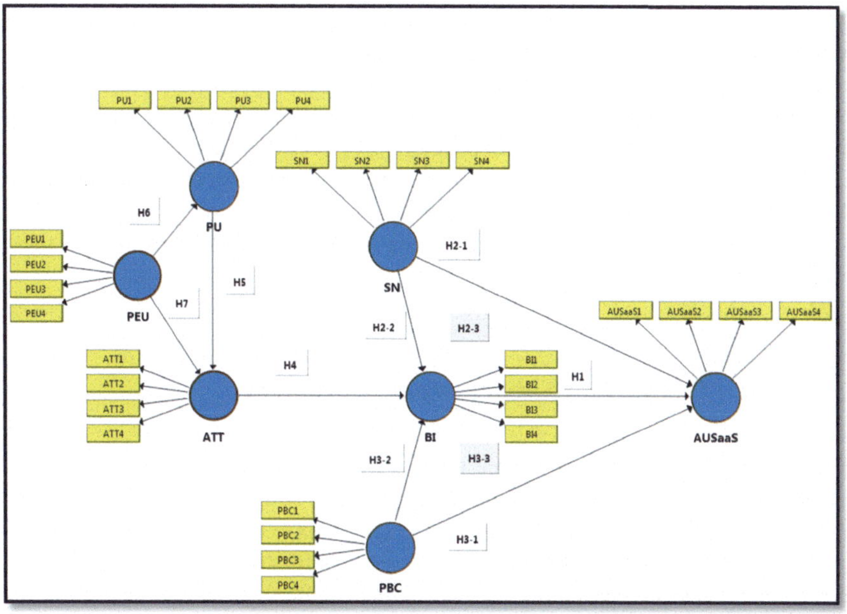

Fig. 1 Conceptual model of the research

2.2 *Cloud Computing*

Cloud computing is in its simplest definition an IT-related capability that is provisioned as a service, which has advantages of higher performance, scalability, availability, and with low cost service that has better features compared with conventional datacentres [17]. These datacentres connect many of its servers together locally and connect with other datacentres together by means of cloud computing to provide IT services to customers that are close to these scattered datacentres without having knowledge of the underlying technology infrastructure. The proliferation of smart mobile devices, increasing need for high performance computing, the high growth of the Internet usage, and the need to reduce cost and energy paved the way for a new and different computing model that provision IT related resources as utility [17].

On another aspect, various studies in the area of cloud computing adopted the definition of U.S National Institute of Standards and Technology (NIST) of its final draft [39, 50, 64]. In this study, the definition of cloud computing is also adopted from NIST in which it defines it as "A model for enabling ubiquitous, convenient, on demand network access to a shared pool of configurable computing resources (e.g., networks, servers, storage, applications, and services) that can be rapidly provisioned and released with minimal management effort or service provider interaction. This cloud model is composed of five essential characteristics, three service models, and four deployment models" [49].

This new paradigm shift to go for SaaS cloud computing services make customers familiar with this novel technology and to practice it from any device, any place, and at any time in a way to Green their practices in using technology. SaaS cloud computing (or a Software as a Service) is the applications on the cloud computing infrastructure that the service provider offers to consumers [49] over the internet in which it has a variety of software packages that can be accessed based on the customer needs [40]. In addition, cloud computing is classified based on two models, i.e., the service model and the deployment model. The following subsections explains more of these topics.

2.3 Cloud Computing Service Model Paradigm

In the service model, there are three types: Firstly, Software as a Service (SaaS), which is the top most layer of this classification and is simply explained as using the applications on the cloud infrastructure of the service provider by the consumer [49]. In this model, the user has no control over cloud infrastructure such as: network, servers, storage, or even operating systems with exception on specific settings of the application on use [49]. Also, SaaS applications are provided through a technology coined as multi-tenancy technology, where the software and data are provisioned to many users at the same time on a single instance of the software [17].

The Second Layer of service model is Platform as a Service (PaaS), which is the middleware layer and programmers can have more control over it and use the programming tools and packages provided in this layer to develop software. Finally, Infrastructure as a Service (IaaS), is the bottom layer of the service level model, where user can have control over the hardware resources (i.e. by means of virtualization) such as memory, disk space, or processors to use or release on the basis of pay-as-you-use. The providers can supply/rent disk storage, RAM, and IP addresses on as needed bases with certain fees paid to the providers according to agreed service contract.

2.4 Cloud Computing Deployment Model Paradigm

The Second Classification of cloud computing is the deployment model with four types: To start with, the Public cloud computing, in which the datacentre can be accessed from the outsider. In Public Cloud, the infrastructure and services are provisioned for open use by the public over the Internet on bases of pay-per-use [73] or even on free-basis to offer services for the public. This model of cloud computing is owned, managed, administered, and operated by organizations in government, academic institutions, or businesses [49]. As an example, Amazon Elastic Compute cloud (Amazon EC2).

The next type in this model is the private cloud computing, in which access is limited to insiders of an exclusive organization that has different business units [49, 73] and is offering more secure resources than the previous one. The Third Type of the deployment model is the Community cloud computing, in which different entities having the same interests such as: Security requirements, shared policy, and common mission interests, are connected in a way to share resources among the members of this community [33, 73]. Finally, the Hybrid Cloud computing, which is a mixture of any of the above-mentioned deployment models.

2.5 SaaS Collaborative and Social Media Services

As explained earlier, SaaS is the topmost service model layer of the cloud. There are many services provided in this layer, and some of these are the collaborative and social media services. These services have attracted the attention of individuals of all occupations, roles, responsibilities, and students are not an exception. They offer key advantages over the traditional computing services provided on the campus or at the premises of any organization such as: accessibility at anytime and anywhere, simplicity in usage, affordability irrespective of the budget, flexibility in expansion, and real time interactions regardless of location, time, operating system or platform.

In addition, the collaborative services offered by SaaS providers overcame the limitations of conventional distributed collaboration services and featured the flexibility, expandability [75], and the accommodation of different operating systems and platforms. Some of the common collaborative services offered by Microsoft Office 365 [75] or Google's apps include a bundle of applications such as: email services, calendar, word processing, presentation and spread sheets, messaging and chatting services, storage, audio/video-editing editing tools, file sharing, and many more. Many of these collaborative services are offered freely for students and lecturers at many Malaysian universities to boost the pedagogical process and evoke the interaction with the latest technologies.

With respect to the social media applications, they are considered another essential aspect of SaaS services that almost every single smart devices possesses an application and a broad scale of individuals use in everyday communication with others. Applications such as Facebook, twitter, Instagram, LinkedIn, etc. are the most common social media apps used for communication purposes, learning, marketing, or entertaining. Furthermore, nowadays social media becomes a virtual platform that attracts people to interact and discuss issues of interest [65]. Consequently, social network's apps and social media sites form a vital impact on the user's decisions in many aspects of life that require a decision. That is, many people consult others by means of social media apps or sites to get feedback, more information, or solutions to many issues encountered in their lives. Lee et al. [41] states that there are three main reasons to use social media. Firstly, it is an easy and quick means of communication through chat or discussion boards. Secondly, people can share their views, issues, queries, and solutions to their problems amongst themselves. Thirdly, getting feedback from others considered one of the top reasons for using social media. Hence, it is logically that the platform of SaaS could-based social networks became popular amongst students inside and outside campus.

3 Hypotheses Development

3.1 Collaborative and Social Media SaaS Cloud Computing Acceptance (AUSaaS)

The adoption or acceptance is referred to in previous works as the implementation, usage, acceptance, utilization, actual behaviour, or use [7]. Usage, acceptance, or adoption behaviour toward technology is defined as the process that has many phases and needs to extend over time and under the individual's will to perform/not to perform it [58]. Consequently, for the purpose of the current study, the student's usage, acceptance, or adoption of collaborative and social media SaaS cloud computing services (AUSaaS) is defined as the process that consists of many phases that extends over time and under student's will to use or not to use collaborative and/or social media SaaS cloud computing services.

Many universities, in the U.S. and UK, use cloud computing services; however, there are still some issues surrounding and hindering higher education sector to be fully adopting cloud computing services that need further investigation [67, 76]. Likewise, Low et al. [45] and Sabi et al. [59] emphasize the crucial role of cloud computing and its diffusion in academic research; however, they indicate lack of exploratory empirical studies on the diffusion and adoption of this technology especially in developing countries. It is noteworthy that adoption is found in literature as an outcome of behaviour Intention and is supported theoretically by theories such as Decomposed Theory of Planned Behaviour (DTPB) [70], Unified Theory of Acceptance and Use of Technology (UTAUT) [72], TPB [5] among others, and empirically in using and adopting technologies in different contexts [20, 37]. Based on the above, the researchers presume the acceptance and adoption of collaborative and/or social media SaaS cloud computing services are essential component of the model that is measured by the construct AUSaaS and is predicted by the behaviour intention, the social norms influence, the perceived behavioural control constructs, in which will be explained further below.

3.2 Behaviour Intention (BI) Relationship with AUSaaS

Reviewing the literature, there are different definitions found of BI in different contexts. For example, the following definition defines BI as, "The degree to which a student has formulated conscious plans to use or not use cloud services in the future" [13]. Another more comprehensive and adopted definition is found by Taufiq-Hail et al. [69] in which they defined BI as, "The degree in which the individual, student & academic staff, using SaaS cloud computing services has formulated conscious plans to use/accept 'or adopt' SaaS cloud computing services in their academic studies or communications with others inside or outside the campus".

It is worthy to note that BI is widely adopted in literature in different contexts and models and found to be a strong and statistically significant predictor of behaviour, adoption process, utilization, or acceptance of certain services or innovative technologies [8, 13, 21, 36]. More importantly, plethora of services and applications available from different vendors, commercial and non-commercial companies, individual developed apps, and service providers are offered freely or with low rates that encourage, motivate, invoke, and persuade individuals to utilize, use, accept, and adopt SaaS collaborative and social media services and applications. That is, these opportunities make individuals' intention driven toward the utilization of cloud services and applications [36].

Additionally, the inclusion of BI is apparent in many models and theories, for example: TRA [24], TAM [23], and TPB [5]. These theories affirm that BI is the main driver of behaviour. Following this line of reasoning and based on the review of literature, the researchers find it essential to include BI in the model and postulates the following hypothesis:

Hypothesis H1. BI has an empirically positive and statistically significant relationship with AUSaaS.

3.3 Subjective Norms (SN) Construct's Relationships: Relationship Between SN and AUSaaS

Social norms influence is studied in literature thoroughly and considerable work has been done and given special attention in the academia. Social norms has been identified in literature as the context where, "a specific individual expects one to perform or not to perform the behaviour combined with one's motivation to comply with these specific individuals" [14]. Ajzen [6] adds that performing any behaviour is based on the evaluation of a person to be positive and when he/she believes that important others urging him/her to perform it. In the context of this study, SN is defined as the extent in which the SaaS collaborative and social media services' users desire and urge others to adopt or accept (or not) these services, combined with motivation that those others would comply with them.

Previous works reviewed show a paucity in studying the direct relationship between SN and the actual behaviour in general, and in specific in SaaS cloud computing area of research. Yet, there are limited number of studies postulated the statistically significant effect of SN on the actual behaviour. However, their findings emerged with no statistically significant relationship between the two. For example,

Janmaimool [32] postulated that individuals would engage in waste management behaviour (WMB) if others (i.e., referents) engaged in WMB. However, their findings did not show statistically significant direct relationship towards WMB, rather it shows this significance indirectly through organizational norms, which is a category of the social norms.

Similarly, the findings revealed by Mafabi et al. [47] in the area of knowledge sharing show that the relationship between SN and knowledge sharing behaviour (KSB) (i.e., actual behaviour) appears to be nonsignificant, which is against their hypothesis. Their findings draw more ambiguity on the direct relationship between SN and behaviour in what is deposited and what is empirically found. That is, they hypothesized that SN has a statically positive relationship with knowledge sharing behaviour in TPB model and that SN has, at the same time, a statistically positive relationship with BI, where these two hypotheses do not match the criteria of mediation. To explain, the full mediation should be hypothesized when the direct effect of a construct driver (i.e. in their case SN) is not statically significant with the outcome (KSB), while, at the same time, the driver construct SN to BI relationship should be statistically significant in the indirect path towards the outcome. On the other hand, they posited that BI mediates the relationship between SN and the behaviour, which does not match the principle of full mediation in the said hypotheses. Surprisingly, the full mediation occurred in their empirical results and SN to KSB relationship was not statistically significant, where it is in congruent with the full mediation principle. Conversely, one study by Alzahrani et al. [10] found a statistically significant and positive direct relationship between SN and the actual use of online game playing. Hence, these contradicting findings in different contexts warrant the investigation of this relationship in the current study and lead the researchers to hypothesize the following to conform with the mediating effect of BI:

Hypothesis H2-1. SN latent construct is expected to have an empirically nonsignificant positive relationship with AUSaaS.

3.4 Relationship Between SN and BI

Drawing upon the previous works, the researchers argue that the effect of significant others has a crucial influence on the intention to adopt innovative technology especially in societies that have strong bonds among themselves as in Malaysian communities inside the university campuses. Also, this study builds on the explorative nature in getting more insights of different beliefs that affect the behaviour and behaviour intention in the context of the study.

Moreover, the relationship between SN and BI is explicated in literature in different contexts and research areas. However, there are inconsistencies in the results obtained. For instance, some research studies' findings demonstrate the statistically significant and positive relationship between SN and BI [22, 30, 43]; whereas other studies obtained nonsignificant relationship between them [32, 34, 48, 63]. Additionally, it has been proven that SN is the weakest predictor of BI in a meta-analysis conducted by Smith and McSweeney [66] that included 185 studies of TPB and evaluated the role of SN in shaping the intention compared with the double effect of ATT on BI. These inconsistencies raise a flag to further investigate the relationship between SN with BI and the actual acceptance of collaborative and social media SaaS services within the Malaysian higher education sector represented by students. Hence, the following hypothesis is assumed:

Hypothesis H2-2. SN latent construct is expected to have an empirically statistically significant and positive relationship with BI.

3.5 The Mediating Role of BI on the Relationship Between SN and AUSaaS

It is worthwhile to know that there are scarcity of studies focusing on the mediating effect of BI in the relationship between SN and AUSaaS– or the behaviour of interests. Generally speaking, this relationship is overlooked in many research areas [47]. Among these few research studies, the research of Albarq and Alsughayir [9] in the context of Internet Baking in Saudi Arabia. In their study they investigated the mediating effect of BI in the relationship between SN and usage of Internet Banking. Their findings shed light on the mediating effect of BI.

Also, a study's results by Mafabi et al. [47] go in congruent with the previous work and the mediating role of BI in the relationship between SN and the knowledge sharing behaviour is ascertained. Besides, the study of Peng et al. [53] affirms the full mediating effect of BI in the relationship between SN and the actual behaviour. However, they indicated that they cannot affirm or obtain a concrete evidence on the dominant role of BI based on TPB. Also, they indicated that although the mediating role of BI with the three main beliefs of TPB is well recognized in previous works, the empirical evidence does not hold for their study with the three behavioural constructs (i.e., perceived behaviour control, attitude, and subjective norms). Looking at these considerations, the researchers come up with the following hypotheses:

Hypothesis H2-3. BI is expected to mediate the relationship between SN and AUSaaS.

3.6 *Perceived Behavioral Control (PBC) Construct's Relationships*

3.6.1 Relationship Between PBC and AUSaaS

The intention to use any technology is driven and predicted by the attitude, SN, and PBC as stated in TPB. PBC can be said to be, "The perceived ease or difficulty of performing the behaviour and it is assumed to reflect past experience as well as anticipated impediments and obstacles." [5]. For the purpose of this research, PBC can be referred to as the perceived ease or difficulty of performing different academic tasks using collaborative and/or social media SaaS services and applications and it is assumed to reflect the students' past experience in using technology skills they possess as well as the anticipated obstacles to perform such tasks.

Reviewing the literature with respect to the relationship PBC → Actual behaviour, it is found that there is a paucity of literature that study this direct relationship in the area of collaborative and social media SaaS cloud services and applications empirically, although theoretically supported by TPB. However, some studies revealed the positive and statistically significant relationship [10, 20, 26, 31] between PBC and the actual behaviour. On the other hand, Mafabi et al. [47] found a nonsignificant relationship between PBC and the actual behaviour. More importantly, when perceiving the mediating effect of BI in the relationship between PBC and AUSaaS, the hypothesis should fall into assuming a nonsignificant relationship between the two constructs. To explain, in full mediation, the mediator tries to absorb the direct effect between PBC and the actual behaviour. That is, the direct effect must be a statically nonsignificant in order to achieve the full mediation and the direct effect to be statistically significant [27]. Based on the contradicting findings in the previous works, and considering the mediation effect of BI, the researchers inclined to posit the following hypothesis:

Hypothesis H3-1. PBC latent construct is expected to have a nonsignificant relationship with AUSaaS.

3.7 *Relationship Between PBC and BI*

It is noteworthy to understand that PBC is a one driving construct in TPB that is determining both the intention and the interested behaviour. In other words, the more easily, within ability, and under control the behaviour perceived by individual, the more likely that one would perform or have the intention to perform that behaviour or adoption [66]. Furthermore, PBC has received theoretical support in TPB [6, 5] and DTPB [70] theories, respectively, as well as the empirical support [19, 30, 77] in which a statistically significant and positive directional relationship from PBC → BI is established. However, other studies revealed different findings in having a nonsignificant relationship between PBC and BI [31, 54, 61]. These inconsistencies urge more attention to investigate the said relationship in the context of Malaysian universities. Consequently, it is presumed the following directional hypothesis:

Hypothesis H3-2. PBC latent construct is expected to have an empirically statistically significant relationship with BI.

3.8 *The Mediating Role of BI on the Relationship Between PBC and AUSaaS*

There are scarcity of empirical works studying the mediating role of BI in the relationship between PBC and the actual behaviour. Among these few research studies, the research of Mafabi et al. [47] and Peng et al. [53] in two different contexts (i.e., knowledge sharing and internet adoption, respectively). However, these two research papers have inconsistent results. That is, Mafabi et al. [47] found a full mediation role of BI in the respective relationship, whereas the findings of Peng et al. [53] revealed that BI is not mediating the relationship between PBC and the actual behaviour. Hence, these conflicting findings lead the researchers to further investigate the said relationship in the context of the current study and the following hypothesis is postulated:

Hypothesis H3-3. BI is expected to mediate the relationship between PBC and AUSaaS.

3.9 *Attitude (ATT) Relationship*

Attitude is defined as, "The degree to which a person has a favourable or unfavourable evaluation of the behaviour in question." [4]. Lining up with the above definition, and in the context of the study, ATT is defined as the degree of favourability or unfavourability felt by the individual in in relation to the adoption and acceptance of collaborative and/or social media SaaS services. Besides, Ajzen [5] noted that the positive ATT that the individual possesses, forms the future intention to embrace a specific behaviour. Additionally, ATT is found to be a strong and robust antecedent of BI in various studies especially those related to ICT [53] and innovative technologies, and exerts a positive and statically significant relationship with BI [11, 12, 61, 69]. It is worthwhile to notice the positive advantages of SaaS collaborative and social media services and applications to the actual users that would most likely drive their intention towards adopting this technology. To explain, the around the clock access to these services regardless of location, time, or device, make it advantageous compared with other traditional methods such as: Using computers at office hours inside the university, accessing data using USB drives, or the use of limited sharing capabilities at premises. Hence, these tangible benefits presumed arguably to affect university's student ATT and, thus, influence his/her intention to perform the behaviour in question. Following this, a hypothesis is considered:

Hypothesis H4. ATT has a direct empirically positive and statistically significant relationship with BI.

3.10 Perceived Usefulness (PU) and Perceived Ease of Use (PEU) Constructs' Relationships

Rogers [57] defines perceived usefulness (or relative advantage as it is called in Diffusion of Innovation theory, DOI) as, "The degree to which an innovation is perceived as better than the idea it supersedes" [57]. That is, the higher the perceived usefulness of an innovative technology, such as collaborative and/or social media SaaS services, the higher possibility, the faster, and the more likely to have influence on the ATT towards the adoption or acceptance. Building on the previous definitions, the study defines PU as the degree in which the collaborative and/or social media SaaS services' acceptance or adoption are perceived to be better to use, access, understand faster, and to reliably access, as compared with the traditional methods, (e.g., updating/uploading/sharing different data in local hard drive or inside local server), to enhance performance and achieve the perceived advantages.

Besides, in previous works there is a consensus among researchers that PU has a positive and statically significant relationship with ATT in a large body of research [3, 16, 46]. Also, Davis [23] emphasized the pronounced and stronger effect of PU on ATT compared with that of PEU. Davis [23] explained further that adopting an application in a system is at the utmost priority accomplished by its usefulness (i.e., the function it performs for them) and then by how easy or hard to get the system perform these functions (i.e., the ease of use). Therefore, the following hypothesis is postulated:

Hypothesis H5. PU has a direct empirically positive and statistically significant relationship with ATT.

Furthermore, perceived ease of use (PEU) is defined as, "The degree to which a person believes that using a particular system would be free of effort." [23]. In other words, Davis [23] is pointing out to the feelings of being free from worries, difficulties, or effort to achieve the perceived performance when using the innovative system. More precisely, the more easily the innovation perceived by the individual, the more likely that this technology influence the ATT. Drawing on the above, the study defines PEU as the degree in which the collaborative and/or social media SaaS services' adoption or acceptance is believed to be easy and free from worries, difficulties, or effort to achieve the perceived performance by the students and is readily understandable. Further, the PEU latent construct towards ATT has gained theoretical support such as of TAM, DOI, DTPB (i.e., [23, 57, 70], respectively) and empirical support in literature [1, 38] in which results revealed positive and statistically significant relationship with ATT. Hence, the following hypothesis is deposited:

Hypothesis H6. PEU has a direct empirically positive and statistically significant relationship with ATT.

Drawing upon the findings of TAM, PEU influences PU as the easier the technology is to use, the more useful the individual perceives to use it [71]. That is, it is assumed to enhance one productivity in performing a certain task. In other words, when the individual perceives collaborative and/or social media SaaS services easy to use, free of hurdles, this makes him/her presumes the productivity in using it; thus, its usefulness is considered. Moreover, the findings affirm the association between PEU and PU [71].

Furthermore, Chang and Hsu [18] consider ease of use and perceived usefulness as motivational and benefit factors that encourage the use of an information system (IS); thus, changing one's attitude towards the usage. More importantly, the authors noted that before considering the acceptance, use, adoption, or switching to alternative options exposed to anyone, one has to make proper judgment and weighs the pros and cons before taking decision or making intention towards the possible optional alternatives [18]. Therefore, these two constructs are of special focus in the current study.

These arguments are further supported in a large body of research [16, 18, 62] and affirm linkage between these two factors. However, other findings are not lining up with the above-mentioned results. Scholars such as Agarwal and Karahanna [2], for instance, found inconsistent result in their study that demonstrates PEU is not associated with PU. These inconsistencies in former works lead the researchers to investigate this relationship in the context of the current study. and the following is hypothesized:

Hypothesis H7. PEU has an empirically positive and statistically significant relationship with PU.

4 Methodology

4.1 Survey Instrument

In terms of the measurements, the questionnaire was divided into two sections. The first section has two parts. It includes the demographic questions, as part one of the section, and part two consists of answering the question of SaaS collaborative and social media services and applications.

Section two includes the indicators/items of all constructs that were adapted from previous literature. Each construct has four items which are adapted based on an acceptable validity and reliability with total number of 28 items. The items anchored with five-point Likert scale [44], where 1 = Strongly disagree to 5 = Strongly agree.

To measure and assess their internal consistency and reliability, convergent validity, and discriminant validity SmartPLS 3 structural equation modelling (SEM) software [56] was used as a main analytical tool to analyse the measurement model (i.e. the items) used and to assesses the structural model (i.e. the hypotheses of the current research).

4.2 Data Collection and Measurements

The respondents of the study are bachelor's degree students from different colleges at a public university in Malaysia. The questionnaire preparation underwent different stages in which small group of students were pre-tested (i.e. 7 responses) and amendments were considered based on their views, then a pilot test conducted from the same university community respondents (i.e. 35 responses), and finally, the actual deployment of the questionnaire has taken place in four months' time, started in September 2018 and ended in February 2019. This procedure was followed in order to ensure its appropriateness and validity of the questionnaire as well as its content is understandable in the way the researcher aims to get specific and reliable answers.

The minimum sample size suggested by applying G*Power software package was 98 at a medium effect size of 0.15, alpha error probability 0.05, and Power of 0.80. However, obtaining more responses has better results and broader view of the sample investigated as emphasized by many scholars [27, 68]. Based on simple random sampling procedures on the final distribution of the questionnaire, the total number of reliable responses obtained were 345 out of 400 distributed questionnaires. On the other hand, after an initial screening and evaluation due to incomplete answers, and outliers' elimination, the valid responses obtained were 289, where they underwent the main analysis. After obtaining the usable responses, the power of analysis was 0.999, which reflect adequacy of the sample obtained.

4.3 Mediation Assessment Procedure

Regarding the analysis of the mediation effect, a large body of research investigating the salient factors influencing the acceptance or adoption of innovative technology has employed mediators in their models drawing upon Baron and Kenny [15] approach. Nevertheless, this approach has been criticized of its drawbacks to explain the mediation effects [28]. Therefore, the current study employs the recent and concrete procedure to investigate the mediating role of BI in the model under investigation. That is, the approach by Zhao et al. [78], which further was supported by Hair et al. [27].

The guidelines of this approach are summarized by Taufiq-Hail et al. [69] as follows: (a) the Direct-only non-mediation is accomplished if the direct effect is statistically significant whereas the indirect is not; (b) In case the direct and indirect effects are not statistically significant, the mediation is referred to as No-effect non-mediation; (c) When both, the direct and indirect effects are statistically significant with the same sign (− or +), this can be called "Complementary mediation"; (d) If the direct effect and the indirect effect are statistically significant but with opposite directions (i.e. one is + and the other is –), in this case we assume a "Competitive mediation"; and finally (e) when the indirect effect is statistically significant and the direct effect is not statistically significant, we then assume the full mediation or Indirect-only mediation [27, 78].

Based on the above, the research applied the concurrent approach as a rigorous assessment of the mediation effect of BI on the relationships estimated. This is to ensure that a complete analysis is taken place to draw more insight on the phenomena under investigation.

5 Conclusion

Cloud computing is a nascent research area that has captured the researchers' attention in recent decades. However, the research in the academia focused on the business side and not the individual side, who is the end-user of the technology and the building block to its diffusion. This study in its first part formulated the conceptual framework of the study taking into account two theories, i.e. TAM and TPB with modifications to suite the objectives of the study. The literature review gave a background on the underlying theories, the cloud computing and its taxonomy, the services and applications of collaborative and social media SaaS services, along with the constructs' definitions and relationships justifications. The results obtained and to supplement the suggested framework in this article to the suggest the conceptual model that can be used in other contexts and in different fields in social sciences studies that consider the human factor as the basis of the success of any implementation of innovative technologies that are IT-based either in business, banking, education, healthcare, tourism, or any other social sciences research.

References

1. Agag, G., El-Masry, A.A.: Understanding consumer intention to participate in online travel community and effects on consumer intention to purchase travel online and WOM: An integration of innovation diffusion theory and TAM with trust. Comput. Hum. Behav. **60**, 97–111 (2016). https://doi.org/10.1016/j.chb.2016.02.038
2. Agarwal, R., Karahanna, E.: Time flies when you're having fun: Cognitive absorption and beliefs about information technology usage. MIS Q. **24**, 665 (2000). https://doi.org/10.2307/3250951
3. Ahmed, E., Ward, R.: A comparison of competing technology acceptance models to explore personal, academic and professional portfolio acceptance behaviour. J. Comput. Educ. **3**, 169–191 (2016). https://doi.org/10.1007/s40692-016-0058-1

4. Aizen, I., Madden, T.J.: Prediction of goal-directed behavior: Attitudes, intentions, and perceived behavioral control. J. Exp. Soc. Psychol. **22**, 453–474 (1986)
5. Ajzen, I.: The theory of planned behavior. Organ. Behav. Hum. Decis. Process. **50**, 179–211 (1991). https://doi.org/10.1016/0749-5978(91)90020-T
6. Ajzen, I.: From intentions to actions: a theory of planned behavior. In: Kuhl, J., Beckmann, J. (eds.) Action Control, pp. 11–39. Springer Berlin, Heidelberg (1985). https://doi.org/10.1007/978-3-642-69746-3_2
7. Al-jabri, I.M., Sadiq Sohail, M.: Mobile banking adoption: Application of diffusion of innovation theory. J. Electron. Comm. Res. **13**, 379–391 (2012)
8. Alalwan, A.A., Dwivedi, Y.K., Rana, N.P.: Factors influencing adoption of mobile banking by Jordanian bank customers: Extending UTAUT2 with trust. Int. J. Inf. Manage. **37**, 99–110 (2017). https://doi.org/10.1016/j.ijinfomgt.2017.01.002
9. Albarq, A.N., Alsughayir, A.: Examining theory of reasoned action in internet banking using SEM among Saudi consumers. Int. J. Mark. Pract.—IJMP **1**, 16–30 (2013)
10. Alzahrani, A.I., Imran Mahmud, T., Ramayah, O.A., Alalwan, N.: Extending the theory of planned behavior (TPB) to explain online game playing among Malaysian undergraduate students. Telematics Inf. **34**, 239–251 (2017). https://doi.org/10.1016/j.tele.2016.07.001
11. Arpaci, I.: Understanding and predicting students' intention to use mobile cloud storage services. Comput. Hum. Behav. **58**, 150–157 (2016). https://doi.org/10.1016/j.chb.2015.12.067
12. Arpaci, I., Baloğlu, M.: The impact of cultural collectivism on knowledge sharing among information technology majoring undergraduates. Comput. Hum. Behav. **56**, 65–71 (2016). https://doi.org/10.1016/j.chb.2015.11.031
13. Arpaci, I., Kilicer, K., Bardakci, S.: Effects of security and privacy concerns on educational use of cloud services. Comput. Hum. Behav. **45**, 93–98 (2015). https://doi.org/10.1016/j.chb.2014.11.075
14. Bagozzi, R.: The self-regulation of attitudes, intentions and behavior. Soc. Psychol. Q. **55**, 178–204 (1992)
15. Baron, R.M., Kenny, D.A.: The moderator-mediator variable distinction in social psychological research: Conceptual, strategic, and statistical considerations. J. Pers. Soc. Psychol. **51**, 1173–1182 (1986)
16. Besbes, A., Legohérel, P., Kucukusta, D., Law, R.: A cross-cultural validation of the tourism web acceptance model (T-WAM) in different cultural contexts. J. Int. Consum. Mark. **28**, 211–226 (2016). https://doi.org/10.1080/08961530.2016.1152524
17. Bhardwaj, S., Jain, L., Jain, S.: An approach for investigating perspective of cloud Software-as-a-Service (SaaS). Int. J. Comput. Appl. **10**, 44–47 (2010). https://doi.org/10.5120/1450-1962
18. Chang, Y.-W., Hsu, P.-Y.: An empirical investigation of organizations' switching intention to cloud enterprise resource planning: a cost-benefit perspective. Inf. Dev. **35**, 290–302 (2017). https://doi.org/10.1177/0266666917743287
19. Chang, I.-C., Chou, P.-C., Keng-Jung Yeh, R., Tseng, H.-T.: Factors influencing Chinese tourists' intentions to use the Taiwan Medical Travel App. Telemat.Inf. **33**, 401–409 (2016). https://doi.org/10.1016/j.tele.2015.09.007
20. Cheung, M.F.Y., To, W.M.: Service co-creation in social media: an extension of the theory of planned behavior. Comput. Hum. Behav. **65**, 260–266 (2016). https://doi.org/10.1016/j.chb.2016.08.031
21. Choudrie, J., Pheeraphuttranghkoon, S., Davari, S.: The digital divide and older adult population adoption, use and diffusion of mobile phones: a quantitative study. Inf. Syst. Front. (2018). https://doi.org/10.1007/s10796-018-9875-2
22. Chu, T.-H., Chen, Y.-Y.: With good we become good: Understanding e-learning adoption by theory of planned behavior and group influences. Comput. Educ. **92–93**, 37–52 (2016). https://doi.org/10.1016/j.compedu.2015.09.013
23. Davis, F.D.: Perceived usefulness, perceived ease of use, and user acceptance of information technology. MIS Q. **13**, 319 (1989). https://doi.org/10.2307/249008

24. Fishbein, M., Ajzen, I.: Belief, attitude, intention, and behavior: An introduction to theory and research. In: Addison-Wesley Series in Social Psychology. Addison-Wesley Publishing Company, Reading, Massachusetts (1975)
25. Glavee-Geo, R., Shaikh, A.A., Karjaluoto, H.: Mobile banking services adoption in Pakistan: are there gender differences? Int. J. Bank Mark. **35**, 1090–1114 (2017). https://doi.org/10.1108/IJBM-09-2015-0142
26. Gopi, M., Ramayah, T.: Applicability of theory of planned behavior in predicting intention to trade online: Some evidence from a developing country. Int. J. Emerg. Mark. **2**, 348–360 (2007). https://doi.org/10.1108/17468800710824509
27. Hair, J.F., Hult, G.T.M., Ringle, C.M., Sarstedt, M.: A Primer on Partial Least Squares Structural Equation Modeling (PLS-SEM), 2nd edn. SAGE Publications Inc., London, England (2017)
28. Hayes, A.F.: Beyond Baron and Kenny: statistical mediation analysis in the new millennium. Nat. Commun. Assoc. **76**, 408–420 (2009). https://doi.org/10.1080/03637750903310360
29. Hofstede, G.H.: Culture's consequences: international differences in work-related values. In: Abridged (ed.) Cross-Cultural Research and Methodology Series, vol. 5. Sage Publications, Beverly Hills (1984)
30. Hopkins, N., Tate, M., Sylvester, A., Johnstone, D.: Motivations for 21st century school children to bring their own device to school. Inf. Syst. Front. **19**, 1191–1203 (2017). https://doi.org/10.1007/s10796-016-9644-z
31. Jain, S., Khan, M.N., Mishra, S.: Understanding behavior regarding luxury fashion goods in India based on the theory of planned behavior. J. Asia Bus. Stud. 11, 4–21 (2017). https://doi.org/10.1108/JABS-08-2015-0118
32. Janmaimool, P.: The role of descriptive social norms, organisational norms and personal norms in explaining solid waste management behaviours in workplaces. J. Org. Change Manag. **30**, 184–198 (2017). https://doi.org/10.1108/JOCM-12-2016-0265
33. Jlelaty, M., Monzer, Y.: Factors in cloud computing adoption. Master's thesis. Lund University, Sweden (2012)
34. Kabra, G., Ramesh, A., Akhtar, P., Dash, M.K.: Understanding behavioural intention to use information technology: Insights from humanitarian practitioners. Telemat. Inf. **34**, 1250–1261 (2017). https://doi.org/10.1016/j.tele.2017.05.010
35. Kim (Sunny), J.: An extended technology acceptance model in behavioral intention toward hotel tablet apps with moderating effects of gender and age. Int. J. Contemp. Hosp. Manag. **28**, 1535–1553 (2016). https://doi.org/10.1108/IJCHM-06-2015-0289
36. Kim, S.-H., Kim, J.K.: Determinants of the adoption of mobile cloud computing services: a principal-agent perspective. Inf. Dev. **34**, 44–63 (2018). https://doi.org/10.1177/0266666916673216
37. Kim, E., Lee, J.-A., Sung, Y., Choi, S.M.: Predicting selfie-posting behavior on social networking sites: an extension of theory of planned behavior. Comput. Hum. Behav. **62**, 116–123 (2016). https://doi.org/10.1016/j.chb.2016.03.078
38. Kim, M., Qu, H.: Travelers' behavioral intention toward hotel self-service kiosks usage. Int. J. Contemp. Hosp. Manag. **26**, 225–245 (2014). https://doi.org/10.1108/IJCHM-09-2012-0165
39. Klug, W.E.: The determinants of cloud computing adoption by colleges and universities. Doctoral dissertation, Prescott Valley, Arizona, U.S.: Northcentral University (2014)
40. Lee, S.G., Chae, S.H., Cho, K.M.: Drivers and inhibitors of SaaS adoption in Korea. Int. J. Inf. Manag. **33**, 429–440 (2013). https://doi.org/10.1016/j.ijinfomgt.2013.01.006
41. Lee, W., Tyrrell, T., Erdem, M.: Exploring the behavioral aspects of adopting technology: Meeting planners' use of social network media and the impact of perceived critical mass. J. Hosp. Tourism Technol. **4**, 6–22 (2013). https://doi.org/10.1108/17579881311302329
42. Leung, L., Chen, C.: Extending the theory of planned behavior: A study of lifestyles, contextual factors, mobile viewing habits, TV content interest, and intention to adopt mobile TV. Telemat. Inf. **34**, 1638–1649 (2017). https://doi.org/10.1016/j.tele.2017.07.010
43. Lian, J.-W.: Critical factors for cloud based e-invoice service adoption in Taiwan: an empirical study. Int. J. Inf. Manage. **35**, 98–109 (2015). https://doi.org/10.1016/j.ijinfomgt.2014.10.005
44. Likert, R.A.: A technique for the measurement of attitudes. Arch. Psychol. **140**, 4–53 (1932)

45. Low, C., Chen, Y., Wu, M.: Understanding the determinants of cloud computing adoption. Ind. Manag. Data Syst. **111**, 1006–1023 (2011). https://doi.org/10.1108/02635571111161262
46. Ma, C.-C., Kuo, K.-M., Alexander, J.W.: A survey-based study of factors that motivate nurses to protect the privacy of electronic medical records. BMC Med. Inf. Decis. Mak. **16**, 13 (2016). https://doi.org/10.1186/s12911-016-0254-y
47. Mafabi, S., Nasiima, S., Muhimbise, E.M., Kasekende, F., Nakiyonga, C.: The mediation role of intention in knowledge sharing behavior. VINE J. Inf. Knowl. Manag. Syst. **47**, 172–193 (2017). https://doi.org/10.1108/VJIKMS-02-2016-0008
48. Md Husin, M., Ab Rahman., A.: Do Muslims intend to participate in Islamic insurance?: Analysis from theory of planned behaviour. J. Islamic Acc. Bus. Res. **7**, 42–58 (2016). https://doi.org/10.1108/JIABR-03-2014-0012
49. Mell, P., Grance, T.: The NIST definition of cloud computing. Recommendations of the National Institute of Standards and Technology, National Institute of Standards and Technology (2011)
50. Mitchell, R.B., Meggison, P.F.: Strategies for integrating cloud computing concepts. J. Appl. Res. Bus. Instr. **12**, 1–6 (2014)
51. Mohammed, F., Ibrahim, O., Nilashi, M., Alzurqa, E.: Cloud computing adoption model for e-government implementation. Inf. Dev. **33**, 303–323 (2017). https://doi.org/10.1177/0266666916656033
52. Okai, S., Uddin, M., Arshad, A., Alsaqour, R., Shah, A.: Cloud computing adoption model for universities to increase ICT proficiency. SAGE Open **4**, 215824401454646 (2014). https://doi.org/10.1177/2158244014546461
53. Peng, T.-Q., Zhu, J.J.H., Tong, J.-J., Jiang, S.-J.: Predicting internet non-users' adoption intention and adoption behavior: A panel study of theory of planned behavior. Inf. Commun. Soc. **15**, 1236–1257 (2012). https://doi.org/10.1080/1369118X.2011.614628
54. Picazo-Vela, S., Chou, S., Melcher, A.J., Pearson, J.M.: Why provide an online review? An extended theory of planned behavior and the role of Big-Five personality traits. Comput. Hum. Behav. **26**, 685–696 (2010). https://doi.org/10.1016/j.chb.2010.01.005
55. Püschel, J., Mazzon, J.A., Hernandez, J.M.C.: Mobile banking: proposition of an integrated adoption intention framework. Edited by Heikki Karjaluoto. Int. J. Bank Mark. **28**, 389–409 (2010). https://doi.org/10.1108/02652321011064908
56. Ringle, C.M., Wende, S., Will, S.: SmartPLS 3 (version 3). SmartPLS GmbH, Boenningstedt (2015)
57. Rogers, E.M.: Diffusion of Innovations, 3rd edn. Free Press A division of Macmillan Publishing Co., Inc., New York, NY (1983)
58. Rogers, E.M.: Diffusion of innovations. 5th ed. 5. Free Press A division of Macmillan Publishing Co., Inc., New York, NY (2003)
59. Sabi, H.M., Uzoka, F.-M.E., Langmia, K., Njeh, F.N.: Conceptualizing a model for adoption of cloud computing in education. Int. J. Inf. Manage. **36**, 183–191 (2016). https://doi.org/10.1016/j.ijinfomgt.2015.11.010
60. Sang, Y., Lee, J.-K., Kim, Y., Woo, H.-J.: Understanding the intentions behind illegal downloading: a comparative study of American and Korean college students. Telemat. Inf. **32**, 333–343 (2015). https://doi.org/10.1016/j.tele.2014.09.007
61. dos Santos, R., Miguel, L., Okazaki, S.: Planned e-learning adoption and occupational socialisation in Brazilian higher education. Stud. High. Educ. **41**, 1974–1994 (2016). https://doi.org/10.1080/03075079.2015.1007940
62. Shareef, M.A., Kumar, V., Dwivedi, Y.K., Kumar, U.: Service delivery through mobile-government (mGov): driving factors and cultural impacts. Inf. Syst. Front. **18**, 315–332 (2016). https://doi.org/10.1007/s10796-014-9533-2
63. Shiau, W.-L., Chau, P.Y.K.: Understanding behavioral intention to use a cloud computing classroom: A multiple model comparison approach. Inf. Manag. **53**, 355–365 (2016). https://doi.org/10.1016/j.im.2015.10.004
64. Simamora, B.H., Sarmedy, J.: Improving services through adoption of cloud computing at PT XYZ in Indonesia. J. Theoret. Appl. Inf. Technol. **73**, 395–404 (2015)

65. Singh, P., Dwivedi, Y.K., Kahlon, K.S., Sawhney, R.S., Alalwan, A.A., Rana, N.P.: Smart monitoring and controlling of government policies using social media and cloud computing. Inf. Syst. Front. (2019). https://doi.org/10.1007/s10796-019-09916-y
66. Smith, J.R., McSweeney, A.: Charitable giving: the effectiveness of a revised theory of planned behaviour model in predicting donating intentions and behaviour. J. Commun. Appl. Soc. Psychol. **17**, 363–386 (2007). https://doi.org/10.1002/casp.906
67. Sultan, N.: Cloud computing for education: a new dawn? Int. J. Inf. Manag. **30**, 109–116 (2010). https://doi.org/10.1016/j.ijinfomgt.2009.09.004
68. Tabachnick, B.G., Fidell, L.S.: Using Multivariate Statistics, 6th edn. Pearson Education Inc., New Jersey, NJ (2013)
69. Taufiq-Hail, G.A., Ibrahim, H., Yusof, S.A.M.: SaaS as a means of Green IT acceptance model: a theory of planned behavior model at Malaysian public universities' context. J. Inf. Syst. Technol. Manag. 2, 01–17 (2017)
70. Taylor, S., Todd, P.A.: Understanding information technology usage: a test of competing models. Inf. Syst. Res. **6**, 144–176 (1995). https://doi.org/10.1287/isre.6.2.144
71. Venkatesh, V.: Determinants of perceived ease of use: integrating control, intrinsic motivation, acceptance model. Inf. Syst. Res. **11**, 342–365 (2000)
72. Venkatesh, V., Morris, M.G., Davis, G.B., Davis, F.D.: User acceptance of information technology: toward a unified view. MIS Q. **27**, 425–478 (2003)
73. Wang, W.Y., Chung, A.R., Chuang, H.-M.: Toward the trend of cloud computing. J. Electron. Comm. Res. **12**, 238–242 (2011). https://doi.org/10.4304/jsw.6.7.1230-1234
74. Yadegaridehkordi, E., Nilashi, M., Shuib, L., Samad, S.: A behavioral intention model for SaaS-based collaboration services in higher education. Educ. Inf. Technol. (2019). https://doi.org/10.1007/s10639-019-09993-1
75. Yadegaridehkordi, E., Shuib, L., Nilashi, M., Asadi, S.: Decision to adopt online collaborative learning tools in higher education: a case of top Malaysian universities. Educ. Inf. Technol. 24, 79–102 (2019)
76. Yaghmaei, O., Binesh, F.: Impact of applying cloud computing on universities expenses. IOSR J. Bus. Manag. **17**, 42–47 (2015). https://doi.org/10.9790/487X-17214247
77. Yeap, J.A.L., Ramayah, T., Soto-Acosta, P.: Factors propelling the adoption of m-learning among students in higher education. Electron. Mark. **26**, 323–338 (2016). https://doi.org/10.1007/s12525-015-0214-x
78. Zhao, X., Lynch, J.G., Chen, Q.: Reconsidering Baron and Kenny: Myths and truths about mediation analysis. J. Consum. Res. **37**, 197–206 (2010). https://doi.org/10.1086/651257

FinTech, RegTech and Artificial Intelligence

Fintech: A Pathway for MENA Region

Gopalakrishnan Chinnasamy, Araby Madbouly, and Sameh Reyad

Abstract The game changer of today's economy is coined with the word 'Fintech' the financial technology [19]. Fintech is growing by leaps and bounds in the current financial services through banks and Fintech companies. It provides solution for the businesses in controlling their operational costs in their business activities. Startup businesses use "Fintech" as platform to run their businesses operations with minimum cost [25]. The startup wave has obsessed the Middle East and North Africa (MENA) region and when it is digital startups, the fintech is closely reaching 12% of all transactions in the region followed by e-commerce [9]. The MENA region governments and apex bodies are encouraging fintech industries to develop their businesses along with IT innovation to establish a sound financial ecosystem. As said by COO of Jordan's Arab Bank, this industry is at an earlier stage which requires lots of rules and policies to regulate this industry with the support of banks and fintech industries. It is the crucial task for the regulators to establish technology based steady financial system which requires well established stringent rules and regulations to prevent cyber-crimes. Hence, this conceptual work attempts to understand the main aspects of Fintech, role of financial institutions, attitude of stakeholders towards the new financial system, challenges and opportunities prevailing towards this industry. This has been achieved by reviewing the existing literature to support the growth of the economy in MENA region using Fintech. The outcome of the study indicates that the sustainability, infrastructure, human resources and government supports are the main pillars of the growth of fintech in MENA region.

Keywords Fintech · Economy · MENA

G. Chinnasamy · A. Madbouly (✉)
Business and Accounting Department, Muscat College, Muscat, Sultanate of Oman
e-mail: araby@muscatcollege.edu.om

G. Chinnasamy
e-mail: gopalakrishnan@muscatcollege.edu.om

S. Reyad
College of Business and Finance, Ahlia University, Manama, Bahrain
e-mail: sreyad@ahlia.edu.bh

A. Hamdan et al. (eds.), *The Fourth Industrial Revolution: Implementation of Artificial Intelligence for Growing Business Success*, Studies in Computational Intelligence 935,
https://doi.org/10.1007/978-3-030-62796-6_7

1 Introduction

Financial Stability Board (FSB) defines the Fintech as "technologically enabled financial innovation that could result in new business models, applications, processes or products with an associated material effect of financial markets and institutions and the provision of financial services". Fintech is the integration of financial services offered by companies with modern technology in order to provide easy and effective services to their stakeholders. Fintech is growing rapidly and leads as new business models [6]. This innovative business models have become the trend of businesses since 2010. Although fintech firms are basically small or medium in size and they do not have enough financial resources to manage their businesses, they have a vision of how to innovate or modernize the existing financial services. The fintech companies consist of established financial institutions and technology startup companies. Most of fintech firms are start-ups and the start-ups number is increasing steadily with the various innovative business approaches to attract and fulfill the needs of their customers [22]. These companies are bridging the gap between service providers and the general public to attain the aim of financial inclusion. Some of these advance technologies like Artificial Intelligence (AI), Big Data, Blockchain, Robotic Process Automation (RPA), Machine Learning (ML), Cloud computing (CC) and Internet of Things (IoT) used to enhance the offered services and products by financial institutions to their customers in a fast and beneficial manner. The fintech industry has revolutionized and resized the existing business models for customers to utilize its' advantages. Broadly speaking, a wide range of countries are progressively incorporating and embedding this technology as a part of their infrastructure [25]. This study makes an attempt to identify the way to attain the holistic fintech ecosystem by developing a theoretical model through underpinning the available literature and the statistical information related to this sector for the benefit of the society.

2 Understanding Fintech Categories

Fintech simply refers to the transformation of financial services by changing the existing business models into innovative technology-based services [6]. A wide range of applications can come under this umbrella. Fintech services can be categorized into four groups:

- Business to Consumers (B2C)
- Business to Consumers for small firms
- Business to Business (B2B) for Banks
- Bank's business clients

The followings are the few Fintech services categories which offered by fintech companies in relation to the type of services (Table 1).

Table 1 Types of Fintech Offered Services

Type of services	Core services
Blockchain	It provides private, transparent and secured transactions with the tracking services to know the lifecycle of financial transactions A wide range of other platforms use this technology, such as: Bitcoin, Chain, Bloq, Ethereum and Wirex
Cryptocurrency	The Blockchain technology supports the distributed nature of cryptocurrencies from business and consumers
InsurTech	It provides insurance solutions with the use of digitized financial ecosystem to enhance the customer experience. The tools like smartphone apps, AI, IoT and ML
RegTech	Regulatory technology facilitates the delivery of regulatory norms effectively It is integrated, secured, cost effective, standardized, transparent regulatory processes the whole compliance system become automated Reporting, risk management, monitoring offered with the platforms like Regis – TR, Provenir, Continuity and IMind
LendTech	Lending solutions using accurate and streamlined way Using AI, ML and verify credentials to make error free outcome
PayTech, International Money Transfers	The mobile transactions and management of assets is made secure and easy through Payment technology Transferring money across countries using peer-to-peer and SWIFT and others MenaPay, PayPal, MobiKwik, Google Wallet, Apple Pay are the few payment platform of PayTech
TradeTech	The effective and secured use of technology in supply chain finance and the distribution of finance to support cross-border trading
NeoBanks, BankTech	Consumer Banking use of digital platform to serve their customers in an effective manner. NeoBanks known as Challenger Bank
Equity Financing Crowdfunding	Raising funds or capital using sale of securities through this technology The participants are Kickstarter, Pebble, Aflamnah, My Salaam and others
Wealth Tech, Robo-Advisors, Personal Finance	Portfolio building and stock trading advise through robotic technology Sarwa, Wahad Invest, WeInvest, SmartWealth, WealthFace, Betterment, Ellevest are few examples of Robo advisors and others

(continued)

Table 1 (continued)

Type of services	Core services
AccountingTech	ML, AI, CC, Digital Tax payment, accounting softwares is little technological application in the accounting operations to have smooth and transparent operations

3 Perspectives and Expectations of Stakeholders

The conventional banks are concerned with fintech business model and its' effect on bank deposits, bank lending style and payment system. However, banks are not worried much about cryptocurrencies and Blockchain business models [11]. Further, it is evident that banks are influenced by their peers and each one is ready to follow others without any disinclination. Based on the above results, it can be said that they believe their trust and worthiness are the major sources of doing business. The goal of fintech is to develop novel ideas and new technologies enable financial services to work as alternate to traditional financial system. These are not fully regulated, as it has limitations in terms of crossing the countries in their dealings. Hence, the sustainability is questionable in some occasions where their focus is on novel ideas and may not look it with long term perspectives [45]. The financial industry is highly regulated and the conventional financial institutions are in line with these well-established set up where they are giants in this field is the hindrance for the survival of new startups [38]. Lending is the biggest challenging part for the startups, as the aversion of conventional banks in offering risky loans may not be taken as business opportunity of new companies to lose their sustainable business activities [21].

Fintech platforms have influenced the conventional banking sector related to the payment system. These are at lower cost with innovative easy and faster services compared to the other mode of payment [46]. Tidebrant [43] agreed with this statement that the banks are in an urgency to adopt it to retain the customers in the competitive market. The earlier day's payment is complex and it is challenging for the bankers in terms of operation, whereas the fintech payment has brought the bank into new convenient and effective mode of operation [23].

The reason for adopting fintech in the banking activities is to enhance the user experience and banking operational efficiency. Further, there should not be any cyber threat in using the fintech in their day to day bank transactions. B2B collaborative model is working well with Indonesian Banks and fintech companies [47]. According Moody's report, the early generation people are highly rely on conventional banking system as they are convenient and learned towards to use in the experience [15]. Whereas, the Millennials prefers the fintech business model and the majority prefers to use adoptable, convenient system of Fintech in the banking operations with high security [36]. Hence, it can be observed that the easy use, convenient, time saving and secured financial technology is need of the hour in any economy.

Fintech hypes flourish. In the financial services, Fintech is "disruptive", "innovative", and "removes the barriers" of conventional financial institutions. The ease use of banking system significantly influences the banking customers [2]. The Mobile Enterprise Applications users are significantly influenced with the ease use of technology [27]. The positive influence among the users of online banking is observed [26].

The determinants of new generation people in adopting Fintech is mainly rely on life expectancy and financial awareness attainment of the technology [14]. The use of complex financial technology relatively affects the willingness to accept the technology through portable devices [37]. If the fintech is easy, convenient and time saving are more willing to adopt it. Taylor and Todd [42] also stated that the ease of use has positive impact on technology adoption.

The Technology Acceptance Model (TAM) has been proposed by Davis [16] to elaborate the impacts variations on consumer intention and willingness. The TAM has been shows the significant influence on the user friendly Fintech solutions [15]. Competitive advantage is an important predictor of the adoption of new technology. There are studies show positive association between intention to use of product or services and competitive advantages [18, 24, 29, 34]. The above evidences represents that there is a significant association between competitive advantage and user's attitude towards the technology oriented system.

Though there is plenty of business opportunities associated with Fintech, there are some digitalization vulnerability is existing and it is to be addressed by regulatory measures [44]. This unlawful, illegal and other type of frauds needs to be addressed through exclusive regulatory framework. The cyber security initiatives are taken as essential to prevent and mitigate those grey areas through an innovative ICT solution [8]. The responsibility of the regulators is to provide some protective covenants to manage the institution specific micro risks and system wide macro risks by providing parameters and regulations through common global framework [22]. The regulatory sandbox has to be structured with robust governance framework to all the jurisdiction compliances to have dedicative Regtech framework.

As like as conventional banking and fintech companies, there are challenges are associated in Islamic finance in terms of operational risks, capital risks, cyber risks, operational frauds and shariah compliance issues in using Fintech oriented services to their customers [19]. Basel Committee and the Islamic financial services board suggested that Islamic finance institutions need sufficient and clear regulatory framework to deal the risk associated with their Islamic finance businesses. The product and service innovation has to be taken as priority to handle the current needs of the society without compromising shariah requirements. These are the major requirements from Shariah Supervisory board and the Islamic financial services board [20].

Fintech sector faces some other global challenges in terms of merging policies and practices. The regulatory challenges are associated more with consumer protection in the larger retail financial market. The MENA region is also brought into the threat of cyber security, threat assessments, vulnerability assessments and security standards with the external business ventures [44]. Crypto assets also facing some security issues, operational challenges, tax related problems at certain extent [8]. The lending processes related to corporate loans are still customer base oriented in GCC. Where there is no strong customer base, banks will lose their businesses due to this new competition came through the innovative fintech strategy. This may create troubles to the conventional banks which may lead to collapse the entire financial system of a country and it may exacerbate to be global. The RegTech can support the sector in managing the process to gain both short-term and long-term yields [17]. The fintech service delivery relies on electricity and internet where these basic amenities cannot be expected at all parts of the nation including MENA region. There are countries still facing some unreliable electricity and internet can lead to service interruption at each point of the operations in both financial institutions and customers operations. There are challenges associated with Fintech business processes [28] especially the technical challenges associated with traditional financial institutions and managerial challenges associated with fintech startups in most of the occasions [9].

Overall, there is a positive impact on the innovative financial services in the industry with many factors are accepting (TAM) towards the use of Fintech. The influencing factors are plenty, particularly ease use of banking system, rely on life expectancy, convenient and time saving. Fintech platforms have been influenced the conventional banking sector related to the payment system. The reason for adopting fintech in the banking activities is to enhance the user experience and banking operational efficiency. Further, there should not be any cyber threat in using the fintech in their day to day bank transactions. The majority of challenges are associated in financial services by conventional banks, fintech companies and Islamic finances in terms of financial regulations, operations, raising funds from investors, cyber security, hiring and retaining talents, geographical expansion and building trust and confidence.

4 Global Emergence of Fintech

The financial services industry indicates the flourishing of the transformation phase especially through technology innovation. The main pillar of financial services is conventional banking. Financial institutions also undergoing an extensive change with an increased competition through many venture capitalists which is building market edge, innovative offerings to the society very faster and easy manner [9]. The global banking sector has some uncertainty due to the recent crisis from pandemic and it requires major change in economies though there is a projected climb in compound average growth rate (CAGR) of 8% over the next five years. The recent business shift has made lot of changes in this sector and making the business processes through technology platform. Recently, there are many Fintech companies doing business along with conventional banks [28].

According to the KPMG recent report 2019, US is the top with 18 companies among top 100 Fintech companies taken into account from 36 countries. The UK holds the second level with 12 Fintech companies and followed by 11 from China. Among these are majorly Neo banks which continue to accelerate their contribution to

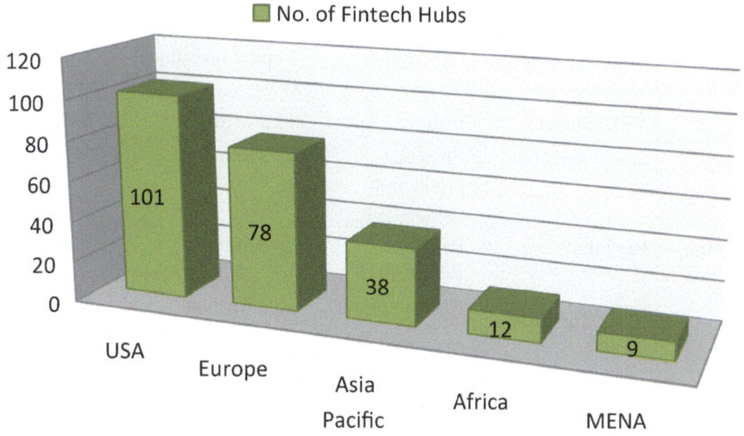

Fig. 1 Number of Fintech Hubs—2019. *Source* Global Fintech Index (2020)

become leaders in the market. Further, payment and lending services became leading next to digital banking services. Conversely, wealth management companies are stands next with 15 companies and Insurtech stands next with 12 fintech companies. These companies have raised venture capital over 53 billion of USD in the last year. This is more than three times increase with 360 percent is the evidence of the success and future development of this industry. According to Global Findex 2020, the 60% of global GDP will be from digitalized economic activities by 2022. The following Fig. 1 is detailed about the number of hubs operated Fintech majorly in each region of the world.

5 Emergence of Fintech in MENA Region

The startup wave has passionate the Middle East and North Africa (MENA) region and when it is digital startups, the fintech is closely reaching 12% of all transactions in the region followed by e-commerce [9]. The fintech business venture came to MENA region predominantly in 2017 only. Though the region has been slow in adopting this technology, the policy makers started implanting the new regulations and procedures to support and motivate the growth of the sector in this region. This has helped the sector to grow well with the sustainable futuristic business model. The expected annual growth is to reach 125 million USD by 2022 because of many new companies launching their business each year in MENA region. The MENA region governments and apex bodies are encouraging fintech industries to develop their businesses along with its' innovation to establish a sound financial ecosystem. The followings are the Major Fintech companies' deals in MENA region (Table 2).

Start-ups in general have been doing quite well in the region—in 2019 with 564 new startups in Middle East and North Africa (MENA) which received $704 million

Table 2 The Fintech Establishments in MENA region

Country	Company name	Description	Capital investment ($)	Year of establishment
UAE	Yallacompare	Insurtec	17.2 Million	2011
UAE	Souqalmal.com	Financial production comparison	15 Million	2012
Saudi Arabia	PayTabs	Payment & Remittances	26 Million	2013
UAE	Eureeca	Crowdfunding	12.3 Million	2013
UAE	BayZat	Insurance & HR services	25 Million	2014
UAE	Beehive	Wealth Management	15.5 Million	2014
Tunisia	Expensya	Multi-platform	5.7 Million	2014
UAE	Beehive	Wealth Management	15.5 Million	2014
Jordan	Liwwa	P2P lending platform	8.55 Million	2015
UAE	Democrance	Insurtech	1.3 Million	2015
Egypt	Paymob	Digital financial enabler	No Data	2015
Lebanon	Nymcard	Payment & Remittances	4 Million	2016
Jordan	POSRocket	Point of sale solution	2.36 Million	2016
UAE	Sarwa	Robo wealth advisor	1.5 Million	2016
Lebanon	Risk + solutions	Financial Intelligence	1.45 Million	2016
Egypt	Vapulus	E payment gateway	1.06 Million	2016
Egypt	Moneyfelows	Crowd Funding	980 K	2016
UAE	Monami Tech	Financial services	1 Million	2016
UAE	Zbooni	Chat commerce app	1.4 Million	2017
UAE	Point checkout	Payment gateway	600 K	2017
UAE	Aqeed	InsurTech	18 Million	2018
UAE	Foloosi	Payment & Remittances	500 K	2018

Source Magnitt [31]

in funding, from 354 deals, which was 40 per cent higher than in 2018, according to a MENA Venture Investment Report. Though there high worthy startups came into this region especially in UAE and Saudi Arabia, the regional and fintech ranking appears in the following way. Overall, in fintech companies' business operations Israel holds the first rank among MENA countries and overall 18th rank achieved globally. The following table represents the ranks of each country according to their volume of operation compared with global and regional level.

Figure 2 present the score achieved by the MENA countries in terms of their performance in Fintech businesses. Out of 19 countries, there are about half of the countries doing well in terms of their digital platform. Table 3 and Fig. 3 reflect the same regarding their fintech performances. Other MENA countries also make an effort to achieve the digital transformation in their business operations.

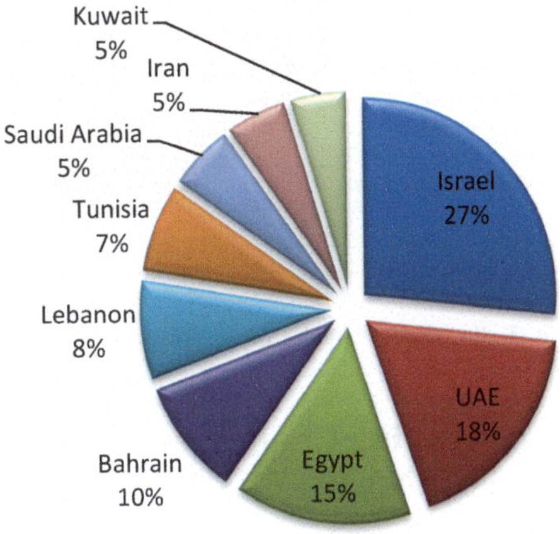

Fig. 2 Rank onto Score. *Source* Global Fintech Index (2020)

Table 3 MENA Regional Ranking of Fintech

Country	Region rank	Global rank
Israel	1	18
UAE	2	58
Egypt	3	106
Bahrain	4	153
Lebanon	5	175
Tunisia	6	196
Saudi Arabia	7	232
Iran	8	234
Kuwait	9	236

Source Global Fintech Index (2020)

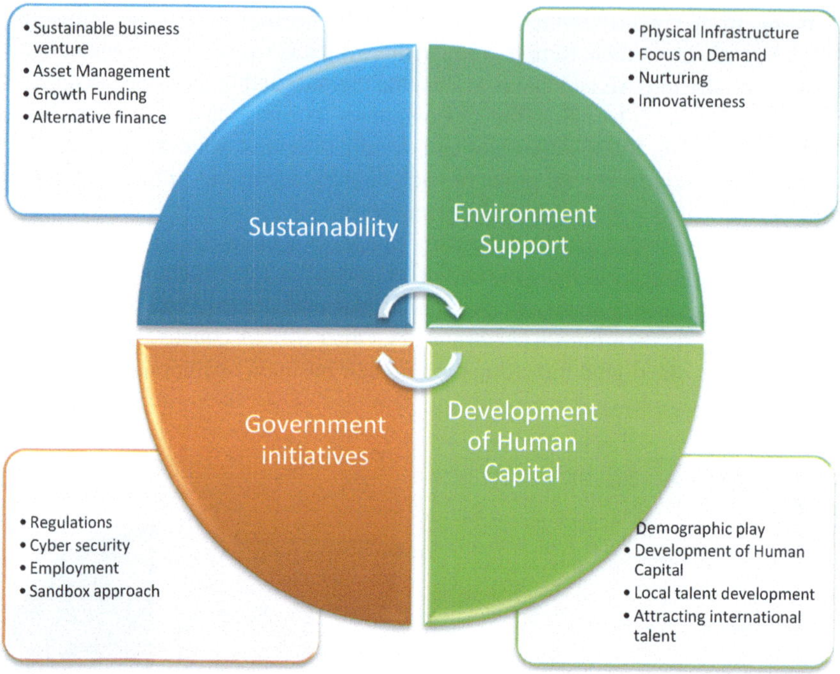

Fig. 3 Proposed Fintech Ecosystem. *Source* Developed by the authors

6 Prevailing Challenges Towards Adoption of Fintech in MENA

The MENA region gradually deploy fintech activities though the industry. There are about 75% of fintech companies started their business activities majorly in four countries such as UAE, Egypt, Jordan and Lebanon in MENA region. Further, the fintech activities have been emerging in the countries like Saudi Arabia, Bahrain and Iran [1]. Alhough there is an emergence of the fintech activities in this region, still there are gaps is existing in infrastructures, rules and regulations, security provisions, adequate funding and human resource management [5]. The major limitation in adopting the fintech activities is network system and the availability of network data as well as the customers' ability in using the advanced technologies. On the other hand, majority of customers are convenient to use the traditional cash transactions rather than moving towards the technology oriented financial activities. Companies must realize this mindset of the people and try to provide the services and products at their convenient [7]. The fintech companies are facing some challenges related to regulatory aspects on setting up the business and fund raising opportunities. There are about 85% of new starts are willing to relocate themselves to other region due to this regulation issues. The cyber threat is the major concerns for the stakeholders of fintech operations. The recent survey states that there is about 55% of policy makers believes that the cyber threat is their major concern than other issues exists in fintech sector. This has been evident that there is 17% increase in the cyber-attack in the year 2019 in MENA countries when compared to the previous year [4]. The sandbox approach will help the companies as well the policy makers to develop a holistic policies and regulations to the growth of fintech activities in MENA region. The initiatives of developing new regulations and policies started implementing in many countries in this region. Notably, the launching of UAE sandbox and the mobile money regulatory policy of Egypt is the evidence of the policy makers' involvement in this sector. The another major challenge for the fintech companies in MENA region is hiring and retaining talent to achieve their aspirations and interest in attracting their customers. There are about 25% of the startups shutdown their businesses due to the shortage of getting talent IT qualified people in MENA region. On the other side, over 40 international fintech companies have started their businesses recently in MENA region [1]. It is very clear that this sector will make massive impact in the global economic activities and the way of doing transactions. Fintech services are the necessity of the current scenario and further developing this sector is the major responsibility of the peoples who are involving in this sector.

7 Looking Ahead—Fintech in MENA

The financial services industry has dramatically changed after the technological revolution imparted in the financial services i.e. Fintech. This revolution has created new culture and customs among the financial institutions. There are many venture capitalists involving in this business processes (Haddad and Hornuf 2016). The increased use of gadgets and internet facility has brought the people rely on technology based financial services through online platform. Nowadays, majority of customers are looking for online transactions rather than going conventional method. This mainly due to that conventional approach is time consuming activity when compared to Fintech platform. The Fintech business processes are flourishing well all over the world due to many reasons like easy way of doing, fast and convenient in handling the financial transaction [32].

Most of the countries considering the financial inclusion is the key driver in reducing poverty and enhancing the growth of any nation. This brings the general public into a common point in accessing the services provided by the financial sector. The financial inclusion is needed for all and especially in the informal sector [3]. It helps poor people to make daily payment reliably, keep the savings for future investments and access the credit supports to survive with certain confidence. The success of the monetary policy of any country highly relies on the above factors. But surprisingly, only the half of the total population is having bank account. The MENA region is also not exceptional in this aspect, where there is only 50% of the population in MENA region have bank accesses and there are about only 30% of women only having bank accounts. Hence, it is essential for all the countries to bring the people into the pool of financial services. This may not be achieved in a moment because the general public needs to be used to it and they should feel that it really helps them. It is quite possible for governments to bring their civilians under one roof called financial services through the effective implementation of fintech based services [32]. If the Fintech is implemented effectively by the policymakers in MENA region, the financial inclusion will be achieved through this innovative service.

Based on the MENA Fintech venture report 2019, the MENA region is 2nd highest worldwide increase in internet penetration between 2014-2018 and in GCC it is found there are about 21% increases in internet penetration during this period. Further, overall there are about 31% higher in payment made using online platform [3]. The use of mobile phone with the internet provision is causal and it is available with most of the people who are having certain extent of educational background [30]. According to the World Bank data 2020, the literacy rate is at 92.68% in MENA region and the users of internet in the year 2016 was at 47.62% and it is growing at an annual average rate of 11.47%. The same source of data states that users of internet are rapidly increasing in MENA region. Hence, this is the step stone to lay the foundation of Fintech startup in MENA region. There are many studies stated that the fintech performs well in MENA region and it is growing phase with certain challenges associated with its' operations. Hence, the present study made an attempt to understand the gap existing and how to overcome these gaps. Since, the study is conceptual in nature, it has taken into account the previous literatures related to fintech businesses in MENA and other regions carefully and drawn a solution to this existing challenges through the powerful word 'Fintech Ecosystem'.

The present study tries to form an holistic approach in building the Fintech Ecosystem for the further growth of this sector. The proposed approach is developed through underpinning the existing literatures available in this field of research. Each literature discusses the issues based on their specific objectives framed for their studies. Further, this study presents the statistical information collected from various agencies and the organization's annual report. The objective of the current study is to bring all the outcomes into one holistic perspective for the benefit of stakeholders of this sector.

7.1 Sustainability

The sustainability is the pillar of any business processes such as environmental, conceptual, financial and social demands and concerns which ensures the ethical, responsible ongoing success. The main strategic risk associated with the fintech services sectors is maintaining or gaining the market share [13]. There is a stiff competition among the rivals in this sectors needs to be managed by maintaining innovative initiations, fulfilling the customers' expectations, efficient operations, substantial profit, delivering the quality services with low cost.

The main asset of the fintech sector is knowledge oriented, innovative information as well technology base. The IT dependence between market players and market resources are limited whereas the development of the business is highly relying on IT development and implementation. If the fintech companies do not concentrate on upgrading or innovating approach in developing their own new venture, IT crisis may arise. Further, the new innovative advanced method of doing business may have complex approach or brutality in complex approach [10]. Hence, the companies should have a proper plan to mitigate those issues to have trustworthy business activities. This can be achieved through training and development. Basically learning platform needs to be used effectively [13].

The fintech firms' needs to take better economic decision to diversify risks associated within the investment avenues. The provision of financial protection can be achieved by reducing cost, transferring risk associated with the business, building real economy assets and develop good financial base. This can be achieved through alternative financing and growth funding via crowdfunding, mezzanine lending and grants [35]. During their initial stage the fintech companies should manage the requirement of funds actively through dedicated funds proliferation, FDIs, angel investors and government supports. Majority of the funds are aligned with the initiatives set by regulators to promote and grow the fintech businesses.

7.2 Environment Support

The physical infrastructure like internet provisions, speed of the net connectivity, network coverage and worthy gadgets are the important support for the growth of fintech business processes [40]. The availability of these basic needs is not that much attained by most of countries in MENA region. This is the major challenge for this sector as well the government and they need to concentrate to resolve this issue on an urgent basis. This may largely affect the new venture capitalists who is involving in fintech business if there is no proper infrastructure facility. Though, the internet penetration looks high in MENA region, only few people use the online tools for their financial transactions (Simanis, [39]. The main issues associated with the fintech businesses is getting target customer with high amount of trust. Further, it is the responsibility for the companies to introduce novelty applications or business models by ease of use. Further, the companies need to provide an assistant to familiar with the new applications in their financial transactions.

The business environment has to support by funding and encouraging the advanced technologies like AI, data management, CC, Blockchain and cryptocurrencies in MENA region. The threat of hidden charges, privacy and security issues in terms of sharing the customers' credentials are the important problems for the companies. However, they need to create trust among the users of the services to have regular business with them. The innovation is the matter of gaining momentum of these types of businesses. Hence, it is essential to look on the success factors in their innovative business models as a key. The innovation should be customer centric and it should solve the real problems and wants of the real end users of the business models.

7.3 Development of Human Capital

The development of organizational culture is highly relying on Organizational focus on innovation. The ecosystem of organization consists of culture, behavior and innovation [12]. Hence, it is inferred that the innovation is the important element for the development of culture in any organization. The success of any organization is highly relying on the development of knowledge and skill based performance [33].

The Millennials are the biggest early users of technology across the world [41]. There are about half of the total population in MENA region is occupied by the Millennials which is the average year of 25. This is really supportive in terms of using the technology. It is also considering as technology-savvy for employees of the fintech companies. The talented peoples can be pooled by proper education and training to be given to enhance or impart the technology skills to associate with the fintech companies as future leaders and developers. The colleges and schools can initiate this step in giving proper education to develop their local peoples to serve in this sector. The World Bank Human Capital Index is the evident for the availability of the qualified human resources in MENA region with the index value of 0.57. Further, the skilled peoples can be recruited from various parts of the world to excel the business to the next level.

7.4 Government Initiatives

Government bodies are supporting the fintech industry in MENA region to grow well with an innovative approach. The government bodies are even associating with private venture capitalist as partner of the business to support the startups to explore further. On the other hand the regulators are viewing in a different perspective is that this sector can bring huge growth opportunities. But be aware that this is equally destructive if it is not regulated well. The Governor of the Central Bank of Kuwait said that the modern regulations and policies is the main factor for a safe, state-of-the-art fintech ecosystem.

There are some 20 countries in this region and harmonizing the regulations and establishing common policy is highly challenging tasks. But this can be achieved by bringing common kind of business processes. Regulators are allowing sandbox regulation approach to operate the fintech businesses in a supportive regulatory environment to test the reality and challenges associated in the real time operations [48]. The same time government bodies needs to look the holistic approach in handling this regulatory issue across the border in MENA region. The sharia requirements and compliances needs to be included in the Islamic fintech ecosystem.

The stakeholders are to be empowered with the information exists regarding the fintech business process through a strategic policy documents. These documents can be disseminated through a common webpage to encourage the required research to fill the existing gap. This approach will lead to some employment opportunities for the local talents [49]. The conferences and workshops can be arranged to ensure the proper mix of skill sets among the local community. Further, the risk register must be maintained at all the levels of the operations to track and mitigate those risks. The major risk are associated with this business is system hacking, money laundering, cyber-attack and data breaches. These issues can be solved by formulating the regulation and follow up through advanced technologies.

8 Conclusion

'Fintech' the financial technology brought the advanced technology tools, web platform and ecosystem together to serve the society with the financial products and services in an efficient manner. The current situation after fintech has impacted a lot in the payment options, funding possibilities, investments, financial services and digital currency based operations. MENA region is also witnessing its contribution to this sector by encouraging the venture capitalists to do business in their region. The government bodies also taking lots of initiatives to develop the financial technology based business processes for the benefit of their economy. The present study clearly highlights the current situation in MENA region as well as forward looking approaches towards Fintech with proposed ecosystem. The study concludes that the sustainable business activities, proper infrastructure facility, the efficient learned human resources and the government supports are the main pillars of the growth of fintech, the revolutionary business model in MENA region.

References

1. Abu Amuna, Y., Abu-Naser, S., Shobaki, M., Mostafa, Y.: Fintech: creative innovation for entrepreneurs. Intell. Syst. Acc. Finan. Manag. **3**, 5–15 (2019)
2. Al-Fahim, N.H.: An examination factors influencing the intention to adopt internet banking among SMEs in Yemen: using an extension of the technology acceptance model (TAM). J. Internet Bank. Comm. **21**(5), 1–23 (2016)
3. Alhassan, A., Li, L., Reddy, D.G.: The impact of formal financial inclusion on informal financial intermediation and cash preference: evidence from Africa. Appl. Econ., 1–18 (2019). https://doi.org/10.1080/00036846.2019.1593316
4. Ali, I., Chandra, G., Sharma, B.: UAE's Strategy Towards Most Cyber Resilient Nation (2019). https://doi.org/10.35940/ijitee.l3022.1081219
5. Ali Ghiyazuddin, M.: Creating Enabling Fintech Ecosystems: The Role of Regulators. Alliance for Financial Inclusion (AIF) (2020)
6. Alt, R., Beck, R., Smits, M.T.: FinTech and the transformation of the financial industry. Electron Mark. **28**, 235–243 (2018)
7. Arner, D., Barberis, J., Buckley, R.: FinTech, RegTech and the reconceptualization of financial regulation. Northwestern J. Int. Law Bus. **37**, 373–415 (2017)
8. Association of Certified Fraud Examiners (ACFE): What is fraud? (2016). www.acfe.com/fraud-101.aspx
9. Begenau, J., Farboodi, M., Veldkamp, L.: Big data in finance and the growth of large firms. J. Monetary Econ. **97**, 71–87 (2018)

10. Buckley, R., Arner, D., Barberis, J.: The evolution of Fintech: a new post-crisis paradigm?. Georgetown J. Int. Law **47**, 1271–1319 (2016). https://doi.org/10.2139/ssrn.2676553
11. Bunea, S., Kogan, B., Kund, A.-G., Stolin, D.: Fintech and the banking bandwagon. J. Finan. Transf. Capco Inst. **47**, 25–33 (2018)
12. Buschgens, T., Bausch, A., Balkin, B.D.: Organizational culture and innovation: a meta-analytic review'. J. Prod. Innov. Manag. **30**(4), 1–19 (2013)
13. Camilleri, M.A.: Corporate sustainability and responsibility: creating value for business, society and the environment. AJSSR **2**, 59–74 (2017)
14. Carlin, B., Olafsson, A., Pagel, M.: Technology adoption across generations: financial fitness in the information age. Working Paper Series; National Bureau of Economic Research: Cambridge, UK (2017)
15. Chang, Y., Wong, S., Lee, H., Jeong, S.: What Motivates Chinese Consumers to Adopt Fintech Services: A Regulatory Focus Theory. ACM, New York, NY, USA (2016)
16. Davis, F.D.: Perceived usefulness, perceived ease of use, and user acceptance of information technology: MIS Q. **13**, 319–339 (1989)
17. Ernst and Young: Innovating with RegTech: Turning regulatory compliance into a competitive advantage (2016). https://www.ey.com/Publication/vwLUAssets/EY-Innovating-with-RegTech/$FILE/EY-Innovating-with-RegTech.pdf
18. Fenwick, M., Vermeulen, E.P.M.: Banking and regulatory responses to FinTech revisited: building the sustainable financial service 'Ecosystems' of tomorrow. Lex Research Topics in Corporate Law & Economics Working Paper no. 4 (2019)
19. Gomber, P., Kauffman, R., Parker, C., Weber, B.: On the Fintech revolution: interpreting the forces of innovation, disruption, and transformation in financial services. J. Manag. Inf. Syst. **35**, 220–265 (2018). https://doi.org/10.1080/07421222.2018.1440766
20. Grais, W., Pellegrini, M.: Corporate Governance and Shariah Compliance in Institutions Offering Islamic Financial Services (2006)
21. Haas, P., Blohm, I., Peters, C., Leimeister, J.M.: Modularization of Crowdfunding Services: Designing Disruptive Innovations in the Banking Industry. Association for Information Systems (2015)
22. Haddad, C., Hornuf, L.: The emergence of the global fintech market: economic and technological determinants. Small Bus. Econ. **53**, 81–105 (2019). https://doi.org/10.1007/s11187-018-9991-x
23. Jaksic, M., Marinc, M.: The future of banking: the role of information technology. Bancni vestnik: Banking Sector at the Crossroads: Challenges for the Future (2015)
24. Krishnan, D.: Electronic finance—recent developments. Manag. Finan. **43** (5), 614–626 (2017). https://doi.org/10.1108/mf-02-2017-0028
25. Laahanen, S., Martikainen, M., Lehner, M.: FinTechs: their value promises and disruptive potential. ACRN Oxford J. Finan. Risk Perspect. Special Issue Digital Accounting, 59–70 (2019)
26. Lee, M.C.: Factors influencing the adoption of internet banking: An integration of TAM and TPB with perceived risk and perceived benefit. Electron. Commer. Res. Appl. **8**(3), 130–141 (2009)
27. Lee, S.: User behavior of mobile enterprise applications. KSII Trans. Internet Inf. Syst. **10**(8), 3972–3985 (2016)
28. Lee, I., Shin, Y.: Fintech: Ecosystem, business models, investment decisions, and challenges. Bus. Horizons **61** (2017). https://doi.org/10.1016/j.bushor.2017.09.003
29. Lee, Y., Kozar, K.A., Larsen, K.: The technology acceptance model: past, present, and future. Technology **12** (2011). https://doi.org/10.17705/1cais.01250
30. Lenka, S.K., Barik, R.: Has expansion of mobile phone and internet use spurred financial inclusion in the SAARC countries? Finan. Innov. **4**, 5 (2018)
31. Magnitt: MENA Fintech venture report, Abu dhabi Global Market (AGDM). https://www.adgm.com/documents/publications/en/adgm-mena-fintech-venture-report-2019.pdf
32. Meyer, E.: Getting to si, ja, oui, hai, and da. Harvard Bus. Rev. (2015). https://hbr.org/2015/12/getting-to-si-ja-oui-hai-and-da

33. Minbaeva, D.B.: Strategic HRM in building micro-foundations of organizational knowledge-based performance. Hum. Resour. Manag. Rev. **23**(4), 378–390 (2015)
34. Ng, A., Kwok, B.: Emergence of Fintech and cybersecurity in a global financial centre: strategic approach by a regulator. J. Finan. Regul. Complian. **25** (2017). https://doi.org/10.1108/jfrc-01-2017-0013
35. Nuccio M.V., Loewy R.R.: Basel III: Impact on Asset-Based Lending. abfjournal **11**(1), 28–29 (2013). abfjournal.com/articles/basel-iii-impact-on-asset-based-lending
36. Priem, R.L., Li, S., Carr, J.C.: Insights and new directions from demand-side approaches to technology innovation, entrepreneurship, and strategic management research. J. Manag. **38**, 346–374 (2012)
37. Riquelme, H.E., Rios, R.E.: The moderating effect of gender in the adoption of mobile banking. Int. J. Bank Mark. **28**, 328–341 (2010)
38. Schueffel, P.: Taming the beast: a scientific definition of Fintech. J. Innov. Manag. **4**(4), 32 (2016)
39. Simanis, E., & Hart, S.: Innovation from the inside out. MIT Sloan Manag. Rev. **50** (2016)
40. Sironi, P.: FinTech Innovation: From Robo-Advisors to Goal Based Investing and Gamification. Wiley, West Sussex (2016)
41. Swart, J., Kinnie, N.: Organisational learning, knowledge assets and HR practices in professional service firms. Hum. Resour. Manag. J. **20**(1), 64–79 (2010)
42. Taylor, S., Todd, P.A.: Understanding information technology usage: a test of competing models. Inf. Syst. Res. **6**, 144–176 (2015)
43. Tidebrant, P.: Disruptive Innovation in the Swedish Payment Market: A Supply-Side Perspective. KTH Industrial, Engineering and Management, Stockholm (2013)
44. Treleaven, P.: Financial regulation of fintech. J. Finan. Perspect. **3**(3), 1–14 (2015)
45. Varga, D.: Fintech, the new era of financial services. Budapest Manag. Rev. **48**(11), 22–32 (2017)
46. Vives, X.: The impact of FinTech on banking. Eur. Econ. Banks Regul. Real Sector **2**, 97–105 (2017)
47. Zavolokina, L., Dolata, M., Schwabe, G.: The Fintech phenomenon: antecedents of financial innovation perceived by the popular press. Finan. Innov. **2**, 16 (2016)
48. Zetzsche, D., Buckley, R., Arner, D., Barberis, J.: Regulating a revolution: from regulatory sandboxes to smart regulation. SSRN Electron. J. (2017). https://doi.org/10.2139/ssrn.3018534
49. Zetzsche, D.A., Buckley, R.P., Arner, D.W., Barberis, J.N.: From FinTech to TechFin: The Regulatory Challenges of Data-Driven Finance (2017)

RegTech and Regulatory Change Management for Financial Institutions

Hazik Mohamed and Ramazan Yildirim

Abstract Financial reform in the last few decades has transformed the global operations of finance watched over by a reactive regulatory framework, in a concerted effort to enhance the resilience of institutional structure, recapture market trust and improve the critical supporting infrastructure to the economy. Many financial institutions are struggling with ever-changing regulatory requirements and the increased burden of intricate rules. Such rules can be conflicting, and subjected to local implementation and jurisdictic interpretation. Non-compliance or the inability to meet the deadlines subject institutions to hefty penalties and multimillion-dollar fines for failures in controlling nefarious and over-the-counter trading activities. In this chapter, we map out ongoing regulatory milestones globally to understand the main supervisory priorities by territory and propose how financial institutions can successfully administer regulatory change through an AI-driven regulatory management model within their organizations. We discuss the evolution, challenges and recommendations for financial institutions (including Shariah compliance) in the change management for regulations using a structured approach that addresses regional priorities.

Keywords Anti-money laundering · Customer due diligence · Data privacy and protection · OTC reform

H. Mohamed (✉)
Stellar Consulting Group, Singapore, Singapore
e-mail: hazik@stellarcg.com

R. Yildirim
Upsite Consulting, Muharraq, Bahrain
e-mail: ramy@upsiteconsulting.com

A. Hamdan et al. (eds.), *The Fourth Industrial Revolution: Implementation of Artificial Intelligence for Growing Business Success*, Studies in Computational Intelligence 935,
https://doi.org/10.1007/978-3-030-62796-6_8

1 Introduction

An effective and conducive regulatory system is key to the success of jurisdictions in order to remain competitive in the global market with the use of FinTech. Persistent financial crises in the last few decades have transformed the global regulatory framework radically, in a concerted effort to enhance the resilience of financial institutions, recapture investor confidence and improve the overall financial system. Failure to comply to new regulatory requirements incur hefty fines which have resulted in deferred prosecution agreements by major financial institutions [1].

International cooperation will be crucial in view of the harmonies and shared dimensions of many global financial activities. "There is potential for international bodies, like the IFSB, FSB, the GPFI and SSBs—such as the BCBS, IAIS, IOSCO and CPMI—to provide avenues for authorities to get together to share experiences on FinTech implications for financial markets" [8]. Especially needed is increased cooperation which will be able to "mitigate the risk of fragmentation or divergence in regulatory frameworks, which could impede the development and diffusion of beneficial innovations in financial services and limit the effectiveness to promote financial stability" [7].

The regulatory changes continue to take place and its implementation continue to prioritize cybersecurity, data privacy and OTC (over-the-counter) derivatives reform [2]. It is observed that European regulators are sorting out the implementation of its 4th EU Money-Laundering Directive as well as priming Brexit's impact, while the U.S. regulators are wrestling with regulatory reform. Meanwhile, in Asia-Pacific, regulators adopting the "wait-and-see" approach are now catching up in anti-money laundering (AML), fraud and identity data regulations.

In this chapter, we map out ongoing regulatory milestones globally to understand the main supervisory priorities by territory and propose how financial institutions can successfully administer regulatory change within their organizations through an AI-driven regulatory management model. We also discuss the evolution, challenges and recommendations for financial institutions (including Shariah compliance) in the change management for financial regulations using a structured approach that addresses regional priorities.

2 Financial Regulations

The pace of regulatory reform did not wane in the last few years. Several key deadlines concluded from January 2018 onwards, one of which was the introduction of MiFID II,[1] which utilized the legal entity identifier (LEI)[2] for transaction reporting, and the mandatory application of the General Data Protection Regulation (GDPR), harmonizing identity data security across Europe. As for the U.S., "FinCEN's Final Rule (CDD) introduced significant new customer due diligence requirements for financial institutions".[3] Generally, the concerns for financial regulations are as follows in the sub-sections.

2.1 Regulation and Supervision

In effective regulations, clarity and critical parameters have to be determined through a sound consensus-led framework (for all, including digital business models) to ensure ease of adoption and improve compliance. In the digital era, market supervisors have to make sure that financial institutions have vigorous governance structures and reconnaissance that are leveraged on data-driven supervision.

Other risk management concern is the operational risk within which cybersecurity, fraud and theft, data privacy and legal issues fall under [7]. Similar to the Basel Committee on Banking Supervision (BCBS), the Islamic finance standard-setting body known as Islamic Financial Services Board (IFSB) prescribes a capital regime for operational risk, which does not address risk-related operational issues. "While regulatory instruments such as BCBS/IFSB capital requirements can create incentives to address certain operational risks, such as business continuity, capital is not sufficient to restore operations if a financial institution suffers a cyber-attack" [4]. In addition, there are serious risks involving money laundering and concerns on terrorism financing (now typically known as AML/CFT) that financial institutions including banks need to incorporate adequate measures for [3].

Specific to Islamic Finance, the innovative solutions for Islamic Financial Services should be consistent with Shariah rules and principles, and it takes adequate knowledge in the relevant economic, financial and technical (AI, blockchain, IoT, machine learning) areas. This is a huge concern in the role of the Shariah Supervisory Board (SSB) in overseeing the product innovation at the institutional level, which impacts Shariah compliance issues for enhanced supervision. The assessment of Shariah-compliance too should include the procedural processes from creation to the result of any crypto-assets and its mechanisms. In this respect, CIBAFI has suggested that "Islamic banks will need to consider how they can safeguard the end-to-end transactions according to the Shariah, including the rights and ownership at each

---

[1] https://www.fca.org.uk/markets/mifid-ii/legal-entity-identifier-lei-update.

[2] https://www.emissions-euets.com/internal-electricity-market-glossary/839-lei.

[3] https://www.fincen.gov/resources/statutes-and-regulations/cdd-final-rule.

stage" [10]. While other bodies, such as the IFSB, have also highlighted the challenges that FinTech innovations cause, specific guidance to address all forms of FinTech operations in Islamic finance would require time and dedicated resources, and such expectations need to be reframed. The appropriate regulations will have to be created (or adjusted) as the technology itself evolves. The assumption that regulations, once crafted, will then remain in place, unchanged, for significant periods of time, has been overturned in today's environment. As new business models and services emerge, such as sharing services (e.g. Airbnb and Grab) and tokenization of businesses, government agencies are challenged with creating or modifying regulations, enforcing them, and communicating them to the stakeholders, while working within legacy frameworks and striving to foster innovation.

2.2 Financial Stability and Consumer Protection

The purpose of market regulations is to safeguard financial institutions from events that could threaten their stability and the larger macroeconomic stability due to their nature of services. This means that banks have to put in place robust plans for scenarios that may undermine financial stability. For example, the increasing use of algorithms in robo-advisory services, or automation in recommending investment portfolios based on predictive client preferences, need to be within approved products or companies offering such services need to be compelled to disclose more information when required. Correspondingly, external vendors which provide third-party service to financial institutions, such as cloud computing, storage, server and data services are becoming more common and an essential part of their *modus operandi*. Such third-party technology providers may need to be regulated in order to manage related operational risks, which may impact financial stability indirectly.

Within the scope of financial stability, one of the areas that are highly focused on is personal data protection, which also falls under consumer protection. The need to protect sensitive consumer and corporate financial data is crucial for financial services to function safely. The danger of cyber-attacks and identity hacking is very real with recent incidents involving online fraud and digital theft. Such incidences include violations of personally information, which raises many issues including data privacy, digitally identifiable rights, cybersecurity protocols in systems management, and legal liability.[4] In order to circumvent any undesirable situations, resolute steps need to be taken to enhance cyber-security and mitigate cyber risks.

[4]Data Protection Law: An Overview https://fas.org/sgp/crs/misc/R45631.pdf.

2.3 Domestic and Cross-Border Transactions

Domestic and cross-border transactions are equally important for market supervisors. The development of FinTech has posed a significant reassessment of traditional regulation since the use of advance technologies warrants a shift in regulating them. With invisible interconnections in cyberspace and faster transactional speeds, regulators struggle to track and keep up with questionable transactions when they cannot even understand how it is done and its legality. As such, there has to be stronger synchronization among the supervisory agencies, and harmonization in FinTech regulations in order to regulate financial institutions better in light of the digital revolution. A common approach and strategy need to be in place to address AML and counter-terrorism financing (CFT) issues at the national and also international levels. Hence, regional cooperation will be integral in regulating FinTech that scale beyond borders. Further to these concerns, digital innovations in financial services utilize smart contracts (for blockchained systems) and their usage raise questions about the cross-jurisdictional enforceability issues and dispute resolution where there are no consensus and recognized legal recourse.

3 RegTech—Digital Reporting, Audit and Compliance

Around the world, regulators are dealing with digital transformation challenges with the rise of new FinTech technologies and non-traditional market entrants (BigTech firms) at extraordinary swiftness. Authorities are "faced with the task to develop regulatory approaches that do not hamper development and innovation while still limiting risks to consumers and financial stability" [1].

RegTech is largely seen as a "category that focuses on technologies that may facilitate the delivery of regulatory requirements more efficiently and effectively" than existing capabilities. However, Arner et al. [1] view "RegTech represents more than just an efficiency tool but rather a pivotal change leading to a paradigm shift in regulation". To them, holistically "RegTech represents the next logical evolution of financial services regulation and should develop into a foundational base underpinning the entire financial services sector".

The application of technology to monitoring and compliance offers "massive cost savings to established financial companies and potentially massive opportunities to emerging FinTech start-ups, IT and advisory firms" [11]. RegTech will enable the option of "continuous monitoring that would improve efficiency by both liberating excess regulatory capital, decreasing the time it takes to investigate a firm" [5], fostering competition and upholding their directives for financial stability (both macro and micro) and market integrity from the regulator's perspective.

3.1 Regulatory Priorities by Region

Looking at the planned regulatory priorities in the different regions, the key upcoming regulatory milestones identified over the next eight quarters are: AML and CDD, data protection and OTC reform, with slightly different priorities for different regions for advanced and developing nations. However, data protection and privacy are high on regulatory priorities[5] globally. The movement towards "GDPR has created a ripple effect of new and revised data protection and data privacy regulations worldwide in order to ensure the continuation of data flows between jurisdictions" [2].

In the U.S., "regulatory reform, conflicts of interests and cybersecurity feature prominently on the list of regulatory priorities".[6] There is genuine concerns for the regulatory impacts of Brexit on U.S. investors and the overall securities markets. Related parties and companies do not seem to satisfactorily disclose the likely impact of Brexit to their firms according to the Securities and Exchange Commission (SEC).[7] To subdue the anxiety of anticipation of undesired repercussions, they have dedicated personnel to record and monitor the exact disclosures on Brexit that companies are making.

In January 2020, the financial prudential regulatory requirements came under closer scrutiny, especially when the 'Current Expected Credit Loss' regime came into effect.[8] The U.S. is also "under increased pressure from the EU to appoint a Data Privacy Ombudsman, as per Privacy Shield requirements".[9]

The European Union concentrated on REFIT (regulatory fitness) for 2019, by revising their legislations and procedures, particularly those relating to reporting measures. "Continued development of the Capital Markets Union and securitization supervision have been called out by the European Securities and Markets Authority (ESMA) as key regulatory priorities".[10] There is an increased emphasis on accountability, specifically making senior managers accountable for violations and irregularities of the firm and non-compliance of the staff, including retail transgression and wholesale agency wrongdoing.

The work culture aspect of organizations have come under the scrutiny to ensure the objectives of compliance pervades through every level of the organization. The

---

[5]https://www.pdpc.gov.sg/-/media/Files/PDPC/DPO-Connect/March-20/Seven-Global-Personal-Data-Protection-Priorities-for-2020.html.

[6]https://blogs.thomsonreuters.com/answerson/top-10-concerns-for-u-s-compliance-officers-in-2019/.

[7]https://tax.thomsonreuters.com/blog/the-sec-is-watching-part-2-brexit-disclosures/.

[8]https://www.occ.treas.gov/news-issuances/bulletins/2020/bulletin-2020-27.html.

[9]https://www.euractiv.com/section/data-protection/news/us-to-appoint-permanent-privacy-shield-ombudsperson-following-eu-pressure/.

[10]https://www.esma.europa.eu/sites/default/files/library/esma80-199-332_confidential_supervision_ar_2019_wp_2020.pdf.

European Banking Authority (EBA) published draft "guidelines on internal governance in the 2017".[11] Also, outsourcing is another concern raised by European regulators to look into the outsourcing of IT security and IT risk management in order to be conscious of data integrity and prevention of unwanted data events. Preliminary standards have been created by the EBA to establish a "framework for the due diligence process of institutions to ensure that functions are only outsourced to reliable service providers so that the ongoing provision of services and compliance with regulatory requirements is ensured" [2].

Cybersecurity is high on the agenda of the APAC regulatory priorities list. Many regulators have already issued "specific cyber-risk management and information security guidance, including on the importance of effective cybersecurity risk management"[12] (Hong Kong SAR) and on "early detection of cyber intrusions"[13] (Singapore).

3.2 Anti-money Laundering (AML) and Counter-Terrorism Financing (CFT)

In the U.S., "compliance with the Bank Secrecy Act (BSA) remains a key area, with an increased focus on stress testing and risk testing programs to ensure they are able to stand up to risk".[14] Digital currencies are still being actively discussed, but it is "unlikely there will be any formalized rules around this yet".[15] In its place, the urgency in 2019 has been to monitor the cryptocurrency space.

In Asia-Pacific, "AML and beneficial ownership regulatory reform progressed with the Anti-Money Laundering Bill taking effect in Hong Kong, along with the removal of a two-tiered approach to ultimate beneficial ownership to align with FATF (Financial Action Task Force) standards".[16] Historically, a "dual-threshold was required where 25% was applied as standard, with the threshold reducing to 10% where there is an increased risk. Several financial institutions have elected to retain the dual-threshold approach or apply 10% across all entities" [4].

[11] https://eba.europa.eu/regulation-and-policy/internal-governance/guidelines-on-internal-governance-revised.

[12] https://www.hkma.gov.hk/media/eng/doc/key-information/guidelines-and-circular/2015/20150915e1.pdf.

[13] https://www.mas.gov.sg/regulation/circulars/srd-tr-012015-circular-on-early-detection-of-cyber-intrusion.

[14] https://www.fincen.gov/news/news-releases/federal-bank-regulatory-agencies-and-fincen-improve-transparency-risk-focused.

[15] https://www.loc.gov/law/help/cryptocurrency/world-survey.php.

[16] https://www.fatf-gafi.org/media/fatf/documents/reports/mer4/MER-Hong-Kong-2019.pdf.

Embedded in the FSB SIFI[17] framework are macro-financial issues related to systemic importance, which recommends that "financial institutions identified as systemically important should have more intense supervisory oversight, higher loss absorbency as well as recovery and resolution plans" [4]. The "majority of regulatory changes and clarifications have been made in the areas of payments, capital raising, and to a lesser extent investment management as many of these economic functions naturally fit within existing regulatory regimes.[18] Only a few regulatory changes to include FinTech innovations in insurance and market support were mentioned" [4].

3.3 Customer Due Diligence[19] (CDD) and Ultimate Beneficial Owner (UBO)

CDD continues long after establishing a relationship with the customer. Regulations require transactions and account activities to be scrutinized for money laundering or terrorist financing risks on a regular basis. Also, accurate records-keeping allow financial institutions to understand its entire relationship with its customer, and help them to meet regulatory reporting obligations.

Within Europe, AML legislation has gone through many changes and undertakings. However, there are some countries that have not adopted into law these rules even when the 4th EU Money Laundering Directive has long past its implementation date [2]. By convention, Member States are required to transpose the Directive by January 20th, 2020. Related to that Directive, the "Sixth AML Directive was adopted in October 2018 and must also be transposed by the Member States by December 3rd, 2020".[20]

Meanwhile in the US, the "FinCEN Final Rule (CDD) came into effect for all new accounts opened on or after May 11th, 2018, implementing a dual-prong approach to

---

[17]Systemically important financial institutions (SIFIs) are financial institutions whose distress or disorderly failure, because of their size, complexity and systemic interconnectedness, would cause significant disruption to the wider financial system and economic activity.

[18]Financial Stability Implications from Fintech. Supervisory and Regulatory Issues that Merit Authorities' Attention, 27th June 2017.

[19]CDD collects different information depending on different criteria for different categories:
e.g. for (Individual) Customer Profiles: Full name, including any aliases, residential address, mailing address, contact numbers, email addresses, place of birth, date of birth, marital status, nationality, race, government-issued identification number, government-issued tax identification number, occupation, specimen signature, parental consent form (where the individual is a minor).
e.g. for Customer Profile (Entity): Name of corporation, type of corporation, date of incorporation, place of incorporation, board resolution on authorised signatories, certificate of incumbency, constitution, Articles of Association, Certificate of Incorporation, annual report, rirectors, shareholders, senior management, Ultimate Beneficial Owners.
e.g. for Enhanced CDD (ECDD): Politically Exposed Person (PEP), customer who are positively identified to have adverse profiles on watchlists, terrorists, non-face to face account opening, correspondent accounts, customers located in high-risk locations.

[20]https://ec.europa.eu/commission/presscorner/detail/en/qanda_20_821.

beneficial ownership".[21] Under the FinCEN CDD Rule, collecting, maintaining and reporting of beneficial ownership information is now a requirement for financial institutions: "Covered financial institutions must collect from the legal entity customer the name, date of birth, address, and social security number or other government identification number (passport number or other similar information in the case of foreign persons) for individuals who own 25% or more of the equity interest of the legal entity (if any), and an individual with significant responsibility to control/manage the legal entity at the time a new account is opened".[22] [2].

In general, there are four elements the FinCEN considers crucial when performing due diligence[23]:

i. customer identification and verification, along with timely updates after onboarding,
ii. understanding the nature and purpose of customer relationships to develop a customer risk profile,
iii. ultimate beneficial ownership identification and verification,
iv. ongoing monitoring to scrutinize transactions for money laundering or terrorist financing risks on a regular basis.

3.4 Cybersecurity

Cybersecurity is another high-priority program in the list of APAC regulatory agenda. A report by the World Economic Forum [13] has rated the large-scale cyber-attacks as one of the five most dangerous attacks on this planet. It is also estimated that the loss from cyberattacks and loss of data will reach US$6 trillion until 2021, double that of 2015 [9]. As mentioned earlier, several regulators have already issued specific cyber-risk management and information security guidance in Hong Kong and Singapore. In Australia, the "Australian Prudential Regulation Authority (APRA) sought to implement enforceable standards around information security"[24] (CPS 234), which commenced July 1st, 2019. The standards on third-party outsourcing are being reviewed, including if 'outsourcing' remains relevant, given that banking supply chains have evolved and are much more complicated.

[21] https://www.fincen.gov/sites/default/files/2018-04/FinCEN_Guidance_CDD_FAQ_FINAL_508_2.pdf.

[22] Ibid.

[23] https://www.trulioo.com/blog/customer-due-diligence/.

[24] https://www.apra.gov.au/sites/default/files/cps_234_july_2019_for_public_release.pdf.

3.5 Data Protection and Data Privacy

The final draft of the ePrivacy Regulation still remains to be published.[25] The "Regulation is intended to increase the effectiveness and level of protection for privacy and personal data in electronic communications. The final version of the Regulation on a framework for the free flow of non-personal information, known as the fifth freedom of the European Union, was published in November 2018 and has been directly applicable from May 2019".[26] The Regulation will prohibit non-personal data localization. Member States continue to implement national legislation to account for permitted derogations. "Given the increased scrutiny on data breaches of late, it would not be too long before national authorities begin imposing enforcement measures—although this has already started in some Member States" [2]. Member States expect a final text of the e-Privacy Regulation so as to prepare for the implementation of the 'Free Flow of Non-Personal Data Regulation'.

Countries in APAC have "begun reviewing—and, in some cases, introducing—legislation relating to data protection in the wake of the implementation of GDPR".[27] In Asia-Pacific, "Thailand, India, Indonesia and Vietnam have all published draft data protection bills as several jurisdictions begin to seek adequacy status".[28] In contrast and contrary to the EU's new data localization prohibition, China is implementing strict data protection regimes with data localization provisions, with India and Vietnam following suit. "New Zealand has also begun reviewing current data protection provisions ahead of a likely review of its EU whitelist status".[29] 'Conduct and culture' are also a critical supervisory focus to align with the process of reform in the lending culture and the way it conducts financial service throughout the industry in the region.

3.6 Remodeling the OTC

The year 2019 marked ten years since the G20 committed to reform and remodel the OTC derivative markets following the financial crisis. Some of the measures taken are the introduction of initial margin requirements globally in phases. "September 2018 was the most recent deadline (becoming applicable for Phase 3 entities), bringing a larger number of counterparties into scope for those requirements. Most recently, Singapore Central Clearing came into effect in October for Singapore and US class

[25] https://www.lexology.com/library/detail.aspx?g = 5e41c7c1-691b-4179-87f1-553d8d638c7a.

[26] Ibid.

[27] https://www.regulationasia.com/managing-regulatory-change-as-a-business-as-usual-activity/.

[28] https://www.dataprotectionreport.com/2020/01/reflecting-on-apac-data-protection-and-cyber-security-highlights-for-2019-and-what-lies-ahead/.

[29] https://mlexmarketinsight.com/insights-center/editors-picks/area-of-expertise/data-privacy-and-security/new-zealands-privacy-law-revamp-to-come-under-scrutiny-in-eu.

currency accounts, while Canadian Central Clearing was phased in from August 2018" [2].

OTC reform was an important agenda on the EU regulation reforms in 2019. Particularly, the "Securities Financing Transaction Regulation (SFTR)—the reporting obligation for investment firms and financial institutions—was expected to come into effect in Q3 2020 (following the European Commission's extension of the scrutiny period of the RTS by three months)".[30] There are also "additional changes under global investor protection in Switzerland, highlighted in the Regulatory Outlook Calendar under January 2020, with the introduction of the Financial Services Act (FinSA) and the Financial Institutions Act (FinIA)".[31]

3.7 *Shariah Audit, Compliance and Monitoring*

The main risk that "Islamic financial institutions (IFIs) face which is unique to them is the Shariah compliance risk. In addition to managing the risks faced by conventional banks, such as credit, market, operational risks, an Islamic financial institution also has to ensure that it complies with Shariah rulings as this carries a significant reputational risk to the institution" [8]. Fintech products (including cryptocurrencies and tokens) and services need to be treated accordingly to its use cases and the fiqh understanding of the Shariah as well as the regulations in place.

The assessment of Shariah-compliance too, should include the procedural processes from creation to the result of any new financial products or digital assets and its mechanisms. For compliance and legitimacy, the regulations for them can be coded as per fiqh guidelines (from Shariah committees) and automated as an algorithm-driven AI module. Such rulings or fatwas that have been previously determined as sound, can be accumulated in a database for future cross-referencing and limit additional resources and fees required for Shariah compliance assessments. Through the use of new regulatory audit mechanisms and validation technology (i.e. RegTech), the compilation of various schools of thought (mazhabs, etc.) can be organized productively and Shariah compliance can be reconciled in a structured and efficient manner.

4 Managing Regulatory Change

Policy-makers and regulatory authorities have a monumental task in confronting rapidly transforming financial systems in the coming years. "Regulatory change management has the potential to impact divisions beyond the compliance function,

[30] https://www.esma.europa.eu/policy-activities/post-trading/sftr-reporting.

[31] https://www2.deloitte.com/content/dam/Deloitte/ch/Documents/financial-services/deloitte-ch-fs-financial-markets-regulatory-outlook-switzerland-2020.pdf.

including data management, operations, client-facing teams, client experience and time to revenue" [2]. Regulatory change has to be managed and precisely calculated, as wide-ranging changes can ultimately use up the allotted discretionary budget and resources. The continuing regulatory changes embroil uncertainty in operating environments, and anticipated regulatory change has already impacted today's business decisions and existing business operations [12].

Under AML and KYC requirements, detailed requirements of a client's identity, income, source of funds, etc., are reviewed at a substantial level. As such, for large financial corporations with global operations in many countries with thousands of customers, this is without a doubt, a tediously demanding process. Especially so when not only the general global requirements needs to be addressed, but the specific requirements of individual markets, and individual regulatory authorities within those markets [1] need to also be complied to.

Operationally, financial institutions face daily challenges from constant modifications of the innumerable amount of regulations with various requirements. They are also increasingly pressured to achieve categorical compliance with limited budgets and substantial resource limitations. Such constraints impact both the financial and operational performance of financial institutions, as they are subjected to the continual adjustment of resources to meet compliance deadlines. Financial institutions have traditionally responded by deploying resources to reassess and gather information, increasing the total cost of compliance at the expense of customer experience.

To overcome this anticipated revolving door of regulatory changes, banks will necessitate the increasing use and reliance on technology to build the necessary infrastructure to support the changing regulations. The "development of RegTech so far has primarily been driven by the financial services industry wishing to decrease costs, especially in light of the fact that regulatory fines and settlements have increased 45-fold" [6]. The next stage of regulatory supervision is likely to increase their capacity by automating compliance and regulatory surveillance through AI and machine learning. Likewise, financial institutions will need the help of AI technology and systematic strategies within a regulatory framework to adapt to these changes.

4.1 Tracking Regulatory Changes

The financial sector is probably one of the few industries that is highly regulated and policed [12]. Financial institutions not only subjected to monitoring and audits, but they also have to comply with the national as well as the international regulatory demand that has steadily increased over time. As such, their challenge is two-fold: first, to remain compliant, financial institutions need to extract and process all relevant information and implement the regulatory requirements on time. Second, to stay competitive, they need to fulfill the regulatory requirements in a way to keep the optimum synergy and balance between people, processes and technology while holding on their corporate strategic objectives. Before completing with one change application, there are already new regulatory requirements waiting in line. The pace of change and volume of regulatory changes is overwhelming and financial institutions are required to develop a framework that provides not only a modern approach to regulatory change management but also ensures organizations' agility and flexibility to future regulatory demand. Therefore, it is essential to understand the key objectives of the regulatory changes and implement an AI framework that supports financial institutions to cater to those requirements for regulatory compliance.

4.2 Compliance Readiness to Regulatory Changes

As mentioned above, financial institutions are exposed to national and international financial regulations. To ensure compliance and operational sustainability, the following steps must be performed:

i. Assess new regulatory changes against the related and applicable 'as-is' implemented business processes.
ii. Design and implement the related and applicable 'to-be' processes to adapt to the new regulatory changes.
iii. Ensure the implemented regulatory changes are adequately adapted and integrated across the organization.
iv. Monitor the implemented regulatory changes continuously to ensure operational sustainability.

The integrated in-house regulatory system ensures that financial institutions have transparent compliance, for instance, during and post-client-onboarding in terms of statutory tax capturing and reporting obligations, i.e. FATCA (Foreign Account Tax Compliance Act) and CRS (Common Reporting Standard).

4.3 A Regulatory AI-Driven Model

Many regulations, which include the EMIR and MiFID II, consists of important and complex requirements. Therefore, the need for RegTech is vital for the management of regulatory reporting and monitoring, including management of large digital datasets and audits, which require thorough design thinking, coding and product development, micro-financial risk management and macro-prudential supervision. As such, financial institutions need a cohesive model for assessing and implementing prioritized regulatory requirements that allow the buy-in of all management levels within an organization.

The Regulatory Model shown in Fig. 1 offers a structured and consistent management approach to allow automation and compliance maintenance of regulatory requirements and consist of the following key components:

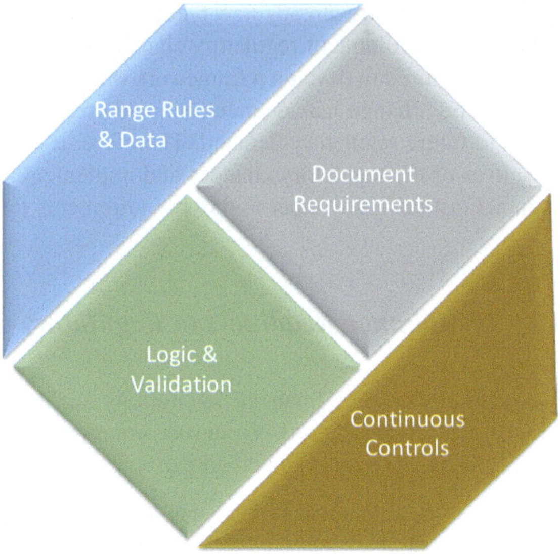

Fig. 1 Components of the Regulatory AI-Model (*Source* Author's Own)

Range the 'Regulatory Rules' and 'Data Requirements': Defining the extent of regulatory rule helps to create a clear picture of when and where the financial institution's compliance will meet the regulatory requirements. All the rules in the new regulations must be interpreted and understood well before a gap analysis can be conducted. This is an analysis to identify how well current practices of the financial institutions meet the new regulatory requirements. There are four main options—either:

(a) the current practices and capabilities meet the new regulatory requirements;
(b) the current practices and capabilities are insufficient, therefore need to be adjusted;
(c) new practices and capabilities are required;
(d) the financial institution does not need to comply with the requirements.

Once the option (b) or (c) applies, it is important to determine whether the scoped requirements (i.e. rules and their characteristics) should be applied through the business processes or the existing regulatory rules engine of the financial institution. The other important aspect during scoping is the data requirements that require thorough assessment primarily related to sourcing. Usually, there are two aspects to be considered—either:

(a) data is available within the organization; whether directly or indirectly
(b) data is not available; hence it must be sourced from third-party service providers.

Document Requirements: The regulations may require documents to be generated by default during the execution of a business process (e.g. client onboarding). In some instances, when a specific rule applies, conditional documentation may be required from the client. Depending on the data captured within the regulatory datasets, the regulators can request ad hoc information if issues of concern arise.

Logic and Validation: The rule-based algorithms are used to extract knowledge (in the form of rules of the data classification) and calculate the compliance values. These computed values drive the validation warnings and ensure that the controls are in line with the regulations.

Continuous Controls: Compliance monitoring mechanisms are essential to ensure ongoing compliance. Following the initial client onboarding, for instance, certain client events such as changes in marital status, political affiliation, citizenship, source of funds, etc. may trigger regulatory monitoring for updated documentation or declarations based on new assessments due to a change in status.

5 Conclusion

Regulators may provide innovation-friendly policies and an environment that "gives incentives to digital platforms to test and refine their innovative ideas, and financial institutions may provide financial services or access to their internal sources and financial expertise" [8]. Stakeholders must ensure their organizations can adapt to market shifts, even making difficult operational changes if the market turns. The capability to adapt to new regulations and managing those changes internally will foster a prosperous and dynamic financial ecosystem, towards better business conduct and service culture.

In the coming years with the upcoming regulatory changes, ensuring compliance will be exigent for global firms, and the continuing review has become the new normal in an uncertain economy plagued with black swans[32] and grey rhinos.[33] Although the "pace of regulatory reform has slowed in Europe and the US, geopolitical factors, such as Brexit, combined with new regulatory priorities and ongoing supervision will increase the regulatory burden on global organizations" [2]. In Asia-Pacific, its regulators continue to trudge forward with the accomplishment of proposed reforms according to internationally agreed standards. Improving the financial industry's culture and business conduct will be a vital supervisory priority for the subsequent years ahead across jurisdictions, emphasizing on individual accountability within institutions and organizations. Reform and remodeling OTC trading and data protection will endure being prevailing premises, as will cybersecurity, where essential service institutions will confront mounting pressure to build resilient systems for economic security and resilience.

[32]Black swans are unexpected events of large magnitude and consequence, first theorized by Nassim Taleb.

[33]Grey rhinos are highly probable, high impact yet neglected threats, coined by Michele Wucker.

References

1. Arner, D.W., Barberis, J.N., Buckley, R.P.: FinTech, RegTech and the reconceptualization of financial regulation. Northwestern J. Int. Law Bus. **37**(3), 371–413 (2017)
2. Dunphy, J., Glynn, L.: Global Regulatory Outlook 2019-2021. Fenergo, Dublin (2018)
3. Financial Action Task Force: Guidance for a Risk-based Approach: The Banking Sector (2014)
4. Financial Stability Board: Financial Stability Implications from FinTech: Supervisory and Regulatory Issues that Merit Authorities' Attention (2017). Accessed 27 June 2017
5. Gutierrez, D.: Big Data for Finance—Security and Regulatory Compliance Considerations. Inside Big Data (2014). http://insidebigdata.com/2014/10/20/big-data-finance-security-regulatory-compliance-considerations/
6. Kaminski, P., Robu, K.: A Best-Practice Model for Bank Compliance, McKinsey, Exhibit 1 (2016). http://www.mckinsey.com/business-functions/risk/our-insights/a-best-practice-model-for-bank-compliance
7. Mohamed, H.: The future of FinTech in ASEAN. In: Anshari, M., Al-Munawar, M.N., Masri, M. (eds.) Financial Technology and Disruptive Innovation in ASEAN, pp. 63–79. IGI Global (2020)
8. Mohamed, H., Ali, H.: Blockchain, Fintech and Islamic Finance—Building the Future of the New Islamic Digital Economy. De|G Press, Boston/Berlin (2019)
9. Morgan, S.: Cybercrime Report. Cybersecurity Ventures, sponsored by Herjavec Group (2017)
10. Raden Aji Haqqi, A.: Strengthening Islamic finance in South-East Asia through innovation of Islamic FinTech in Brunei Darussalam. In: Ordoñez de Pablos, P., Almunawar, M.N., Abduh, M. (eds.) Economics, Business, and Islamic Finance in ASEAN Economics Community, pp. 202–226. IGI Global. (2020)
11. Shedden, A., Malna, G.: Supporting the Development and Adoption of RegTech: No Better Time for a Call for Input. Burges Salmon (2016), https://www.burges-salmon.com/-/media/files/publications/open-ac-cess/supporting_the_development_and_adoption_of_regtech_no_better_time_for_a_call_for_input.pdf
12. Turki, M., Hamdan, A., Ajmi, J.A., Razzaque, A.: Regulatory Technology (Reg-Tech) and money laundering prevention: exploratory study from Bahrain. In: Hassanien, A., Bhatnagar, R., Darwish, A. (eds) Advanced Machine Learning Technologies and Applications. AMLTA 2020. Advances in Intelligent Systems and Computing, vol. 1141 (2021)
13. World Economic Forum: The Global Risks Report, 12th edn (2017)

Artificial Intelligence and Financial Technology FinTech: How AI Is Being Used Under the Pandemic in 2020

Haochen Guo and Petr Polak

Abstract The chapter is about the development of Artificial Intelligence (AI) technology in finance, especially under the case of the COVID-19 pandemic in 2020. It does not only present the applications, but also the regulation of AI and Financial Technology FinTech. An innovation regulatory framework at the regulation level and compulsory restrictive guidance and supervision for AI-based technology to allow sustainable growth will promote the accelerated growth of AI in finance. The AI in the financial industry itself focuses on the main characteristics of "digitalization", "onlineization", "remoteization", "visualization", and "intelligence", building a multi-functional, all-process end-to-end system based on data, enabling multi-user multi-terminal concurrent office, intelligently assisting in dealing with problems and giving solutions. The advent of AI and its ever-broader effects on other industries demands an assessment of its influence on achieving sustainable development goals.

Keywords Artificial intelligence (AI) · Financial technology (FinTech) · Finance and Treasury 4.0 · Finance 4.0 · COVID-19 pandemic · Regulatory technology (RegTech)

H. Guo
Faculty of Economics, VSB-Technical University of Ostrava, Sokolská Třída 33, Ostrava, Czech Republic
e-mail: haochen.guo@vsb.cz

H. Guo · P. Polak (✉)
Faculty of Business and Economics, Mendel University in Brno, Brno, Zemědělská 1, Brno, Czech Republic
e-mail: petr.polak.68@gmail.com

P. Polak
UBDSBE, Universiti Brunei Darussalam, Jalan Tungku Link, Gadong BE1410, Brunei

A. Hamdan et al. (eds.), *The Fourth Industrial Revolution: Implementation of Artificial Intelligence for Growing Business Success*, Studies in Computational Intelligence 935,
https://doi.org/10.1007/978-3-030-62796-6_9

1 Introduction

The COVID-19 pandemic—as a serious public event nowadays—has affected the adjustment of a series of social operation models and also the modification of behavior models—"quarantine" and "home office" became frequently used words, to raised reply to the present "physical quarantine" demand, the money trade has promoted on-line services to guide enterprises through online channels.

Meanwhile, the COVID-19 pandemic creates uncertainty, on several fronts many FinTech companies are under stress. However, as the broader economy shifts from responding to recovery, COVID-19 may create new opportunities for some FinTech companies that have taken hold worldwide with AI technology, such as social distancing, in the use of digital financial services and e-commerce has grown tremendously. AI can evolve into autonomous intelligent systems. At present, an excellent deal of analysis within the field of computing focuses on deep learning, however, there is a limitation of deep learning it needs loads of human intervention. Therefore, the researchers try to reduce the synthetic intervention of the autonomous intelligent ways and improve machine intelligence's self-learning ability, so in the future, computing can improve operational potency of assorted industries, promote mixing and upgrading of computing technology, and build a variety of benchmark application situation innovations to attain affordable, high-efficiency, wide-ranging inclusive intelligent society see (Deloitte, n.d.).

Governments around the world have recognized the value of this new generation in manufacturing through aggressive programs (see Table 1), including awareness-raising, action strategies, funding, expenditure in services, sponsorships and tax advantages to promote the adoption in corporations.

Table 1 Main countries' industrial plans

Country	Industrial plan
Germany	High-Tech Strategy 2020
France	La Nouvelle France Industrille (The New Industrial France)
UK	Future of Manufacturing
US	Advances Manufacturing Partnership
China	Made in China 2025
Singapore	Research, Innovation and Enterprise
South Korea	Innovation in Manufacturing 3.0
Italy	Impresa 4.0

Source Büchi et al. [2]

2 Literature Review

The thought of industry 4.0 was first introduced in 2011 to boost the aggressiveness and potency in manufacturing industries. There is higher property within the system using different sensors—the Internet of Things (IoT), AI, and other innovative technologies. In production lines, their technologies area unit accustomed give acceptable data and decision throughout the complexness. This revolution strengthens the smart factory and smart manufacturing system to manufacture any customized products. Industry 4.0 creates interconnected advanced manufacturing and information system which could adequately communicate, analyze, and provide useful information in the physical world.

Industry 4.0 has had a significant impact on the debate on the discourse on manufacturing process digitalization, despite being blamed for missing a clear concept. However, industry 4.0 may provide an unlimited amendment to align the goals of sustainable development with the continuing digital transformation in industrial development [1]. The conception industry 4.0, which suggests applying principles and technologies on the manufacturing industry from the IoT, was launched in 2011 by a German government council composed of representatives of scientists and industry. Because of this context, the term was not only defined as a technical advancement but was also meant to have a political connotation with the AI to supporting Germany's "position as a leader in the manufacturing engineering industry". The definition has been widely disseminated and gained significant international attention. Consequently, the amount of publications taking up the idea has risen both in the academic community and private sector. Kagermann and Wahlster [6]. The Industrial Age where humanity has entered an extended time with steam, series has caused primitive mechanization. Nowadays, with the development of internet and mobile technologies, electronics, nanotechnology, advances in medicine, health and digital applications, and so on accelerate mechatronics studies. Every day we tend to are witnessing tremendous news and articles in business pages about these topics and obviously, corporate life and professionals can't resist these changes any longer. Dynamic business terms and workforce shape, the way business is done through the use of new technologies will have severe impacts on the daily business life and countries and the world economy as a result. Many items and headlines such as jobless ratio, Philips Curve, performance, management, CRM Analytics, customer relationship management, sales, strategic planning, mass production, Purchasing Power Parity, GDP, inflation, money, Central Banks, Banking System, coaching, training, accounting, taxes, etc. with business and economics, the AI and robotics must face significant risks, impacts, shifts, losses as well as incentives and benefits see Dirican [3].

Büchi et al. [2] builds an operationalization of Industry 4.0 and performance openness principles. Basic terminology for Finance and Treasury 4.0, as well as an overview of the artificial intelligence's influence on this area, was first implemented on Polak et al. [10]. The authors discuss the corporate treasury management move to e-BAM, e-billing, e-procurement, future e-services, common enterprise-wide XML standard, and "follow-the-sun" cash management/cash pooling solutions. There will be no spare cash hanging around ever again. Treasurers should look perfect in all situations in overseeing financial risk and find increasing viewpoint to fuse the impacts of interaction decorrelation, and preferably oversee opportunity. The way to handle financial risk should once ever be a fragmented handheld solution, but a nationwide and organized process of tools that assist in simple leadership and BI (Business Intelligence) studies covering the concerns of the Boards of Directors and Audit Committee. And the corporate treasurers should again consider their position at the top of the organizational hierarchy and strong reasons for their presence owing, in return, to greater anticipated participation in the competitive planning and decision-making of the company. The incremental change from manufacturing financial reporting to a process of assessment, preparing and promoting decision-making, while often subtle, still seems to be a major challenge and role for Treasury 4.0. Polak et al. [11] advanced work in the field is the integration of AI information and language within finance and treasury management 4.0. We may also find several articles discussing dramatic developments in the structure of finance and treasuries. The changes began with the complete centralization of all finance and treasury activities—some at the global level (financial risk management), some at the level of regional finance and treasury centers (cash management) as a prerequisite for automation and digitalization.

Roszkowska and Prorokowski [13] investigate the current position of bank treasuries and show how the Treasury's work in the banking sector is being changed. They use a sample of international bank surveys that represent emerging and advanced markets alike. Analyze the present and potential challenges that treasuries pose, too. The study provides for a review of how treasury departments vary in their quest for sales diversification under tougher funding and liquidity constraints, and what are the issues involved with the selling of treasury goods as part of the updated strategies.

Polak et al. [9] discuss the position of the new corporate treasurer in a global company and its evolution in the light of existing problems confronted by corporations and treasurers. The most significant incident driving change in the role of the corporate treasurer is the credit crisis that occurred in 2007–2011 according to this research. Ramiah et al. [12] report the activities of corporate treasurers engaged in the capital, inventory, accounts receivable, accounts payable, and risk management decision-making phase during the global financial crisis. It used a survey questionnaire and sought to figure out whether working capital managers are susceptible to such heuristic-driven prejudices, such as loss avoidance, the high degree of trust, anchoring, and self-serving prejudices. Its findings show that these professionals show signs of biased behavior. While the prejudices in some fields of working capital management (WCM) contribute to sub-optimal choices, in certain facets of WCM they can still be attractive qualities. We are suggesting a summary of effective management working capital. Moosa and Ramiah [8] provide a succinct overview of philosophical differences and their consequences for decision taking in financial matters. It argues that the corporate treasurer's narcissism is found to be pervasive behavior in the finance sector and that it has contributed to the global financial crisis erupting.

Zeidan and Shapir [16] show that, according to the state of the art, corporate treasurers overinvest in working capital as defined in the literature and that such expenditures are economically inefficient. In their analysis, in the context of x-inefficiency, they decompose working capital expenditures in the cash exchange process and the development consequences. It forecasts that the cash conversion period decreases will improve the shareholder value. It implies the cash-oriented operation of the economy, accounting for impacts on operating profits, culminating in improved asset values and productivity, and enhanced cash flow. The theoretical context and the findings complement the literature and offer a basis for inefficiencies in overinvestment and working capital management. In the era of Industry 4.0, digitalization and virtualization control the character of finance and exchange between market participants. Most financial institutions are moved to the virtual plane allowing alternative finance to be created-raising funds and making borrowings across online platforms. Accompanied by an overview of the impact of factors on alternative finance growth, the paper formulates five hypotheses on the dependency of the amount of financial capital attracted by online platforms on the quantitative characteristics of the factors concerned. This uses the approach of connection study to analyze the causes of the internet finance sector as an alternative to conventional financial intermediaries (including banks). The usage of generally agreed indices and specific relative metrics is recommended, as the quantitative parameters of each component [15].

Since the COVID-19 pandemic occurs as a serious public event at the moment, the adjustment of a series of social operation models has been affected, and also the modification of behavior model -"quarantine" and "home office" have become hot words. To adapt to the increasing demand for "digital quarantine," money trading has enabled internet platforms to direct companies across electronic networks. Hence, computation Reveals the pattern of activities in the years to come. AI can grow into autonomous, smart systems. A great deal of research within the computer sector is based on deep learning, but the drawback of deep learning is that there is a need for lots of human input. Therefore, researchers began to listen to the reduction of the synthetic intervention of the autonomous intelligent ways, improve the machine intelligence's self-learning ability supported the setting. within the future, computing can improve the operational potency of assorted industries, promote the mixing and upgrading of computing technology and every one walks of life in society, and build a variety of benchmark application situation innovations, to attain affordable, high-efficiency, wide-ranging inclusive intelligent society.

The COVID-19 eruption in China has had significant impacts on enterprises and industries. Zhu et al. [17] summarized China is responding to minimize the impacts, respond to the shifts, and making new prospects available for good. Under the semi-permanent cycle of technical progress and capital market change, investors may exploit opportunities. The New York Stock Exchange had to shut the floor briefly, switch to online trading following successful COVID-19 results. Li et al. [7] through the years, the stock exchange has often stopped, such as after World War II and in the wake of 9/11, but that could be the only time the Big Board's physical trading floor has completely shut down as electronic trade occurs.

The rise of AI and its ever-broader effect on other industries requires an assessment of its effects on achieving sustainable development goals. Vinuesa et al. [14] to allow sustainable growth, the accelerated advancement of AI must be accompanied by the necessary restrictive perspective and supervision of AI-related technology. Failure to try and do so could lead to gaps in transparency, safety, and ethical standards. AI has influenced every "office"—front, middle, and back. Gossett [4] presents if you may not know the way your organization employs, claim, sophisticated AI to ward off money launderers or sifting through volumes of data for fraud-related irregularities, you are likely to have a lack of encounters with your customer service chatbot working on AI. You communicated at least with their customer support chatbot, which operates on AI.

3 Applications of AI and FinTech in Finance

FinTech helps securities brokerage online transformation—based on facial recognition, optical character recognition (OCR), public security network verification and other technologies, many securities companies through intelligent audit support staff to improve the efficiency and quality of customer identity audit, to ensure the efficiency of off-site service; 24 h account opening the application to accept, based on natural language understanding, machine learning, knowledge mapping, multi-wheel algorithms and other technologies, intelligent customer service platform effectively support 7×24 h multi-channel online business consulting, divert artificial customer service pressure.

For example, in China, Suzhou Cheersson Precision Metal Forming Co., Ltd. (002976.Schenzen) on 3rd March 2020 in Suzhou should take a part in the Shenzhen Stock Exchange listing ceremony, but because of the need to prevent the pandemic, the Shenzhen Stock Exchange temporarily canceled the on-site listing ceremony held in the listing hall of the exchange, with the help of the Shenzhen Stock Exchange and Panorama Network jointly set up the cloud live room held an online listing ceremony, which is also the first Shenzhen Stock Exchange held a "cloud ring opening bell" enterprises.

Also, in the United States, the New York Stock Exchange moved temporarily to fully electronic trading on 23rd March 2020. This is the first time that the Big Board's physical trading floor has ever turned itself off as electronic trading proceeds.

Customer needs to satisfy and improve service efficiency through intelligent smart investment—brokers promote intelligent investment, account diagnosis, and other online services, based on user portraits and intelligent algorithm technology to meet customer broad-spectrum wealth management demands. There are also companies relying on the online live broadcast and video conferencing platform to organize a series of conference calls, and through the mobile financial terminal retransmission promotion, to guide investors to a rational and objective view of the impact of the pandemic, to help investment advisory services in an orderly manner.

Explore the whole process of managed outsourcing business of managed products—the securities company upgrades the management and outsourcing business system of the managed products, provides the online processing service of the whole life cycle of the product from the aspects of product preparation, raising and setting up, investment operation, product liquidation, etc., and provides the quality and efficiency for the remote office. In the product preparation link, the use of manager service platform online processing, for fund managers assembled financial technology and data services, support 7 × 24 h a day business processing, to achieve the management of outsourcing business online. In the collection and establishment process, promote paperless and electronic signature application, for managers to provide electronic signing services. To meet the manager's investor identity authentication, investor appropriate management, as well as contract electronic contracting and other collection processes of all aspects of the functional requirements, and reduce human interaction, conducive to the prevention and control of the new crown pandemic. In the daily operation, in advance to develop an emergency management plan for pandemic prevention and control work, from personnel management, site, and system arrangements, operational problems troubleshooting, logistics support, and other aspects to make relevant preparations, to ensure that the pandemic prevention and control of all the daily operation of all management products such as valuation, requisition redemption, dividends, deductions, value-added tax, payment, etc. can be normally handled. In the product liquidation link, after the termination situation agreed in the contract, the manager may initiate a liquidation process on the manager's service platform to end the product operation.

Intelligent control system to ensure the smooth and orderly development of business—based on the AI technology such as big data and knowledge map and natural language processing, in combination with theoretical methods of business intelligence, the securities companies have constructed an intelligent control application system. Such a system includes the risk management leadership cockpit, the group control index and table management system, the intelligent risk early warning system, the internal rating system, etc., and through the internal and external big data analysis. Combined with the application of AI risk analysis engine and innovative tools and algorithms, it can deeply excavate the potential risk transmission mechanism, intelligently identify and warn the source of hidden risk, visually and dynamically display the overall risk situation and risk analysis data, and ensure that the management of risk management of securities companies is carried out normally and efficiently.

Through remote access to the risk management system through VPN and other tools, the securities company realizes the remote operation and operation of the risk management system, ensures the dynamic monitoring and early warning of all kinds of business risks remotely during the pandemic prevention and control, supports the efficient development of risk management, and ensures the smooth operation of the securities company's various businesses. The securities company uses the investment bank electronic draft system, all the drafts and declaration documents are audited through the system multi-line defense and realize the systematic and real-time approval, retention and filing of all the drafts and processes of the investment banking business. Some companies use risk measurement engine RiskMetrics, the use of big data technology to analyze market macro factors and historical market, stress testing, to measure the company's proprietary investment in extreme market profit stakes, for management decision-making. Some companies use the remote adjustment method of working to carry out stock pledge business. Using video interviews, video recording access to information process, video viewing to understand the financing party's office or assets, etc., require the stock pledge business customer to remotely sign all written materials, should record video, confirm the customer's identity and customer's wishes, and record the signature process, through APP to obtain the scanned, and through courier to obtain the original, the original, scanned copy and video match ingress with the submission. There are also companies combined with years of accumulated derivatives business, risk management experience, independently developed the Global Derivatives Trading System (GDTS), through real-time valuation, T-0 day-by-day market, multi-scene analysis, and stress testing and other functions, to achieve the Group's domestic and foreign, subsidiary over-the-counter derivatives business of the entire process of risk control.

The practice of AI in investor education—in the aspect of investor education, online education, represented by the course broadcast, has ushered in an explosive increase in user base and frequency, and securities companies have dug deep into the customer's demand for information services in this special period, exploring the various possibilities of AI in investor education.

AI self-written, some securities companies have been through the "AI writing" "AI and investment" of the human-machine combination, the machine completes the report of the market objective data part, manual investment analysis, and supplementary views, to achieve high efficiency, high-quality investor education content, but also to strengthen the customer's position-related content personalized production.

To carry out single-player/multi-person online live service, securities companies business units have assembled the research force of research institutes, headquarters investment to take into account external professional institutions, the use of telecommuting and online service systems, and on the macro-economic impact of investors concerned and investment strategies under the pandemic, etc., invited experts to invest, fund managers, and so on to carry out online investor education and related services, investors can participate at home around the clock.

Realize the accurate distribution of investor education content, through big data to customer investment preferences, needs, and key active scenarios to classify and intelligent identification, comprehensive use of scenario delivery, active reach, intelligent investment assistant, etc., the content accurately distributed to customers.

3.1 Application of Treasury Technology

When the COVID-19 sent shockwaves across the economy, managers are seeking to rapidly determine the damage amount. Regardless of the value of cash flow to the company, treasury departments are increasingly reviewing cash control, finance, and risk reduction to better handle corporate financial balances as a consequence of the pandemic disruption. Although business vital issues would understandably need the most attention, tax concerns and law enforcement will have a huge effect on the result of decisions that treasury teams take.

Treasury creativity relies on four innovations that are now being implemented in treasury operations. Application program interfaces (APIs), robotics, AI, and cloud systems are also available. Growing invention streamlines and automates manual processes thus saving resources, reducing delays, and quickly adapting to changing needs of an enterprise see Isaacs [5].

APIs function behind the scenes to have a more flexible end-user interface and smooth connectivity. The fuel efficiency by accelerating the changeover from batch to process in real-time. APIs can also enable batch processing of payments across multiple banks and assist with received information in reconciling payment. APIs would create cash receipts in the system when reconciling payments, reduce the time spent, and increase efficiency. This can also increase the chances of an erroneous or fraudulent payment being successfully spotted, stopped, and recalled.

Robotic technology is most important for high-volume, manual operations, and processes automation. This approach will improve the precision, synchronize data through different channels and reduce reporting time when implemented accordingly. As far as treasury functions are concerned, robotic technology frees up skilled professionals to focus on other activities.

A computer program utilizes algorithms to process large amounts of data with AI, or machine learning. It then begins to develop new rules for task performance or problem fixing based on what has worked in the past. AI is expected to have a huge impact on liquidity management and the processing of payments. This technology might learn from payment background, use counterparty and value data to help detect possible fraud, or to recognize possibilities for lower-cost networks and forms of payments. Liquidity management, especially liquidity forecasting, will benefit from the capacity of AI to analyze large volumes of historical data and predict potential cash flows.

Third-party cloud computing systems that are efficient at handling FX risks and predicting cash needs—time-intensive processes that are potentially vulnerable to human error—including Enterprise Resource Planning (ERP) software, Treasury Workstation (TWS), or Treasury Management Systems (TMS), FX trading platforms. The new iterations of these devices will improve performance whilst delivering possible reductions in time and expense. Since they are cloud-based, in the future they will need fewer hardware updates.

Example of on-demand treasury management software provider TreasuryXpress introduces the COVID-19 treasury development relief system that provides free access to a patented newly built treasury management application for self-service administration. It seeks to help existing treasury consumers who have endured loss to manual and physical processes since the pandemic. The self-service treasury management system facilitates the monitoring of unified currency and financial instruments, liquidity forecasts and placement, and reconciliations of bank accounts. The COVID-19 pandemic accelerates the introduction of emerging technologies and the software intends to simplify treasury systems by delivering scalability and self-service setup.

3.2 Application Scenarios of AI and FinTech in the Financial Industry

There are four application scenarios of AI and FinTech in the financial industry.

- Voice recognition and natural language procession applications—smart customer service, voice data mining

Smart customer service—integrating different customer support networks within the community, offer electronic smart customer care with multi-mode integration; incorporate market intelligence solutions internally, such as voice recognition and customer service assistants. Provide agents with an additional tool to support agents to overcome customer issues rapidly. The customer support assistant captures consumer desires through real-time speech recognition and semantic awareness in real-time and instantly drives user apps, knowledge bases, and other material.

Through telephone customer service channels, online customer service, APP, SMS, and intelligent robot terminals, interact with customers in voice or text, understand customer business needs, reply to business queries put forward by customers, and navigate to designated business modules according to customer's voice. Transform the traditional push-button menu, users use natural voice to interact with the system, flatten the menu, improve user satisfaction, reduce the pressure on manual service, and reduce operating costs. Customer support by telephone is no longer restricted by the guide, you may conduct a full support voice navigation service.

Voice data mining—this will automatically arrange large phone calls and numerous consumer documentation based on voice and semitone technologies, attach different marks, evaluate useful knowledge, and provide data and decision support for services and marketing.

Voice semantic analysis dynamically provides essential knowledge clusters, blends associative data sets, retrieves keywords, and summarizes hot terms to uncover the new business trends and consumer attention. Around the same period, the request condition of the customer support and clients in the finance sector makes company consultation hotspot problems to be filtered out and data are automatically obtained by the system, and the information question and response database is filtered out and used as a research base for future computers to automatically address consumer questions.

- Computer vision and biometrics recognition application—portrait monitoring and early warning, monitoring of employee violations, security monitoring in the core area

Portrait monitoring and early warning—the use of branch and ATM cameras provide a portrait recognition feature to anticipate criminals, trigger suspicious activities, and recognize VIP customers.

Identify the features of suspect employees in the network region, such as: if there is a mask covering the forehead, carrying unusual items, irregular level of activity, workers dropping to the ground, staffing, etc.

Monitoring of employee violations—using the camera within the branch's counter to increase the suspicious behavioral recognition and tracking feature of the employee, report and identify the suspicious activity and notify the context surveillance workers to further evaluate and alert them.

AI can monitor and track employee behavior and determine whether employee behavior is compliant and safe. Such as the use of graphics and video processing technology, real-time monitoring of bank tellers outside the prescribed actions, to remind background staff to pay attention. Through paper text reading technology, check all transaction documents, and establish keyword promoting technology. Or answer the visitor service question and answer, counter dialogue records, establish a risk model, and promptly discover suspicious transactions. Identify and mark video clips of employees' suspicious behaviors found in video surveillance, prompt background personnel to check them; at the same time, play a psychological deterrent effect on front-line operators.

Security monitoring in the core area—add a face recognition camera in the core area of the bank. Personnel entry and exit must be through face recognition and the identity of the document before entering. At the same time, portraits are registered for all persons entering and exiting to prevent strangers from trailing in and out of the relevant area. Such as a unified operations center, control space in the data center, etc.

Centralized operation centers, computer rooms, safes, vaults, and other important places can use face access control to improve internal security control. Through face recognition verification, the bank's internal security management can be implemented to effectively prevent illegal intrusions by criminals and conduct multiple people at the same time. Face recognition to achieve intelligent recognition and achieve the goal of security.

- Machine learning, neural network application and knowledge—financial forecasting and anti-fraud, the decision on financing and credit, intelligent investment advisor

Financial forecasting and anti-fraud—large-scale application of machine learning, the importation of vast volumes of financial transaction data, and the usage of deep learning algorithms to automatically detect patterns from financial data, such as examining credit card data, detecting fraudulent transactions and forecasting activity trends in advance and taking necessary countermeasures. Creating a financial information graph centered on deep learning technologies, risk management focused on big data requires to bring data from multiple sources (structured, unstructured) together, it can identify data anomalies, evaluate upstream and downstream, business collaboration and competitiveness Adversaries, parent firms, innovation, benchmarking, etc.

*The decision on financing and cre*dit—score by screening, modeling, and predicting the data, and separately classify and deal with different assets. For instance: bad assets may be explicitly identified as "judicial litigation" to alert related workers to start the litigation phase. Through collecting data from individuals and businesses on their homepages, social networking, etc., it can determine the impact of the business or its goods in society, such as analyzing the number of device downloads; however, the danger points of the investment may even be estimated after the data is organized. Using machine learning to complete the lending process that traditional financial companies are unable to achieve, real-time monitoring of the borrower's ability to repay the loan so that those who may not be able to repay the loan can intervene promptly to reduce the losses caused by bad debts.

Intelligent investment advisor—according to Markowitz's modern asset portfolio theory (MTP), the combination of risk appetite and financial goals for individual customers, the use of AI algorithms and internet technology to provide asset management and online investment advice to customers to achieve bulk investment advisory services for individual customers.

The intelligent investment advisory system continues to learn according to AI technology to use a multi-layer neural network to collect all important economic data indicators in real-time. It adopts an effective investing policy for asset diversification and may carry out a wide range of tailored investment advisory services for various entities, with the goal of not chasing short-term fluctuation returns but seeking steady long-term returns. The operation level of wealth management is lowered to an average family community utilizing an insightful investment advice approach.

- Application of service robot technology—computer room inspection and branch intelligent robots

Deploy 24 h monitoring robots in key areas such as computing rooms and servers to identify and fix possible threats in time and supplement or assist with manual surveillance. Try to set up intelligent robots in the lobby of the branches to give the robot's anthropomorphism and give them human images and corresponding emotions and actions. Branches for business customers advice, auxiliary aid diversion, collect customer data, carry out big data marketing efforts to complete the inquiry, the auxiliary handle to open the card, card sales, and other business.

Using robotic technologies, smart robots are installed in the branches or machine rooms to execute the automated cruise role in the specified field, which is capable of welcoming customers, performing interactive voice contact and performing regular business consultation and questioning and addressing based on customer knowledgebase information, decreasing the reproducibility of lobby manager jobs. Around the same period, predictive targeting is carried out by the front-end processing of consumer data. In turn, the advancement of technical progress and existing financial industry expertise has introduced additional variables into the transition and modernization of banking services.

3.3 Application Cases of AI Technology in Chinese Financial Industry

Alibaba—Alibaba's Ant Financial Services Group has a special team of scientists, specializing in cutting-edge research in AI, such as machine learning and deep learning. A series of innovations and applications under the business scene of Ant Financial Services Group, including internet loans, insurance, credit, intelligent investment, customer service, and other fields.

Bank of Communications (BOCOM)—in 2015, BOCOM lunched an intelligent robot—Jiaojiao—is a physical robot, using voice recognition and face recognition technology to communicate with people and machines, familiar customers can also be recognized, branch customer guidance can be provided and different types of banking services introduced. The robot will address numerous consumer questions during the language dialog process, relieve the possible feelings of bank customers waiting for the company, share the job of lobby executives, distract customers and save customer service time.

Ping An Insurance (Group) Company of China—Ping An Group has a Ping An Technology AI Laboratory to develop AI financial applications, which could use face recognition technology to conduce overall monitoring in designated bank areas, identify strangers, suspicious persons, and suspicious behaviors, and improve the security of the bank's physical area. Also, it could provide intelligent customer service in insurance, funds, banks, and securities.

4 AI Technology Changes of Financial Business

Nowadays, AI changes the main two aspects of financial business.

- *Improve financial business processing efficiency and user experience*

The continuous maturity of perception technologies (computer vision, speech recognition, natural language processing, etc.) and their use in financial business processes have pushed the automation level of business processing in the financial industry to new heights and greatly improved the customer experience.

Typical usage cases of highly advanced AI technologies in the finance sector include finance customer support chat robots, intelligent computer recognition, and OCR bill smart entry. The financial customer support chat robot will pursue the consumer journey's normal route, use machine learning algorithms to monitor the dialog and recognize the meaning of the conversation, return the hardening question to human processing when confronted with difficulties, and learn the manual solution to enhance customer service. The position of quality, and cost of service reduction. Automated authentication of computers is to use automated speech recognition or automatic facial recognition to authenticate users and to examine the expression, eyes, and facial features of the person to validate identity, remove the original security query or password. This approach is more about raising questions about the proof. The approach is also easier, and because the user does not need to recall the answer, the consumer service will significantly increase. Intelligent bill entry is managed by errors and the entry output has a big benefit over the manual entry of conventional consolidated operating centers.

- *Improve the accuracy of intelligent analysis and decision making*

The continuing development in data processing and deep learning technologies will significantly enhance the quality of intelligent research and decision-making, thus generating or improving the market appeal of financial goods and financial services.

In the past, business intelligence and quantitative research have mostly stayed at the stage of pattern analysis, trigger mining, data mining and simulation, and AI may enhance the validity and accuracy of advice and achieve "personalized insight" through continuous learning and development. Provide tailored research and decision taking in fields such as management, promotions, and services. For example, AI can achieve credit scores based on social networks, optimize existing scores, or generate scores for people without credit records; carry out personalized marketing based on customer and product DNA, provide unique personalized products; and carry out dynamic fraud mode detection To detect fraud from real-time complex transaction patterns.

5 Regulation—Technology in Finance Must Be Strictly Regulated and Standardized

Financial institutions have partnered with FinTech over the past few years on diverse business aspects. While the main focus was on customer-facing areas, businesses have now also begun to devote attention to backend aspects. RegTech[1] as a niche segment within the FinTech ecosystem has gained importance with much pressure from regulators on overall data compliance and governance. One of the objectives of new financial regulations is to prevent criminal activity FinTech may provide the opportunity for. At present, the rapid development of the FinTech industry, rather than financial enterprises themselves, take the initiative to use technology to empower the industry. Hence, it should clarify the boundaries of FinTech. Before the Chaos of FinTech, the reason for the technology company did not possess the qualifications for business integration, risk identification, and management capabilities, but involved the high risky areas in the financial industry. Therefore, such as so-called FinTech companies were not real FinTech, but in turn, affected the dynamic environment of FinTech development. Hence, support and protect the development of the real FinTech industry is to understand the boundaries of FinTech, adhering to the principle of consistency in supervision.

[1]Regulatory Technology.

In addition to this, it is necessary to pay attention to the safety and specification of data usage. Big data is the fundamental of FinTech development, but the process of data collection, analysis, and application involves many issues, such as personal privacy and data security. The legal institution needs to set up the personal information protection law to deal with personal information security issues, and also, to determine the appropriate specifications for how institution to collect and use the data, to avoid the issues caused by data leakage. To order to fulfill the standards of regulators, financial intermediaries need to store, view, and process data on a scale never expected before. New regulations are pushing the financial sector to tackle data infrastructure more front-to-back. Data sets must be comprehensive and precise, and their contents must be far more flexibly analyzed, interpreted, and abused than ever, with varying latencies and for specific purposes.

FinTech itself focuses on the main characteristics of digitalization, onlineization, remoteization, visualization, and intelligence, building a multi-functional, all-process end-to-end system based on data, enabling multi-user multi-terminal concurrent office, intelligently assisting in dealing with problems and giving solutions, while drawing graphics visual methods such as images are presented to people. Under the pandemic, the application of FinTech has revealed some shortcomings.

5.1 Disadvantages of the Regulation

First, some financial companies are ill-prepared. Although FinTech has developed rapidly at the technical level and developed many applications and solutions, some financial enterprises have previously habitually followed the existing regulatory processes out of inertia and require artificial completion of all aspects of the work. For the unexpected outbreak caused by the demand for online office, some financial enterprises to apply FinTech to solve practical problems are not prepared, many employees on the FinTech itself and its application using is relatively limited. From the practice of the war "pandemic" period, some financial enterprises for the application of science and technology are still more stuck in the online meeting, VPN connectivity stage, from the use of FinTech intelligent transformation of business processes, still have a certain distance.

Second, there are still shortcomings in the combination of FinTech and various types of financial services. On a technical level, FinTech already has a certain amount of technology accumulation, but how to combine existing technology with various types of financial business, better design in line with the actual business needs of products, but also need more accumulation and exploration. At present, all kinds of products can be directly applied to the financial industry is still relatively limited, in the financial industry war "pandemic" period, it is difficult to form the financial technology and financial industry during the outbreak of the situation more rapid integration.

5.2 *Advantages of the Regulation*

For the financial industry, the outbreak has made regulators and financial institutions fully aware of the importance of FinTech and has gradually developed a strong consensus on the use of FinTech to transform existing business models and business processes. For all kinds of FinTech enterprises, the rapid increase in demand will certainly promote the rapid growth and development of the industry, if some enterprises in the FinTech industry can seize this opportunity keenly, there are conditions in this wave to promote the excellent company, just as the SARS period has contributed to some excellent e-commerce enterprises into a business. For traditional financial enterprises, closer integration with FinTech also means a broad space. In the traditional financial industry market competition pattern gradually formed, especially the strong and strong trend and background, the combination with financial science and technology can be some industry enterprises to consolidate their industry status of the moat, but also some small and medium-sized financial institutions to achieve the development of characteristic opportunities and magic weapon. For those in the financial industry, on the one side, with the stronger convergence of FinTech and the financial industry, the potential need for manual labor at the financial sector's executive level would be significantly diminished, and on the other side, the need for professionals with a hybrid technical system in the financial industry will be greatly enhanced.

6 Conclusion

The advent of AI and its ever-broader effects on other industries demands an assessment of its influence on achieving sustainable development goals. The rapid development of AI in Finance needs to be supported by a state-of-the-art regulatory environment and mandatory restrictive insight and supervision of AI technologies to enable sustainable development. Failure to attempt to do so can lead to gaps in transparency, safety, and ethical standards. One area of FinTech that will be positively affected by the social distancing of the COVID-19 crisis is the trend towards non-cash payments and perhaps digital currencies. In February 2020 the European Commission implemented its Digital Strategy in the European Union. 'This policy is gaining renewed urgency in the face of the coronavirus pandemic, as the Commission deploys digital resources to track the spread of the virus, study and improve vaccinations and therapies and ensure that Europeans can remain informed and healthy online' see (European Commission).

The financial sector has always been associated with technology. In the 1970s it was the arrival of ATMs and credit cards, in the 1980s debit cards and the 1990s internet banking and e-commerce. Thanks to the huge increase in the popularity of smartphones, the pace of technological progress has also accelerated, laying the foundations for the development of mobile banking. The ubiquity of mobile phones is breaking down barriers between customers and banks. Customers expect personalized, interactive, and instant service at their fingertips, virtually anytime. This change has become a catalyst for a new generation of financial startups that offer services on digital platforms tailored to specific parts of the market. Technology giants, the 'big tech' companies have also taken some bold steps in the financial sector.

As described in the chapter, with the COVID-19 pandemic being exacerbated, there are signs that the smallest banks and FinTech companies have shown an extraordinary capacity to adapt their business model and react to drastic changes in the marketplace. From the capacity to create and execute on APP loans to the rapid delivery of innovative digital banking solutions, the smallest organizations have demonstrated a commitment and flexibility that suggests the potential to withstand the COVID-19 pandemic 's effect. FinTech and traditional banking, the biggest and most capitalized, have reacted well to adversity. Such companies were not only the furthest down the road of digital change but were still able to adapt rapidly to customer needs. Meanwhile, midsize FinTech firms and legacy regional banks are experiencing the worst of both worlds. Companies there enjoyed the prosperity before the COVID-19 pandemic but did not have the culture to see the risks of complacency. If it's not providing a competitive approach or not accepting the need to be new, all these companies don't have the funding or the ambition to deal with the innovative community banks or the well-funded grants.

References

1. Beier, G., Ullrich, A., Niehoff, S., Reißig, M., Habich, M.: Industry 4.0: How it is defined from a sociotechnical perspective and how much sustainability it includes—A literature review. J. Clean. Prod. (2020)
2. Büchi, G., Cugno, M., Castagnoli, R.: Smart factory performance and indutry 4.0. Technological Forecasting and Social Change, 150 (2020)
3. Dirican, C.: The impacts of robotics, artificial intelligence on business and economics. In: World Conference on Technology, Innovation and Entrepreneurship, pp. 564–573 (2015)
4. Gossett, S.: AI Bankability: 10 Ways Artificial Intelligence is Transforming Banking (2020, March 19). Fintech News https://www.fintechnews.org/ai-bakability-10-ways-artificial-intelligence-is-transforming-banking/
5. Isaacs, K.: 4 Technologies Driving Treasury Transformation (2019, November 14). J.P.Morgan https://www.jpmorgan.com/commercial-banking/insights/technologies-driving-treasury-transformation
6. Kagermann, H., Wahlster, W., Helbig, J.: Umsetzungsemofehlungen für das zukunftsprojekt Industrie 4.0 (2013). https://www.bmbf.de/files/Umsetzungsempfehlungen_Industrie4_0.pdf
7. Li, Y.: NYSE to Temporarily Close Floor, Move to Electronic Trading after Positive Coronavirus Tests (2020, March 18). CNBC: https://www.cnbc.com/2020/03/18/nyse-to-temporarily-close-trading-floor-move-to-electronic-trading-because-of-coronavirus.html
8. Moosa, I.A., Ramiah, V.: The Financial Consequences of Behavioural Biases: An Analysis of Bias in Corporate Finance and Financial Planning. Palgrave Macmillan (2017)
9. Polak, P., Robertson, D.C., Lind, M.: The new role of the corporate treasurer: emerging trends in response to the financial crisis. Int. Res. J. Finan. Econ. **78**, 48–69 (2011)
10. Polak, P., Masquelier, F., Michalski, G.: Towards treasury 4.0/ The evolving role of corporate treasury management for 2020. Manag.: J. Contemp. Manag. Issues **23**(2), 189–197 (2018)

11. Polak, P., Nelischer, C., Guo, H., Robertson, D.C.: "Intelligent" finance and treasury management: what we can expect. AI & SOCIETY (2019)
12. Ramiah, V., Zhao, Y., Moosa, I., Graham, M.: A behavioural finance approach to working capital management. Eur. J. Finan. **22**(8–9), 662–687 (2016)
13. Roszkowska, P., Prorokowski, L.: The changing role of a bank's treasury. Asia Pac. J. Finan. Stud. **46**(6), 797–823 (2017)
14. Vinuesa, R., Azizpour, H., Leite, I., Balaam, M., Dignum, V., Domisch, S., Felländer, A., Langhans, S., Tegmark, M., Nerini, F.F.: The role of artificial intelligence in achieving the sustainable development goals. Nat. Commun. Retrieved from Nature (2020)
15. Yuriy, B., Pavlo, R., Tetiana, V., Serhiy, L.: The influence of Industry 4.0 on financial services: determinants of alternative finance development. Polish J. Manag. Stud. **19**(1), 70–93 (2019)
16. Zeidan, R., Shapir, O.: Cash conversion cycle and value enhancing operations: theory and evidence for a free lunch. J. Corp. Finan. **45**, 203–219 (2017)
17. Zhu, N., Tse, E., Tang, N., Wu, T., Xin, L., Dai, Q.: Coronavirus in China—insights on the impacts and opportunities for change (2020, March 04). World Economic Forum: https://www.weforum.org/agenda/2020/03/coronavirus-china-opportunities-change

Conceptualising the Corporate Governance Issues of Fintech Firms

Khakan Najaf, Alice Chin, and Rabia Najaf

Abstract The board of directors plays a vital role in reducing the divergence of interest, which effectively alleviates agency problems. This study aims to highlight the current corporate governance issues which Fintech firms are facing. This line of enquiry is necessary as a majority of Fintech firms are collaborating with the listed firms, and quality corporate governance is vital to protect the shareholders' rights. Fintech governance structure faces four key challenges: lack of anti-misconduct policy, CEO duality, over-boarded directors, and the inability of audit firms to detect fraud. Organization for Economic Co-operation and Development (OECD) provides a guideline for corporate governance, suggesting that anti-misconduct policy is a sign of better governance. The agency theory postulates that concentration of power weakens corporate governance mechanism; thereby, CEO duality and over-boarded directors are the indicators of weak governance. Also, the supervisory structure of Fintech firms is different than counterparts, which hinder the audit firms to deduct any financial misstatement. Based on the literature, we find that the Fintech firms lack anti-bribery policy and prone to have CEO duality, over-boarded directors and audit firms' failure to detect fraud. It implies that the Fintech firms have embedded weak governance mechanisms, which will expropriate the shareholders' right after the listed firms' collaboration. The governance issues are not merely a technical issue of the Fintech firms, as it may cause financial instability across the world.

Keywords Fintech · Anti-misconduct · CEO duality · Over-boarded directors · Corporate governance

---

K. Najaf (✉) · R. Najaf
Taylor's University, Subang Jaya, Malaysia
e-mail: khakannajaf@sd.taylors.edu.my

R. Najaf
e-mail: Rabianaajf@gmail.com

A. Chin (✉)
IPE Management School Paris, Paris, France
e-mail: alicechin861@gmail.com

A. Hamdan et al. (eds.), *The Fourth Industrial Revolution: Implementation of Artificial Intelligence for Growing Business Success*, Studies in Computational Intelligence 935,
https://doi.org/10.1007/978-3-030-62796-6_10

1 Introduction

Fintech is considered as an epicenter of innovation and rapidly evolving via its economy of scale and technological transformation capabilities. However, in the past few years, the Fintech firms have shown a slowdown in growth rate. In 2019, the Fintech organization's market value was around USD 127.66 billion, with a growth rate of 24.8%, which is 12.6% less than last year [1]. The weak corporate governance remains a crucial concern for all Fintech firms as it hinders the growth rate. The academic research shows that the spike in the number of Fintech firms prevails in weak regulatory enforcement countries [2]. According to the Fintech governance census, almost 25% of respondents suggest that quality corporate governance is a principal challenge for Fintech management [3]. This issue is crucial to investigate as poor corporate governance is a barrier to the Fintech emerging growth. In this study, we argue that Fintech firms face more corporate governance issues relative to others. For example, the taxi service Uber is mostly in the news due to the multiple corporate misconducts: such data breaches, allegations of sexual harassment, and anti-competitive behavior. In addition, the Revolut corp., a famous digital bank provider recently involved in a money laundering case [4]. In just 18 months, Rovolut has fired the third head of compliance due to corporate negligence. It seems that the international regulatory authorities have already circled Fintech firms as a rebel from the code of corporate governance.

According to the Global Fintech Index report, the Fintech firms are enlarging their ecosystem by collaborating with the listed firms [5]. The Fintech collaborated firms are trying to achieve economies of scope with the combination of other services such as big data analytics and sharing-economy businesses for the value-addition of the economy; while, the threats of poor quality governance embedded with the Fintech firms inhibits future growth of the partner listed firms. This study categorizes four broader areas of governance issues currently related to Fintech firms. First is the anti-misconduct policy, which means the Fintech firm has corporate policies to stop the misconducts of board of directors and employees. Also, the anti-misconduct policy limits all types of corrupt business practices. The practical examples of anti-bribery policies are corporate fraud, nepotism, cartels, collusion, preferred patronage, and price-fixing for the products. Second is over boarded directors, which means when a board member of Fintech firm is holding multiple directorships in other firms. Third, is CEO duality means when the CEO is holding the office of chairman as well. Fourth, the inability of audit firms to detect fraud or any misstatements from the Fintech firms' accounts. We argue that recent collapse of Wirecard (Fintech firm) is an evidence of audit firms' inability to detect any anomaly from Fintech financial statements. In the end, this study emphasizes that the embedded weak corporate governance risk of Fintechs becomes more prevalent after the collaboration with listed firms, as it endangers the interest of the shareholders of the listed firm.

We provide a comprehensive look at the traditional corporate governance issues related to the Fintech firms while contributing to the prior research [6–8]. We make an effort to address a central research question: *What are critical corporate governance issues faced by Fintech firms?* Understanding the weak corporate governance of Fintech-traditional firms' will facilitate the policymakers to identify channels via which policy adjustments strength the economic welfare. Subsequently, this study sheds light on significant aspects of anti-misconduct policy, CEO duality, over-boarded directors of Fintech firms, and audit firms' failure to audit, which will be helpful for policymakers to form the legislation, which Fintech partners with the listed firms. This ensures public interest is protected by allowing government intervention in corporate governance-related matters. The rest of the paper is organized as follows: the theoretical background and proposition development are outlined in Sect. 2. Furthermore, we precisely conclude the issue in Sect. 3.

2 Literature Review

Organization for Economic Co-operation and Development (OECD)[1] advocates that quality governance of the firm's principles reinforce a relationship between a shareholder and manager, which helps in the realignment of interests of the manager-owner. Effective corporate governance practices create an environment where it persuades the board and management for the mutual interest of shareholders and firm. Corporate governance practices induce and articulate an effective monitoring mechanism. Although OECD provides a benchmark framework for best corporate governance practices, there is hardly any research that elaborates on the theoretical aspects of these principles. The sound quality corporate governance principles promote transparent and efficient markets, where not only the primary shareholders' rights are protected and facilitated but also promote an equitable treatment for stakeholders. OECD emphasizes on the existence of the anti-misconduct policies, under-boarded directors, CEO holding one position, and quality external audit process in the organization is a sign of quality governance. These four factors promote transparency and timeliness in the disclosure of all material facts to the shareholders. The literature related to these three factors is as follows.

2.1 Fintech Firms' Corporate Policies

The literature suggests a linkage between corporate policies and corporate governance mechanisms [9–12]. Where, Jiraporn and Chintrakarn [10], and Schellenger et al. [12] suggest that there is a positive correlation between corporate governance and stagger boards with more generous dividend payout policies. In other words, corporate policies alleviate corporate governance. Davila and Penalva [11] depicts

[1]OECD was founded in 1961, it operates for the world trade and economic growth. Currently, OECD has 36 countries as member world-wide.

that the weak corporate governance enhances the in cash and in-kind compensation of CEO. Chakraborty and Sheikh [9] suggest that corporate investment policies are negatively affected by the managerial entrenchment. Chan et al. [13] document that there is an association between corporate social responsibility policy and corporate governance. Nevertheless, the researcher ignores the effect of corporate policies on the corporate governance that are specifically made to control managerial. The presence of anti-misconduct policy is more likely to establish a quality corporate governance mechanism, and consequently.

In corporate governance literature, signaling and agency theories have significant value. Morris [14] posits that the agency and signaling theories are substitute of each other. Precisely, under sufficient circumstance, signaling theory is consistent with the agency theory. Basically, signaling theory addresses the issues for agency problems rooted in information asymmetry [15]. To investigate this issue of whether signaling theory substitutes or complements the agency theory or vice versa. In this study, we assume corporate governance as agency argument and high-quality firm policy on market valuation as signaling argument. Li et al. [16] purpose that a firm with a better anti-misconduct policy has more market valuation as compared to those firms which have low-quality anti-misconduct policies. It implies that anti-misconduct policies affect the market valuation of a firm because it serves as a good signal for the investor.

A firm establishes a set of specific and strict policies to prevent opportunistic behavior by its managers and staff, which is an internal signal to a firm's staff and managers [16]. For the institutional investor, the existence of these anti-misconduct policies is an indicator that the firm is making earnest efforts towards the prevention of any managerial misconduct with the help of a quality corporate governance system. Spence [17] introduces the signaling theory to solve the information asymmetry problems in agency theory, particularly for those market investors who are unable to separate the good firms from the bad firms and thus, they have to give the same market valuation to all the companies. The literature shows an association between various types of corporate signals and market reactions by the investors [18–21]. In other words, a firm has a quality anti-misconduct policy that gives signs of quality corporate governance practices to institutional investor. Furthermore, prior literature also articulates that taking the point of view of investors, the firms' market value depends on the better anti-misconduct policies [16]. It means that the directors in a company with anti-misconduct policy trends to disclose the standard procedure of practice for avoiding any misconduct at managerial level. Also, transparency in the disclosures serves as a good signal for the investors which differentiates good firms from the bad one.

The proponents of non-executive directors (NED) argue that the policy of the inclusion of non-executive directors on boards improves the corporations' compliance with corporate governance which is beneficial to enhance the quality of the disclosures. Forker [22] posits that disclosure quality is crucial to know that information about the proportion of non-executive directors in the financial report that can increase the quality of corporate governance. His argument does not entirely give support to the presence of the negative relationship between NEDs and firms disclosure quality. It means corporations are not disclosing complete information on NEDs.

Among various roles of NED, shareholders expect to monitor the board policies and more ensure the accuracy of all the information provided by management [23]. In order to enhance the performance of corporate boards, the code of corporate governance committee passes the corporate policy about fixing a certain numbers of NEDs within the listed corporations [24]. Thus, the high quality of anti-misconduct policies affects many corporate governance decisions. The intervention of non-executive directors in the decision-making process depends upon the corporate policy of the firm.

In summary, the presence of anti-bribery policy in the organization is a signal of quality governance. Also, the anti-bribery policy facilitates enhancing the corporate performance, number of NEDs, the market value of the firm, and serves as a quality corporate social responsibility policy. Prior literature depicts that a departure of corporate social responsibility policies exists among the Fintech firms [25]. In the future, the Fintech and traditional firm's collaboration will face the embedded issue of the absence of anti-misconduct policy.

2.2 *Fintech Firms' Over-Boarded Directors Are Good Advisor or Bad Monitor?*

In the research field of accounting, economics, and finance, the monitoring and controlling role of a corporate board are always subject to the interest [26–28]. The board of directors is the agent of the shareholders in the company, and their main motive is to supervise the management behavior for the shareholders explicitly. The governance structure of a board defines the role of the board of directors in the entity. A director needs to be a good monitor of management and as well as better advice, which generates value for the firm. The question arises from here that *whether the over-boarded directors influence the corporate governance*? Some studies suggest that this phenomenon affects the transparent decision-making power of the corporate board and alleges that personal relationships between directors reduce the corporate board independence [29, 30]. While, some studies report the theoretical evidence that a firm seemingly manages better earnings when directors are over-boarded [31]. This study discusses the dimensions of the impact of the board unlocks with the board's independence of the firm. An enormous board of literature has analyzed the ramifications of board unlocks on the corporate governance of a company but unable to determine the core relationship of board networks with regard to corporate governance and board independence.

The literature is indecisive about the cost and benefit analysis of over-boarded directors. Some studies imply board interlock adds more value for the firm as the over-boarded directors are has diversified experience and also, they have quality advisory skills, which will bring reputation capital for the firm. Such as, Lennox and Yu [32] articulate that there is enough proof to believe that board interlock and director networks can lead to the best business practices. While, some studies articulate that a firm with more board interlocks unable to utilized monitoring skills as he is a busy director, which will be detrimental for the firm [33]. On one end, the board interlock seems beneficent for the firm as many directorships mean too much experience, whereas, on other end, it appears to be value-reduction in terms of corporate governance lost. The issue of board interlocks affects the firms' monitoring skills and enhancing the consulting experience of directors. The board of directors is the agents of the shareholders, and their decisions are supposed to the unbiased and transparent.

The current issue is necessary to investigate because the manager of an entity needs to serve the interests of its shareholders with individual board independence. A board member who has more than one executive position to perform, then that person may be too busy to monitor the management [33]. In conclusion, it can drive from the literature that over-boarded directors face more conflict of interest than other firm counterparts. It is because the under-board directors have no obligation or responsibility in other firms, and their main focus is entirely on the current company operations. A recent study shows evidence of over-board directors among the Fintech firms [34]. Therefore, the Fintech over-boarded directors are facing more governance issues due to the over-boarded directors relative to a company which board of directors has fewer interlocks.

2.3 Fintech Firms' CEO Duality

In the Fintech corporate governance literature, a key controversy is the relationship between CEO duality and board independence. The CEO duality means a position where the CEO is holding two positions first holds an office as a CEO and also serves as a chairman of the board of directors. The prior body of corporate governance literature develops a great extent over the last two decades and show theoretical and empirical work regarding the association between CEO duality and the corporate governance [35–41]. However, a critical theoretical gap still exists as the studies are inconclusive about the direction of the relationship between the Fintech CEO Duality and board independence.

Agency theory is unquestionably among the prevailing theories of management and economic organization. In the global context, agency theory assists in identifying the agency problem and way to resolve it [42–44]. The agency problem probably rises when shareholders appoint the board of directors for wealth creation. One of the essential features of agency theory is the divergence of interest between the shareholders and the managers. In agency theory, it is assumed that the principal has a lack of information about the agents' contribution. These unique features of agency theory point out the problems, which ultimately results in cost inefficiencies born by the principal. Although agency cost is difficult to measure still, it is significant [45]. The literature documents that large manufacturing firms have to bear 0.2 percent of revenue and small manufacturing firms have to pay five percent of their total income as an agency cost every year [46, 47]. As per agency theory, the shareholders try to encourage the managers to reduce the agency costs. Eisenhardt [48] argues that agency theory has three fundamental assumptions (1) all managers are self-interested, (2) agents monitor all the managers rationally, and (3) the managers are more risk-averse than the shareholders. Still, there are mixed theoretical and empirical results about the agency theory. One of the main issues is the duality of the CEO, where the CEO holds two offices at the same time. Standard agency assumption indicates that when a firm pays the CEO with a stock option, it will help in the alignment of interest of the CEO with the shareholders, but some studies show that the losses of CEO duality and higher payment outweigh its benefits [49].

As per agency theory, independent boards have significant incentive and inspiration to monitor CEOs' activities. Jensen and Meckling [50] argue that a vigilant board should not favor CEO duality because by providing CEO with two positions, it can lead to inefficient and opportunistic behavior and promotes CEO entrenchment that diminishes principal wealth. In corroboration with this line of argument, Fama and Jensen [51] posit that when the board of directors grants dual positions for CEO then the CEO duality can influence the decision-making process of the corporate board and it enhances the likelihood of deviation from the interest of shareholders' interest which is wealth creation. The basic board purpose is to protect the shareholders' interest, and extensive management participation upsets this basic board purpose [52]. Hence, CEO duality is a signal of a lack of board independence because of the central decision-making power [51]. A CEO who holds two offices at a time can become dominant in the content and agenda of board meetings and influences corporate decision-making. Nevertheless, CEO with dual positions get a chance to control the directors' nomination process and facilitate only those directors who are faithful to the CEO-chairperson [53]. Thus, the monitoring role of the board of directors over the CEO activities becomes prejudice in the presence of CEO duality. In fact, when a CEO establishes the overall strategic management and as well as monitors the effectiveness at the same time, then it leads to a conflict of interest in the CEO role [54]. In conclusion, the agency theory suggests that CEO duality cause divergence of interest between the board and shareholders which exacerbate agency problem.

In summary, the literature depicts that the CEO duality presents weak corporate governance, which leads to the negative firms' performance. The facts reveal that the CEO is most typical in the Fintech form of governance structure, where less human resource is required. For example, Revolut corporation, a leading UK-based Fintech firm, has a CEO who holds the chairman [55]. A similar trend is observed in other Fintech firms listed in the Nasdaq Fintech index [56]. It implies that the presence of CEO duality spikes within the Fintech firms, which deteriorates the corporate governance.

2.4 *Role of Audit Firms in the Fintech Corporate Governance*

Recently, one of the re-known Fintech firms, "Wirecard" in Europe, is declared insolvent. The insolvency of Wirecard raises dwarfs relating to the governance of the Fintech firms. After a proper financial investigation, the head of the regulator describes Wirecard insolvency as a "total disaster" [57]. Following the same premises, the German finance minister categories this event as the "Fintech scandal" and suggests reviewing the internal regulation system of the Fintech firms. The collapse of the Wirecard Fintech firm has shaken Germany's economy [58] and leaves the authorities with a question "*Is FinTech governance structure different than others*?". The failure event like Wirecard is an indicator of a bad monitoring system and seems like a wake-up call for adequate governance legislation for Fintech firms.

The role of external auditors has significance the corporate failure. Ernst and Young (EY) is the external auditor of Wirecard for the last two years, and it is one of the "big four" accountancy and auditing firms in the world. EY is a traditional symbol of credibility, and shareholders prefer to invest in the firm, which has to appoint big four firms as external auditors [59]. The history of EY is full of facing an enormous amount of litigations after the disastrous oversight of Arthur Andersen's event of Enron energy company, US. On behalf of bondholders and shareholders, Wirecard also files class action against EY. The management of Wirecard argues that in the past two years, how the firm operates without being questioned by the EY auditors. In a nutshell, the traditional "big four" firms, which are considered as benchmark audit firms, unable to detect any anomaly within the Fintech firms' accounts because the Fintech firms operate differently than non-Fintech firms and face more governance issues. Therefore, the collapse of Wirecard serves as an evident that Fintech firms have the potential to bring the next financial crisis due to weak governance structure.

3 Conclusion

This study sheds light on the Fintech firms' implications for financial inclusion and collaboration with financial institutions, and its embedded weakness of corporate governance. Indeed, Fintech-banks partnership has many advantages like operational cost-saving, low cost for customers, and faster service; however, this article emphasizes on the governance challenges faced by the Fintech firm along with the collaborated listed firms. Based on the prior literature, we conclude that the Fintech firms face four key governance issues: the absence of anti-bribery policy, CEO duality, over-board directors, and failure of audit firms to detect any fraud. Therefore, corporate governance is a prime concern for Fintech firms. Due to the inherent governance issues of Fintech firms, the partner firms' shareholder rights can be expropriated. One of the possible reasons for Fintech weak corporate governance is that it is more focused on financial growth than the internal governance system. Therefore, the government needs to redesign the supervisory structure for the Fintech firms.

This study has few limitations that can be addressed by future research. First, Fintech firms do not disclose corporate governance information. The future researchers can gather the Fintech collaborated firm's corporate governance data and compare it with non-collaborated firms. Second, future research can examine the usefulness of the Fintech corporate governance on the corporate stability of partner firms. Third, this study can be extended by the future research by using a systematic econometric model such as the ordinary least squares model or probit model.

References

1. Prnewswire: Global Fintech Market Value is Expected to Reach $309.98 Billion at a CAGR Of 24.8% Through 2022 (2020). https://prn.to/3ghUu6Z
2. Cumming, D.J., Schwienbacher, A.: Fintech venture capital. Corp. Gov. An Int. Rev. **26**(5), 374–389 (2018)
3. Barclay, S.: FinTech: Growth versus Governance (2019)
4. Times, F.: Revolut Reports Suspected Money Laundering on Its System (2018)
5. Findeable: The Global Fintech Index 2020 (2020). https://findexable.com/wp-content/uploads/2019/12/Findexable_Global-Fintech-Rankings-2020exSFA.pdf
6. Peng, M.W., Tan, J., Tong, T.W.: Ownership types and strategic groups in an emerging economy. J. Manag. Stud. **41**(7), 1105–1129 (2004)
7. Ding, Y., Zhang, H., Zhang, J.: Private vs state ownership and earnings management: evidence from Chinese listed companies. Corp. Gov. An Int. Rev. **15**(2), 223–238 (2007)
8. Yu, X., Zhang, P., Zheng, Y.: Corporate governance, political connections, and intra-industry effects: Evidence from corporate scandals in China. Finan. Manag. **44**(1), 49–80 (2015)
9. Chakraborty, A., Sheikh, S.: Antitakeover amendments and managerial entrenchment: new evidence from investment policy and CEO compensation. Q. J. Finan. Acc., 81–104 (2010)
10. Jiraporn, P., Chintrakarn, P.: Staggered boards, managerial entrenchment, and dividend policy. J. Finan. Serv. Res. **36**(1), 1–19 (2009)
11. Davila, A., Penalva, F.: Governance structure and the weighting of performance measures in CEO compensation. Rev. Acc. Stud. **11**(4), 463–493 (2006)
12. Schellenger, M.H., Wood, D.D., Tashakori, A.: Board of director composition, shareholder wealth, and dividend policy. J. Manag. **15**(3), 457–467 (1989)
13. Chan, M.C., Watson, J., Woodliff, D.: Corporate governance quality and CSR disclosures. J. Bus. Ethics **125**(1), 59–73 (2014)
14. Morris, R.D.: Signalling, agency theory and accounting policy choice. Acc. Bus. Res. **18**(69), 47–56 (1987)
15. La Porta, R., Lopez-de-Silanes, F., Shleifer, A., Vishny, R.W.: Law and finance. J. Polit. Econ. **106**(6), 1113–1155 (1998)
16. Li, C., Li, J., Liu, M., Wang, Y., Wu, Z.: Anti-misconduct policies, corporate governance and capital market responses: international evidence. J. Int. Finan. Mark. Inst. Money **48**, 47–60 (2017)
17. Spence, M.: Competitive and optimal responses to signals: An analysis of efficiency and distribution. J. Econ. Theory **7**(3), 296–332 (1974)
18. Lourenço, I.C., Callen, J.L., Branco, M.C., Curto, J.D.: The value relevance of reputation for sustainability leadership. J. Bus. Ethics **119**(1), 17–28 (2014)
19. Veenman, D.: Disclosures of insider purchases and the valuation implications of past earnings signals. Acc. Rev. **87**(1), 313–342 (2012)
20. Cohen, B.D., Dean, T.J.: Information asymmetry and investor valuation of IPOs: top management team legitimacy as a capital market signal. Strateg. Manag. J. **26**(7), 683–690 (2005)
21. Hribar, P., Jenkins, N.T.: The effect of accounting restatements on earnings revisions and the estimated cost of capital. Rev. Account. Stud. **9**(2–3), 337–356 (2004)
22. Forker, J.J.: Corporate governance and disclosure quality. Acc. Bus. Res. **22**(86), 111–124 (1992)
23. Wheeler, S.: Non-executive directors and corporate governance. N. Ir. Leg. Q. **60**, 51 (2009)

24. Xie, B., Davidson III, W.N., DaDalt, P.J.: Earnings management and corporate governance: the role of the board and the audit committee. J. Corp. Financ. **9**(3), 295–316 (2003)
25. Nicoletti, B.: Nicoletti, and Weis, Future of FinTech. Springer, New York (2017)
26. Bhagat, S., Black, B.: The non-correlation between board independence and long-term firm performance. J. CorP. **27**, 231 (2001)
27. Bushman, R.M., Smith, A.J.: Financial accounting information and corporate governance. J. Acc. Econ. **32**(1–3), 237–333 (2001)
28. La Porta, R., Lopez-de-Silanes, F., Shleifer, A., Vishny, R.W.: Legal determinants of external finance. J. Finan. **52**(3), 1131–1150 (1997)
29. Rosenstein, S., Wyatt, J.G.: Outside directors, board independence, and shareholder wealth. J. Finan. Econ. **26**(2), 175–191 (1990)
30. Hermalin, B.E., Weisbach, M.S.: Endogenously chosen boards of directors and their monitoring of the CEO. Am. Econ. Rev., 96–118 (1998)
31. Chiu, P.-C., Teoh, S.H., Tian, F.: Board interlocks and earnings management contagion. Acc. Rev. **88**(3), 915–944 (2013)
32. Lennox, C.S., Yu, Y.J.: The role of director and executive interlocks in mitigating uncertainty in auditor hiring decisions (2016)
33. Shivdasani, A.: Are busy boards effective monitors. J. Finan. **61**(2), 689–724 (2006)
34. Galasso, A., Simcoe, T.S.: CEO overconfidence and innovation. Manag. Sci. **57**(8), 1469–1484 (2011)
35. Lin, Y.: Corporate governance, leadership structure and CEO compensation: evidence from Taiwan. Corp. Gov. An Int. Rev. **13**(6), 824–835 (2005)
36. Abdullah, S.N.: Board composition, CEO duality and performance among Malaysian listed companies. Corp. Gov. Int. J. Bus. Soc. (2004)
37. Boyd, B.K.: CEO duality and firm performance: A contingency model. Strateg. Manag. J. **16**(4), 301–312 (1995)
38. Rechner, P.L., Dalton, D.R.: CEO duality and organizational performance: a longitudinal analysis. Strateg. Manag. J. **12**(2), 155–160 (1991)
39. Donaldson, L., Davis, J.H.: Stewardship theory or agency theory: CEO governance and shareholder returns. Aust. J. Manag. **16**(1), 49–64 (1991)
40. Chaganti, R.S., Mahajan, V., Sharma, S.: Corporate board size, composition and corporate failures in retailing industry. J. Manag. Stud. **22**(4), 400–417 (1985)
41. Berger, A.N., Humphrey, D.B.: Bank scale economies, mergers, concentration, and efficiency: The US experience. Federal Reserve Board (1994)
42. Ghoshal, S.: Bad management theories are destroying good management practices. Acad. Manag. Learn. Educ. **4**(1), 75–91 (2005)
43. Dalton, D.R., Hitt, M.A., Certo, S.T., Dalton, C.M.: The fundamental agency problem and its mitigation: independence, equity, and the market for corporate control. Acad. Manag. Ann. **1**(1), 1–64 (2007)
44. Hill, C.W.L., Jones, T.M.: Stakeholder-agency theory. J. Manag. Stud. **29**(2), 131–154 (1992)
45. Bosse, D.A., Phillips, R.A.: Agency theory and bounded self-interest. Acad. Manag. Rev. **41**(2), 276–297 (2016)
46. Ang, J.S., Cole, R.A., Lin, J.W.: Agency costs and ownership structure. J. Finan. **55**(1), 81–106 (2000)
47. Dobson, J.: Agency costs in US manufacturing: an empirical measure using X-efficiency. J. Econ. Finan. **16**(1), 1 (1992)
48. Eisenhardt, K.M.: Agency theory: an assessment and review. Acad. Manag. Rev. **14**(1), 57–74 (1989)
49. Sanders, W.G., Hambrick, D.C.: Swinging for the fences: the effects of CEO stock options on company risk taking and performance. Acad. Manag. J. **50**(5), 1055–1078 (2007)
50. Jensen, M.C., Meckling, W.: Theory firm Manag. Behav. agency costs Ownersh. Struct. En J. Finan. Econ. **3** (1976)
51. Fama, E.F., Jensen, M.C.: Separation of ownership and control. J. Law Econ. **26**(2), 301–325 (1983)

52. Riordan, M.H., Williamson, O.E.: Asset specificity and economic organization. Int. J. Ind. Organ. **3**(4), 365–378 (1985)
53. Berg, S.V., Smith, S.K.: CEO and board chairman: a quantitative study of dual vs. unitary board leadership. Dir. boards **3**(1), 34–39 (1978)
54. Finkelstein, S., D'aveni, R.A.: CEO duality as a double-edged sword: How boards of directors balance entrenchment avoidance and unity of command. Acad. Manag. J. **37**(5), 1079–1108 (1994)
55. Revolut: One app for all things money (2020). https://www.similarweb.com/website/revolut.com/#referrals
56. Nasdaq FIntech: Fintech Index (2020). https://index.co/list/22259-nasdaq-fintech
57. FinancialTImes: Wirecard collapses into insolvency (2020). https://www.ft.com/content/ac949729-6167-4b6c-ac3f-f0aa71aca193
58. Reuters: 'The money's gone': Wirecard collapses owing $4 billion (2020). https://uk.reuters.com/article/uk-wirecard-accounts/the-moneys-gone-wirecard-collapses-owing-4-billion-idUKKBN23W17C
59. FinancialTimes: Wirecard: the timelinel Free to read (2020). https://www.ft.com/content/284fb1ad-ddc0-45df-a075-0709b36868db

Implementation of Artificial Intelligence in Healthcare and Public Sector

Artificial Intelligence in Public Sector: A Framework to Address Opportunities and Challenges

Sertaç Yerlikaya and Yaman Ömer Erzurumlu

Abstract In the fast-paced technological change environment, Artificial Intelligence (AI) is among the fastest evolving policy areas. The opportunities promised by AI technologies are beyond the imaginations of scientists and innovators. Therefore the risk which may impose is not easy to predict. AI entails major challenges, such as gender and race biased discrimination, non-transparent nature of decision process or abuse of our privacy. Major developed and developing countries prepare AI national strategies to enhance national competitiveness and maintain national security, through policies enabling power in the international and technological competition. Large technology (platform) companies and governments are two major entities which have the big data. During Covid-19, governments and large technology companies work together on artificial intelligence enabled solutions to take immediate actions. As a side effect, this has accelerated the acceptance of digital transformation both in public and private sectors. In this paper, we have analyzed governments' strategies and policies to foster AI enabled solutions, and proposed an approach to policymaking by identifying challenges and opportunities.

Keywords Artificial intelligence · Technology policy · Innovation

S. Yerlikaya (✉) · Y. Ö. Erzurumlu
Bahçeşehir University, Yıldız, Çırağan Cd., 34349 Beşiktaş/İstanbul, Turkey
e-mail: sertac.yerlikaya@tubitak.gov.tr

Y. Ö. Erzurumlu
e-mail: yamanomer.erzurumlu@eng.bau.edu.tr

A. Hamdan et al. (eds.), *The Fourth Industrial Revolution: Implementation of Artificial Intelligence for Growing Business Success*, Studies in Computational Intelligence 935,
https://doi.org/10.1007/978-3-030-62796-6_11

1 Introduction

1.1 Why Technology Policy Is Required for Artificial Intelligence?

In the fast-paced technological change environment, Artificial Intelligence (AI) is among the fastest evolving policy areas. AI comes with many opportunities such as improving healthcare services (i.e. diagnostics to prevention of diseases and management of major public health issues such as spread of epidemics), higher output in agriculture and better climate change mitigation. On the other hands, AI entails major challenges, such as gender and race biased discrimination, non-transparent nature of decision process, or abuse of our privacy. European Commission (EC) has mandated a comprehensive AI Strategy in 2018 to address the risks and the opportunities of this shift. EC under the leadership of Ursula Von Der Leyen published a 2020 white paper to propose a framework with the aim of a ensuring a well-coordinated Europe wide approach both to mitigate the possible risks on human and ethical consequences and to identify the opportunities of AI for improving the lives of Europeans and ensuring their rights.

Technology policy environment is characterized by society's distrust of big businesses and policymakers' increasing eagerness to regulate them. Large technology companies are accused of having escaped regulatory oversight for many years to accumulate shareholder wealth while being short of meeting societal expectations at the expense of short term consumer welfare. Research conducted between March 23–26, 2020 indicates that 61% of the general public does not believe that government understands emerging technologies enough to regulate them effectively [1].

This declining trust escalated further in the post Covid-19 pandemic as they are perceived as being the main beneficiaries of lockdown measures.

1.2 Covid-19 Will Redefine the Relations Between People, Business and State Actors

Covid-19 pandemic will increase the ripple effects of reputation change on public policies and regulations. The general public's expectations from companies to play their part in solving economic and social problems are on the rise. The same research shows that 90% of people believe that brands should partner with governments to solve the crisis, and 71% say that brands that put profits before people will lose trust forever.

Representing the most affluent populations on earth, Europeans are among the most valuable customers of the entire technology industry. Yet, the European Union (EU) technology companies are lagging behind US and China based companies. The EU has a unilateral ability to regulate the global marketplace, and this phenomenon is called "The Brussels Effect". This effect arises from a set of enabling conditions sustained by markets rather than from the EU's efforts to export its regulations. However, European norms are becoming the gold standard of the technology regulation. A research conducted on digital regulation in 16 jurisdictions across the globe in the first half of 2019 shows that 49% of total 452 legislative proposals originate from Europe, either at EU level or coming from EU Member States, compared to 28% from the U.S. and 8% from Asia. This makes the EU a regulatory avant-garde in other legislations and at international organizations such as the OECD. The Brussels Effect makes the EU the most influential regulatory power through its ability to set the global standards. General Data Protection Regulation (GDPR) has been an excellent example of how Europe can define the global playing field and set the rules.

Over the past 2 to 3 years, many countries published their policy positions and started the implementation of competitive AI ecosystems. The USA published AI report in December 2016, France published an AI strategy in January 2017, followed by a detailed policy document in March 2018; Japan released a policy document in March 2017; China published the AI strategy in July 2017; and the UK released its industrial strategy in November 2017; India released a white paper in March 2018. Although there are common approaches applied, no two national strategies are similar, as national priorities and strengths vary. Implementation of these documents outlining either a strategy or policy requires billions of dollars of investments as a starting point. Lately, Covid-19 triggered the growth of AI technology. They escalated its impacts on various sectors to levels that would not be possible with the policy implementations. AI creates challenges for the governments in many areas such as data protection, ensuring ethics in decisions, accountability and possible unemployment due to the change in the labor force environment as a result of digitalization.

Governments are responding to these challenges by similar but original mechanisms in order to get a competitive rank in the artificial intelligence era [2]. World Economic Forum (WEF) has published a framework in Oct 2019 [3], to help government officials and/or experts develop AI strategies and policies for the governments. This framework defines strategic objective settings in four dimensions: *Target for capacity*: human resources and digital infrastructure, *target for investments*: research and development, grants, target for adoption: socioeconomic sectors, industrial sectors, *target for regulation*: enabling regulation such as privacy and ethical standards for the use of data.

Government AI index reported by Oxford Insights classify capacity, implementation and impact of OECD countries' AI strategies and policies. There are three main pillars of country-specific strategies according to their capabilities. These pillars also identifies the initial starting point of policymaking: attracting talent (UK, France, India), providing data capability (UK, India, South Korea) and boosting technology innovation (USA, UK, Germany, France, Israel, South Korea, Japan).

We have reviewed strategy and policy papers that are recently published within the above and other frameworks. Our analyses and interviews with various countries policymakers have also revealed that there is a need to clarify process flow, ensuring feedbacks through pilots on prominent use cases.

In the second section, we discuss why countries prepare AI strategies, what their upcoming opportunities and threats are, and how they plan to address their challenges. In the third section, we define pillars for AI policymaking.

1.3 Why Do Countries Prepare AI Strategies, and What Are Their Challenges?

By 2030, AI is expected to contribute 15.7 trillion USD to the global economy. Global income from AI technologies and businesses is expected to reach 126 billion USD in 2025, which was just 10.1 billion USD in 2018. China and the UK estimates that 26% and 10% of their GDP in 2030 will be sourced from AI-enabled technologies and businesses. AI is expected to deliver an additional 939 USD billion in value across the public sectors of 16 major developed economies by 2035 [4].

Every day, from education to health, from cybersecurity to finance, many sectors integrate AI technologies like machine learning, computer vision, natural language processing into their products and technologies. Therefore, it seems not unlikely that the global AI market will reach a compound growth rate of 30% annually from 2017 to 2025 [5].

According to indicators such as research, innovation, investments and patents in 2017, the USA is the leading country in the development and implementation of AI. In 2016, the USA made about 66% of all AI investments. China tries to catch up with the USA through substantial government investments (17% of investments in 2016). Europe is lagging behind. Neverthless, the development of AI technology and practices is increasing in Germany, France and UK [6] which are working hard to lead the policy environment in order to set the rules for the consumption of AI.

Prof. Mariana Mazzucato advocates for "mission-oriented policies and institutions" to foster innovation in her report prepared for European Union. (*The epitome of the mission-oriented concept was the Apollo program, the space program designed to land Americans on the Moon and return them safely to Earth. More than 300 different projects contributed, not only in aeronautics but in areas such as nutrition, textiles, electronics and medicine, resulting in 1,800 spin-off products.*). As missions are framed within challenges that are broadly agreed to be of high societal importance, and clear and ambitious objectives are set, real value to the societies will be achieved by a portfolio projects with the partnership of public and private stakeholders. Therefore setting a missionary goal for each country is essential.

- The USA has set its target to sustain global leadership. Areas of interest include economic prosperity, educational opportunities, quality of life, and national and homeland security. The USA is focusing on growing the AI ecosystem through public spending on contracts, e.g. the US Department of Defense spent over 2.4 billion USD on AI related technology in 2017 ($2 \times$ increase from 2015). The US government has announced to increase the budget by 70%, excluding the defense sector, until 2021.

- China targets the global leadership in research, technology and application of AI by 2030. Areas of interest include education, healthcare, energy, transport, quality of life, smart cities. China has revealed the necessity of "establishing an open and collaborative artificial intelligence technology innovation system". China is focusing on developing a talent pipeline, using AI for delivery of public services and financial support, and leveraging international cooperation. To become the "World Leader", China aims to increase the value of the core AI industry by 1 trillion RMB, and to drive the scale of related industries to exceed to 10 trillion RMB [7].
- Japan is focusing on moving from the "Industry 4.0" paradigm to "Society 5.0" through the development of AI use cases for delivering public services. Areas of interest include industrial productivity improvement, healthcare, medical care and welfare, mobility and information security.
- In this competition, Germany has allocated a budget of 3 billion Euros by 2025 to protect Europe's competitiveness in the future by making Europe a leading center in the field of artificial intelligence. In the strategy, 14 goals are set for being a center of excellence and attracting talent in artificial intelligence to the country. Germany aims to establish a national network consisting of at least 12 centers and implementation centers, and to open more than 100 staff in the higher education system.

  - France government announced 1.5 billion Euro allocation on AI by the end of 2022, in order to develop AI education and training systems to educate and withdrawal of the best AI talent [8]. Areas of interest include healthcare, environment, transport mobility, defense. The government is planning to support AI startups through data availability, public spending and talent reskilling.
  - UK signed an "Artificial Intelligence Strategy Agreement" with the private sector, and announced to invest 1.5 billion USD together with the private sector's contribution by 2025. The UK government targets to increase its total R&D investment to 2.4% of GDP by 2027. Areas of interest include services, life sciences, agriculture and public-sector. The government focuses on growing innovative technology firms and making deals with the private sector to provide better public services thru AI solutions.
  - South Korea targets to become one of the world's leading AI countries, will invest approximately 1.7 billion Euros by 2022. Areas of interest include autonomous cars, smart factories, drones, and smart cities with smart infrastructure and green energy [9].
  - Israel targets to be one of the top five countries in AI and announced to fund AI projects 500 million USD per year. Areas of interest, autonomous cars, precision medicine, defense and drones [10].
  - India targets to be "CERN of AI", by opening unique data resources to global academia and research centers, investing on common computation platforms

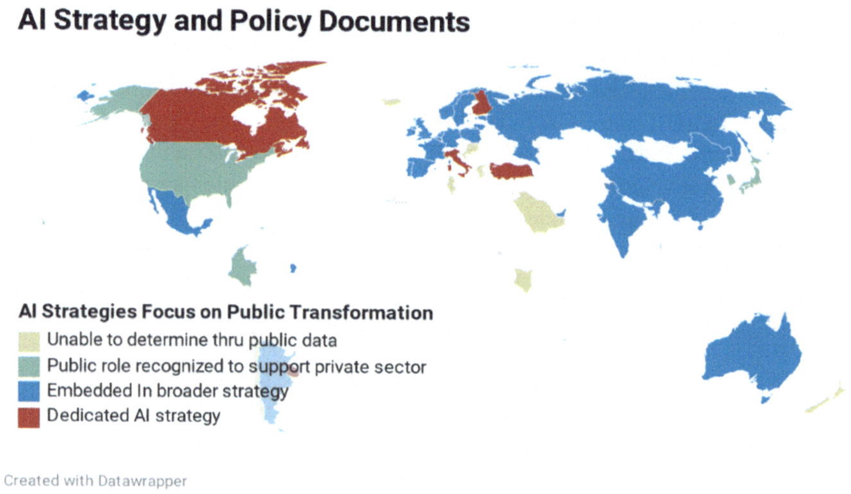

Fig. 1 Countries published their AI strategy and/or policy document

to address nationwide problems. India targets to ensure inclusivity of AI technology and solutions worldwide [11].

- Turkey announced that "National AI Strategy" will be published in 2020, with focuses on various use cases where public and private sector can collaborate. Turkey also established an "AI Institute" to coordinate technology readiness from level 1 to level 9 with relevant stakeholders through fostering talent pipeline, and an "OpenSource Platform" to develop AI and digital projects through communities within an open innovation ecosystem [12].

Countries have identified their strategic focus areas in AI development and deployment, according to their economic and political powers. Some countries deploy a dedicated strategy for AI, some embed AI into a broader digitalization strategy (Fig. 1). Another group has defined the role of public to support private sector.

Strategies that address to the real problems of countries, open collaboration between government-business-academia, and then transform the economic power into global benefit will be successful.

1.4 What Are the Real Problems of the Society and How to Prioritize Them?

The United Nations Sustainable Development Goals are among the best-known and most frequently cited social challenges, and all use scenarios in the chart match each of the 17 development goals. Mapping AI-enabled application use cases to the

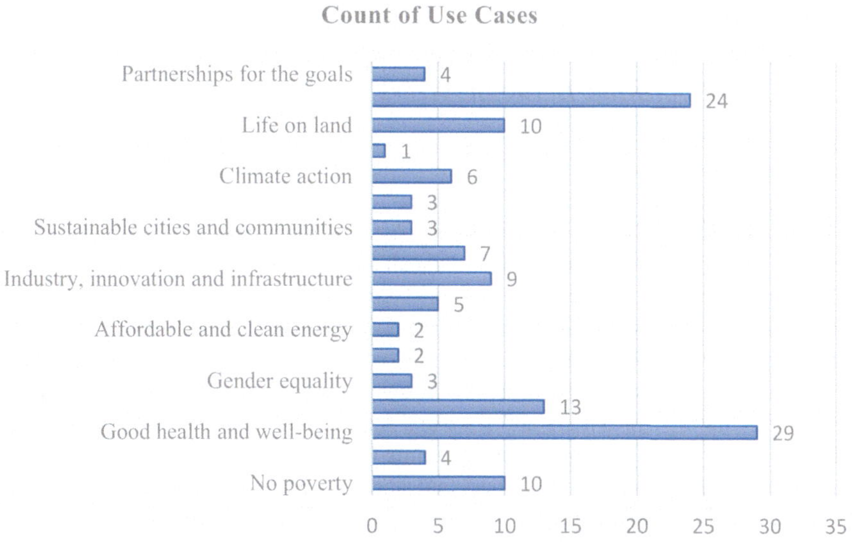

Fig. 2 AI uses cases that support UN Sustainable Development Goals [6]

Sustainable Development Goals of the United Nations for 2030, shows that number of use cases in health is higher, and followed by use cases in peace, justice and strong institutions [13]. The number of use cases in the fields of industry, innovation and infrastructure and qualified education is above average. Each use case highlights a kind of meaningful problem that can be solved by an AI-enabled technology or some combination of AI capabilities (Fig. 2).

Prioritizing real problems of the society depends on recent developments in the research and innovation. Today, computer vision techniques beyond facial recognition, have gained importance for medical imaging, agricultural monitoring and autonomous navigation. Natural language processing is becoming widespread on chat-bots, autonomous speech recognition other than translations and searches. Other important use cases in AI-enabled technologies are examples such as cyber threat hunting, drug discovery, network optimization and advanced health biometrics.

In the 2018–2019 period, start-up companies have attracted 77.8 billion USD investments, which could be the pioneer in the ecosystem for disruptive innovation. The top 3 areas where the total investments made are autonomous vehicles (9.9%), drug and cancer studies (6.1%) and facial recognition (6.0%), respectively. The appetite of investors, as highlighted in section one, was not always in line with social challenges. Therefore, governments, through policymaking have to play a substantial role in identifying real problems of the society.

1.5 According to Our Analysis 90% of 48 Strategy or Policy Documents Have Emphasized Importance of Collaboration Between Actors

The main actors of the AI strategy are the policymakers, research institutes, private sector and technology providers (Fig. 3).

Countries like France and Singapore established "innovation sandbox" or "testbed" in their strategy, focused areas such as transport, education, smart cities, safety and security, to solve countries' problems and to identify policies enabling these solutions. Access to data is crucial for problem setting and solving. To ensure inclusiveness and ethics, various scenarios have to be applied, which means new sets of big data are necessary. Therefore, countries have to collaborate on data sharing as well. Technology providers and start-ups require rich data sets to validate their solutions. India set its AI strategy as becoming CERN of AI, which implies opening its enormous data to global researchers and startups, Indian policymakers expecting to attract tales as a side benefit.

Talent is scarce. All countries' (and companies') primary focus is to educate a new generation in AI, train workforce to upskill and attract global talent pool. There will be a fierce competition, which may require a new type of cooperation among countries with the support of digital transformation.

Research is not a new area of cooperation. OECD analysis shows AI publications with international cooperation in Fig. 4 below. There is extensive collaboration between USA, China, EU(27) and UK. Australia, Canada, Japan, India, South Korea

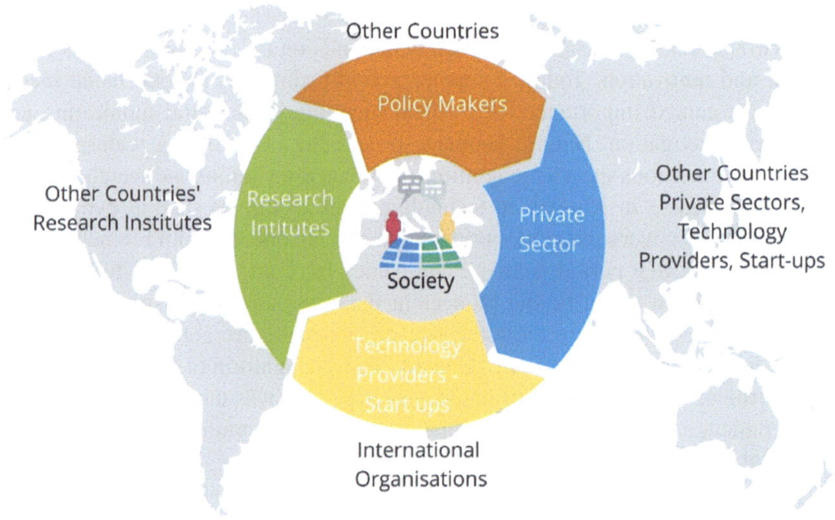

Fig. 3 Main Actors in AI Strategy for Collaboration

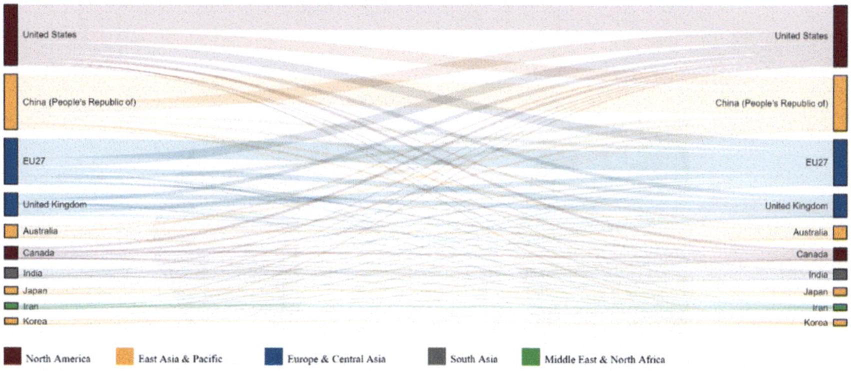

Fig. 4 Distribution of Publications Published by International Cooperation in the Field of Artificial Intelligence in 2019 by Countries *Resource* OECD.AI (2020), visualizations powered by JSI using data from MAG, version of 29/05/2020, accessed on 12/7/2020, www.oecd.ai

and Iran. They are the top 10 at international collaboration. Increasing and diversifying number of cooperation is essential to achieve the most critical AI principles in inclusiveness and ethics. The thickness of a connection represents the number of joint AI publications between two countries for the selected period.

2 Pillars of Government Strategies and Challenges

In this section, we propose a process approach to policymaking where three main processes are defined as capability, implementation, and impact on society. Any of them will not be successful without implementation of the others. However, instead of a full coverage of all strategic areas, starting from most prominent use cases and ensuring links between actors and their roles are defined step by step is recommended (Fig. 5).

2.1 Capability: Talent, Infrastructure, Investment and Governance

Providing talent, implementing infrastructure for communication, data gathering, data sharing, attracting investments and ensuring governance to sustain all capability requirements are the main challenges of the governments to develop robust policies, compatible with other countries and allowing cross-border collaborations. There is a fierce competition between countries to attract the capability components where each one triggers the others.

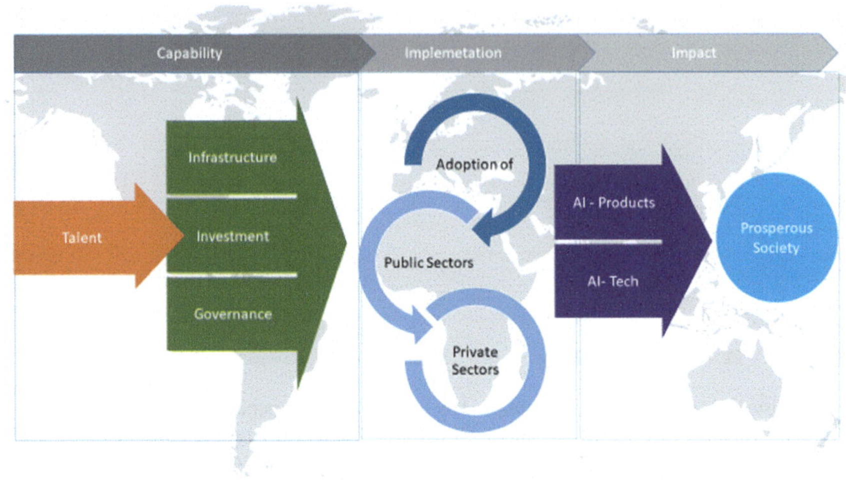

Fig. 5 Pillars of Government Strategies of Challenges on AI Policy Making

Capability building for talent: AI talent is scarce. There is a high demand both from universities, and platform companies for AI talents. Especially, USA platform companies such as Amazon, Google, Facebook or Microsoft attract talents, including researchers all over the world. The current education system, designed to address the needs of first industry revolution, is not good enough to educate talents for unprecedented development of technology. This system is based on problem solving; however, AI requires questioning rather than trying to find out solution first. AI also requires multidisciplinary approach to analyze the requirements and deliver the solutions for application areas.

OECD working paper [14] indicates that 14% of jobs are at high risks of automation and 32% are requires significant changes, which also highlights wage polarization and inequalities. In UK, about 1.5 million jobs [15], in Turkey, 7.6 million jobs [16] are at high risk of being at least partially-automated. These threats require redesigning social security nets, and up-skill and/or re-skill workers. Meanwhile, the young generation must be inspired to study technology and to work in the related and/or adjacent fields. The young generation is purpose-driven rather than career and benefits.

Governmental policies significantly focus on allocation of resources on increasing the number of AI talent, and developing the ecosystems for collaboration. Education must be specialized to provide re-skilling and up-skilling for the new and existing workforce and address the demand/supply mismatches between employers and employees. Governments invest on increasing number of AI researchers and teachers. AI education requires a robust STEM-A starting at k12, a multidisciplinary approach on undergraduate and graduate levels to solve human centric issues. Philosophy and ethics are two main courses for technology and application developers to ensure the proper use of AI. On the other side, public awareness have to be built on understanding impact of AI applications and their proper usage.

Although demand is high, currently, there are 300,000 AI talent including students, academia, entrepreneurs and professionals all over the world and 1 million required according to studies [17]. Each group requires specialized educations on top of core AI education. On the other hand, policy and lawmakers who will define the regulation and resolve the disputes have to be aware of technological benefits and ethical risks of AI. Decision-makers and technology buyers also have to be able to evaluate cost-benefit analysis of AI properly. Therefore there is need to invest in redesigning education systems in law schools, and re-skilling training programs for decision-makers.

On top of education and trainings, multinational research and innovation collaborations are key to develop AI technology and solutions. All these efforts and investments requires high collaboration between governments and the private sector. USA plans to grant 200 million USD for STEM-A education matched by 300 million USD by the industry. UK targets to increase R&D spending to 2.7% of its GDP by 2027, to invest 42 million GBP in the training of teachers, and 64 million GBP in re-skilling of current workforce in digital technologies. The UK also targets to build over 1,000 government- supported PhD institutions by 2025 with a Turing fellowship program for AI [15]. Japan has founded a "national consultative body" with three universities and private sector representatives to develop education programs for re-skilling [18]. France targets to triple the number of AI graduates in three years and to double starting salary of researchers in public universities [8]. China has launched an "AI + X" program with 50 teaching and research institutes to train 500 teachers and 5000 students, 50 nationwide online open courses and 50 AI faculties by 2020 [17].

Governments want to attract international talent as well. UK and France plan to ease work-permit applicants working on AI technologies, and to reduce administrative formalities and offer high payments and quality of life.

Capability building for infrastructure: AI strategies can be prioritized according to the availability of big data, investment appetite of public and private sectors, and readiness of computing and connectivity (5G and fiber optic network roll-outs) infrastructure, cybersecurity, and collaboration systems for joint research and innovation which are the components of infrastructure capacity.

Data is the oxygen of AI technologies and the fuel of forth industrial revolution. To build a data ecosystem, governments have taken significant actions. Open data policies aim to ensure the interoperability of data. Governments issue policies to open public data and share with academia and private sectors, and also to enable cross-border data flow. Each phase requires policymaking on data communication standards and data privacy and cybersecurity. Roll-out planning on data connectivity and investments on computing has to be planned accordingly. Other than investments, governments are working hard on ensuring privacy, safety, transparency, accountability and inclusivity of AI-enabled systems without disrupting innovation and the potential economic impact.

UK, in line with "Artificial Intelligence Strategy Agreement" signed with private sector, is exploring the feasibility of creating data trusts where the government underwrites the process of data sharing and storage. The UK also focuses on defining data rights for potential participants of the agreement. Around 1 billion GBP investment planned to upgrade digital infrastructure, including 5G and full-fiber networks roll-outs. France has built an open platform called "innovation sandboxes" and offered resources for testing use cases. AI research requires supercomputers designed explicitly for AI usage and devoted to researchers and their public or private sector partners during their co-initiatives. The USA, China and Turkey are developing open source platforms and software libraries, toolkits where AI ecosystem actors can contribute, and actors can define the rules where necessary. China has already defined funding policies for 5G roll-outs and supercomputers, high performance semiconductor chips for AI use.

Capability building for investments: Countries investing in research and innovation ecosystem are attracting funds and becoming pioneers of AI. The USA, China and Japan universities and research institutes lead the publications on AI topics between 2010 to 2016. The USA universities, Carnegie Mellon, Massachusetts Institute of Technology, and Stanford are the top leading universities in terms of paper publications. These universities have also established common infrastructure and industry collaborations such as IBM, Microsoft, Amazon and Google. On top of research in academia, Chinese universities, Peking and Tsinghua universities also build a strong relationship with industry partners such as Tencent, Alibaba and Baidu.

Japanese universities and research centers, though, historically machine-oriented, have rapidly adapted themselves to collaborate on theoretical and applied research on AI with the industry. Japan's ecosystem focuses on developing sector-specific platforms for public and private partnerships on AI-enabled technologies, and inter-linking these platforms. To achieve, Japan is planning to increase its science and technology fund to 900 billion JPY by 2020 and already announced an R&D tax exemption including AI and big data and subsidies for building new robots with integrated AI. The UK is planning to establish over 1,000 government-supported PhD institutions by 2025. It announced a Turing Fellowship to support initial cohort. It increased R&D spending to 2.7% of its GDP, and establishing reforms on enterprise investment schemes and venture capital trusts to unlock 7 billion GBP over 10 years. France is planning to spend 1.3 billion EURO to develop AI-enabled innovations. Academia-industry fellowship programs are mostly used tools to convert outputs of researches into outcomes.

Large technology companies dominates the AI investments. Table 1, listing market cap of the top seven companies investing on AI, indicates that their funding capacity is even over many countries except the USA, China and total of EU. They invest intensely on AI areas such as autonomous cars, precision medicine, and robotics.

Venture capital and private equity companies are also funding startups and scaled ones. According to AI100 index conducted by Stanford University, the number of active AI startups in the USA increased 14 times since 2000. Consequently, the volume of Venture Capital funds increased six times at the same period.

Table 1 Companies Market Cap investing to AI

Companies	Market Cap	Country
Amazon	1 trillion USD	USA
Apple	1 trillion USD	USA
Alphabet	Almost 1 trillion USD	USA
Microsoft	1 trillion USD	USA
Facebook	1/2 trillion USD	USA
Alibaba	1/2 trillion USD	China
Tencent	1/2 trillion USD	China
TOTAL	6 ½ trillion USD	

Machine learning, computer vision, natural language processing, autonomous vehicles and smart robotics are attractive areas for investors. AI100 index results are limited by USA data, which may be representing most of the current AI universe. However, to ensure a globally prosperous society through AI-enabled economy, measuring performance indicators and ensuring sustainability are the governments' core responsibility.

Technology transformation from academia to public and private sectors was always challenging. Ensuring facilitation and collaboration thru government policies and incentivizing accordingly is the area to focus. Funding from seed to market requires public-private partnership, on top of cost of research and proof of concept, injection of five times more investments required for commercialization, and reaching to the market. Therefore, in collaboration with investors and trade bodies, Venture Capitals are essential for the functionality and profitability of new businesses. Governments mostly build mechanisms to support and scale up small and medium businesses. Corporates and funding institutions' roles differ as investing thru funding, capital injection and order placement to mature business. In addition to these, public policies required to ensure a sustainable and predictable market environment for foreign and local investors.

Capability building for governance: Allocation of resources and monitoring performance for AI-enabled strategies requires a well-established governance model. Multidisciplinary task forces/councils are required to realize the strategic pillars within government organizations, private sectors and academia. Most of the governments have established their own central governance councils, typically consist of ministers, representatives from private sectors and academia. U.K. has a dedicated department "Office of AI" to collaborate with multiple departments, ministries and other stakeholders to deliver AI project. France has a shared specialist center of 30 members from governments departments to provide input and resources on implementation of the projects. In China and Japan, individual ministries and departments are responsible for implementing AI solutions across different sectors. Singapore established a program office to coordinate the strategic projects and use cases.

Local governments have also established their governance bodies to solve local challenges. London has a Smart City board and a Chief Digital Officer to apply best practices in smart infrastructure and AI. Over 19 Chinese cities, including Beijing, Shanghai, Hangzhou, Zhejiang, Tianjin, have been mandated to develop their own city-level AI agenda. In Turkey, Istanbul and Gaziantep have established AI Offices to provide open data and AI enabled application developments. Challenging real, local issues have the potential to democratize and sustain the inclusivity of AI-enabled systems use.

2.2 *Implementation: Adoption of Public and Private Sectors*

Rapid technological developments are leading not only structural changes but also redefinition of roles of actors in the public and private sectors. In addition to applications such as data-analytics, speech recognition, filtering unsolicited e-mails, credit risk calculations, machine learning-enabled decision making is increasing both in public and private sectors. Since many of these techniques are data-intensive or compute-intensive, performance metrics have to be defined to understand impact and efficiency which will not be easy. Data-driven scenario analysis are required instead of setting the rules through policies and ensuring the compliance, and governments have to be prepared to be agile. Accountability, transparency, inclusiveness and ethics are the primary metrics to be achieved. Rather than a single task traditional policy-making approach, all these capabilities require evaluation of multi-tasks where the importance of various sub-tasks have to be weighted for assessing overall progress.

The roles of central and local governments may vary from country to country. Still, b education, health, infrastructure are mostly their focus areas to sustain and improve the quality of their citizen's life. Governments are working on data acquisitions, data distribution and access to foster innovation to solve the society's major problems. Thanks to COVID-19, digital transformation is progressing faster than expected. Online education, remote health services with the availability of data, are expected to cause decrease in cost of public services. Start-ups will develop customized solutions as a spill-over effect.

Companies already generating big consumer data are ahead in this competition, others need to collect data from their products and production area. Likewise multi-task orientation in the public sector, the private sector is shifted to a data-driven decision making processes. Adoption will be slower but fast adapters with responsible implementations will win their customers and shareholders' value.

Governments and public institutions are getting prepared for next-generation technologies mostly thru responsible AI procurement processes. Only in EU, more than 250,000 public authorities deal with procurement of external services, works and supplies, software, cloud storage and IT support around EUR 2 trillion annually. Governments consciously using their economic power, could support AI innovation and set technological and ethical standards for their citizens welfare.

Thru policymaking, governments will set the playing rules for controlling large technology companies and making space for startups while ensuring data privacy and ethical use for owner of the data. Adaptation of the public and private sector to an AI-driven ecosystem has to be designed as an iterative process where scenarios are tested, and best ones are implemented.

2.3 *Impact on Society*

From problem identification and defining strategic use cases for the country to delivery of an AI technology or AI-enabled solution, only impact on the society will present the success. This success may not directly correlated with the solution as enormous spill-over effects are possible. Anyhow, success could only be achieved by the development of a human-centric ethical framework for transparent and fair use of AI applications. As the prosperity of the society measures impact, evaluations may also end-up by not to implement one of the AI-enabled solutions. Actors must be aware of these possibilities and ready to accept to move on for a better solution.

Governments have already started to use AI technologies such as chatbots used to effectively manage overwhelming inquiries from the general public, tracking systems to identify people infected COVID-19. However, this increased concern about the misuse of sensitive information about people in the future. Similarly, applying predictive policing helped law makers to fight crime, but couldn't ensure discrimination and biased decision making yet. Therefore, establishing standards for audit, setting performance measures on fairness, openness and accountability, ensuring that there is human involvement in the course of decision making will foster the impact on society.

3 Conclusion

It is not easy to define what is an accountable and ethical deployment of AI should look like. Policymaking is being restrictive and prescriptive but AI policymaking is only possible by enabling openness and innovation. On top of them, AI policies have to sustain flexibility as technology matures. Current systems are focused on too many rules on process in order to benefit from technology we have to focus on outputs.

AI-enabled technologies may be useful for the society overall. All actors have the responsibility to design a human-centric ethical framework to ensure continuity of the humanity in the world. Today, we know what the capacity requirements are, such as talent, data, computation power, investment. But answers on how to foster adoption and how to evaluate impact are yet to be responded. For that reason, the last sections may require revisions and additions as AI-enabled solutions evolve.

References

1. Edelman, R.: Global report edelman trust barometer, online survey report, pp. 15. (2020)
2. Miller, H., Stirling, R.: Government Artificial Intelligence readiness index. Oxford Insights and the International Development Research Centre. (2019)
3. Madzou, L., Shukla, P.: A framework for developing a national Artificial Intelligence strategy, White Paper. World Economic Forum. (2019)
4. Le Masson, B.: Transforming public sector with AI, A Govtech Ecosystem Approach Report, pp. 5–7. Accenture (2019)
5. Lam, T., Li, F., Han, Z., Chung, R.: Global artificial intelligence industry whitepaper, pp. 33–36. Deloitte (2019)
6. Bughin, J., Hazan, E., Ramaswamy, S., Chui, M., Allas, T., Dahlström, P., Henke, N., Trench, M.: Artificial Intelligence: The next Frontier?, Discussion Paper, pp. 6–42. McKinsey Global Institute (2017)
7. The New Generation of Artificial Intelligence Development Plan: State Council, China. http://www.gov.cn (2017)
8. Villani, C.: For a meaningful artificial intelligence towards a French and European strategy, pp. 10–98. http://www.aiforhumanity.fr (2017)
9. Groth, J. O., Nitzberg, M., Zehr, D., Straube, T., Kaatz-Dubberke, T.: Comparison of national strategies to promote artificial intelligence, pp. 39–41. (2019)
10. The Israeli Innovation Authority: AI R&D framework and activities, OECD policy observatory information. http://www.oecd.ai (2018)
11. National Strategy for AI: Discussion Paper, pp. 6–91, NITI Aayog, Government of India. http://www.niti.gov.in (2018)
12. 2023 Industry and Technology Strategy, pp. 73–77. Ministry of Industry and Technology, Turkey. http://www.sanayi.gov.tr (2019)
13. Shoham, Y., Perrault, R., Brynjolfsson, E., Clark, J., LeGassick, C.: Artificial Intelligence Index. Annual Report, Stanford Institute of Human-Centered Artificial Intelligence. (2019)
14. Nedelkoska, L., Quintini, G.: Automation, skills use and training. OECD Report, pp. 6–10. https://doi.org/10.1787/2e2f4eea-en (2018)
15. Policy Evidence and Analysis Team Office for National Statistics: Which occupations are at highest risk of being automated?, Office for National Statistics UK. http://www.ons.gov.uk (2019)
16. Singh, A., Korkmaz, B., Kendi, K., Cenudioğlu, C., Demirdağ, E., Dandona, G. S., Chui, M., Kokal, Ö., Tanrıkulu, Ö., Gökler, P.: Future of work, Turkey's talent transformation in the digital era report, pp. 26–38. McKinsey & Company Turkey in cooperation with the McKinsey Global Institute (2020)
17. Vincent, J.: China and the US are battling to become the world's first AI superpower. Article, The verge, August 3, 2017. http://theverge.com (2017)
18. AI Strategy 2019 AI for Everyone: People, Industries, Regions and Governments, Innovation Strategy Promotion Council, Japan. http://www.cao.go.jp (2019)

Big Data for Healthcare: Opportunities and Challenges

Fatima Lalmi and Laadjal Adala

Abstract The implementation of Big Data Analytics (BDA) in healthcare has significant contribution in ameliorating patient care with effective cost, predictive analysis of diseases, improving value in healthcare organizations and accelerating medical research with low costs. The paper provides a general survey of recent advances in BDA in healthcare, BDA tools and roles. Also, a healthcare framework that offer miscellaneous data analytical capabilities is proposed. Additionally, various opportunities and challenges inside this field are explored. This study reached several results, mainly: the application of BDA in healthcare provides opportunities like: enhancing the quality of care, decreasing waste and error, decreasing the charge of care, facilitating of medical organization activities and making decisions and enhancing personalized healthcare service. While security and privacy remain, major challenges facing the application of this technology in healthcare. Therefore, basic pillars should be adopted to take advantages of BDA and overcome its challenges. BDA has potential positive effects and global implications in healthcare; however, it must overcome some obstacles and design an effective system based on BDA.

Keywords Big Data · Big data analytics · Tools · Applications · Healthcare · Opportunities · Challenges

1 Introduction

The Information and Communication Technologies (ITCS) have seen significant developments in last years, the most recent one known as Industry 4.0, it's the latest trend in manufacturing technologies, it relies on three pillars of technologies: Internet of Things(IoT) which is characterized by the presence of a variety of cooperating objects like mobile phones, sensors and actuators; Cloud and Fog Computing which provides virtual unlimited computing storage and communication sources; and Big Data Analytics(BDA), it extracts value from a huge amounts of

F. Lalmi (✉) · L. Adala
Abdelhamid Ibn Badis University, Mostaganem, Algeria
e-mail: lalmi.fatima@yahoo.fr

A. Hamdan et al. (eds.), *The Fourth Industrial Revolution: Implementation of Artificial Intelligence for Growing Business Success*, Studies in Computational Intelligence 935,
https://doi.org/10.1007/978-3-030-62796-6_12

data [1]. This advancement in ITCS have positively influenced in several fields the most important one is the healthcare [2]. Hence, the expression eHealth, intended as a result of the application of ICTS to healthcare. Also, healthcare organizations have adopted electronic health records (EHR_S) which reduce administrative activities, ensure data availability, enable faster time to treatment and reduce costs [3].

The use of BDA has become necessary in the healthcare field as a result of several factors such as rapid demographic growth, increasing demand of healthcare and increase the number of chronic diseases [4]. Since the domain of healthcare always had a massive amount of data, coming in different forms (structured, unstructured, semi-structured) and from various sources [5], which they are difficult to analyze and manage with traditional software or hardware [6]. The use of BDA with innovative tools is essential in order to improve patient care and facilitate decision-making [7].

Today, the decision makers in the health system are able to guide the stage coming towards development in the healthcare sector by taking advantages of progress in data science to generate valuable visions from huge, and complicate datasets that health systems contained [8]. BD science is already contributing to a quantitative transformation in some other main sectors such as transportation and financial services.

Car automation is a prominent example of the application of data science. Today, cares are programmed to activate the brake or deviation to avoid collision alongside self-parking feature. These systems use computer vision technologies, machine learning algorithms, and despite the potential error presented, it is expected to lead to significant improvement in road safety [9]. The same approach can be applicated to many complicated healthcare operations. Some new hospitals, such as Humber Hospital in River, have already automated the dispensing of medicines [10]. As such clinical research organizations use BDA to determine the optimal mix of sites in clinical trials. They implement predictive risk algorithms that include natural language processing and machine learning techniques in order to choose sites that are most likely in recruiting eligible participants as well as achieve the goals of the experiments in a timely manner and with appropriate data quality [11]. This technic has contributed to a 15% increase in registration time, reduced patient visits 10%, in addition to improve targeting quality by 19.40%, and similar improvement are also expected for comparable areas in healthcare research, which means the ability to do more research with limited resources [12].

The potential benefits adopting BDA in healthcare have attracted many scientists and academics to research in this field. Nevertheless, actual contributions to clarifying the various positive and negative effects are still few. The objective of this study therefore is to gain a comprehensive understanding of current outlook on this technology in healthcare field. It purposes at answering the study question on: How "Big Data Analytics" contributes to enhance healthcare value in the light of the opportunities and challenges of BDA in this field? Accordingly, this study highlights the conceptual aspect of implementing BDA to healthcare and its importance in improving care delivery. It also discusses the opportunities as well as threats associated with adoption of BDA in healthcare and proposes solutions to overcome them. Additionally, it discusses a proposal framework to turn into a healthcare system based on BDA.

In the same context, this paper's objectives are:

- Determination of different definitions and characteristics of BA in healthcare;
  - Exploration of both traditional and novel sources of BDA in healthcare;
  - Identification of BDA tools and their role in healthcare;
  - Discussion of the potential benefits BDA within healthcare;
  - Analyze of most important challenges faced BDA in healthcare and proposition of some solutions to overcome them;
  - Proposition of framework to turn into a healthcare system based on BDA.

By achieving these goals, this paper will make a valuable contribution in comprehending BDA and its impacts on healthcare. The paper is structured as follows: previous studies are mentioned in Sect. 2, background of Big Data Analytics in healthcare is illustrated in Sect. 3, both opportunities and challenges discussed in Sect. 4, basic pillars should be adopted to take advantages of BDA and overcome its challenges are demonstrated in Sect. 5. the conclusions are shown in Sect. 6.

2 Related Works

Several studies have contributed in different ways to the understanding of BDA in healthcare. Pashazadeh et al. (2018) [4] and Bora (2019) [13] are literature reviews of the big data meaning and applications in the healthcare literature. They gave an overview about BD existing concepts in healthcare and its applications in different categories, including machine learning and cloud-based, etc. They concluded that these factors have positive impact on healthcare domain for many reasons: ease of use; scalability; improve efficiency, speed, accuracy, etc., Thus, these factors can easily analyze the BD of health care.

The study of Belle et al. (2015) [14] presented an insight into how BDA tools can be used to manage large volume of data in order to aid in the decision making and performance of healthcare personal and patients. They focused on three areas: Medical Signal Analytics (MSA), Medical Image Processing (MIP) and genomic data. For this work, they concluded that these areas address several benefits including, ability to use many sources of data to increase the accuracy of diagnostic, reducing costs and improving the accuracy of processing methods. Karen et al. (2017) [15] discussed the challenges related to manipulate large-scale next-generation sequencing (NGS) data and diverse clinical data from the EHRs for genomic medicine. Also, they introduce solutions for these challenges in manipulating, managing, and analyzing genomic and clinical data to implement genomic medicine.

Some studies focused on discussing drawbacks of BD in a specific area of medicine, Peek et al. (2014) [16] presented technical and methodological weakness of BD research in biomedicine and health. They concluded that the most important challenge is to transform analytical methods that are used in the biomedical and health domain, to fit the distributed storage and processing model that is required to handle BD, while ensuring confidentiality of the data being analyzed. While Fillinger et al. (2019) [17] reviewed the state of BD in bioanalytical research and provided an overview of the guidelines for its proper usage. However, Cheung et al. (2019) [18] illustrated the use of BD research on gastrointestinal and liver diseases and concluded that BD in this field faces important weakness including, data validity, missing data, incomplete data capture due to the unavailability of diagnosis codes for certain clinical situations, and individual privacy.

Other studies interested in vital roles of BDA implementation in various sub-sectors of healthcare, among them the study of Liang et al. (2016) [19] showed the contribution of BDA in human genomics, they concluded that BDA have positive impact on the biomedical and health science fields such as improved understanding of human life, health, diseases, and behavior possible. The studies mentioned above include important contributions and innovative ideas which provide understanding BDA technology, its tools, its adoptions, as well as the benefits and weaknesses in healthcare.

The current study differs from previous studies in that it proposes practical solutions to overcome the challenges related to BD in healthcare, and it also suggests the pillars adopted to ensure an effective transformation of healthcare based on BDA.

3 Background: Big Data Analytics in Healthcare

3.1 Big Data Analytics Definition

Data are increasing at an exponential rate. This huge growth in both volume and type of data is leading a sea change in data management that is encapsulated under the designation of Big Data. Big data offers exciting possibilities in several domains including business, manufacturing, health care, and scientific research [20].

Despite the frequent use of the term "Big Data" there are various definitions for this term, all above them mentioned that Big Data is a large volume of high velocity, complex, variable and visible data that requires advanced techniques to enable the collection, storage, distribution, management and analysis of information in order to facilitate decision making. It refers to the six basic characteristics of Big Data, also known as 6 V_S, which can be described as following (see Fig. 1) [21]:

- Volume, means a large scale of BD, which requires innovative tools for their collection, storage and analysis;
- Velocity, represents the speed in which data created or updated, pointing to the real-time nature of Big Data;
- Variety, signifies the variation in types of data. Big Data comes in diverse and dissimilar forms from multiple sources;
- Veracity, implies the uncertainty of data;

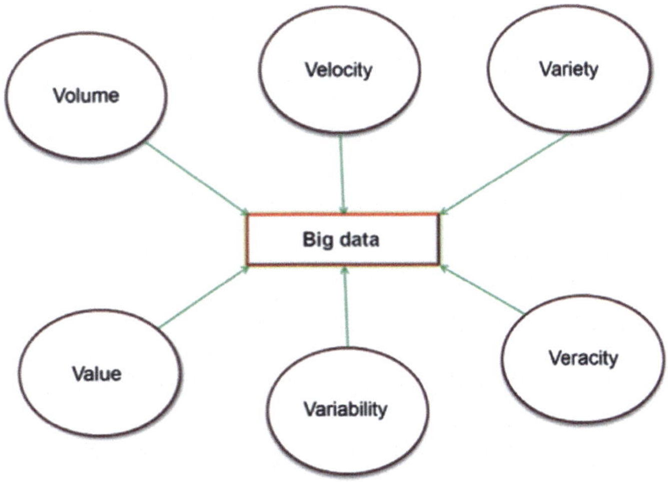

Fig. 1 Big Data and its 6V's [21]

- Value, means the core of Big Data that addresses the cost/benefit proposition;
- Variability, means that data may change during the processing period or any stage in its life cycle.

For describing BD in healthcare, Auffray et al. (2016) defined as: "high volume, high diversity biological, clinical, environmental, and life style information collected from single individuals to large cohorts, in relation to their health and wellness status, at one or several time points". This definition concentrated on the kinds of healthcare data, however Karen et al. (2017) focused on the tools used to analyze the data, they defined BD as follows: "Big Data refers to novel technological tools delivering scalable capabilities for managing and processing immense and diverse data sets" [15]. However, Liang and Kelemen (2016) described BD as: "tools used to exhibit abundant potential to support a wide range of medical and healthcare functions such as clinical decision support, diseases surveillance and population healthcare management" [19].

From the above definitions we conclude that BD with its characteristics and tools aims to facilitate data management, improve healthcare from depiction and record to prediction and optimized decision-making.

Ultimately, the value of BD is not about the BD, it's about how to turn big data into good research problems/questions/hypotheses, then transform into valuable solutions which benefit society. This is rendered simpler by their tools. As that BDA in healthcare involve the methods and techniques of analyzing the massive amount of electronic patient care data [22], for the purpose to summarize useful and important information and focus on the insight gained from historical data in order to determine patterns and predict future outcomes and trends. So, the methods of BDA refer to techniques like forecasting, optimization, simulation, visualization, machine learning and others which provide insights to managers and policy-makers and assist decision-making.

3.2 Sources of Healthcare Data

Healthcare field uses an enormous increase data collected from different resources (see Fig. 2), which can be classified as follows [3]:

- **Clinical data**: it's the main resource of data, it contains patient data such as electronic health record (HER), medical images, and medication history, etc. These data are mostly unstructured;
- **Patient behavior and sentiment data**: these are the data collected from social media, which gives the information about behavior and lifestyle of individuals;
- **Omics data**: data includes genomics, micro-biomics, proteomics, and metabolomics, which helps to understand the structures of diseases and speed the personalization of medical theraps;
- **Administrative and external data**: these are data collected through different external sources such as insurance companies, hospitals. It describes costs, bills, reimbursement categories and other patient characteristics;
- **Health publication data**: these are data obtained from different publications like journal articles, clinical practice guidelines, medical research materials, health reports, etc. [13];
- **Pharmaceutical R&D**: it's a new concept of analytical pharmacy solutions to enhance care coordination. It contains drug therapeutic mechanisms, R&D data from target behavior in the body like effects of toxicity.

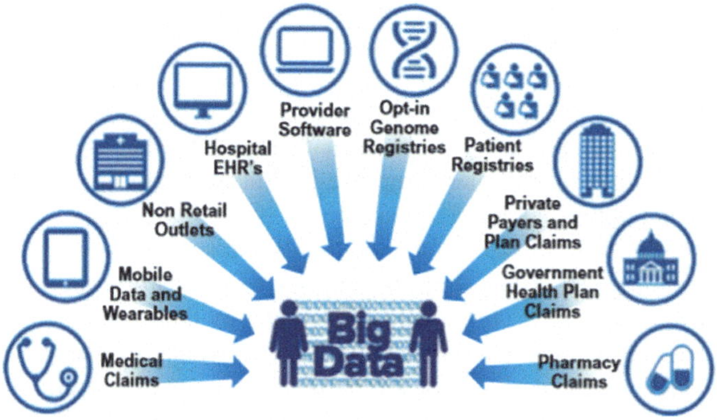

Fig. 2 Types of healthcare data resources [21]

3.3 Big Data Analytics Goals in Healthcare

BDA in healthcare aims to achieve a group of goals which can be summarized in the following points [24]:

- Enhancement of every detail of healthcare containing professionals in this domain, patents and management;
- Personalization of the patient care with effective cost and specializing in improved effects with smarter selections at any time when required;
- Taking the benefits of BDA for enhancing any disease to right affected person on right time;
- Development of algorithms for predicting the inpatient time in coming years;
- Potentially benefit all the components of a healthcare system i.e., provider, payer, patient, and management.

3.4 Tools Used in Big Data Analytics in Healthcare

Hadoop. It's considered as the most famous platform of BDA. That's why it is the first choice for many analysts and decision makers. It is programmed in Java which supports large data sets in distributing computing. It has several advantages such as: it is linearly scalable and reliable, a fault-tolerant system, acceptance of hardware failure and fast data transfer. Hadoop has two main components namely Map Reduce and Hadoop Distributed File System [25].

Hadoop Distributed File System (HDFS). Hadoop distributed file system is the prime data storage system used by Hadoop; it is a file system which extends all nodes in Hadoop clusters for data storage. It joins all the file system together on local node to make into a large file system. To overcome the node failures HDFS enhances the security by depicting data across multiple sources. It is optimized for large files [26].

MapReduce. It is a programming model for processing large data workloads. It has two roles: mapper and reducer. The first one, it filters and parcels out work to various nodes within the cluster or map and, the second one, it reduces the results from each node into a consistent answer to query [13].

Lucene. It is a high-performance scalable information retrieval library which is implemented in Java. It has several advantages, the first one is Lucene is an open source project so it is available for free, the second one is the speed and high-performance indexing. Efficient and accurate search thanks to algorithms implemented is another advantage [25].

Hive. It is a data warehouse infrastructure tool for processing structured data in Hadoop. It has different storage kinds like plain text, RC file, etc. Built-in user-described functions are used to handle dates and other data mining tools. It is SQL-like Bridge that allows BI application to run queries against Hadoop clusters. Hive lacks few things such as full SQL support [26].

HBase. It is an open-source, non-relational distributed database model. It comes as a handy alternative to Hive which lacks of full SQL support as mentioned earlier. HBase is not only provides row-levels queries but is also used for real-time application processing unlike Hive. It offers both linear and modular scalability and is strictly maintains consistency of read and write which in return helps in automatic failover support. All these features make HBase different and unique and also the go-to tool for horizontal scaling of huge datasets [27].

Pig and Pig Latin. Pig is a data flow handling language. It consists a high-level programming language named Pig Latin for creating MapReduce programs that requires Hadoop for data storage. The Pig Latin code is extended with the help of User-Defined Functions that can be written in Java, Python and few other languages. It is amenable to substantial parallelization, which in turns enables them to handle very large data sets [13].

4 Opportunities and Challenges of Big Data Analytics in Healthcare

The technologies developed in the Industry 4.0 contribute to changing traditional applications in all fields, especially in healthcare field. This technological development provides access to many advantages as well as produces some challenges that will be discussed in this section.

4.1 Big Data Analytics Benefits in Healthcare

The use of BDA in healthcare field provides several benefits that can be summarized as follows:

Improvement of patient care. Big Data (BD) identifies novel data sources such as social media platforms, sensors, wearable devices, etc., in addition to traditional sources that include patient medical history, doctor's diagnostic, etc. [2]. All of these sources provide comprehensive data about the patient, the accurate analysis of which uses BDA tools contributes to the early diagnostic of the disease, regulars patient surveillance,minimizes drug dosage to avoid side effects and reduces costs on the one hand, and offers a valuable source of information for healthcare researcher to find new healthcare solutions on the other hand.

Real-time patient monitoring. BDA provides complete data in real time on the patients and at the same time able to administer medical intervention without any delay.

Predictive analysis of diseases. BDA aid detecting disease earlier, which facilitates the personalization of prediction, reduction of costs, minimize waste in resources and giving recommendation to individuals for maintaining their health [28].

Improvement of treatment methods. BDA refers to the tools and processes that enable the conversion of huge amounts of healthcare data into predictive models to recognize the requirements, services to be supplied, predict and impede health disaster for the benefit of people. In addition, it facilitates both the measurement of efficiency of specific drugs and the process of developing required vaccines by researchers in this field [2].

Smart Healthcare System. Application of BDA in healthcare can transform healthcare providers into smart and efficient delivery of care by enriching various fields in Smart Healthcare System like virtual hospitals, telemedicine and smart healthcare devices [29].

Genomics Analytics. Genomic information plays an important function in keeping the affected person document. Applying genomic and scientific facts together, cancer remedy may additionally turn out to be extra easier than before.

Research & development. The use of BDA allows the provision of comprehensive data on various disease, which aids researchers in studying these diseases and finding new medicines.

4.2 Big Data Analytics in Healthcare: Challenges and Solutions

BDA application in healthcare sector faces several challenges, including [19]:

- The main challenge faces healthcare researchers are still the ability to locate, analyze, integrate and interact with the enormous amounts of data in real time;
- Integration of the huge amount of data in multiple forms (structured, unstructured, semi-structure) from a variety of sources (novel and traditional sources);
- Human errors, EHR_S data are influenced by the personal who entered the patient's data, which can lead to missing data, incorrect data, misunderstanding or wrong interpretation of the original data;
- Privacy and confidentiality are the most important challenges in healthcare;
- Security is the major hindrance in the broad integration of BDA in industries lies in its security, particularly when dealing with patient's health data.

Overcoming the previous challenges and minimizing their damages requires designing an effective system based on the following principles [30]:

- Only collect, store and give access to necessary data, for example, an enterprise that develops systems for improving the mechanism for scheduling surgeries may only to see the dates of operations, diagnoses and results, but not medical records or scan results. However, a patient should be given access to all data related to him;
- Use unique, consistent but pseudonymous patient identifiers in order to link records with their owners throughout the system;

- Consider using internal or cloud-databases instead of one central data bank. This process has two benefits: the first, is to maintain the quality of the data; it contributes to providing access to the original sources and preventing from the risk of introducing errors during the process of sending and copying data; second, in the event of security breach, only a small portion of data would be at risk;
- Use of cryptographic techniques and enterprise-grade security measures;
- Ensure traceability and transparency of all interactions with data.

5 Moving to a Full Healthcare System Based on BDA

5.1 Transformation Phases to Full Health Care System Based on BDA

The transition to a healthcare system based on BDA needs to formulate a deliberate a plan that balances goals and investments in order to achieve the desired results, which consist in moving from the current healthcare system to an integrated healthcare system based on the use of BDA. In this context, the report of the World Innovation Summit for Health (WISH) in 2018 included a proposal for the path of transformation in healthcare, which can be summarized as follows [31]:

- *The current status of most healthcare system*: Today, most of healthcare systems are at the beginning of the transformation. Even high-income countries still combine paper records and digital inputs. Where EHR_S are available only to provider that created them. Also, these systems have several characteristics among them: decisions rely on clinicians' knowledge and judgment, limited use of applications and wearables for self-management, and services targeted at high-level risks.
- *Integrated EHR_S*: It used across settings, it offers many advantages: data available to the clinicians, patient's data driven decision-making, contribution of patients to their own records, and services targeting based on a comprehensive view of risk factors.
- *Using digital algorithms in decision making*: The use of digital algorithms facilitates data analysis of the whole system offline, diagnoses patients and makes decisions. Some online prediction systems used in diagnostics and point of care devices. Thus, these algorithms offer several benefits in supporting decisions, transforming hospitals efficiency and primary care facilities using mobile applications.
- *Personalized treatment and precision medicine*: Using genomic and other personal data collected from many sources like wearables and patients' applications integrated with EHR_S allows to make personal health records (PHR_S) that facilitate providing personal treatment to the patient as well as identifying patients at risk.
- *full healthcare system based on BDA*: This system based on integrated real-time and real-world data sets and Artificial Intelligent are incorporated into all element

of planning and decision-making. It offers a lot of advantages, including: optimal treatments in most cases, portable drug delivery systems with continuous monitoring liberate patients from hospital beds, healthcare services focused on risk management and prevention, research outcomes integrated into health systems using all population data sets.

5.2 *Pillars of Transformation*

To ensure a successful transformation to a full healthcare system based on BDA, it is necessary to develop a strategy included the following pillars:

- Create databases in all the concerned organizations: These organizations must use the same system with the same unified quality standards and the same method of displaying results, in addition to ensuring data interoperability between them in order to maximize the benefit from the data [32];
- Ensure data security and governance, by supporting investment in cybersecurity and validation of technologies, in addition to establishing routes of legal liability for all healthcare services;
- Invest in the development of capacities and skills to provide high qualified personal in this field;
- Use data collected to enhance decision-making and improve healthcare services, because the purpose of collecting this data is to improve healthcare systems.

6 Conclusion

This research paper is considered as a reference aimed at aiding researcher understanding how Big Data Analytics (BDA) can leverage healthcare sector by facilitating access to services and improving their efficiency.

Previous studies were presented in Sect. 2 that included important contributions to the topic of BDA in healthcare, as the present study is similar in that contains various concepts related to BDA, its sources, its tools, and importance in reforming the healthcare system, while It differs in terms of its proposal solutions to overcome challenges, in addition to proposing the basic pillars to be adopted to transform into an effective healthcare system based on BDA. While the theorical framework was included (BDA concepts, its traditional and novel sources of data, its tools and role in healthcare) in Sect. 3.

We have identified a number of advantages detailed (improving patient care, real-time monitoring, predictive analysis of diseases and improving treatment methods, etc.), besides the most important challenges (security, privacy, human errors and integration of huge volume of data), a number of solutions have also been proposed to overcome previous challenges (using unique, consistent but pseudonymous patient identifiers, using internal or cloud-databases instead of one central data bank, using cryptographic techniques and enterprise-grade security measures, etc.) (Sect. 4).

In addition, we suggested five phases of gradual transformation into a healthcare system based on BDA (the current status of most healthcare system contained a combination of paper records and digital inputs, integrated EHR_S, using digital algorithms in decision making, personalized treatment and precision medicine, and full healthcare system based on BDA), as well as the pillars to be included in the transformation strategy (creation of databases in all the concerned organizations, insurance of data security and governance, investment in the development of capacities and skills, and usage of data to enhance decision-making and improve healthcare services) in Sect. 5.

Finally, lessons learned that new technologies in general, and especially BDA in healthcare can add more effectiveness, facilitate access to services, and improve their efficiency, and that will support healthcare practice in diagnosing injury in patients at an early stage, then provide treatment with a faster and more effective use, in addition ensuring patients' participation and providing self-care. This technology will contribute to reforming of the healthcare sector, enhancing the level of healthcare delivery as well as supporting decision-making. However, the application of this technology is accompanied by many challenges, including security and privacy. Therefore, effective measures must be taken to overcome them, such as using unique, consistent but pseudonymous patient identifiers, using internal or cloud-databases. Moreover, maximizing the benefits of BDA requires a gradual transformation of the healthcare system and needs an appropriate environment, and among its pillars are: create an unified databases in all the concerned organizations and ensure data security and governance.

References

1. Aceto, G.P.: Industry 4.0 and health: Internet of Things, Big Data, and cloud computing for Healthcare 4.0. J. Ind. Inf. Integr., 18, 100–129 (2020)
2. Palanisamy, V., Thirunavukarasu, R.: Asurvey on Internet of Things architectures. J. King Saud Univ. Comput. Inf. Sci. **31**, 415–425 (2019)
3. Mehta, N.P.: Transforming healthcare with big data analytics and artificial intelligence: a systematic mapping study. J. Biomed. Inform. **100**, 1–36 (2019)
4. Pashazadeh, A., Navimipour, N.J.: Big data handling mechanisms in the healthcare applications: a comprehensive and systematic literature review. J. Biomed. Inf. **82**, 47–62 (2018)
5. Andreu-Perez, J., Poon, C.C., Merrifield, R.D., Wong, S.T., Yang, G.Z.: Big data for health. IEEE J. Biomed. Health Inf. **19**(4), 1193–1208 (2015)
6. Ristevski, B., Chen, M.: Big Data Analytics in Medicine and Healthcare. J. Integr. Bioinf. **15**(3), 1–5 (2018)
7. Monteith, S., Glenn, T.: Automated decision-making and Big Data: concerns for people with mental illness. Curr. Psychiat. Rep. **18**(12), 112 (2016)
8. Martin-Sanchez, F., Verspoor, K.: Big data in medicine is driving big changes. Yearbook Med. Inf. **9**(1), 14–20 (2014)

9. Liu, X., Shen, D., Lai, L., Le Vine, S.: Optimizing the safety-efficiency balancing of automated vehicle car-following. Acc. Anal. Prev. **136**, 1–9 (2020)
10. Humber River Hospital Foundation: www.hrhfoundation.ca. Accessed 12 June 2020
11. Shameer, K., Johnson, K.W., Glicksberg, B.S., Dudley, J.T., Sengupta, P.P.: Machine learning in cardiovascular medicine: are we there yet? Heart (British Cardiac Society) **104**(14), 1156–1164 (2018)
12. Narasimhan, V.: Three things that will change medicine in 2018. World Economic Forum. www.weforum.org/agenda/2018/01/3-things-change-medicine-2018-big-data-healthcare. Accessed 20 April 2020
13. Bora, D.J.: Big Data analytics in healthcare: a critical analysis. In Big Data analytics for intelligent healthcare management. J. Biometr. Biostat. **7**(4), 43–58 (2019)
14. Belle, A., Thiagarajan, R., Soroushmehr, S.M., Navidi, F., Beard, D.A., Najarian, K.: Big Data analytics in healthcare. BioMed Res. Int., 1–16 (2015)
15. Karen.Y., He, D.: Big Data analytics for genomic medicine. Int. J. Mol. Sci. (IJMS) **18**(2), 1–18 (2017)
16. Peek, N., Holmes, J.H., Sun, J.: Technical challenges for big data in biomedicine and health: data sources, infrastructure, and analytics. Yearbook Med. Inf. **9**(1), 42–47 (2014)
17. Fillinger, S., de la Garza, L., Peltzer, A., Kohlbacher, O., Nahnsen, S.: Challenges of big data integration in the life sciences. Anal. Bioanal. Chem. **411**(26), 6791–6800 (2019)
18. Cheung, K.S., Leung, W.K., Seto, W.K.: Application of Big Data analysis in gastrointestinal research. World J. Gastroenterol. **25**(24), 2990–3008 (2019)
19. Liang, K.: Big Data science and its applications in health and medical research: challenges and opportunities. J. Biometr. Biostat. **7**(3), 1–9 (2016)
20. Janet, L.K.W.: The intersection of Big Data and the data life cycle: impact of data management. Int. J. Knowl. Eng. **3**(2), 32–36 (2017)
21. Natalija, et al.: Big Data in building efficiency: understanding of Bid Data and main challenges. Proc. Eng. **172**, 544–549 (2017)
22. Galetsi, P.K.: Big data analytics in health sector: theoretical framework, techniques and prospects. Int. J. Inf. Manag. **50**, 206–216 (2020)
23. Mehta, N.P.: Concurrence of big data analytics and healthcare: a systematic review. Int. J. Med. Inf. **114**, 57–65 (2018)
24. Pastorino, R., De Vito, C., Migliara, G., Glocker, K., Binenbaum, I., Ricciardi, W., Boccia, S.: Benefits and challenges of Big Data in healthcare: an overview of the European initiatives. Eur. J. Public Health **29**(Supplement_3), 23–27 (2019)
25. Rajkumar, et al.: Big Data: Principles and Paradigms, 1st edn. Morgan Kaufmann, USA (2016)
26. Vijayarani, S.S.: Research in Big Data: an overview. Inf. Eng. Int. J. (IEIJ) **4**(3), 1–20 (2016)
27. Loganthan, A.S.: A systematic approach to Big Data: exploration of the Hadoop framework. Int. J. Inf. Comput. Technol. **4**(9), 869–878 (2014)
28. Khatibi, T., Kheyrikoch, N., Sepehri, M.M.: Analysis of big data for prediction of provider-initiated preterm birth and spontaneous premature deliveries and ranking the predictive features. Arch. Gynecol. Obstet. **300**(6), 1565–1582 (2019)
29. Paik, S.H., Kim, D.J.: Smart healthcare systems and precision medicine. Adv. Exp. Med. Biol. **1192**, 263–279 (2019)
30. Wang, Y., Kung, L., Byrd, T.A.: Big data analytics: understanding its capabilities and potential benefits for healthcare organizations. Technol. Forecast. Soc. Chang. **126**, 3–13 (2018)
31. Colclough, G.E.A.: Harnessing Data Science and Artificial Intelligence in Healthcare: From Policy to Practice. Qatar Foundation, Qatar (2018)
32. Khennou, F.: Improving the use of Big Data analytics within electronic health records: a case study based open HER. Proc. Comput. Sci. **127**, 60–68 (2018)

Artificial Intelligence and Its Impact on Public Management and Decision-Making

Ahmad Yousef Areiqat and Ahmad Fathi Alheet

Abstract Artificial Intelligence (AI) is a high-speed technology that influences our everyday lives. It is traditional to mean an artificial development of an intelligence that enables the learning, planification, perception, or process of natural language to create vast and ethical and socio-economic opportunities. As AI is a technology activated by the Internet, the Internet Society recognizes that an Internet with which people can trust is important to the creation of an opportunity and challenge associated with AI. At the same time, in this dynamic area, AI or machine learning problems are more common and involve concerns when it comes to users' trust in the internet. This happens most frequently in goods and services. A variety of problems, including the socio-economic effects, concerns on openness, partiality, and accountability, new uses of data, health, ethical issues and how AI makes it easier to build new ecosystems, have to be addressed when it comes to dealing with AI. We plan to give policymakers and other players in the broader Internet ecosystem help in this chapter. Nonetheless, if it continues unabated some people find AI to be a threat to mankind. Others claim AI would create the possibility of mass unemployment, as opposed to past technological revolutions. This chapter describes the basics of the AI system, described key technology issues and challenges, examined the AI impacts on the performance of organizations and finally provided some guidelines for coping with this technology.

Keywords Artificial intelligence · Socio-economic challenges · Impacts · Transparency · Bias · Ecosystems ethical · Security and accountability

A. Y. Areiqat (✉) · A. F. Alheet
Al-Ahliyya Amman University, Amman, Jordan
e-mail: ahmadareiqat@ammanu.edu.jo

A. F. Alheet
e-mail: a.alheet@ammanu.edu.jo

A. Hamdan et al. (eds.), *The Fourth Industrial Revolution: Implementation of Artificial Intelligence for Growing Business Success*, Studies in Computational Intelligence 935,
https://doi.org/10.1007/978-3-030-62796-6_13

1 Introduction

In recent years the focus has been on artificial intelligence (AI), in the midst of rising Internet usage patterns that bring AI closer to our day-to-day lives. This puts AI at the center of various contemporary discussions. Industry investment in AI is rising rapidly, and policymakers are seeking to consider the consequences of technology for their countries.

Machine learning AI systems have already entered the realm of public management. It has taken many forms, including the use of catboats in public participation systems. Artificial intelligence is a technology that also influences the experiences of users and how the Internet impacts them. AI has the ability to fundamentally change the way people communicate with, through their working and other socio-economic structures, the digital world as well as with one another. While AI is gaining much attention these days, it is not as new as some would believe.

The area of IA is at least 25 years old but it rapidly spreads and enters government halls because of a convergence of technologies. Like the human brain, machine learning not only simulates how information and experiences are interpreted, but also learns from them. Add robotic technology to computer education and you have machines that can quickly substitute for routine human work. Some economists expect a big workplace upheaval by 2050. Others tend to build new jobs to offset most losses.

The Internet Society has developed a set of principles and guidelines to identify the various issues and to understand the different challenges with regard to what we believe are the main "functionalities" that sustain the interest of the Web. Although AI implementation is not new in internet infrastructure, the current move towards AI becomes more and more relevant in the future growth and usage of the internet. Such guiding principles and guidelines are therefore an initial attempt to direct the debate. These include: ethical concerns for implementation and design; ensuring the 'interpretability' of AI systems; motivating the consumer; responsible use of AI systems; ensuring accountability; and creating an economic and social climate created by the transparent participation of various stakeholders.

2 Definition of AI/Literature Review

"In computer science, artificial intelligence (AI), sometimes called machine intelligence, is intelligence demonstrated by machines, in contrast to the natural intelligence displayed by humans and animals. Leading AI textbooks define the field as the study of "intelligent agents": any device that perceives its environment and takes actions that maximize its chance of successfully achieving its goals, as quoted from Wikipedia. The field was founded on the assumption that human intelligence "can be so precisely described that a machine can be made to simulate it". This raises philosophical arguments about the nature of the mind and the ethics of creating artificial beings endowed with human-like intelligence.

In general use, the term "artificial intelligence" means a program which mimics human cognition. At least some of the things we associate with other minds, such as learning and problem solving can be done by computers, though not in the same way as we do. Kaplan et al. [4] define AI as a system's ability to correctly interpret external data, to learn from such data, and to use those leanings to achieve specific goals and tasks through flexible adaptation. In the twenty-first century, AI techniques have experienced a resurgence following concurrent advances in computer power, large amounts of data, and theoretical understanding; and AI techniques have become an essential part of the technology industry, helping to solve many challenging problems in computer science, software engineering and operations research.

Until recently, AI was primarily taught in computer science, electronics, robotics and computer programming schools. Those who want a degree in public policy or administration of public services are mostly excluded from any technical emphasis, let alone from AI instruction. Several researchers describe the potential of a computer program or Mache to think and understand artificial intelligence (AI). It is also a field of study which tries to make computers "smart". They work on their own without being encoded with commands. The name "artificial intelligence" was given in 1955 (John McCarthy, an American computer scientist and inventor, was recognized by Artificial Intelligence father). John McCarthy's principal contribution to AI was his development in the area of representation and reasoning for information, which in the last 50 years became the key subject of his work. He died in 2011. At present we use the term AI for successfully understanding human speech, competing at a high level in strategic game systems (such as Chess and Go), self-driving cars, and interpreting complex data. Nonetheless, some experts consider AI a danger to humanity if it continues to progress at its current pace.

An extreme goal of AI research is to create computer programs that can learn, solve problems, and think logically. In practice, however, most applications have picked on problems which computers can do well. Searching data bases and doing calculations are things computers do better than people. On the other hand, "perceiving its environment" in any real sense is way beyond present-day computing.

AI involves many different fields like computer science, mathematics, linguistics, psychology, neuroscience, and philosophy. Eventually researchers hope to create a "general artificial intelligence" which can solve many problems instead of focusing on just one. Researchers are also trying to create creative and emotional AI which can possibly empathize or create art. Many approaches and tools have been tried.

3 Types of AI

Referring to the management literature [4] classify artificial intelligence into three different types of AI systems: analytical, human-inspired, and humanized artificial intelligence, Analytical AI has only characteristics consistent with cognitive intelligence generating cognitive representation of the world and using learning based on past experience to inform future decisions. Human-inspired AI has elements from cognitive as well as emotional intelligence, understanding, in addition to cognitive elements, also human emotions considering them in their decision making. Humanized AI shows characteristics of all types of competencies (i.e., cognitive, emotional, and social intelligence), able to be self-conscious and self-aware in interactions with others.

However, some researchers argue that there are four types of artificial intelligence including:

3.1 Reactive Machines

The most basic types of AI systems are purely reactive, and have the ability neither to form memories nor to use past experiences to inform current decisions. The machine that directly perceives the environment and acts on what they see in this form of intelligence. It is not based on an internal world definition. AI researcher [2] argued in a highly significant paper that only machines like that should be designed. He was mostly because people are not really successful in the detailed programming of computer simulated environments, what is called a “representation” of the world in the AI grant.

Today’s intelligent machines do not have such a broad scope, or are rather constrained in terms of their specific tasks. These methods improve AI systems’ ability to play other games better, but they can not be modified or added easily in other circumstances. Such computerized imaginations do not have a wider world perspective—this means they do not function outside the basic tasks assigned to them.

3.2 Limited Theory

In this sort you can look in the past with computers. Cars that drive themselves are already doing some of this. We track the speed and direction of other vehicles, for instance. This can not be achieved in a moment alone, but involves detection and tracking of individual objects over time. Such findings are applied to the preprogrammed depictions of the environment by the self-driving vehicles, which also include traffic lanes, lights and other essential elements such as road curves. They will be used if the vehicle determines when lanes will be shifted and no driver is cut off or struck by a nearby car.

However, these basic bits of past knowledge are just transitory. They are not spared from the way human drivers accumulate experience over years behind the wheel as part of the car’s experience library. So, how are we able to create IA structures that construct full representations, recall experience and how to manage new situations?

3.3 Theory of Mind

We should pause and call this the essential difference between the machines that we have and the machines that we are going to create in the future. However, it is best to address in greater depth the forms and what they must be of representation machines. In the next, more advanced level, computers not only serve the earth, but also other agents or entities worldwide. This is called "mind theory" in psychology—the concept of the idea that humans, animals and objects may have thoughts and feelings that affect their own behaviour.

It is important because they allow us to have social experiences about how we human people developed societies. Working together is at best complicated, at worst impossible without knowing each other's motives and intentions and without taking into consideration what anyone else knows, either of me or of the community. If AI systems are to actually just walk between us, then each one of us has ideas, feelings and perceptions about how we are handled. And their actions will have to be changed accordingly.

3.4 Self-awareness

The final step in the development of AI consists of building structures which can reflect themselves. Ultimately, we IA researchers do not only have to consider understanding, but we will need to construct machines. To a way, this is an extension of the 'thought principle' of the artificial intelligences of type III. Consciousness is often for a reason called 'self-consciousness.' "The aware beings are aware of themselves, know about their inner thoughts, and can foresee other people's feelings." (I want that object "is a very different assertion to" I think I want that object.) We presume that someone behind us is frustrated or irritated in traffic because we feel like that if we honk with others. We do not make these kinds of inferences without a theory of mind. Although we might not be building self-confidence devices, we will concentrate on knowing memory, learning and decision-making based on previous experience. This is an essential step towards an objective view of human intelligence. And it's important if we want to build or grow machines that are more than excellent when classifying what they see.

4 Current Uses of AI

Although artificial intelligence evokes science fiction thought, artificial intelligence already has various applications today, including email filtering: email providers use artificial intelligence to filter inbound emails. By labeling emails as "spam," users can train their spam filters.

Personalization: online platforms will personalize your experience through artificial intelligence. In order to suggest content specific to you, services like Amazon and Netflix "know" from your past photos and purchases by other users.

Detection of fraud: banks are using artificial intelligence to determine whether your account is involved in a foreign crime. The algorithm may be distinguished by irregular behavior, such as international transactions.

Recognition of speech: artificial intelligence systems are used to automate voice recognition functions. Smart personal helpers, for example, are examples. "Alexa" or "Siri" from Amazon. The Internet Society agrees that the opportunities and challenges associated with AI are essential for building a trust network.

This is especially important as the Internet is a key to the technology behind AI and it is the main forum for its delivery. This policy paper offers an summary of the main areas of AI consideration, including a set of guiding principles and guidelines to help make sound policy decisions. Machine learning, a basic AI methodology and the driving force behind recent advances are of particular importance.

5 Machine Learning: Algorithms that Generate Algorithms

A series of instructions used to solve a problem are algorithms. Programmers have developed algorithms to train computers on new tasks and are the building blocks of today's advanced digital world. Based on specific rules and guidelines, computer algorithms organize vast quantities of data into knowledge and services. It is an important principle since algorithms—not computer programmers—build the rules of machine learning. This approach provides the machine with instructions to learn from data without the programmer needing to learn step-by-step instructions instead of programming each step in the process. This means that machines can be used for modern and complex tasks that can not be programmed manually. Issues like apps for visually impaired object recognition or converting images into speech. Not fresh is machine learning. Many of the learning algorithms, including neural networks, have been built on decades of study. AI and machine learning are currently rising in three significant fields:

Data availability: Online with an estimated 17 billion connected devices or sensors are just over 3 billion people. This creates a great deal of data that is readily available for use along with rising data storage costs. Machine training will use it to build new rules for increasingly complex tasks, as training data for algorithms.

Power to use: powerful computers and the capacity to link remote control through the Internet, allow the processing of enormous amounts of data in machine learning techniques.

Algorithmic innovation: new techniques of machine learning have driven new services, especially in layers of neural networks—also called "deep learning"—and also boost investments and research in other sectors.

As the technology progresses over time, it is likely that other innovative applications of AI-technology will be used within the public sector. Public organizations working with AI-technologies will be able to automate some repetitive tasks, freeing up time for civil servants to concentrate their time on more valuable tasks. In addition, the insights given by AI-technologies could support civil servants with relevant knowledge to improve the quality of public services.

In short, the use of Artificial Intelligence is able to provide numerous benefits to society, but could potentially lead to unintended consequences. The AI-Watch will monitor the current uptake and use of different kinds of AI-technologies within the public sector to better assess how governmental organizations are using these new technologies. The findings from the studies will be used for the creation of a roadmap for the development of AI within governmental organizations. This will include recommendations to further stimulate the valuable usage of AI-technologies in public services.

Many factors contribute to the challenges facing stakeholders in AI development, including socio-economic effects, with major socio-economic implications due to the new functions and services of AI. The ability of machines to demonstrate advanced cognisant skills in the encoding, learning, planning and interpretation of natural language enables smart systems to perform new tasks with greater efficiency than humans. New AI technologies could deliver exciting possibilities for more effective medical treatment, healthier products and services and increase massive productivity.

Transparency, bias and accountability: decisions taken by AI may have significant implications for lives of people. AI can discriminate against some people or make mistakes because of biased training data. How AI decides is also difficult to grasp and makes it much more difficult to address bias problems and to ensure transparency.

New data usages: machine learning algorithms have been demonstrated in vast numbers, usually called "big data," to be effective in analyzing and detecting patterns. Big data is used as instruction to boost their efficiency for learning algorithms. This leads to increased data demand, increases data collection and generates risks of information being exchanged at the expense of the privacy of users.

Safety and security: Advances in AI and its implementation will also pose new problems in protection and security. Those include the AI's erratic and damaging actions, but also malicious learning.

Ethics: AI can make choices that may not be ethical but which are also a logical outcome, demonstrating the importance of integrating the AI systems and algorithms in ethical considerations.

New ecosystems: AI makes new apps, technologies and new ways of communicating with the network possible, including the effects of mobile Internet. For instance, this will pose new challenges in the degree of Internet access in the near future through voice and smart agencies.

Taking decisions: clarity and "interpretability." It is important that we understand the decision of an AI agent when artificial intelligence performs tasks from auto driving to insurances. However, openness with respect to algorithmic decisions is often constrained by issues like corporate, state or technological confidentiality. The internal decision logic for the model isn't always understandable, even for the programmer, as machine learning complicates this further.

Automation could lead to higher productivity and cheaper goods for consumers. Furthermore, artificial intelligence can create new jobs or raise competition for certain existing jobs. However, it also means that all existing workers can be replaced in a decade or two. Some estimate that the number of jobs in the US may be as high as 47%. Increasing automation of unqualified and inexpensive jobs, however, would also impact highly qualified jobs, which rely heavily on cognitive routine tasks. This can lead to higher systemic unemployment, depending on the net impact.

The global allocation of labor will also be influenced by automation. Over the past few decades, production and services in some economic sectors have shifted, primarily as a result of relatively lower labor or material costs, from developed economies to emerging economies. The positive and negative impacts of AI and automation are not without their own problems on the labor markets and regional division.

Despite its complexity, the IA best governance approaches should be identified by all stakeholders, including governments, industry and users. All stakeholders will cooperate with market-based regulatory strategies to handle the economic and social effects of technology in the coming years. In addition, technology governance can not eliminate entirely AI's social impacts, but requires efforts to control the effects of technology.

6 What Is the Difference Between Artificial Intelligence and Machine Learning?

Machine Learning (ML) are two very common buzzwords and sometimes tend to be interchangeably used right now.

It is not exactly the same, but often the belief that it is can lead to misunderstanding. So I thought a work to illustrate the difference was worth writing.

The words Big Data, Analytics, and the broader waves of technological change that are sweeping across our planet come together very often.

In short, the best answer is that: Intelligence is the wider idea that machines can perform tasks in a way we see as "smart." Intelligence

Machine Learning is the new AI technology, focused on the premise that machines can only have access to knowledge and know for themselves.

7 Conclusion and Recommendations

At the moment, numerous governmental organizations have already implemented various AI-technologies in support of their processes. While new innovative applications are being developed and used every day. Artificial intelligence will change the entire government. In the next decade, policymakers and senior public managers must make decisions on the role of AI in society that are of great significance.

The literature indicates that there are enormous unanswered questions about and the ethics, prejudice and privacy as well as the legal consequences of such technologies.

Public administrators must consider the capabilities and weaknesses of AI, how and why they can go wrong, and what to do to get their benefits. It is time to focus more attention on the proper application of the AI in the area of public policy and administration.

Systems developers and constructors of AI will use a technology user-centric approach. They must take collective responsibility for building AI systems that pose no security risks to Internet users. In designing artificial materials researchers and industry should be motivated by values and standards of ethical considerations.

Innovation policies should be focused on adherence to ethical principles as a precondition for such matters as funding; decisions made by an AI officer should be understandable, particularly when those decisions are of public safety significance or lead to discriminatory practices, and AI systems should be built with a minimum requirement that designers can account for the actions of an AI agent. In addition to providing accurate information in the event of an emergency incident, certain devices have potentially important consequences for public safety.

Providers of AI services must be able to provide customers with the required clarity as to why a decision was made. While complete transparency in the machine learning techniques and training data of a service is usually not acceptable because of the safety risk, public information must also be accessible to allow people to query the information given.

There is a need to implement human checks on new decision-making strategies in AI system design, particularly where the risk to human life and safety is high and any implementation of an autonomous system should be thoroughly checked beforehand to ensure the AI agent's safe contact with its environment (digital or physical) and that it works as expected. Throughout operation, autonomous systems should be controlled and AI systems shall be responsible for data update or correction if necessary. You can just use and delete what you need ('data minimization') if it isn't necessary. They will encrypt and limit access to approved persons in transit and at rest ('access control'). In line with privacy and personal information laws and best practices, AI systems can collect, use, exchange and store only data.

Data which are biosecure, inexact, incomplete or deceptive should not be trained by AI systems. AI systems connected to the Internet should not only be shielded from malfunctioning or malware-infected AI systems for protection of the Internet, but also to protect them from the next wave of botnets. High standards should be implemented for protection of computers, systems and networks.

Security researchers should be able to verify the safety of AI systems safely without fear of prosecution or other legal proceedings. In addition, researchers and others who identify security vulnerabilities or other design defects must report their findings responsibly to those in the best position to address the issue.

The task of policymakers will ensure legal clarity as to how algorithmic decisions and use of autonomous systems can be implemented in the existing laws and policies to ensure a stable legal climate. It involves focusing on the detection of possible differences and legal frameworks with specialists from all disciplines. Likewise, those who design and use AI will comply with established legal frameworks.

Politicians must ensure that all legislation related to the use of AI systems puts consumer needs at the center of attention. This must include the ability of users to question independent decisions which affect their interests negatively.

Governments working with all parties concerned will now take some tough decisions about who is responsible in the event of a error in an AI program and on how to fix any harm suffered. Stakeholders will form an ecosystem in which AI offers all socio-economic opportunities. Both stakeholders will pursue an ongoing dialog in order to define the approaches needed to capture the enormous socio-economic opportunities of artificial intelligence for all and the their future negative effects. A debate, such as school reform, equal income and social care analysis, may discuss similar issues. Openstewardship is an important means of informing and engaging, in the secure implementation of the governance of AI, different stakeholders—civil society, government, private sector or academia and technical community.

Multi-stakeholder governance Promote an open, transparent, and inclusive approach, including organizations, institutions and processes relevant to AI management. It should be based on four main attributes: inclusivity and transparency, mutual accountability, efficient decision-making, and implementation and collaboration through interoperable and responsive governance.

References

1. Brooks, R.: Intelligence without representation. Artif. Intell. **47**(1–3), 139–159 (1991)
2. Brooks, R.: The Artificial Life Route to Artificial Intelligence: Building Embodied, Situated, 1st edn. Routledge (2018)
3. Kaplan, A., Haenlein, M.: Siri, Siri, in my hand: Who's the fairest in the land? On the interpretations, illustrations, and implications of artificial intelligence. Bus. Horiz. **62**(1), 15–25 (2019)
4. Kaplan, A., Haenlein, M., Tan, C., Zhang, P.: Artificial intelligence (AI) and management analytics. J. Manag. Anal. **6**(4), 341–343 (2019). https://doi.org/10.1080/23270012.2019.1699876
5. McCarthy, J.: Recursive functions of symbolic expressions and their computation by machine. Commun. ACM **3**(4), 184–195 (1960)

Emirates Leading Experience in Employing Artificial Intelligence

Fahad Khaled Alkhaldi and Suad Altaei

Abstract Artificial intelligence has become important today for countries and governments, but it is also an important part of the daily lives of individuals. At present, it is necessary for governments to introduce artificial intelligence in all their fields in order to be up to date with the rest of the world. Numerous studies have demonstrated that artificial intelligence has a prominent role in the renaissance of economics. Through this study, we will present the unique experience of the Emirates in activating artificial intelligence and how it has contributed to its economic rise. In this study, the researchers seek to present the Emirati experience in artificial intelligence and how the rest of the Gulf countries can benefit from it. The researchers used theoretical analysis methodology based on previous theoretical literature such as academic articles, scientific papers for conferences, government reports, and reports from international institutions that dealt with the subject of artificial intelligence from a different perspective. The researchers were able to view The UAE experience and the steps that were taken. It is very important for the Gulf countries to benefit from the Emirati experience and the speedy activation of the role of artificial intelligence in their government and other fields.

Keywords Artificial intelligence · Arab Gulf · GCC · United Arab Emirates · Smart cities · Knowledge based economy · Digital based economy

F. K. Alkhaldi (✉) · S. Altaei
Sustainable Development, College of Science, University of Bahrain, P.O. Box 32038, Sakhir, Bahrain
e-mail: fka.alkhaldi@gmail.com

S. Altaei
e-mail: suad.434@gmail.com

A. Hamdan et al. (eds.), *The Fourth Industrial Revolution: Implementation of Artificial Intelligence for Growing Business Success*, Studies in Computational Intelligence 935,
https://doi.org/10.1007/978-3-030-62796-6_14

1 Introduction

The fairy tales that we read about and see in science fiction movies have turned into reality because of the use of artificial intelligence (AI), and it has become an essential part of our daily lives. We did not arrive at our current level of AI in a short period of time, but it has been the build-up of accumulated and continuous hard work over many years, perhaps more than 200,000 years [1]. It should be noted that AI is not a new topic, but rather it has been around since the 1940s with the beginning of the "Turing test", an old term that was mentioned in 1956. Alan Turing wanted to define practical intelligence, and therefore created the Turing test. In addition, he believed that one day it would be possible to simulate human intelligence [2].

The field of AI is growing very quickly and today it affects the movements and decisions of individuals, societies, companies and government institutions. It is expected that in the year 2020 that AI revenues will reach nearly 50 billion dollars [3].

It is certain that AI is the promising future on which we build our hopes not only for the renaissance of economies, but also to achieve sustainable development goals (SDGs) [4].

The topic of AI asks many questions, where individuals want to know its impact on their daily lives, and employees want to know its effect on their jobs and income, while government and private institutions want to know the AI opportunities to help them get the best outcomes [3].

The aim of this paper is to shed light on the benefits to and experience of The United Arab Emirates from its scientific and technical development, where sources of economic income were diversified through activating AI initiatives, filling the void created by science and drawing attention to the Arab forerunner experience in this field.

Moreover, the researchers used theoretical analysis methodology based on previous theoretical literature such as academic articles, scientific papers for conferences and government reports, reports of international institutions that dealt with the subject of AI from a different perspective, where the researchers were able to view the Emirati experience and the steps that were taken so that future studies can benefit from it.

2 Artificial Intelligence

2.1 What Is Artificial Intelligence?

AI incorporates computing devices; machines or robots that perform tasks and functions as human beings. Being able to imitate human intelligence to achieve the greatest number of goals in many fields with better efficiency comparable to human competence, is what AI should deliver. Another definition of AI is a group of computer systems that operate with amazing capabilities that enable it to detect and perceive the actual environment and simulate it, before analyzing the actual situation and responding to it in innovative ways, which may be better than actual human responses [3, 5, 6].

From a production perspective, AI is an effective factor in increasing productivity, through which the workforce can be empowered and assisted to acquire many of the skills that the labor market requires. AI allows the creation of a virtual workforce, which is called "smart automation", which leads to the promotion of innovation and the renaissance of the economy. Many studies and research have proven that the most important goals that AI has achieved are enhancing national economic growth and increasing labor productivity with higher efficiency and at lower costs. Through AI, companies will be able to increase the workforce and in doing so increase their productivity. As for consumers, goods will be produced and more specialized services will be provided according to their specific needs and consequently increase the demand for the product [3, 4].

2.2 The Importance of Artificial Intelligence

It is said that AI can be a reason for realizing unlimited wealth capable of raising the economy to stages that no country has reached before [6]. AI has achieved impressive results in raising the economy and it has provided opportunities and innovations that boost the economy. AI will certainly have a profound impact on the future economy of many countries around the world. It has already had an effective role in the economies of China, Singapore, India, Japan, Taiwan, South Korea and Hong Kong [1, 7, 8].

Add to that, one of the most important uses of AI is innovation and, there is no doubt it will have an effective role in achieving the SDGs. Its innovations have contributed to solving several issues related to sustainable development [9].

On a personal level, AI has facilitated our modern way of life by providing us with productive answers and solutions as if they were from real human intelligence. In fact, it has become the perfect way in which we learn how to live, learn, communicate and achieve the best results from current opportunities. Moreover, AI has contributed to the improvement and development of several areas, most notably education, where one of the programs or tasks that have been developed for AI is literacy. Some studies have examined the possibility of AI raising the rate of literacy, especially in the Asia-Pacific region. Worldwide speaking, AI will have an effective role in generating new ideas that will help the growth of societies and countries, and these ideas will be based on scientific analysis and facts, and therefore have a positive. It is also worth noting that through AI, predictions are more reliable because they are undertaken by analyzing and measuring very specific data. AI predictions include tsunami forecasts, earthquakes, infrastructure maintenance periods [1, 6, 8].

2.3 *Artificial Intelligence Application Areas*

AI has attained achievements in several sectors, for instance, in the automotive industry, there are driverless cars, and in the areas of communications there are digital devices. As the employment of AI requires considerable investments in financial and human resources, economists believe that it is very important to take into account whether the benefits of AI exceed the costs incurred [8].

2.4 *Does Artificial Intelligence Lead to Unemployment?*

Regarding job losses, AI has, in fact, reduced productivity time and effort. Getting many tasks done takes less time, especially time spent paying bills, collecting accounts receivable and preparing the necessary reports. However, it does not necessarily mean unemployment. Instead, training and development must take place to cultivate alternate talents and teach new skills, also, staff rotation can be undertaken to implement other needed tasks [8].

There is no doubt that AI is a tremendous advancement and its effectiveness has been tested in its application in real life, and therefore many questions and doubts have surfaced about the possibility of replacing the human element categorically. Thus, while AI might eliminate the human factor in some areas, AI can provide solutions that can eradicate poverty [1]. In contrast to this, others believe that AI will open doors for the unemployed in various specializations such as programming, computer systems engineering, and many others, including water, energy, education, health, environment, security and space. Therefore, it should be noted that the educational institutions as a whole need to offer AI specialization to students, as it is the future of economic advancement [7].

Governments and companies should incorporate AI through modern devices or robots to develop the best practices and policies. It is very important to recognise that the use of AI does not necessarily mean dispensing the human element. AI offers predictions and statistics and generates better results that can help humans perform and make decisions more efficiently and effectively [1].

3 The United Arab of Emirates Economy and Their Experience Using AI

Before the discovery of oil, The United Arab Emirates' (UAE) economy (like other Gulf countries) depended on fishing, pearl trade, and agriculture [10]. Throughout history, the oil sector in The Emirates prospered, reaching 80% of its revenue which is used to finance the public sector. Thus, the country has become a rentier state. This applies to other Gulf countries as well. With the passage of time, the economy's dependence on oil has become unstable, as the oil sector has suffered from several problems, most notably the fluctuation in oil prices. However, The Emirates has initiated a process to diversify sources of income and made several investments in other non-oil sectors, which certainly will assist them in achieving its 2021 national agenda to transform into a smart country [11]. The UAE is classified as a wealthy country because it is an oil-rich country, and its population base is considered modest. The country is famous for the Emirati family companies, which have opened its doors to employ a large number of expats [12].

Today, The UAE has gained the respect and the appreciation from all countries around the world, as its leadership over the years, has effectively used its oil revenues, to invest in building the country and profiting its people. The Emirates focused its investments in education and in human resource development, which in turn contributed to building a modern state. The country has won the trust of well-known world leaders and leading world companies and has become the main destination for tourism and investment. Its leaders continue to strive to develop and improve the country's living standards and economic life. They promised in their 2021 vision to give the priority to providing the best standards of living and welfare to its citizens by providing only the best government services in education and health sectors as they know this contributes to social development. The education sector in The UAE has witnessed a remarkable development, as it has become a powerful educational station for high-quality universities with thousands of students from all over the world. This important educational development has strengthened the position of The Emirates globally and has had a prominent role in facilitating the process of importing and exporting knowledge. This has lead to achieving huge technological progress with smart cities and AI [10].

The UAE has recorded an impressive economic growth, and high figures in gross domestic product per capita which reached 30.4 thousand US dollars in 2008, from 18.5 thousand US dollars in 1990. Because of the successful experiences of some countries in the transition to a knowledge economy, (Singapore and the Republic of Korea, both achieved highly competitive positions) The Emirates has sought to transform its own economy into a knowledge-based economy, with its vision 2020 directed to promoting investments in technology, innovation and in R&D [13]. Comparing to the previous figure according to UAE's Federal Competitiveness and Statistics Authority the Gross Domestic Product per capita for 2018 shows that UAE grew more than 44 thousand US dollars [14].

What makes The UAE a developed and a leading country is innovation. Its government always seeks renewal and development from an innovative view. Therefore, The Emirati government services have become of a high quality, is innovative, and is based on competitive strategies and policies, which are not easy for other countries to implement or imitate. Innovation is achieved with the help of the human factor, and the employees of The Emirati government are highly motivated and they always pursue higher education, in addition to receiving ongoing training in the field of innovation to enhance their skills and capabilities. Moreover, the government is concerned with choosing creative leaders, who act as role models for the other employees and, leave a positive impact on the work environment, and therefore management encourages their employees to innovate and create opportunities and solutions that would increase work productivity efficiently and effectively. Of course, it cannot be denied that the private sector as well as entrepreneurs have a prominent role in the country's innovation process [15].

The government of The UAE work in a transparent and competitive way to reach the top in the world in delivering the best government services, and they have succeeded in transforming from providing regular government services, to electronic and finally have become a mobile-smart platform. Today, The Emirates sends a message to the whole world that it will precede everyone. AI impacts many UAE government services. In fact, in 2017 the government created a ministry under the name of AI, through which the foundation for solutions to all technological problems are laid to be applied in all government institutions. The most important desired result from the use of AI has been to reduce government's costs and expenditures to more than 50%, while at the same time improving governmental work and services provided to citizens. Therefore, the government has put into place the necessary strategies that aim to invest in the use of an integrated and smart digital system that makes AI applicable in all sectors and create a new market that will advance the country's economy [16].

4 How Gulf Countries Can Benefit from the UAE AI Experience?

The Emirates has vast experience in the field of diversifying their economic resources rather than relying on oil as a core economic resource [17]. Therefore, this part of the paper reviews the extent of the possibility of other Gulf countries benefiting from The Emirati experience in the field of AI. First, it is necessary to study the Emirates as a system, while highlighting its economic, social and political situation in order to understand it closely. Then only, can the possible benefits to other Arab Gulf states be determined.

From a political point of view, The UAE is part of the Gulf Cooperation Council, which includes Bahrain, Saudi Arabia, Kuwait, Oman and Qatar. In fact, all these countries are similar in their form of political system, as well as their social level. Furthermore, all Arab Gulf states share Arabic as their native language and the official religion is Islam [18].

Economically, like with the other Arab Gulf states, oil is considered as a major resource for the local economy in The UAE [19]. In the year 2014, The World Bank reported that due to fluctuating oil prices, Arab Gulf countries need to take austerity measures to reduce the potential deficit from the collapse of oil prices. However, the same report stated that The Emirates is one of the countries that is expected to not be affected much by oil prices [20].

To get better understand about the nature of GCC states and how these countries could benefit from UAE experience in the field of AI in particular. Authors are suggested that regional integration approach are applied to accept the assumption of the similarity of political, social, and economic natures [21].

the GCC states formed a natural sub regional community, which all have a similar political system, all were made prosperous by the oil trade; and all faced similar development and security challenges. In the other word, other GCC states, their political stability delicately balanced. Moreover, is applied neofunctionalism approach to argue that the GCC states are one sub regional era through these aspects.

so, based on the above economic, social, and political dynamics, one would assume that because all Arab Gulf states are similar in political, economic, and social formation, therefore, theoretically, other GCC states can benefit from the Emirates experience in the field of knowledge, technology, and AI.

Through this study of The UAE and the analysis of the steps taken to achieve its technical renaissance, in particular in the field of AI, it can be seen that these steps did not occur randomly but rather came through painstaking work and strategic plans that were developed with the aim of The Emirates becoming one of the leading countries in the region to keep up with international development. This was done by acquiring and maintaining knowledge as a major resource for the local economy. The UAE activated Knowledge Initiative (concerned with measuring the axis of knowledge economy in the Arab countries) and was prepared to convert from a traditional economy to a knowledge-based economy [13].

To ensure the optimal use of AI, governments must enable education and continuous training to prepare a conscious and an aware generation of the importance of the same [4]. To move forward to achieve its strategic goals, The UAE focused on training, education, higher education, effective governance and identifying potential challenges that it might face, because the Emirates realized that the world today is different from what it was in the past, and countries must keep up with global changes to avoid a decline in their level. AI forms the fourth stage of the output of the industrial revolution, and it includes vital sectors in the Emirates such as military, industrial, economic, medical and technical. The UAE has taken important steps to activate AI, the steps are [22]:

1. Educating the Emirati community through media awareness of the concept of artificial intelligence and its applications.
2. Creating awareness among leaders and officials in the state of the importance of artificial intelligence.
3. Creating working teams that include government managers to study opportunities and challenges to adapt their services by applying artificial intelligence.
4. Developing local scientific competencies.
5. Encouraging government employees to enroll in training courses related to Data science.
6. Launching university educational programs to keep up with the expected changes.
7. Establishing a research center in cooperation with the University of Dubai that serves the roads and transportation sector.

8. Adapting programs in educational institutions through virtual learning technology.
9. Launching the Silicon Park project in Dubai in 2014 as the first smart city.

• **Emirates artificial intelligence strategy**:
UAE government AI strategy was announced in October 2017, aiming to reach the top of the world in all fields, in line with the Emirati 2071 vision. As this strategy comes with a new phase of smart government and includes all services, sectors and the infrastructure of the country, the strategy aims to invest all the available energies, resources, human and material capabilities in a creative way in order to raise The UAE. This will be done through the promotion and support of the government and by registering AI, through which innovative and competitive work environments will be created and real achievements accelerated.

A new market that will boost the country's economy will also be set up. Moreover, the strategy will support the private sector by setting a strong research and development base that will increase productivity by investing in AI tools [23].

The sectors targeted in the strategy	**Transportation sector**
	Health Sector
	Space segment
	Renewable Energy Sector
	Water sector
	Technology Sector
	Education sector
	Environmental Sector
	Traffic sector

5 Discussion and Conclusion

Today, AI has become not only the focus of developed countries and governments, but it has also become an important part of the lives of many people. Reaching this level of AI is the result of hard and accumulated work over the years, as the first appearance of the term AI dates back to 1940s with the emergence of the "Turing test".

AI contributes to advancing the economy for developed and developing countries where income sources are distributed and a continuous search for sustainable economic resources takes place. It has been indicated in more than one study that AI will bring wealth to countries. In fact, AI has recorded large profits, estimated to reach nearly 50 million USD in 2020.

The use of AI has become inevitable in order to achieve development and sustainability of the economies for many countries, and it has also contributed significantly and tangibly in achieving the sustainable development goals where education is one of the most important goals of sustainable development, hence a program has been developed, through AI, to eradicate illiteracy and increase literacy around the world.

One of the comprehensive definitions of AI is being able to imitate human intelligence to realize the greatest number of goals of many fields with better efficiency comparable to human competence and, at a lower cost.

Numerous studies have demonstrated that through AI, work productivity has increased with better efficiency and at minimum cost. Moreover, AI contributes to emancipating employees from time consuming routine tasks which are now being completed at a faster pace. As a result, this has led to motivating employees to use their time in more creative and innovative ways. As a result, we now work in what is called: smart automation, which is an employee-stimulating working environment.

From here, we can answer an important question that many specialists have asked; whether the activation of the role of AI will lead to job losses and thus increase the number of unemployed. As we mentioned earlier, AI will contribute to the accomplishment of many tasks that were using up human effort and consuming a lot of time, and from here employees will be given the opportunity to take care of other organizational and developmental matters that contribute to achieving the strategic goals of governments and countries. In fact, the shift to the era of AI introduces many new job opportunities in areas which will deliver better results with the use of AI.

When we talk about The UAE economy, previously it was similar to the rest of the Arab Gulf states, where its economy depended on fishing, pearl trade and agriculture. After oil was discovered, The Emirati government invested oil revenues in promoting and developing the country in all sectors, including infrastructure, education, health, tourism, industry and many others. The most important sectors that the government has focused on are education, R&D and innovation, and this has contributed effectively to the country reaching the summit in competitiveness, as its economy shifts from an oil-based economy to a knowledge-based economy. Consequently, it has become an educational and economical hub for attracting international investments and knowledge.

The intelligence and wisdom of the Emirati government has positively reflected on the living standards of its citizens and residents, and the high ambition of the government has become a reality and is sought by all Emirati people. Therefore, the Emiratis strive to obtain continuous education and training to keep up with what is new around the world, and that is why the arrival of AI and its application in the UAE was not impossible for them. In fact, the country allocated sections specialized in AI and its application in all fields and areas.

The UAE turned out to be a smart and advanced country and it has become difficult for other countries to reach the high level that it has reached. However, that does not mean they should not try to enhance the spirit of competitiveness, raise the bar of ambition and strive to achieve what the Emirates has achieved.

The UAE government has initiated important steps to activate the use of AI, the most important of which are investments in technology, innovation and knowledge.

In addition, the government has directed to facilitate government and legal actions to ensure the safe and effective application of AI. The government has also encouraged the use of AI in all areas, including the military, health, financial and other important services. This is what must be followed by the rest of the Gulf countries so that they too can benefit from the Emirati AI strategy.

To conclude, to create and develop artificial intelligence capacity, governments have to develop strategic plans, activate programs, initiatives, and workshops in all government sectors and, provide training courses for government leaders and employees, to develop their technological capabilities and skills related to the area.

References

1. Goralski, M.A., Tan, T.K.: Artificial intelligence and sustainable development. Int. J. Manag. Educ. **18**(1), 100330 (2020)
2. Hassan, O.: Artificial intelligence, Neom and Saudi Arabia's economic diversification from oil and gas. Polit. Q. (2020)
3. PWC: The macroeconomic impact of AI (2018). https://www.pwc.co.uk/economic-services/assets/macroeconomic-impact-of-ai-technical-report-feb-18.pdf%0Awww.pwc.com
4. Purdy, M., Daugherty, P.: Why artificial intelligence is the future of growth. Accenture, 27 p (2016)
5. Marwala, T.: Impact of artificial intelligence on economic theory (2015). arXiv preprint arXiv:1509.01213
6. Aghion, P., Jones, B.F., Jones, C.I.: Artificial Intelligence and Economic Growth (No. w23928). National Bureau of Economic Research (2017)
7. Munoz, J.M., Naqvi, A.: Business Strategy in the Artificial Intelligence Economy. Business Expert Press (2018)
8. Haseeb, M., Mihardjo, L.W., Gill, A.R., Jermsittiparsert, K.: Economic impact of artificial intelligence: new look for the macroeconomic assessment in Asia-Pacific region. Int. J. Comput. Intell. Syst. **12**(2), 1295–1310 (2019)
9. Vinuesa, R., Azizpour, H., Leite, I., Balaam, M., Dignum, V., Domisch, S., Felländer, A., Langhans, S.D., Tegmark, M., Nerini, F.F.: The role of artificial intelligence in achieving the sustainable development goals. Nat. Commun. **11**(1), 1–10 (2020)
10. Zaabi, F.A., Awamleh, R.: Determinants of soft power: the case of United Arab Emirates. In: Future Governments, pp. 57–74. Emerald Publishing Limited (2019)
11. Warner, R., Moonesar, I.A.: Diversity management: the case of the United Arab Emirates. In: Diversity within Diversity Management. Emerald Publishing Limited (2019)
12. Youssef, M.A.E.A., Moustafa, E.: Societal institutions and control system characteristics: empirical evidence from the UAE. J. Islamic Acc. Bus. Res. (2015)
13. Ahmed, A., Alfaki, I.M.A.: Transforming the United Arab Emirates into a knowledge-based economy. World J. Sci. Technol. Sustain. Dev. (2013)
14. UAE's Federal Competitiveness and Statistics Authority: National Account Estimates (2018). https://fcsa.gov.ae/_layouts/15/download.aspx?SourceUrl=%2Fen-us%2FLists%2FD_StatisticsSubjectV2%2FAttachments%2F881%2FNational%20Accounts%202018%20-%20English.pdf
15. Aisha, Z., Saeed, H., Spraggon, M.: The role of Mohammed Bin Rashid School of Government in capacity building towards making Dubai a leading innovative city of the future—case-study. In: International Triple Helix Summit, pp. 13–19. Springer (2018, November)
16. Halaweh, M.: artificial intelligence government (Gov. 3.0): the UAE leading model. J. Artif. Intell. Res. **62**, 269–272 (2018)

17. Hamdan, A.: Economic diversification in UAE and its role in economic growth. Ajman J. Stud. Res. **16**(1), 21–61 (2017)
18. Al-Khouri, A.M.: The challenge of identity in a changing world: the case of GCC countries. In: Conference Proceedings: the 21st-century Gulf: The Challenge of Identity (2010)
19. Al-Muharrami, S., Matthews, K., Khabari, Y.: Market structure and competitive conditions in the Arab GCC banking system. J. Bank. Finan. **30**(12), 3487–3501 (2006). https://doi.org/10.1016/j.jbankfin.2006.01.006
20. Hussein, N.: The collapse of oil prices and its impacts on the GCC countries. Ministry of Economy - United Arab Emirates (2016)
21. Murden, S. W. (2009). The secondary institutions of the Middle Eastern regional interstate society. In International Society and the Middle East (pp. 117–139). Palgrave Macmillan, London
22. Majed, A., Alhashemi, N.: Artificial intelligence in the United Arab Emirates (2018). https://www.economy.gov.ae/EconomicalReports في تقرير الذكاء الاصطناعي 2018 الربع الأول-الخطة التشغيلية- المتحدة العربية الإمارات دولة.
23. UAE Government: The Official Portal of the UAE Government. UAE Strategy for Artificial Intelligence (2020). https://government.ae/en/about-the-uae/strategies-initiatives-and-awards/federal-governments-strategies-and-plans/uae-strategy-for-artificial-intelligence

Implementation of Artificial Intelligence in Education and Smart Universities

Smart University and Artificial Intelligence

Dileep Kumar Mohanachandran, Cheng Tat Yap, Zohr Ismaili, and Normala S. Govindarajo

Abstract The concept of smart university and Artificial Intelligence (AI) is gaining enormous attention, specifically during the covid19 impact phase, where the education institutions are searching digital platforms and Internet of Things (IoT) for academic continuities. The application AI is changing the operating models of several academic institutions, extending numerous possibilities for learning and development. Though AI provide several opportunities to academic institutions, it has it disadvantages also. Exploring the structure, system, process, strategy, infrastructure, and skills towards a smart university, this chapter has explored the concepts, tools, opportunities, advantages, and disadvantages of AI, with qualitative content analysis and narratives. The output of the chapter facilitates expansion of body of knowledge integrating contents of smart university, smart infrastructure, smart pedagogue, smart classrooms, smart teachers, smart learners, artificial intelligence, online education, artificial intelligence applications, role of AI in education, and AI applications for education, facilitating theoretical implications. Modern universities with integrated AI applications provide new vistas to students learning experience, teacher's instructional improvement and facilitating effortless academic administration.

Keywords Smart university · Artificial intelligence · Smart pedagogue · Smart education · Smart classroom · Smart infrastructure

D. K. Mohanachandran (✉)
Gopal Narayan Singh University, Jamuhar, India
e-mail: prof.dr.dil@gmail.com

C. T. Yap
Help University, Kuala Lumpur, Malaysia

Z. Ismaili
Mohamed V University, Rabat, Morocco

N. S. Govindarajo
Xiamen University, Sepang, Malaysia

A. Hamdan et al. (eds.), *The Fourth Industrial Revolution: Implementation of Artificial Intelligence for Growing Business Success*, Studies in Computational Intelligence 935,
https://doi.org/10.1007/978-3-030-62796-6_15

1 Introduction

In order to upkeep with the challenges of educational environment, universities are continuously evolving novel pattern of academic—infrastructure integration. With creative amalgamation of state-of-the-art concepts, effective software and hardware organizations, smart teaching space with advanced technologies and machine-driven platforms a new concept of university is emerging, which is termed as 'Smart University'. Making use of artificial intelligence—the technology based intelligent platforms—new generation universities are supporting scholars and educators in avail of better teaching learning experience. Artificial intelligence space support is integrated into academic learning environment with the purpose of constant diffusion of knowledge and cultivating level of ingenuity and increasing competence. Smart universities are integrating Smart Pedagogy oriented academic strategies making use of artificial intelligence-based learning analytics and academic analytics not only in engineering disciplines, but also social, arts and several interdisciplinary areas. Around the world university leadership is sharing about the role of artificial intelligence in learning and development process, with the support of Smart technology, Smart Classrooms, and Smart Pedagogy, making students smart and the teachers, smart educators. This chapter discusses the concepts of smart university, smart classrooms, smart infra structure, smart pedagogue, tools and equipment's artificial intelligence, and the role of smart leadership in leading modern universities.

2 Smart Education—Global Scenario

Scenario of education has been changed drastically with innovation and technology applications. From 1990s onward several educational institutions started smart education projects. Malaysia during 1997 started the Malaysian Smart School Implementation to meet the challenges of the twenty-first century [1]. To restructure the instruction system and infrastructures, South Korea had initiated the SMART education project [2]. Several Middle East Countries, especially UAE initiated smart learning program in 2012 to extend better learning experiences to the new generation students. Technology based schooling is widely applied in Singapore, with the initiation of Intelligent Nation Master plan since 2006. Quiet vividly, with the partnership from IBM Australia also followed the technology driven schooling patterns making the education multi-disciplinary and student oriented [Intelligent Nation, 2015]. With technology advanced learning culture New York' Smart School program also had its history to integrate smart classrooms. The directions and policies of Finland nonetheless focused into a systemic smart educational process with the support of pedagogical academic institutional relations (value network) to apply user driven and inspiring learning explanations that facilitate advancement of twenty-first century [3, 4]. Last but not the least, when look into the smart school scheme in Iran, which was initiated in 2004, with the goal of preparing institutions with computer hardware, software, network connectivity, and refining scholars' academic achievements [5].

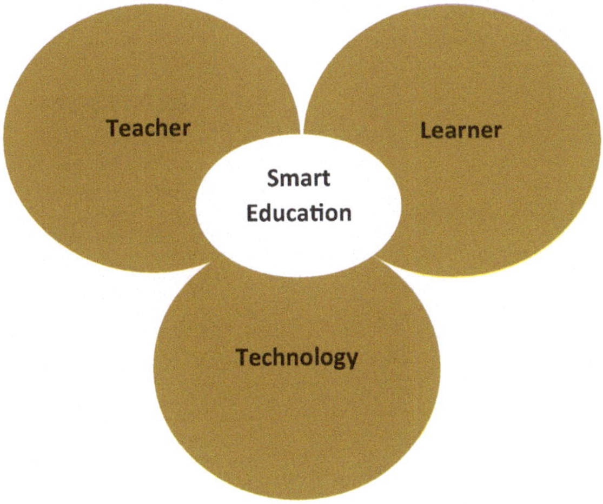

Fig. 1 Zhu et al.'s smart education framework [50]

2.1 *Smart Education Framework*

In the smart education framework, Zhu et al. [4] indicates that smart education is the byproduct of an intelligent environment where the teachers and learners are related to technology. Such intelligent system facilitates, customized education services and enable students to be as smart learners. The smart pedagogue will significantly influence the technology driven learning environment improves the learning goals to cultivate smart learners (Fig. 1).

2.2 *Smart Learning Environment*

Learning environment is a broader concept. Such environment includes several aspects like the learner's, teachers, or a learning system, learning establishment, staff service, technical experts, as well as the culture of the school room, curriculum, organization, and community. Several factors are incorporated into to the learning environment viz., Information, Task assistance, Student understanding, Context understanding and Contemplation and feedback.

Digital innovation has brought opportunity link all those factors in the academic culture since it encourages broader application of new indeterminate requirements of contemporary living. Digital innovation has bought opportunity for digital learning. A learning environment which make use of technology is termed as smart and which involves the change in instruction to deliver students with innovative knowledge and competences. The smart learning ecosystem for teaching and learning is digitally developed that allows access to various kinds of resources, gives shared tasks, and can be effortlessly tailored for effective learning and development (Fig. 2).

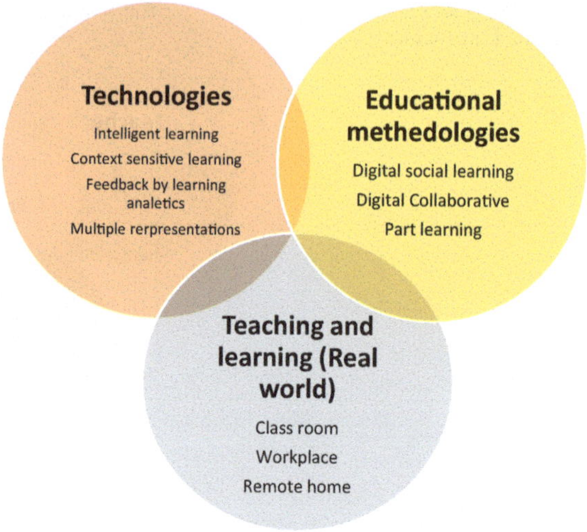

Fig. 2 Components of a smart learning environments [6]

Smart learning environment is a combination of application AI based technologies in the educational methodologies for effective teaching and learning. The framework of technology includes the integration of intelligence learning system, context driven, tailor made sensitive learning system, providing automated feedback for continuous improvement and it facilitate multiple representations. While the educational methodologies facilitate the collaborative, social, part, and digital learning, allowing the students continuous improvement. The last feature of this smart learning environment is the facilitation of learning at workplace, home environment and even with modern classrooms. Such systems are genuinely differentiated with the traditional educational system.

Contemporary schooling system is mainly depending on technology integrated, viz., AI enabled learning and development platforms. Instead of entirely depending on the instructor's time, automated tools and systems facilitate the students to access the tutorials, home assignments, feedbacks, and continuous improvement. The students are integrated with self-directed academic goals in a tech-driven learning management system, which primarily be dependent on visual and AI applied way of learning. The students are accessible to online based data bases, rather into a crowded library, with lot of distractions, flexible enough to create his own learning atmosphere that is most beneficial to their learning. In a smart learning environment system, the virtual school is not duty-bound by deep space or time restrictions. With An open system structural design AI based learning and development system provide better assistance in the assimilation of growing interfaces, smart machines, and various learning data. One of the major features of such smart learning environment is pedagogical network of learning institutes called "value network" that is the crux of AI integrated smart learning programs.

3 Difference Between Traditional and Smart Education

Traditional pattern of learning in schools and colleges are meeting the infrastructural requirement of a physical classroom using the tools of a blackboard, chalkboard, learning books, formal examinations, and home assignments. They must be intime in the classroom, attend the lecturers, and the education is designed for the group. The study mode will be teacher and student centered. However, online and AI based education is more with adaptive learning, depending up on the interest of the students, their timing and it is more customized learning and development. The students can access to the modules by sitting anywhere in the world and can read and retrieve any material related to their studies. They can attend the classes and there is automated clarifications and continuous feedback on their performance. Integration of AI platform thus enhances the teacher's competencies as well as students learning opportunities. Cloud based infrastructure and online education with the support of digital textbooks and educational contents encourage more and more regular and working professionals to avail of the learning opportunities. Expansion of time, education methods, educational contents, and space with the support of AI and technology applications this paved better learning and development opportunities to the world (Fig. 3).

4 What Is a Smart University

Several definitions are available to understand the concept of a 'Smart University"—which delivers a platform for initial data to drive the analysis and facilitate absolute quality enhancement in instruction and education environment. The concept of Smart University is clearly defined by [8] clearly involves an all-inclusive transformation of all educational progressions which integrate the smart classrooms, teachers, pedagogue, students, and technology. Tikhomirov and Dneprovskaya [9] offers a smart university knowledge which includes a wide-ranging apprise all instructional procedures as a part of information and technology applications, students, campus, and teachers which are unified to attain a comprehensive value of actions, outcomes of institutional research, commercial aspects and allied institutional proceedings and activities. Stressing the significance of technology integration in teaching and learning in academic institutions it is stated that the smart system is capable enough to recommend students to acquire in the real life by entering digital resources [10] (Fig. 4).

Making use of technology in every realm of academic administration is the principle behind smart university concept. With cutting-edge technologies and automated administrative plans, smart universities will execute learning and development, in addition to regular administrative work. Staffs belong to all hierarchies of administration thus would be able to engage them into work and realize targets, with the

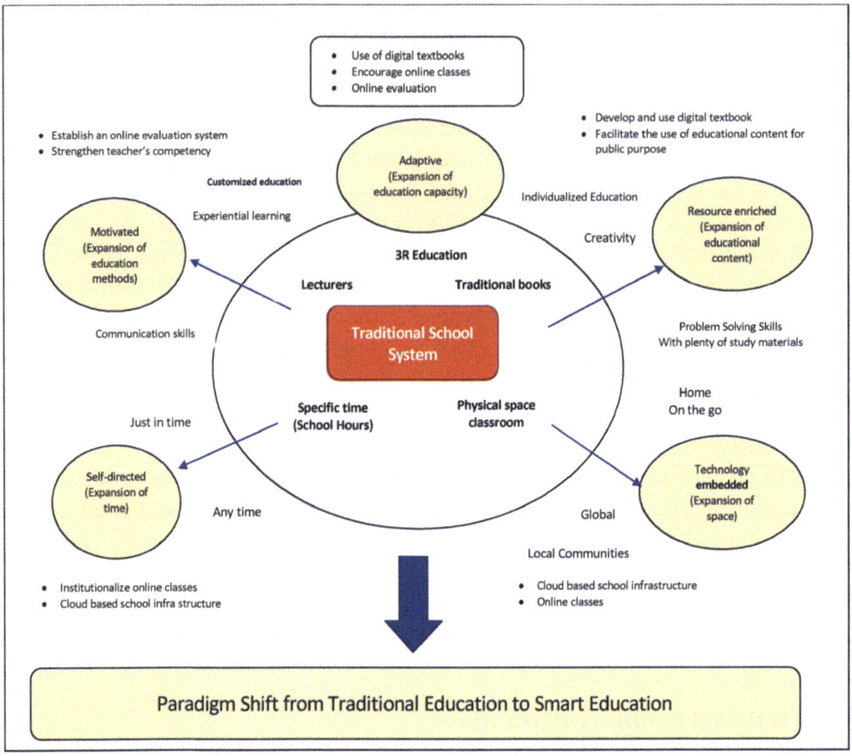

Fig. 3 Paradigm shift from *Source* [7]

support of automated processes. Such technology engagement will reduce the operating expenses when compare to the traditional work organization and execution [11]. The students, teachers and supportive staff members can make use of network and automatic tools in technology driven academic and administrative environment [36]. By revealing extremely sophisticated feature of technology, structure and processes in learning and development, smart university facilitate the usage of state-of-the-art software/hardware structures, use of innovative technical platforms and pioneering teaching and learning strategies [4]. Extending the importance of collaborative digital learning, Elmonem [13] defined smart university as an academic organization which uses smart systems technology that permits teachers to promote participation in growth and promotion of the excellence in campus instruction.

4.1 Characteristic Features of a Smart University

- Use of "Smart" features of smart systems.
- Use of smart instrumentation between the used structures.

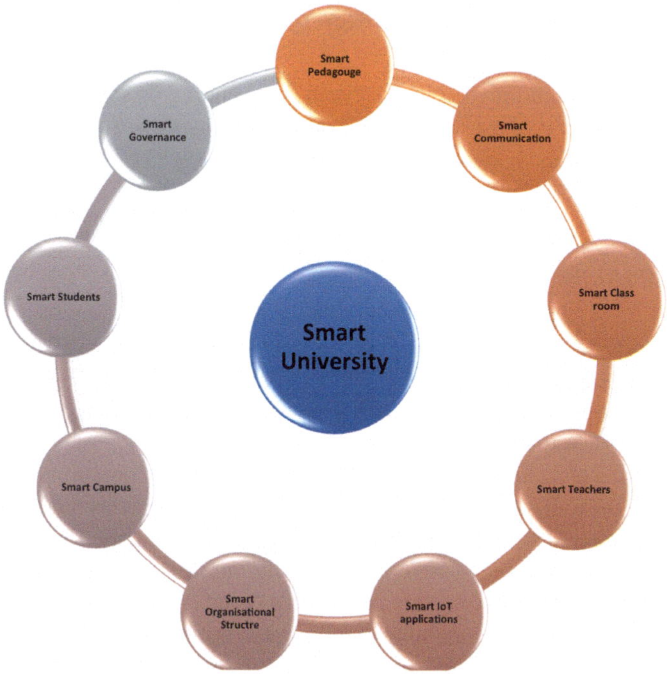

Fig. 4 Conceptual Model—Smart university [Own source]

- Innovation in educational processes.
- Attains and distributes foundational data.
- Technology enhanced learning system.
- Aimed to meet quality of academic operations.
- Using 'a technology enhanced learning system.
- Data accessibility from everywhere.
- Capable of access to digital resources.
- Automated administration plan with advance technologies.
- Automated teaching and learning strategies.
- Interactive digital learning environment.
- Stakeholders use of network and automatic tools.
- Reduced operation expenses.
- Emphasize, state-of-the-art software/hardware features.
- Permits progression of the excellence in university education.

4.2 Advantages and Disadvantage of Smart University

Whether to adopt smart university ecosystem is depending on several pros and cons associated with perceived use and perceive ease of user aspects. Some the advantages and disadvantages can be summarized as follows.

4.3 Advantages

- Some smart academic advantages can be summarized as follows:
- Traditional universities can widen the scope of the academia's day to day tasks and responsibilities without expanding the capabilities through communication and machine learning technologies.
- The platforms of technology and its effective integration allows highest efficiency and productivity in academic and administrative deliveries through continuous data processing, dissemination, and management of information.
- Smart universities which are equipped with machine learning technologies provides the possibility to broaden the awareness and appreciation of the programs by assimilating numerous learning methods at several intensities.
- Enable collaborative learning between students either internally or externally and provide the opportunity to work with students from other countries.
- Effective interaction between teachers, learners as well as management which offer stability and easiness in managing system and processes
- Can eliminate several problems associated with the traditional learning systems through technological empowerment.
- Replacement of traditional staff with smart staff which extend who can integrate their work with appropriate technology and intelligence.
- Sustainable academic deliveries and developmental possibilities through partnership with the community and consortium of universities [14].

Disadvantages. Though such advantages are well illustrated and widely appreciated, there are several drawbacks. Traditional universities which are willing to accept the smart university education and administrative intelligence system should investigate all such drawbacks.

- A total dependence on e-learning platform for academic deliveries
- Any breakdown of technology can disturb the smooth flow of teacher—learner learning process.
- the interaction between the learner and the outer sphere is limited to faculty members, learners, and associates on the Internet.
- Serious psychosomatic and social consequences if there are no alternative means of communication.
- Absence of the real identity of the learners can lead to numerous social challenges.

- Permanent renewal of the smart infrastructure is needed for the efficient assistance and development means of the scholastic operations, workforce preparation and interaction with all stakeholders [14].

5 Artificial Intelligence (AI)

AI is described as the competence and advancement of an information technology-based computer systems or other devices to accomplish the responsibilities that typically involve individual intelligence and plausible reasoning [15]. AI has come to be a vital component in a variety of fields of science and industry. It has set out to produce swift improvement in the academic sector. It has the greatest role in supporting students. Employed in the educational arena, artificial intelligence systems have effectively computerized fundamental and essential activities, as observed in appliances' capability to grade students' assignments and examination scripts etc. AI tools for learning functions is the development of what is currently described to as the 'smart classrooms' or 'intelligent classrooms.'

5.1 Smart Campus

An ecosystem whether physical or digital environment, where people are closely open, connected, and coordinated with the support of technology-enabled systems is called as smart campuses. Several factors are included in this detention viz., people, processes, services, technology which contribute the development an ecosystem where the staff, teachers, and stakeholders of a university is experienced with technology enabled learning and development activities. Radically changed from traditional human interaction system, Smart Campuses adapt the instant future with technology driven platforms and produce innovative patterns on human interactivity. The output of such technology driven human and administrative interaction is the generation of intelligent ecosystem, which support the mechanization of procedures, remote control, and decision making in their setting.

A university ecosystem is where lot many individuals study or work every day. While the impact of intelligence system driven by information and communication technology platforms improves educational institutions and the quality of life of the people around it [16]. A better control over the students learning and development as well as better interactional environment between the actors of the ecosystem is the advantage of ICT enabled university environment. This will charm the co-occurrence amongst the university inhabitants and its surrounds, effectively bring about the resources inside the university, and offers satisfactory places for learning.

In addition to making provision a safe, steady, environmentally friendly, energy-saving and technology driven ecosystem for teaching and learning, the IOT applications facilitates allied activities like scientific research, management, and campus life together.

Smart campus integrates the autonomous business systems and Artificial Intelligence (AI) with the support of data fusion, cloud computing, data mining and other information technologies, which is transforming the traditional pattern of human interactions to extremely sophisticated technology driven interactive experiences in learning and work. Major factors associated with the smart campus thus include infrastructure, operations, and people. These three factors to be integrated with proper intelligence system to make the ecosystem work in a unified way to use resources competently [17]. The smart campus is considered as a modern trend in academic realm which make use of intelligent system and process for effective coordination and implementation of academic administrative system [18], which are and are characterized by the ability to adapt and flexibility.

5.2 *Smart Infrastructure*

A smart infrastructure is characterized as an interrelated detecting system that delivers instantaneous digital data regarding the status of the system [19]. Based on this description, intelligent infrastructure is connoted as self-scrutinizing capability of the network and the system, integrating possibilities of tangible holdings and digital machineries [20, 21]. As it is detailed the data thus received from various sources would be extended to the stake holders through processing, storing and delivery, with adequate digital technologies and administration of infrastructure properties [22] CSIC, [23] describes smart infrastructure as 'the outcome of blending physical and digital infrastructure, delivering better knowledge, which enable the decision making, faster and cheaper' [24, 25].

A smart university infrastructure is linked to two main parts, viz., the hardware component and the software; The former takes in the form of smart buildings, sophisticated wire mesh system, smart gadgets, innovative labs, photographic camera, storage space, transmission systems, liquid-crystal display (lcd), and intelligent board. While the software is in the form of learning management systems, governing systems, safety measures and security systems, community networks, digital library, and automated pages [26, 27]. To perform the system effectively a good quality supportive infrastructure is an essential. The concept of Smart Infrastructure is further effectively defined by CSIC, [23] as the blend of physical and digital infrastructure offering enhanced knowledge to facilitate better decision making, quicker and less expensive.

5.3 Smart Pedagogic

An amalgamated teaching and learning with smart applications are termed as Smart pedagogy [28]. A pedagogue associated with a discipline which link theoretical concepts and practical learning methods and establishes the relationship between teaching and learning. Pedagogy exemplifies the association between instruction and studying. In general pedagogy symbolizes the components of subject matter, and instructional and technical knowledge [29]. The connection of the pedagogic elements offers unparalleled prospects for instruction and education. However, the new technology wave has perpetually transformed the role of teachers, must reconsider the changing aspects of the learning method to accelerate, organize, empower, and direct education into the digital age [30].

A smart instruction facilitates integration of digital platforms in learning and development, where AI will deliver smart instructional agents, and such digital platforms needs to ensure technologies effective integration with perceptions of actuality, intellect, and societal collaborations. A smart pedagogy for digital renovation, where AI will offer smart instructive drivers, requires thinking about how machine and tools affect perceptions of actuality, intellect, and societal collaborations. Converging on smart pedagogy provides the prospect to emphasize the significance of assimilating technical tools, technological development, and education arena into the syllabus [31]. Collaborative cloud is considered as one of the best examples of smart pedagogy.

5.4 Smart Classrooms

A classroom which is outfitted with supercomputers and audiovisual instruments which allow the instructor to use a variety of digital tools in learning and development as well as administration. Figure 5 provides an illustration of the smart classroom.

The idea of smart classrooms lays on the juncture of:

- interacting in the design and use of the learning space.
- layout and its ergonomic; and
- operational, unseen, rationalized and rigorous assimilation of technology.

The concept of smart classroom makes use of technology driven educational methodology, suitable to that space, which provides better satisfaction to teachers and learners in collective learning, project-based education and self-sufficiency, educational co-responsibility [32, 33]. Smart classrooms are mainly technically improved surroundings which are supposed be having the capability to enhance the prospect learners have to vigorously involve and contribute in the instruction and knowledge exercise, through technological tools such as: dedicated software's, super computers, assertive listing devices, audience reply know-how, interacting equipment's and audio-visual proficiencies. The idea of smart pedagogue

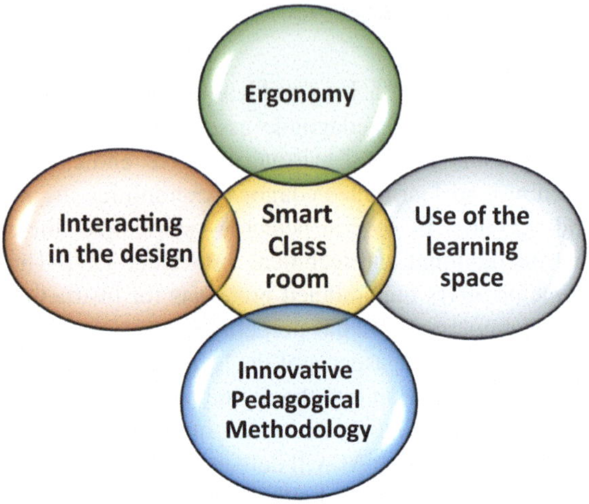

Fig. 5 Concept of smart classrooms

and effective learning and instruction productively occur in smart classrooms have deliberated among modern teaching and AI intellectuals [34–37].

5.5 Barriers to Smart Classroom Implementation

Several researches conducted in the past but emphasized the impediments for the effective incorporation of the technologies in classrooms [38–42].

- Not getting proper support from the institutional leadership.
- Did not make effective planning.
- Inadequate access to the information and communication technology.
- Firewall option created by the university.
- Educators' low-level of confidence in using technology.
- Opposition to shift to smart tools application.
- Absence of time.
- Absence of adequate technical assistance.
- Educators' skill, mindset and values, students' ability.
- Gain access to information and communication technology.

5.6 Advantages and Disadvantages of Smart Classroom

See Fig. 6.

ADVANTAGES	DISADVANTAGES
• Enhance teacher-learner collaboration and communication. • Real-time combined instruction and studying. • Greater conceptual clarity to the students. • Integrate teacher and learner in technology driven platforms. • Enriches visualization and innovation. • Offers improved classroom experiences to the students. • Fosters e-learning and computer-generated Classroom. • Enhances educational performance of students. • Virtual web-based knowledge acquisitions. • Students learn knowledge at specific phase. • Perceived ease of use and ease of access.	• Difficulty in making presentations, displaying videos and courses. • Costly and complicated to implement. • High upkeep expenses. • Lack of easiness in operating computers. digital boards, etc., which are delicate. • The classes dependent on electrical energy. • Necessitates appropriate network connectivity viz., LAN, WAN, Internet, etc.

Fig. 6 Advantages and disadvantages of Smart Classroom

5.7 Vital Technologies that Support the Vision of AI

AI is viewed as a key technology nowadays which is applied in many fields including high-quality education. The application of machine learning in instruction and learning is a form of customized education that extend each student a personalized learning experience. The scholars are getting personalized learning and with which they can follow own pace of doing the learning task and cake a decision on what to learn. While deep learning establishes an adaptable ecosystem with high dimensional state and activity areas. The algorithms of deep learning, which is in tune with the interaction and rewards are skillful of changing one's own result of activities based on pilot-based trial and error [49]. On the other hand, NLP supports the analysis of textual or linguistic data, such as documents or publications, make use of computational techniques, is providing insights from linguistics [50]. Finally, IPA application make use of user's voice, vision (images), and circumstantial data to offer support by resolving queries in natural language, making recommendations, and performing actions (Table 1).

Table 1 Technologies that support the vision of AI

Technologies	Purpose
Machine Learning (ML)	The purpose of ML for prediction. In such a learning structure, the device is trained, by analyzing information, to release a mission provided by a user [43]
Deep Learning (DL)	DL, a comparatively new technical know-how in the dominion of ML, includes the multifaceted effort to unravel perception and cognition of human beings. The speedy embracing of DL, where a person level of accuracy has been grasped with neural systems that utilize big data compilation [44]
Natural Language Processing (NLP)	Hirschberg and Manning [45] describe NLP as computational semantics, as a part of computer science which facilitate utilizing computational methods to understand, comprehend, and deliver human language matter. This require the consolidation of person's understanding and machine interpretation. The technique intends to understand individual-given oral or scripted commands that necessitate automated response, content interpretation and speech generation
Intelligent Personal Assistants (IPAs)	IPAs reassures hardware and software creators to focus on AI technologies such as natural language processing and machine learning. IPAs application has been prevailing on modern mobile phones since Siri's introduce in October 2011, Google Now's arrival in 2012 and Cortana's launch in 2013 [46]. Johnson et al. [47], in this context stated that educational representatives like STEVE (Soar Training Expert for Virtual Environments) and ADELE (Agent for Distance Learning: Light Edition) are "started to do a several type of jobs in astonishingly realistic ways", and these small prototype systems have quickly become practical (p. 31)

Source Goksel and Bozkurt [48]

5.8 AI Versus the Role of the Teacher

It is argued that the teacher's role is redefined with the augmentation of artificial intelligence. The burden of teachers in grading, coordinating forums, submission of assignments, access to library resources, taking sessions face to face standing in a classroom, etc. could largely take away with the intelligent systems/software. With the application of AI in education, the teachers can focus on constructing better relationships with pupils, understanding them, direct them in getting skills for appropriate employment, that will support young generation to develop as responsible human being. AI is on the quest of giving a aiding hand with preparing, customizing, envisioning, and enabling the learning process. Make use of Microsoft Azure and Machine Learning, Grade scope like technology drive teaching tools can support in arranging scholar educational activities or submissions, producing and distributing comprehensive analytics. AI associate can support educators to adapt education projects corresponding to various learner's needs. AI can effortlessly turn out to be a learner's and instructors' finest collaborator delivering customized education and support. But what AI cannot accomplish is to appear as a source of motivation for studying, unlike a human educator.

5.9 Artificial Intelligence and Online Education

Another use of artificial intelligence is widely observed in online education. The quest to expand the frontline of education and to bring in schooling unrestricted and inexpensive to all, regardless of the person's geopolitical locations or time zone is one of the main aims and purposes of Massive Open Online Courses' (MOOC) suppliers. AI tools transformed online learning in the twenty-first Century. Major academic institutions which has adopted such application include Harvard, Stanford, and Massachusetts Institute of Technology (MIT). One of the advantage of AI based online education is the application of cutting-edge text conversion and machine learning build deep-learning systems which can convert English discourses into the learner's local language. Machine learning algorithms know how to analogously be used with a course syllabus to identify fields of prejudice, difficulty, and vagueness for tighter evaluation by the teacher (Fig. 7).

5.10 AI Application in Education

1. Make Simpler Managerial Tasks
   One of the major functions of AI application is automate the mission of managerial responsibilities for teachers and academic establishments. The time teachers used to spend for marking exams, evaluation of homework, and offering useful replies to their learners will be reduced with the application of AI and the teachers can focus into quality time with students in their grooming. Administrative departments like admission, registration marketing, HR, etc. can make us of AI platform. AI is letting for mechanization of organization and dispensation of traditional mode of paperwork.
2. Smart Content
   Machines can provide digital content of comparable quality in comparison with any other means of delivery. Smart content in the form of video conferencing, video lectures etc. has already reached to the smart classroom setting. AI systems are making use of traditional syllabuses to generate customized textbooks, in the form of digitized, learning interfaces, which support academic grading. An easy navigation is extended with the support of chapters, flashcards, and functional tests. An AI cutting edge Netex Learning which allows teachers to produce

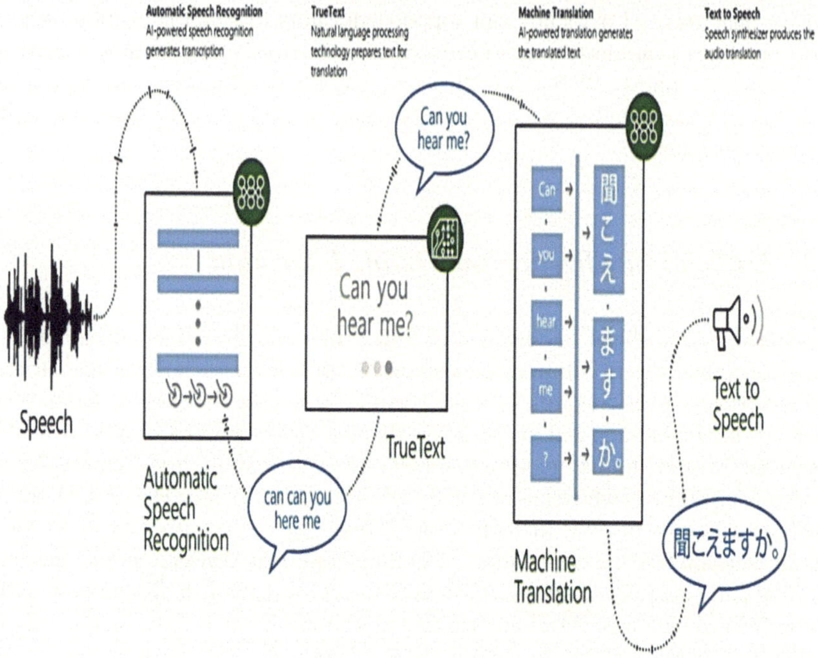

Fig. 6 AI-powered speech recognition and translation. *Source* Microsoft Education Blog

automated syllabuses and educative data throughout machines. Automated seat allocation, traffic management, manage movement between classes, manage classrooms temperature. humidity sensors, stock management etc. can be easily manageable with AI.

3. Personalized Learning
   AI facilitate better performance of teachers by offering personalized suggestions to every scholar. Major tasks like in-class assignments, final exams, instant and customized feedback to students etc. are properly integrated in AI. Smart teaching methods, like Carnegie Learning, can offer quick feedback and work directly with students. AI technology may be used to personalize educational content to suit individual learners. Such features of AI support learners with several scholastic needs. As it is evidenced, the new generation students started looking at all possibilities of AI driven academic experiences which are extending better academic experience suit their strengths.
4. Global Learning
   One of the advantages of technology integration into academic deliveries is the drastic transitions in enabling the learning of any program, though sitting anywhere and at any time. AI-powered learning prepares scholars with vital information technology skills. With more developments, there will be a broader

choice of programs accessible online and with the assistance of AI, pupils will be studying from wherever they like to study beyond locational boundaries. Not only does this advance effectiveness in more geographical places, but access to greater level of information will essentially assist in improving the AI's performance.

5. New efficiencies
   AI support and advances the processes and administrative effectiveness. The academic institutions can govern the suitable means of averting scholars from getting lost in crowds when they run in corridors. AI can likewise be utilized in the displaying of multifaceted data to empower the processes and procedure of department to generate data-driven predictions. AI lets appropriate preparation for the upcoming activities like allocating seats during institutional events or arranging food from local lunchrooms. The institutions henceforth can evade a lot of wastages instigated by over-ordering, which leads to saving expenses. Though, AI come with expenditures for machinery and instruction, nonetheless, the costs will come to be insignificant.

5.11 Roles for Artificial Intelligence in Education

Education is the sector where the artificial intelligence is swiftly making drastic changes. The following are roles for artificial intelligence which details the surprising educational experience of the future.

1. AI can automate fundamental actions in learning, like grading
   The usual complains of educators like 'marking brings up a significant amount of time', can be evaded with the support of AI applications. Automated tools are capable enough to grade all kinds of multiple choice and fill-in-the-blank testing. Though, essay-grading software is still in its early stages and yet it will allow the educators to extend attention more on in-session activities and scholar interaction than marking. AI advances adaptive learning opportunities.
2. AI can draw attention to areas where programs need to get better
   Some of the AI software's can be used to eliminate the gaps or any confusions in the lectures and educational materials extended by the teachers to the students. Coursera, a massive open online program supplier, is extending better support in resolving this issue. For example, when lot of scholars are given into the incorrect answer to an exercise or assignments, the AI approach notifies the instructor and provides potential learners a tailored letter that extends clues to the appropriate answer. This form of approach supports all students in developing better conceptual foundation. Scholars get instant advice that supports them to comprehend an educational content and recall way of doing the assignment correctly.
3. Scholars could get added assistance from AI teachers
   Education curriculum made on AI can support scholars through elementary mathematics, writing, and additional topics. Such curriculum can impart students' basics, though are not replacing the role of teachers who facilitate high order

thinking and creativity to the students through traditional classrooms. Nevertheless, in future, there will be every possibility of AI's role in inducing creativity and analytical skills.

4. AI-driven curriculum can give scholars and teachers helpful feedback
   AI not only help educators and scholars to customize the programs deliveries, but also provide response to the achievement of the whole course. AI systems are used to monitor scholar's advancement in learning and to inform professors when there will be a student performance issue. AI systems permit scholars to avail the support that they need and for instructors to identify the improvement areas for scholars.
5. AI is changing the pattern of interface with data
   Several interactive AI applications are existing which support the search and preference of people. These kinds of intelligent systems, for example Google familiarizes outcomes to users, by identifying the location; Amazon extends suggestions followed by prior consumptions; Siri acclimatizes to the requirements and orders enable the desirable selections and play a major part in interacting with data. Such intelligent systems also backing schools and academia as well, in acquiring, analyzing, and managing data for excellence.
6. AI is shifting the educator's role in academia
   The role of teachers turned to be more efficient with the application of AIs in assisting and supplementing lessons with practical experiences. In countless options, AI is pouring radical changes in the teaching and learning space, which has embraced the flipped classroom model.
7. AI can extend trial-and-error learning less unapproachable
   During learning situation, a trial and error part is very much critical. Students are finding difficulty in understanding the content of learning and failing to answer properly turned the traditional educational system paralyzed. An intelligent computer system could offer scholars options to experiment and learn in a comparatively decision-free atmosphere, particularly when AI instructors can provide answers for improvement.
8. AI can modify universities' prospects in find, teach, and support students
   Several changes are supported by intelligent computer systems in teaching and students learning interaction. This intelligent system is extending is extending better learning experience which are custom-made to scholar's desires and academic objectives. Data mining structures are performing an essential part in education scene.
9. AI is altering student's way of learning, who educates, and obtaining basic skills.
   Using AI systems, software, and support, scholars can learn from anyplace in the world at any time. Academic courses driven by AI are helping students to learn basic skills.

5.12 *Some Application Developed for Education*

Though several applications and its use has been incorporated into the Table 2. its customisation needs to be examined before the section.

Table 2 Application developed for education

Application	Features	Specialization
NUANCE	• A speech recognition software • The software facilitate transcription up to hundred and sixty words per minute. • Improves spelling aptitude and word identification. • Teachers may apply the software to dictate class session	Documentation and collaboration with email system
KNEWTON	• An adaptive education technology. • Program, called ALTA, assist in identification of knowledge, gaps in scholars. • Delivers pertinent academic assignments. • Put scholars on track	Adaptive learning - Managing and monitoring student's performance
COGNII	• Virtual education associate. • Employees interaction interface with technology. • Open format replies. • Improve critical-thinking skills. • Delivers real-time response, individual tutoring and tailored to separately to scholar's need	Virtual learning - Collaboration and communication hub
QUERIUM	• AI for STEM teaching. • AI provides instructors acumens into a scholar's knowledge behaviors and suggests areas for improvement	Assessment - Monitoring of students' performance

(continued)

Table 2 (continued)

Application	Features	Specialization
CENTURY TECH	• Utilizes cognitive neuroscience and data analytics. • Produce individual learning strategies. • Decrease work assignments for teachers • Following student development, recognizes knowledge gaps and suggests individual learning comment. • Decreases time spend on preparation, grading and handling home assignment	Cognitive learning - Automated process using analytics research
KIDSENSE	• Speech-to-text tool uses. • Students talking into text that facilitate note taking. • Practice vocabulary. • Practices firm algorithms to interpret exactly	Early learning - AI tools
CARNEGIE LEARNING	• Application of AI and machine learning • Assist scholars in evolving a profound theoretical knowledge of mathematics. • AI which studies a scholar's behaviors and tailormade the mathematics learning experience	Exploration - AI application specialize in mathematical
KIDAPTIVE	• Support academic institutes to gather information and enhance student involvement. • Used optimally to test scholars perceived strengths and weaknesses. • Foresees upcoming educational performance based on the fundamental outlines and relationships	Early learning - Adaptive learning platform with games

(continued)

6 Conclusion

Higher education has developed more accessibility with the integration of technology and artificial intelligence in teaching, learning and administration. Students, who are accessible to internet and technology prefer technology integrated learning in comparison with traditional classrooms learning. Smart classrooms with smart

Table 2 (continued)

Application	Features	Specialization
BLIPPAR	• Integrate intelligence technology and augmented reality. • Improve learning in the class space. • Carry collaborating resources amongst subjects like geography, biology, and physics to a graphic space. • Displays scholars a virtual 3-D model	Exploration - Augmented Reality (AR) Application
THINKSTER MATH	• A tutoring program integrate laptops, tablets, and desktops. • Integrate student's interaction with artificial intelligence. • Deliver scholars with custom learning opportunities. • Distinguish wrong and right answers	Assessment - AI learning and problem-solving tools
QUIZLET	• Online smart study resources. • Provides adaptive plans. • Apply machine learning and data from millions of studies sessions. • Select most pertinent education materials	Assessment - Online adaptive learning tools
OSMO	• Assimilates online and hands-on learning	Early learning—interactive game
I-READY	• Adaptive reading and math software	Adaptive learning - mathematical application
COUSERA	• For learning business analytics, graphic design, Python, etc. • AI and cloud engineering. • On-demand learning and development session	Online learning - Self-development learning hub
UDACITY	• Application of AI and Machine learning. • Real-time projects supporting through practice. • Other hands-on exercises that lead to real skills mastery	Online learning - Self-development learning hub with real-time projects
VIPKID	• Make use of translation services • Including machine learning	Language learning application

(continued)

Table 2 (continued)

Application	Features	Specialization
IBM WATSON EDUCATION	• By means of ai to expand learning outcomes. • Apply explanations that will help all students succeed. • Apply digital trends. • Provide educators the platform to assist students to perform	Content analytics - Personalized online tools
ALEKS	• Online artificially intelligent valuation and education system. • Assessment of student's level and integrating learning contents. • Re-evaluates the scholar to guarantee that themes learned are also retained	Learning analytics—Adaptive Online AI assessment tools
ST MATH	• Online visual educational program. • Influences the brain's inborn spatial-temporal cognitive ability to resolve mathematical puzzles, edifying feedback. • Build deep conceptual understanding. • Established repetitive outcomes	Learning analytics—Game based Online visual application solving mathematical concept
MANGAHIGH	• Support in excelling at math and coding. Adaptive puzzles, innovative games, and social competitions. • Real-time analytics with AI support	Learning analytics -Game based Analytics and AI in math and coding
TURN-IT-IN	• Plagiarism recognition result. • Succeed possible academic misbehavior by extending resemblances	Writing, feedback and scoring systems powered by machine learning
IBM-SPSS	• Cost-effective use of SPSS for teaching and learning purposes. • To meet the data science needs to students, faculty, and researchers	Process intelligence tools - Analytics reporting on large data

pedagogue, is extending unending opportunity to teachers and is further expending better students experience and feedbacks for their development. A smart university, with cutting-edge technologies and artificial intelligence applications thus changes the concept of old university model into centers of learning and academic excellence in higher education sector and such smart campuses today resemble smart cities of the future. This chapter thus provide detailed understanding on the opportunities of smart education under smart universities with the support of artificial intelligence application.

References

1. Annaswamy, A.M., Hussain, A., Chakrabortty, A., Cvetkovic, M.: Foundations of infrastructure Cps. In: Proceedings of American Control Conference. IEEE, Boston, MA (2016)
2. Balakrishna, C.: Enabling technologies for smart city services and applications.In: 6th International Conference On Next Generation Mobile Applications, Services And Technologies (NGMAST), Paris, France. IEEE (2012). https://doi.org/10.1094/pdis-11-11-0999-pdn
3. Bingimlas, K.: Barriers to the successful integration of ICT in teaching and learning environments: a review of the literature. Eurasia J. Math. Sci. Technol. Educ. **5**(3), 235–245 (2009)
4. Borawska-Kalbarczyk, K., Tołwińska, B., Korzeniecka-Bondar, A.: From smart teaching to smart learning in the fast-changing digital world. In: Daniela, L. (ed.) Didactics of Smart Pedagogy, pp. 23–40. Springer Nature, Cham (2019)
5. Boulton, A.: Applying data-driven learning to the web. In: Leńko-Szymańska, A., Boulton, A. (eds) Multiple Affordances of Language Corpora for Data-driven Learning, pp. 267–295. John Benjamins, Amsterdam (2015). https://doi.org/10.1075/scl.69.13bou
6. Brynjolfsson, E., Tom, M.: What Can machine learning do? Workforce implications. Science **358**(6370), 1530–1534 (2017)
7. Copeland, M.: What is the difference between artificial intelligence, machine learning, and deep learning? Nvidia (2016). https://blogs.nvidia.com/blog/2016/07/29/whats-difference-artificial-intelligencemachinelearning-deep-learning-ai/. Accessed 26 June 2020
8. CSIC (Cambridge Centre for Smart Infrastructure and Construction).: Smart Infrastructure: Getting More from Strategic Assets. CSIC, Cambridge, UK (2016). http://www-smartinfrastructure.eng.cam.ac.uk/files/the-smart-infrastructure-paper.Accessed 20 May 2020
9. Davison, M. (n.d).: AI and the classroom—machine learning in education. In: Sticca, J. (ed.) True Information Architect: Education (2020). https://www.trueinteraction.com/ai-and-the-classroom-machine-learning-in-education/. Accessed 20 May 2020
10. Dillon, T., Potdar, V., Singh, J., Talevski.: A cyber-physical system: providing Quality of Service (QoS) in a heterogeneous systems-of-systems environment. In: 5th IEEE International Conference on Digital Ecosystems and Technologies (IEEE DEST), pp. 330–335. IEEE, New York, USA (2011)
11. Diwan, P.: Artificial intelligence: Changing the education paradigm. Chatbotslife (2017). https://chatbotslife.com/artificial-intelligence-changing-the-educationparadigma26fac41dd5. Accessed 02 March 2020
12. Elmonem: Towards Smart University in the light of 21st Century Skills. Online document (2019). file:///C:/Users/V3670/Downloads/smartuniversnajah%20(2).pdf. Accessed 22 May 2020
13. Eric, E.: The Art of SEO: Mastering Search Engine Optimization—O'Reilly Media, Inc. Online Document (2017) file:///C:/Users/V3670/Downloads/11849-ArticleText-14753-1-1020191127.pdf

14. Gaffney, M.: Enhancing teachers' take-up of digital content: Factors and design principles in technology adoption (2010). http://www.thelearningfederation.edu.au/verve/_resources/Enhancing_Teacher_Takeup_of_Digital_Content_Report.PDF. Accessed 28 April 2020
15. Goksel, N., Bozkurt, A.: Artificial intelligence in education: current insights and future perspectives. In: Sisman-Ugur, S., Kurubacak, G. (eds.) Handbook of Research on Learning in the Age of Transhumanism, pp. 224–236. IGI Global, Hershey, PA (2019)
16. Gros, B.: The design of smart educational environments. Smart Learning. Environment **3**(15) (2016). https://doi.org/10.1186/s40561-016-0039-x
17. Gros, B.: The dialogue between emerging pedagogies and emerging technologies. In: The Future of Ubiquitous Learning, pp. 3–23. Springer, Berlin Heidelberg (2016)
18. Hagen.: The changing nature of participation in design: a practice-based study of social technologies in early design research. A thesis submitted for the degree of Doctor of Philosophy in Computing Sciences. University of Technology, Sydney (2011)
19. Hammer, M., Champy, J.: Reengineering the Corporation: A Manifesto for Business Revolution. Harper Business, New York (1993)
20. Han, Y., Chunyan, M., Cyril, L., Timothy, J.W.: Towards AI-powered personalization in MOOC learning. npj Sci. Learn. **2**(15) (2017). doi:https://doi.org/10.1038/s41539-017-0016
21. Hirschberg, J., Manning, C.D.: Advances in natural language processing. Science **349** (6245), 261–266 (2015). https://doi.org/10.1126/science.aaa8685
22. Hramiak, A., Boulton, H.: Escalating the use of Web 2.0 technology in secondary schools in the United Kingdom: barriers and enablers beyond teacher training. Electron. J. eLearn. **11**(2), 91–100 (2013)
23. Hwang, G.J.: Definition, framework, and research issues of smart learning environments—a contextaware ubiquitous learning perspective. Smart Learn. Environ. **1**(1), 4 (2014). https://doi.org/10.1186/s40561-014-0004-5
24. James, J.: Artificial intelligence & future warfare: implications for international security. Defense Secur. Anal (2019). https://doi.org/10.1080/14751798.2019.1600800
25. Kaimara, P., Deliyannis, I.: Why should I play this game? The role of motivation in Smart Pedagogy. In: Didactics of Smart Pedagogy, 2019. pp. 113–137. Springer, Cham. Kankaanranta, M., M€akel€a, T. Valuation of emerging learning solutions. In: Viteli, J., Leikomaa, M. (eds.) Proceedings of Edmedia 2014—World Conference On Educational Media and Technology, pp. 168–172). Association for the Advancement of Computing in Education, Tampere, Finland (2019)
26. Konnikova, M.: Will MOOCs be flukes? The New Yorker (2014). https://www.newyorker.com/science/maria-konnikova/moocs-failure-solutions2020/06/01
27. Kumar, K.A., Gupta, M.P.: How to make a Smart Campus - Smart Campus Programme in IIT Delhi. Technical Report IIT, Delhi (2015). file:///C:/Users/V3670/Downloads/SmartCampusReportV12.pdf
28. Kwok, L.F.: A vision for the development of i-campus. Smart Learn. Environ. **2**(2) (2015)
29. Marzano, G., Martinovs, A., Usca, S.: Mechatronics education: needs and challenges. Environment, technology, resources. Rezekne. In: Latvia Proceedings of the 12th International Scientific and Practical Conference, vol. II, pp. 214–217 (2019)
30. Microsoft Education Blog.: AI technology helps students who are deaf learn. Online document (2020). https://blogs.microsoft.com/ai/ai-powered-captioning/. Accessed 22 June 2020
31. Morimoto, R.: Estimating the benefits of effectively and proactively maintaining infrastructure with the innovative smart infrastructure sensor system. Socio-econ. Plann. Sci. **44**(4), 247–257 (2010)
32. Oblinger, D.: Bootstrapped learning proposer information pamphlet for broad agency announcement (2006). http://fs1.fbo.gov/EPSData/ODA/Synopses/4965/BAA0704/BLPIPfinal.pdf. Accessed 07 April 2006
33. Pelgrum, W.J., Doornekamp, B.G.: Indicators on ICT in primary and secondary education (2009). http://ec.europa.eu/education/moreinformation/doc/ictindicrep_en.pdf. Accessed 29 April 2020

34. Pishva, D., Nishantha, G.G.D.: Smart classrooms for distance education and their adoption to multiple classroom architecture. J. Netw. **3**, 54–64 (2008)
35. Potter: Artificial intelligence and swarming in maritime systems. Commentary. Marine Technol. Society J. **53**(5) (2020)
36. Radhakrishnan, T.K., Balasubramanian, S.: Modeling and control of multivariable process using intelligent techniques. Sens. Transducers **121**(10) (2010)
37. Riley, T.: Artificial intelligence goes deep to beat humans at poker. Sci. Mag. (2017). http://www.sciencemag.org/news/2017/03/artificial-intelligence-goes-deepbeat-humans-poker
38. Rudd, T., Morrison, J., Facer, K., Gifford, C.: Futurelab: What if…? Re-imagining learning spaces. HAL Telearn Archives (2006). https://telearn.archivesouvertes.fr/file/index/docid/190334/filename/rudd-2006-Learning_Spaces.pdf SAS. Artificial Intelligence for Marketing: Practical Applications. Wiley and SAS Business Series. Accessed 20 April 2020
39. Schroer, A.: Companies using AI in education to enhance the classroom. Bullettin.Com. Artif. Intell. (2018). https://builtin.com/artificialintelligence/ai-in-education. Accessed 15 May 2020
40. Shahzadi, I., Amin, S., Chaudhary, K.M.: Drivers of supply chain performance enhancing organizational output: an exploratory study for manufacturing sector. Facilit. Manag. **5**(14), 53–64 (2013)
41. Siau: How will technology shape learning.: education in the age of artificial intelligence. High. Educ. (2018). https://www.Researchgate.net/publication/324006126Education_in_the_Age_of_Artificial_Intelligence_How_Will_Technology_Shape_Learning. Accessed 17 May 2020
42. StrontBytes.: Best Artificial Intelligence Solution (2019). https://www.carnegielearning.com/. Accessed 30 April 2020
43. Taleb, Z., Hassanzadeh, F.: Toward smart school: a comparison between smart school and traditional school for mathematics learning. In: Proc.—Soc. Behav. Sci. **171**, 90–95 (2015)
44. Torres, S.J., Avariento, J.P., Rambla, D., et al.: Enhancing integrated indoor/outdoor mobility in a smart campus. Int. J. Geogr. Inf. Sci. **29**(11), 1–14 (2015)
45. Van Laar, E., van Deursen, A.J.M., van Dijk, J.A.G.M., de Haan, J.: The relation between 21st-century skills and digital skills: a systematic literature review. Comput. Hum. Behav. **72**, 577–588 (2017). https://doi.org/10.1016/j.chb.2017.03.010
46. Vander Ark.: The Shift to Digital Learning: 10 Benefits (2018). https://www.gettingsmart.com/2015/11/the-shift-to-digital-learning-10-benefits/. Accessed 30 May 2020
47. Wogu, I.A.P., Misra, S., Assibong, P.A., Apeh, A., Awogu-Maduagwu, E., Maskeliunas, R., Damasevicius, R.: Artificial Intelligence, MOOC's and the dilemma of teaching and learning in tertiary institutions: A critical appraisal. In: International Conference on Information Systems and Management Science (ISMS), University of Malta, 22–23 Feb 2018. http://www.proceedings.com/38672.html
48. Wogu, I.A.P., Olu-Owolabi, F.E., Assibong, P.A., Apeh, H.A., Agoha, B.C., Sholarin, M., Igbokwe, D.: Artificial intelligence, alienation and ontological problems of other minds: a critical investigation into the future of man and machines. In Proceedings of the IEEE & International Conference On Computing, Networking, and Informatics (ICCNI 2017), Covenant University, 29–31 Oct 2017. https://doi.org/10.1109/iccni.2017.8123792
49. Wogu, I.A.P., Olu-Owolabi, F.E., Atayero, A.A., Sholarin, M.A., Ugbogu, P.C., Ogbuehi, U.K., Akoleowo, V.O.: The changing face of education and the dilemma of Massive Open Online Courses (MOOCs) in Nigeria's tertiary institutions: Implications for development. In Conference Paper Presented at International Conference On African Development Issues (CU- ICADI) (2019). https://scholar.google.com/citations?user=J5h7gSwAAAAJ&hl=en
50. Zhu, Z.T., Yu, M.-H., Riezebos, P.: A research framework of smart education. Smart Learn. Environ. Springer **3**(4) (2016). https://doi.org/10.1186/s40561-016-0026-2

A Review on Smart Universities and Artificial Intelligence

Mohammad Al-Shoqran and Samer Shorman

Abstract Smart universities and artificial Intelligence play an important role in the educational operation and they are helpful for all educational operation components. Smart universities structure was studied in multiple research papers that studied artificial intelligence and how to apply it in smart universities. This study addresses critical issues on smart universities characteristics and components using artificial intelligence. Likewise used the content analysis to determine the strengths and advantages for smart universities. In addition to explain the role of artificial intelligence and smart universities in developing the educational process and its benefits. Smart universities have an expert system to improve the educational environment by deploy information and communication technology (ICT) and smart systems. Therefore, it is suggested that universities take serious steps to becoming smart universities because of its great benefits on all educational process components and to improve the education quality and to use of resources effectively. Moreover, to increase the use of artificial intelligence in teaching, learning, and scientific research and to embed it significantly in university curricula.

Keywords Smart University · Artificial intelligence · Teaching · Learning · Educational process

1 Introduction

Nowadays many universities around the world seek to turn into smart universities, this is because of many factors such as: the huge technological progress, The impact of artificial intelligence and its role in the development and improvement of education

M. Al-Shoqran (✉)
Department of Mathematical Sciences, Ahlia University, Manama, Kingdom of Bahrain
e-mail: malshoqran@ahlia.edu.bh

S. Shorman
Department of Computer Science, Applied Science University, Eker, Kingdom of Bahrain
e-mail: samer.shorman@asu.edu.bh

A. Hamdan et al. (eds.), *The Fourth Industrial Revolution: Implementation of Artificial Intelligence for Growing Business Success*, Studies in Computational Intelligence 935,
https://doi.org/10.1007/978-3-030-62796-6_16

and the diversity of all educational process components needs, in this period especially. Therefore, Universities are using smart technologies in their infrastructure to develop the educational operation and to be more effective. The smart university also seeks to raise the level of scientific production, reduce production costs, improve the quality of education, and help students to teach and learn under any conditions and at any place. [20] defined smart universities as "Smart education is a concept that involves a comprehensive modernization of all educational process" and he explained this definition in terms of education area that smart university includes the emergence of technologies such as smart boards, smart screens and wireless Internet access everywhere. [13] he explained the meaning of the smart campus by new term which is the intelligent campus by "A campus can be considered traditionally as a piece of land on which buildings of an educational establishment are constructed" and these smart universities contains many facilities such as libraries, classrooms, student centers, and recreation areas. and "intelligent campus can be interpreted that a campus which has the ability to response to new situations occurred on campus in its daily operation" [6] gave a definition of smart learning environment by "smart learning environment is able to offer instant and adaptive support to learners by immediate analyses of the needs of individual learners from different perspectives". The huge technological progress and the using of computers and many modern technical tools play an important role in the development of the educational process, there are many examples on that such as: using of smart boards, educational platforms, distance learning programs and various advanced infrastructure, which became a basic need for the transition from traditional education to digital education or transformation from universities Traditional to smart universities.

Therefore, decision-makers at the universities must take some decisions regards with transform their universities into smart universities gradually, which provides new forms of learning and teaching. Which leads to saves time, effort, and money in terms of accessing and producing the knowledge. In addition to the need to apply this transformation on the administrative aspects, university buildings and engineering designs, which makes them more suitable for this important transformation. Which positively effects on the educational process and development of societies and to service humanity in general.

This study present a review for ten articles related to smart universities, to highlight on the main method and techniques that are used and implemented in the current universities to be as smart, with advantages and disadvantages of those articles.

This study is organized as the following: the current section is introduction which contains the literature review. In the section two contains the research methodology to articulate main objectives and how to achieve them. In section three is smart university components; section four is characteristics of smart university. Section five to clarify the artificial intelligence definition and its features. Section six is smart universities systems and what are main systems should be included in smart university. In section seven to review of articles and discussion, finally is conclusion.

2 Research Methodology

Research methodology aim to clarify the method that conducted in this research to achieve the objectives. Include role of technological progress in serving the educational process and its role in smart universities. Demonstrate the mechanisms of transformation universities from traditional universities into smart universities. Articulate the benefits of Artificial Intelligence and Smart universities on all components of the educational process on the economic and computerized systems. This study is based on a descriptive analytical study based on content analysis of many previous studies related to smart universities and artificial intelligence. In addition to some graphical presentations and some associated interpretations.

3 Smart University Components

The components of the smart university are the main pillar on which the idea of smart universities is based. They are the elements that do its work smartly using artificial intelligence systems. These smart components must communicate with each other to exchange information and experiences to solve problems, to improve the work environment and to enhance the level of students in the scientific field. The following Fig. 1 shows the smart University Components based on [21].

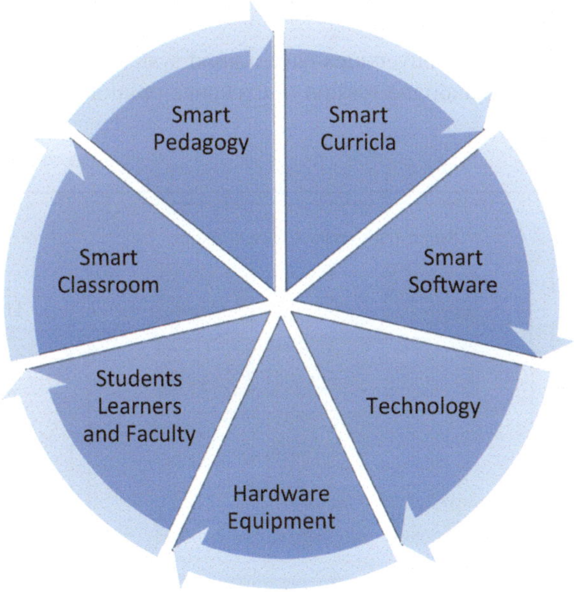

Fig. 1 Smart university components in [21]

4 Characteristics of Smart University

The smart university characteristics are discussed to spotlight on the differences between of smart universities and others (see Fig. 2) in [18].

Effectiveness: smart universities results or desirable outputs, preferably it is given a good expectation level comparing with tradition universities.

Efficiency: smart universities must have better performance in cost saving and speed of data exchanging.

Scalable: smart universities scale extent to be unlimited in campus, to cover outsider bounders.

Autonomous: smart universities able to be autonomous to deal with real situations with students and professors. It has ability to organize the learning plan to help learners to gain the proper knowledge in the proper time that reflect improvements on their outcomes.

Engaging: smart universities able to motivate the learners to participate and interactive with smart universities systems better than other.

Flexible: smart universities have a high-level response to new changes or updates, deal with new resources, courses, and objectives.

Adaptive: smart universities capable to deal with learners in terms of their needs, interests, learning model for each of them.

Personalized: Smart universities are leaders to consider the individual differences between learners, while a student who suffers from difficulties in educational attainment needs support exercises and special duties (assignments). Moreover, the distinguished and progressing rapidly student needs to increase his experience and academic level.

Conversational: Smart universities have a good system and facilitates to make learners create a group dialogue session to exchange opinions and information on relevant subjects.

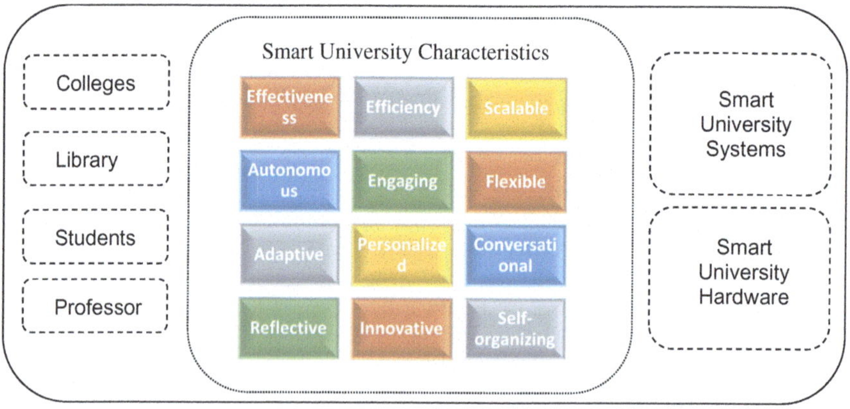

Fig. 2 Characteristics of Smart University and its environment

Reflective: Smart universities able to produce and conduct a self-assessment using its systems, to evaluate the student progress and employees and suggest plan to enhance the effectiveness.

Innovative: Smart universities keep abreast of scientific and technological progress and creativity.

Self-Organizing: Smart universities must rearrange and control overall resources to improve the productivity of students and employees in varied aspects [18].

5 Artificial Intelligence

In 1956 John McCarthy defined the artificial intelligence (AI) as "the science and engineering of making intelligent machines". And it also defined as a branch of computer science which deals with the study and design of intelligent agents that perceives its environment and takes actions which maximize its chances of success. Artificial Intelligence has many applications in the fields of research such as: the using of it in designing and the mechanical elements on basis of size limitations, in electronics and electrochemical systems, to diagnose the software development process, to identify chemical compound structures and chemical compounds and in medical diagnosis. This in [9].

Artificial Intelligence started to be an integral part of various environments such as workspace, education, health, military, etc. AI plays an important role to assist and enhance our lifestyle, beside to do hard work and danger work of it. AI has been proven through many proofs its ability to implements the work in formal and professional manner [11].

Artificial Intelligence has various characteristic and components, those properties make the AI systems being able to work in different environments. Each of AI application or machine has features and some of common characteristic and components such as data collection, data processing, create new prediction rules, decision-making and exchange the data with its environment or other devices. There are diverse of techniques that apply and based on AI such as machine learning, deep learning, Internet of Things (IOT) [17] etc. Most of AI used a multiple algorithm such as genetic algorithm, neural networks [16] Such as to find an optimal solution for different problems which contributed effectively to many fields such as industry, education, health and more.

Artificial Intelligence based on software and hardware components at the same time, each smart hardware managed by smart software, those systems that called an expert systems which means a real life simulation for human experts in particular field see [5, 12].

6 Smart Universities Systems

By rise of smart building, smart cities, and smart classroom, Down to an integrated process of digital interconnection between parts of smart campus. A vision of smart campus to digitally interconnected learning buildings and public facilities such as

Colleges, classrooms, library, parks, etc. Digitally interconnected means send and receive messages between all smart campus parts, through exchange the data and information. Furthermore, the universities need to enhance their services to keep abreast of scientific developments in university ranking and students' levels. Most of scientific developments need to collect a good information from real environment, this idea will be implemented in smart campus [1].

To create a smart campus with high efficiency and reliability, should contains a sufficient communications infrastructure able to serve the requirement to follow the progress in education process, such as The smart grid which means a good electric power grid infrastructure monitored via automated system, energy management, special power converters, and network management and availability for a long time. This is in [8]. Moreover, Smart universities relied on robust systems that execute and manage the transformation to be smart university. Some of universities started to use the systems to learn the students, such as University of Southern California (USC) which has a project called Designing a Personal Assistant for Life-Long Learning (PAL3) to observes techniques that used in classrooms in [19].

Some universities must have varied of computerized systems to manage the integration and control the transactions and communications between the smart campus parts, in the following section will show what the expectations from smart universities systems are.

Follow-up the students on campus.

Managing the relationship between the student and the lecturer. Follow-up of the delivery of duties on time, automated corrections for assessments.

Mentoring of discussions related to lessons.

Analysing the information using artificial intelligence algorithms to discover hidden relationships.

Assess and predicting student grades.

Take decisions in a timely manner to address problems before they occur, determine the extent of students' strengths and weaknesses by analysing data.

Propose mechanisms to improve student learning, individually or in groups.

Communicate with students or teachers when needed.

This trend prompts serious questions related determine the roles between teachers and students as main players in educational process. How does the AI effect or change teacher and student's role? And how to create a standard to achieve balance between them.

7 Review of Articles Discussion

Illustrates some articles that are discussed some topics that are related to artificial intelligence and smart universities (see Table 1).

8 Conclusion

Smart universities have an integrated system to manage and process data flow, controlling, monitoring and decision-making systems, etc. Smart universities using an Artificial Intelligence to enhance all educational process and students' achievements. This research has concluded some findings, based on the content analysis of some articles that are related to Artificial Intelligence and Smart universities. Smart universities have significant influence in improving the educational process, therefore, it is suggested that universities take serious steps and show the ways to becoming smart universities based on Artificial Intelligence techniques that are the future. Because of its benefits on educational process and to improve the education quality and to use of resources effectively. It is necessary to encourage universities around the world to increase the using of artificial intelligence in their teaching, learning, and scientific research systems and to embed it significantly in university curricula. Finally, based on these results, we expect that there will be a significant change in teaching methods and it will depend on the use of smart and expert systems (Artificial Intelligence) and universities will gradually transform into smart universities, which will make reaching knowledge easier. The future work will focus on the best practice on smart universities and how implemented.

Table 1 Literature review of Artificial Intelligence and Smart universities

Articles	Advantages/strengths	Disadvantages/weaknesses
[3]	This paper discussed an approach that plays a role in designing and developing of smart universities and smart classrooms. The paper presents the outcomes of research project aims to analyze how disable students can get benefit from using smart software and hardware systems, identify the characteristics of students with various types of disabilities. And to identify what are the smart technologies to help the disable students. The paper gave classification of levels of "smartness" of a smart system. It discussed Impact of SMART boards on academic difficulties. In addition to discussion of Technologies that help disable students in using of smart technology regards with language disabilities, students who are visually impaired or blind, hearing impaired or deaf. The paper concluded that there is a significant impact of using smart technology on the students of disability	A shortage in research projects that are studying the impact of using smart technology on the desirable students. Some disable students' needs some special devices in their studying. Therefore, there is a need to offering theses special devices
[21]	This paper presented the outcomes of a continuous research Project, the main goal of it is to find a taxonomy, features and components of smart universities that exceeds those in a traditional university. The authors vision depends on that a smart system must implement and demonstrate high significant at many "smartness" levels, including adaptation, sensing, inferring self-learning, anticipation, and self-organization The paper concluded that there is a need to find a taxonomy of a smart university. Identification of the main components of smart universities	Some rustles were given based on the authors vision, without convenience evidence like doing surveys or using some clear scientific research tools
[20]	This paper proposed the system of key features of smart education as a possible way of finding a well-rounded paradigm of smart education The article was done based on conceptual analysis of smart, Education as a research method. The theoretical framework depended on outcomes-based education approach, instructional design aspects, e-learning theories, socio-cognitive paradigm. The paper denoted the most important dimensions of smart education which are educational outcomes, ICT, and organizational dimensions. In addition to the Key features of smart education and how implement them. The outcomes of educational dimension include varies types of cognitive skills such as: different learning aspects and critical thinking	The authors stated that ICT is not the main dimension of smart education component. However, they mentioned that there is another approach concerning ICT as the most important of smart education component The authors think that there is a probability of facing some problems in the implement of smart education components in real educational projects

(continued)

Table 1 (continued)

Articles	Advantages/strengths	Disadvantages/weaknesses
[4]	This paper discussed university guidance counsellors' activities, providing an appropriate environment that helps first-year students to succeed. The authors gave a description of some innovative smart technology's applications in the guidance counselors activities. The study stated that the main goals of student guidance counselling is to demonstrate the role of university facilities in helping students to succeed and to develop their personal growth and professional development. The study concluded that there is an effect between effectiveness of active learning methods and electronic educational resources. Also, there is an effect of collaboration between teachers, psychologists, and students on university guidance counseling model	This university guidance is theoretical approach without real life application
[6]	This article conducted a survey to find the possibility of using smart devices in the university education. This study discussed two main factors; the first one is student learning outside of classroom concerning the suitable time and location. the second factor is using of smartphone in student learning regardless of anxiety of Internet use such as computer virus, Leakage of personal information, Tampering or eavesdropping on personal data, etc. This study used a survey contains 18 questions which concentrated on several parts such as: Anxiety of Internet use, using situation of Smartphone to items related to lesson, Effectiveness of using smartphone for various learning, Using situation of access tool. The result of this study approves that there is no effect of the Time zone or place on learning outside of classroom. Moreover, there is no effect of the anxiety about the use of the Internet on using smart phones for students learning outside of classroom and activities	Network connectivity is important to be active and secure Some skills for teachers and student should be considered in the training sessions to control of learning activities and the assessment of educational effectiveness

(continued)

Table 1 (continued)

Articles	Advantages/strengths	Disadvantages/weaknesses
[7]	This article discussed the Smart education features as an example of future Smart cities. The author believes that smart education is the education future and it is one of the main factors of Smart cities. The article said that smart education development must include Using different educational techniques, creating a new technological environment that allows access to the Internet of Things and Providing a new learning environment for all ages Moreover, the integration of the student model in the smart learning environment supports educational activities within smart cities. The author presented some smart learning environments models such as Hwang Framework of a smart learning environment. The model put student in the center; it contains four levels, which are learning experience, support technology, learning scripts and basic principles of teaching and learning. Active student model is very important to quality implementation. The paper concluded that there is a need to transform the traditional learning environment into a smart learning environment., it concluded the changing in ICT infrastructure effect on educational approaches and educational technology. Therefore, the development of Smart Education environment is very important to the development of Smart cities. So that today's education systems and its institutions, with their current capabilities, are not sufficient for the future requirements of smart education	It would have been better if the study presented a new vision of how to design the active student model
[14]	This investigation focused on smart parking, smart rooms and how to deliver smart education to students. Moreover, smart classrooms and campuses shaping using IOT applications. This paper proposed a model of developing a university smart campus via IoT technology. The model defines a wide range of Standards to make universities smart and urges a new interaction between things and people. The paper discussed the higher education and internet-of-things (IOT) by clarifying the idea and giving a solution to some problems related to it. For example, IOT enable universities to obtain big data from sensors and devices to do an important action. In addition to clarifying the role of Internet of things in this field. This paper concluded that higher education institutions have changed dramatically because of IOT applications and big data and computing technologies can measure Many educational parameters. These computing technologies made a link between the students and the educational environments	The study mention some of good factors that are effective in smart university campuses and missing some others. The study focused only on the role of the Internet of Things in smart universities campuses

(continued)

Table 1 (continued)

Articles	Advantages/strengths	Disadvantages/weaknesses
[10]	This paper presented a criteria and definition of smart learning environments and framework to address the development and the design of smart learning environments regards with supporting learning activities In addition to features and criteria of smart learning are addressed and some research issues related to smart learning such as: Learning behavior and learning pattern analysis, Learning and assessment strategies for smart learning and Learning behavior and learning pattern analysis. The study predicted that new learning concepts might raise many disputed Issues, such as ethics and moral principles when applying technologies in education. They mentioned that smart learning environments is expected and exciting. Moreover, this paper also concluded that there is a potential positive effect of smart learning	It would have been better to suggest some solutions to some future problems that related to the use of smart technology in education
[2]	This paper presented the components of smart university and an interpretation of the research on smart university. In addition to the collaboration in smart university regards with Smart collaborative learning, motivations, and collaborative smart university services. The paper concluded that "smart collaborative learning", uses smart interactions to enhance the new methods of collaboration between learners and smart teams and collaborative learning is crucial for acquiring and sharing knowledge smartly via smart interactions between the team of learners where the implementation of this concept requires a smart service architecture. As future work, the authors plan to define a mechanism of the collaborator via the most suitable collaborator recommendation. This mechanism depends on the description of the service structure and functionality at developed collaborative learning process	This university guidance is theoretical approach without real life application

(continued)

Table 1 (continued)

Articles	Advantages/strengths	Disadvantages/weaknesses
[15]	This conceptual study explored the potential of artificial intelligence systems in the university's teaching, learning and knowledge, with globalization and rapid technological change and the impact of the internet on universities There is a need for a new university model where communications and computers replace roads, buildings. The paper presented the idea of a Hyperclass based on Hyperreality, an exciting form of artificial intelligence and human intelligence overlapped and interacted to provide anyone, anywhere, anytime. Moreover, this paper came to suggest possible philosophical basics for the future of virtual university model. The paper concluded that there is a need to do more research all around the world that can deal with problems from various aspects of different cultures. To include its environmental impact and its effect on cultures and communication	It would be better if the paper present more philosophical basis for the concept of smart universities or virtual education with some applications

References

1. Abdulrasool, F., Turnbull, S.: The Role of IT governance in enhancing the performance of smart universities. In: Joint European-US Workshop on Applications of Invariance in Computer Vision, pp. 708–720. Springer, Cham (2020, April)
2. Akhrif, O., Benfares, C., El Idrissi, Y.E.B., Hmina, N. : Smart University services for collaborative learning. In: The Proceedings of the Third International Conference on Smart City Applications, pp. 131–142. Springer, Cham (2019)
3. Bakken, J.P., Uskov, V.L., Penumatsa, A., Doddapaneni, A.: Smart universities, smart classrooms, and students with disabilities. In: Smart Education and e-Learning, pp. 15–27. Springer, Cham (2016)
4. Belskaya, E., Moldovanova, E., Rozhkova, S., Tsvetkova, O., Chervach, M.: University smart guidance counselling. In: Smart Education and e-Learning, pp. 39–49. Springer, Cham (2016)
5. Castillo, E., Gutierrez, J.M., Hadi, A.S.: Expert Systems and Probabilistic Network Models. Springer Science & Business Media (2012)
6. Chen, N., Yamashita, Y.: The possibility for the active use of smart devices in university education. In: 2017 6th IIAI International Congress on Advanced Applied Informatics (IIAI-AAI). IEEE. pp. 99–104 (2017, July)
7. DUMANĈIĆ, M.: Smart Education in Smart City and Student Model. In Conference proceedings of "eLearning and Software for Education" (eLSE), vol. 2, No. 15. Carol I" National Defence University Publishing House., pp. 64–71 (2019)
8. Gungor, V.C., Sahin, D., Kocak, T., Ergut, S., Buccella, C., Cecati, C., Hancke, G.P.: Smart grid technologies: communication technologies and standards. IEEE Trans. Ind. Inf. **7**(4), 529–539 (2011)
9. Gyanendra, S., Ajitanshu, M., Dheeraj, S.: An overview of artificial intelligence. SBIT J. Sci. Technol. **2**(1), 2277–8764 (2013)
10. Hwang, G.J.: Definition, framework, and research issues of smart learning environments-a context-aware ubiquitous learning perspective. Smart Learn. Environ. **1**(1), 4 (2014)
11. Jarrah, M.I., Jaya, A.S.M., Alqattan, Z.N., Azam, M.A., Abdullah, R., Jarrah, H., Abu-Khadrah, A.I.: A novel explanatory hybrid artificial bee colony algorithm for numerical function optimization. J. Supercomput., 1–25 (2020)
12. Jarrah, M.I.M., Jaya, A.S.M., Azam, M.A., Alqattan, Z.N., Muhamad, M.R., Abdullah, R.: Application of bat algorithm in carbon nanotubes growing process parameters optimization. In: Intelligent and Interactive Computing, pp. 179–192. Springer, Singapore (2019)
13. Kwok, L.F.: A vision for the development of i-campus. Smart Learn. Environ., **2**, 1–12 (2011)
14. Majeed, A., Ali, M.: How Internet-of-Things (IoT) making the university campuses smart? QA higher education (QAHE) perspective. In: 2018 IEEE 8th Annual Computing and Communication Workshop and Conference (CCWC), pp. 646–648 (2018)
15. Rajasingham, L.: The Impact of Artificial Intelligence (AI) Systems on Future University Paradigms. Victoria University of Wellington Wellington, New Zealand (2009)
16. Shorman, S.M., Pitchay, S.A.: Significance of parameters in genetic algorithm, the strengths, its limitations and challenges in image recovery. Faculty of Science and Technology, Universiti Sains Islam Malaysia (USIM), Bandar Baru Nilai, Nilai, Negeri Sembilan, Malaysia. (2015)
17. Shorman, S.: Internet of Things Application to development of smart classroom system. Int. Rev. Comput. Softw. (IRECOS) **14** (1), 22–26 (2019). https://doi.org/10.15866/irecos.v14i1.17374
18. Spector, J.M.: Conceptualizing the emerging field of smart learning environments. Smart Learn. Environ. **1**, 2 (2014). https://doi.org/10.1186/s40561-014-0002-7
19. Swartout, W.R., Nye, B.D., Hartholt, A., Reilly, A., Graesser, A.C., Van Lehn, K., Wang, L.: Designing a personal assistant for life-long learning (PAL3). In: The Twenty-Ninth International Flairs Conference (2016)
20. Tikhomirov, V., Dneprovskaya, N., Yankovskaya, E.: Three dimensions of smart education. In: Smart Education and Smart e-Learning, pp. 47–56. Springer, Cham (2015)

21. Uskov, V.L., Bakken, J.P., Pandey, A., Singh, U., Yalamanchili, M., Penumatsa, A.: Smart university taxonomy: features, components, systems. In: Smart Education and e-Learning, pp. 3–14. Springer, Cham (2016)

Artificial Intelligence and Smart Universities

Abeer AlAjmi

Abstract "Smart University" is the practical application of smart intelligence in the field of education. The adaption of IoT (Internet of Things) and smart technologies has created new learning platforms and emphasizes the value of learning. Smart campuses are equipped with smart class rooms and smart infrastructures. They are staffed by smart individuals who implement smart strategies. The results will enhance advanced learning outcomes and facilitate acquisition of information. This chapter will examine the transformation which occurs in traditional universities/colleges where artificial intelligence is taking place, thus reshaping the future of current learning models.

Keywords Smart University · Smart campus · Smart home · Artificial intelligence · E-learning · IoT · Covid-19

1 Introduction

The concept of "Smart University" evolved through the twentieth century when the educational system started to become more dependent on computers. It was impossible not to change with such innovations as the world started to network and educational systems became more functional and improved in efficiency. As a matter of fact, the evolution process of smart education is still taking more time to adapt to new methodologies. Computers and internet, in other words artificial intelligence, have become more convenient for everyone. There are smart buildings, artificial intelligence applications, elements used in the health sector and an advanced navigational system, just to name a few.

The idea of developing traditional education and converting it to a digital and smart model has indeed been delayed in application. In his article, Mason is asking an important question, 'how is smart learning different from traditional learning?' Presenting a bigger challenge is the question, 'how can standardization work be best

A. AlAjmi (✉)
Box Hill College Kuwait, 29192, 13152 Safat, Kuwait
e-mail: Abeer.AlAjmi@bhck.edu.kw

A. Hamdan et al. (eds.), *The Fourth Industrial Revolution: Implementation of Artificial Intelligence for Growing Business Success*, Studies in Computational Intelligence 935,
https://doi.org/10.1007/978-3-030-62796-6_17

scoped in today's innovative-rich, networked, cloud-based and data-driven learning environments?' [1].

To answer this question, one has to identify the standards of the conceptual learning framework to promote innovations such as in smart universities. In other words, exploring the new dimensions of smart education.

This chapter will highlight the significant trend of smart education, especially in higher education and the evolvement of traditional education with new artificial intelligence applications.

In order to understand the involvement of artificial intelligence involved in the smart learning process, it is important to first understand what artificial intelligence is and how it started.

John McCarthy was an American computer and cognitive scientist who is considered the first to coin the term Artificial Intelligence, in 1955 at the Dartmouth conference. In a Q&A style article he conducted for the Computer Science Department at Stanford University in 2007 he defined Artificial Intelligence as "It is the science and engineering of making intelligent machines, especially intelligent computer programs. It is related to the similar task of using computers to understand human intelligence, but AI does not have to confine itself to methods that are biologically observable" [2]

The wide use of AI started to take off in the late twentieth century in different sectors as it has been a useful assistant in different agencies and institutions. As many think that using AI in the educational field is still an evolving idea, it has proven its importance as a tool to enhance educational experiences. There may not be robots to teach at higher institutions and universities, yet today there is a greater dependency on machines and computers to function efficiently. Today, many highly ranked universities around the world have been offering the opportunity to major in AI. Several have even started researching and studying AI. Stanford University has been involved in this since the 1960s. The University of California in Berkeley, MIT and Carnegie Mellon are other institutions currently expanding in this major.

2 Artificial Intelligence and E-Learning

Smart education provides e-learning technologies and internet based software and hardware systems for users. Many campuses around the world have successfully implemented smart education systems and facilitated learning tools. This shall be examined later in this chapter as case studies will be presented, along with the challenges smart educational systems are presenting.

As e-learning has become one advantage of distance education, it has created exceptional opportunities to endorse technology, which in turn exposes the beneficial use of artificial intelligence. Such innovations, when implemented, will have many advantages. As an example, teaching strategies will require smart tools and devices to present academic materials.

Internet and cloud computing technologies provide the opportunity for many online classes, online libraries and websites, virtual labs and self-study programs. These allow for flexible attendance and supervisions.

Although artificial intelligence facilitated electronic learning and presented many advantages, just like any other invention, it can also present disadvantages. This might vary from different perspectives, but one cannot deny the fact that e-learning has eliminated all geographical barriers. It has allowed students from anywhere to enroll in different majors. This was not even an option until the internet was discovered and implemented in educational facilities. E-learning has provided users like students, academic staff and researchers with a new vision. Learning to learn has become a common feature of e-learning. Students can chose their area of interest and decide the time frame that they wish to enroll. Also, being a user of e-learning platforms can enhance ones computing potentials and skills. Today everything depends on computers. For some tasks to be accomplished, it will be hard not to go digitally to save time and money. Almost all books, resources, libraries, historical archives have digital copies online that can be accessed at any time and free of charge.

In the e-learning environment, the routine of a standard class environment can be virtually replicated. Artificial intelligence today can design online classes where student's attendance can be taken, instructors can use a whiteboard to teach, with the entire lecture being recorded and uploaded online. Even tests can be done online, providing instant feedback with the results. One important feature of e-learning is the information updates. Now the world is strongly connected with increasing numbers of internet networks and digital webs. This means that any type of information can be located, uploaded, downloaded or updated in seconds. Such features abbreviate time and cost, allowing for more research and inventions to be accomplished.

Futurist and environmental activists appreciate the green approach of such invention. E-learning depends on wireless connections, so students and instructors are not driving or taking transportation to schools and universities. This assists in minimizing the pollution problems. Since most tests and exams are taken online, paper usage is completely eliminated. This will result in the reduced consumption of trees. To make 1 ton of standard office paper it takes about 24 trees that's billions of trees every year [3] (Fig. 1).

On the other hand, e-learning can have many disadvantages. First, many think that allowing critical majors, that require traditional learning environments, to be offered online or as a distanced learning course, is a major mistake. Many universities and colleges are offering, for instance, medical classes on line. Such a major, which is a very serious and sensitive one, requires direct interaction with face to face meetings either with instructors or patients. Although artificial intelligence can format and support many applied practices, it still has some limitations. This is especially obvious in regards to anatomy classes, where practical interaction is required. Specialist's performances will be absolutely out of comparison as people will prefer doctors who have graduated from highly ranked universities with a rich history. Thus, the artificial intelligence might still be considered as inferior, in regards to several specific fields. This problem may be overcome in the future.

Fig. 1 Acacia plantation in Sumatra. The impact of paper production on forests can be substantial. (Photo by R. Butler)

Second, e-learning is more of a "loss" platform where control is almost lost and it depends on personal commitment from all participants. In the traditional class environment, it is attendance and class assessments which impact greatly on the grade. Instructors are motivating and supervising students on a regular base to guarantee their understanding and commitment. Counter to this, in a virtual class environment, technology assures online communication channels but not student commitment or perseverance. It diminishes the student's responsibility and obligation to pursue their studies. It is expected that students with low adherence will lack the motivation to pursue with e-learning.

Third, Smart universities are investing in technology, high speed internet; efficient computer laptops and updated systems and software necessary for e-learning education. Students pursuing via the e-learning system must possess laptops or computers, printers, internet and protected authentic software. This technology dependency requires long term maintenance as devices have to be certified and servers have to be protected. It is common for systems to crash but it has to be of a minimum standard and not frequent. Thus maintenance is required periodically and this will add a cost to the e-learning process.

Fourth, for standard learning methods, instructors' face to face discussion will strengthen the determination of students who possess limited skills or potentials. The use of in class technology can help students with physical or mental disabilities. For example, e-learning today cannot serve students with major physical disabilities such as blindness or deafness. Not all materials can be delivered with available technology today for such cases. This puts the e-learning in a recessive situation that can't serve a large and important segment of society.

Artificial intelligence has played a big role in e-learning education and has progressed beyond face to face learning methods. That being the case, credit goes to wireless networks which provide the learning environment platforms necessary, anywhere and at any time.

Education around the world is becoming accessible to almost everyone, anywhere and offered in a wide range of different majors. University students might major in art, science, economics, social studies, history or other more computer dependent fields. Also, for majors that can't be taught online, lectures, books, references and information need to be digitally accessed. Regardless of the studies that students are majoring in, computers and internets are the accessible tools of modern education with high demands on periodic maintenance.

When technology was applied to education, in the late twentieth century, it was not necessary in all the different fields of education. It was considered an obstacle since it was supposed to link two different eras together. Individuals who did this were considered the early learners, going on to become pioneers of knowledge. However, the evolvement of higher education facilities took slow steps, compared to other facilities such as health and care facilities, airlines and navigations, and military. Thanks to technology and continuous research, there have been rapid developments for machines and equipment. For example, today's technology in hospitals, airlines industry, and military weapons are advanced and more energy and cost efficient. Priority had to be given to education, traditional teaching and learning methods can no longer correspond to today's requirements.

Many universities and colleges around the world are now distinguished e-learning platforms, offering creative solutions for skilled potential and leadership. The objective of the e-learning industry adds value and advantages for professional life. As the domain of education is expanding, information becomes more available, thus research is stimulated and more data is being served on a silver platter for scholars, students, instructors and investigators. Books, articles and scientific materials, for example, are being offered on many websites for free, or in many cases, for an affordable cost.

3 Smart University

A smart university is a smart campus environment, smart learning system, smart staff and students and smart researchers. The concept of "Smart University" allowed education to be available anywhere. Such innovation has its own less risky and less challenging circumstances as information became available and updated regularly. Smart Universities took a step toward empowering individuals' potentials and enhance the quality of education by using the artificial intelligence.

The smart university is an emerging and rapidly evolving institution that creatively integrates innovative concepts, smart software and hardware systems. Smart classrooms have state-of-the-art technologies, using technical platforms and smart pedagogy based on modern teaching and learning strategies. Smart learning takes place in academic analytics; as well as various branches of computer science and computer engineering.

Smart campuses are adapting artificial intelligence as fundamental tools to operate the system such as advanced technology infrastructures for the best economic, social and environmental deliverables and impacts. In smart campus class rooms, laboratories, libraries and different facilities are monitored and secured. It is essential to secure intelligent campus' network, security, communication and management.

Recently, smart universities are witnessing continued development which vitally affects society. One of the most prominent goals of smart universities is the ability of student's perceptions on how to receive information, process and develop it. Students are shifting from being a consumer of knowledge to a producer of it. This has opened many doors and opportunities for people with different backgrounds and different abilities.

Smart Universities are the translation of information technology to enhance situations for individuals with special needs. Readable materials can be heard or touched electronically as with Braille language in the case of the visual impaired. Special electronics and computers can encourage the case of special needs to embrace their skills and better motivate them to be an effective member in society. This I accomplished by offering jobs that allow these students to increase their motivation and inspire others. Yet, as previously mentioned, one disadvantage that remains an obstacle in the face of technology for many special need students is the narrow domain of majors.

Long term objectives are to create more effective methods to assist all individuals. Other special needs individuals will find advanced networking and operation system solutions available to meet their requirements. Smart universities are employing smart intelligence to help individuals with what could be considered hopeless cases to get educated. Machines can read lips, help cases of paralysis in reading and writing and even facilitate their mobility on campus. Such employment of artificial intelligence has created more personalized courses in the context of smart education.

In the pursuit of advancing the human race and the obstacles facing the learning mechanism, the smart university directs individuals to invest in industrial artificial intelligence and information technology to develop research. One of the most important goals of educational research is to reach radical solutions that may assist education to bridge gaps. Technology helps speed up the process of finding, listing, and solving sub problems without affecting the university's public order.

To reach this goal, the infrastructure systems used in the smart universities must be examined. Interactive institutions compete in employing the information revolution, in building a strong infrastructure and developing sophisticated software methods. These may take a long time and effort if they are to replace the human component. Therefore, smart system data that has durability, flexibility and ability to store information and use through a high-speed network is required. The methodology of work is to conserve energy and the environment.

The idea of a Smart University is based on several interconnected and indivisible elements. To create a Smart Campus essential elements that are required including Smart IT Infrastructure, Smart Buildings, Smart people and Smart Education System.

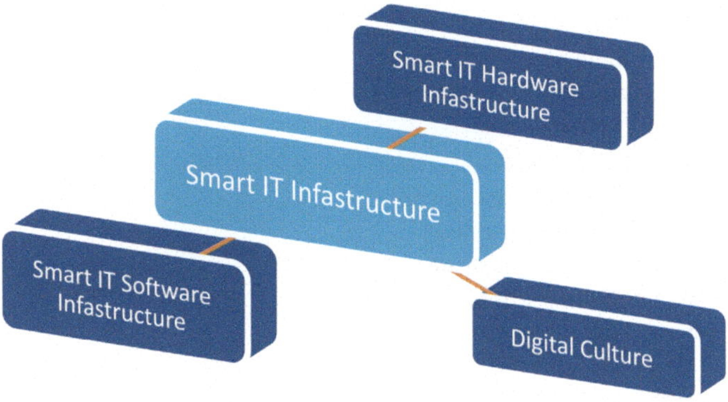

Fig. 2 Smart IT Infrastructure

4 Smart Campus

4.1 Smart IT Infrastructure

Many researchers in the field of technical information have proposed three primary requirements for a smart IT infrastructure (Fig. 2).

4.2 The Smart IT Hardware Infrastructure Includes

1. High Speed Wired and Wireless Facilities.
2. Advanced Tablets and Portable Computers.
3. Wired and Wireless Photocopy Devices and Printers.
4. Cameras and Sensors.
5. Storage Devices and Data Centers with High Quality Specifications.
6. Smart Boards and Projectors.
7. Intelligent Controlling and Monitoring Systems.
8. Security and People Identification Systems.
9. Green Power Systems and Storages.
10. Advanced Scientific and Digital Laps and Classrooms.

4.3 The Smart IT Software Infrastructure Depends on What It Has to Do with All Applications and Controlling Systems Such as

1. Management Systems.

2. Social Network Systems.
3. Enterprise Management Systems.
4. Safety and protection Systems.
5. Maintenance Systems.
6. Smart Electronic Library.
7. Network database Systems.
8. Cloud Abandonment Systems.
9. Intelligent educational software.
10. Electronic laboratory software.

4.4 Digital Culture

It basically concerns the basic skills needed by learners, instructors, and support members in a digital learning environment.

Artificial intelligence is the basic ingredient required to operate a smart IT infrastructure. It would be impossible to activate the previous component of the structure without functioning use of advanced technology that facilitates human work, as well as and in substituting human operations in complex and vital operations. Furthermore, artificial intelligence quickly diagnosis and track software problems and analyze patterns of problems and solutions without affecting the overall entity or disrupting the other digital functions of the system.

4.4.1 Smart Campuses

In his open journal, Kwok defines intelligent campus (i-campus) as "… a new paradigm of thinking pertaining to a holistic intelligent campus environment which encompasses at least, but not limited to, several themes of campus intelligence, such as holistic e-learning, social networking and communications for work collaboration, green and ICT sustainability with intelligent sensor management systems, protective and preventative health care, smart building management with automated security control and surveillance, and visible campus governance and reporting". [4]

Smart pedagogy needs the implementation of smart components that are linked together and interact together as a unit. Smart classrooms, Smart Libraries, Smart Environment, Smart Infrastructures, Smart management, and Smart teaching staff. Traditional class rooms require much equipment that operates manually or separately and has to be connected to electricity. For example, projectors, computers, paper slider, TV etc. today, the Smart class room is considered to be bifurcated cells that links to one remote control, or perhaps can be activated by sound sensors or movement sensors. Smart classrooms are active, fixable and multipurpose learning environments. The class purpose can change depending on the taught subject. Sound systems and lighting systems can be controlled wirelessly and smart technologies can take place to engage with distant students. Also, Smart classrooms have contextual awareness thanks to IoT and AI implications. Classrooms can analyze and adjust its status; temperature can be adjusted accordingly, computes can be activated by touching, lights turn on be movements and other human stimulation of human behavior.

Furthermore, AI can enhance and support teaching methods by helping instructors with certain tasks like face recognitions for door access, attendance and exams. Also, it can anticipate classroom damages, crimes of danger, and serious gas leakages, sending an instant email or messages to high safety and security maintenance.

The same idea is applied to different Smart Campus facilities like laptops, Libraries, courtyards, students housing, cafeteria…etc.

Smart Learning Environments have many interacting requirements. It has to be connected via internet network or cloud computing system to ensure communication of all devices and secondary networks. Also, all information and data has to be updated and accessible. The Smart Learning Environment has to be adjustable, addressable and shareable. Which means all students can access such smart environment anytime, from anywhere and for as long as they require. The virtual learning environment is considered an example of a Smart Learning Environment, especially for online taught courses which demand high utilization of technology and AI approaches.

In the Smart University, instructors, staff and students are facing a paradigm shift in the history of contemporary education. This quantum leap involves speeding up of modernization and normalization with modern technological methods. As AI requires different potentials, it's a must to expand all knowledge and experiences to keep pace with this development in the mechanism of education.

5 Smart Buildings

The online oxford dictionary has defined "Smart Home" as "a home equipped with lighting, heating, and electronic devices that can be controlled remotely by smartphone or computer. People can contact their smart home on the Internet to make sure the dinner is cooked, the central heating is on, the curtains are drawn, and a gas fire is roaring in the grate when you get home" [5].

Surely, the previous definition reflects a simple meaning of remotely controlled inhabitation, yet it does not show where artificial intelligent took place in such development. In another word, a house that is already programed for certain number of tasks but would not predict emergency updates and such invariables is not really a smart home. Today, Smart homes are those engineered systematic dwellings that are programmed to take action stimulating in it natural human behaviors and predict future decision making. In this case it is definitely known that a smart home has efficiently applied artificial intelligence that replaces human presence. For example, instead of a timing light system to turn on automatically at 6 pm every day and turn off by 6 am, the house will sense human presence and through sensors can efficiently turn the light system on or off.

The previous definition of a smart home is only an example that can pave the road before us to understand a greater meaning of Smart Buildings. We know now that AI will upgrade the tasks and nature of smart building; however we need to know the way it's upgrading it to understand the difference and also the benefits of AI.

Thus, what is needed to function a smart building? In his paper "The distinctive features of "smart" buildings", Eugeny I. Batov suggested that there are three important components that would feature a building as a smart building. There must be a hardware system and software system and a network system.

Eugeny I. Batov explains that the hardware systems are devices such as cameras and sensors which are capable to understand human behavior and interact with the surrounding environment. AI located in smart buildings today, has maximized capacity to allow the building to change its status depends on the situation. The building can sense heating system status, lighting system status, air conditioning system status etc.

As for the Software system it is represented in the programming of devices by analyzing data and making decisions accordingly. It is even possible to anticipate events and simulate human reactions in many cases. For example, light can be programmed to turn on in corridors once movement has been detected. The building can detect failures such as damages in the water system. The water system stops in the broken pipeline circuits without stopping the whole plumping system.

Last, network acts as a tie between the two previous systems in a way that allows efficient and successful communication. Network systems link different devices and machines together such as cameras to the mother computer and triggers alerts resulting from software glitches.

Although it might have an essential cost and high maintenance requirement yet one cannot deny the long term benefits of Smart Buildings. These benefits vary depending on the way the smart building functions. Below are the main and general benefits:

5.1 Safety

Buildings are monitored and secured as AI can be applied to recognize faces and sounds further than finger prints. This is considered important to ascend to the level of crime today. Also, smart buildings can detect fire and predict the least hurt sections of the building to be isolated or shut it down to minimize damage. It can predict gas or water leakage and contact maintenance departments or fire stations for rapid intervention.

Artificial intelligence can provide high standards of security in smart universities. Devices like high efficiency digital cameras and sensors are required to ensure public safety and prevent crimes.

Miami University in Oxford, Ohio, is like many other American Universities, that have successfully employed technology on campus to prevent crime. The university has distributed many emergency phone towers that in case of emergency it broadcasts messages to all phones on campus and the police as well. The information technology enhances students and faculty staff safeties as these messages are not only related to crimes but also natural disasters such as earthquakes, fire, rains storms etc.

In Smart Universities, campus safety has become an important public concern for students, families and employee. The artificial intelligence can provide smart and fast solutions to ensure a maximum safety environment, especially on large campuses and campuses with student housing. A new emerging concept of IoT (Internet of Things) has become an essential requirement in smart universities. Classrooms, car parking, student housing, libraries and other campus facilities need to be monitored 24 h 356 days. The Wi-Fi serves in connecting the cameras and sensors to a mother board or what is considered the brain of campus and records that help to prevent or reduce crime.

> Another important aspect of AI is the ability to predict behavior. In contrast to the imaging and identification of criminal activity in progress, the University of Houston has used NIJ funding to develop algorithms that provide continuous monitoring to assess activity and predict emergent suspicious and criminal behavior across a network of cameras. This work also concentrates on using clothing, skeletal structure, movement, and direction prediction to identify and reacquire people of interest across multiple cameras and images [6].

Many universities are researching technology of algorithms working on "predictable crime programs". Such researches can increase students' tendency to enroll in universities and campuses that are ranked to be highly secure and safe. Furthermore, the application of "smart" would extend from Smart Campus to a Smart City.

It is important to understand that the adaption of AI does not necessarily convert the campus to a smart campus. If the system is too weak to analyze and enhance the learning environment Information, the technology must then be purred in the correct mold.

5.2 *Cost Effectiveness*

Smart buildings can save time by allowing AI to intervene in the event of human absence. Action can be taken and remotely recorded, enabled or disabled. This way a substantial amount of money can be saved by this enhancing feature rather than paying for electronic programming errors.

> While many universities have no way of knowing when a building is in use or not, smart campus IT managers can use Internet of Things sensors to detect whether a building is empty and adjust power to the area accordingly to conserve electricity or gas. This can cut building maintenance and energy costs and is also a more environmentally friendly choice [7].

5.3 *Energy Saving (Green Building)*

In Smart buildings, energy production depends to a large extent on renewable energy sources or in many cases vernacular. Solar panels are one example to feed the Smart Building with energy. This does not produce waste and requires low cost and demands of maintenance. This classifies as a sustainable green energy source. The advanced development of the solar panel industry has created a branch for energy or power banks that can store excessive energy for emergency cases and for a long time. This has a high demand in many buildings such as hospitals and military structures.

One of the most important features of green buildings is its environmental and social impact. Many smart buildings are taking huge positive steps to be able to react to many natural issues such as water pollution, air pollution and global warming.

The term eco is a recent term that revolves around human behaviors for a better ecological system, in other words, the care of the environment and nature.

Many associations are active to create eco systems that are seen nowadays in many smart buildings and especially, on smart campuses. Uses and application are different yet the objective is creating a better habitat and environmental impact for future generations. Many universities adapted this terminology along with sustainable and green building to minimize human waste, recycle materials, and use biodegradable materials. This is one of the main purposes of the smart building. The University of Adelaide is an example of "Eco- University". According to the university website, in August 2019 the university started working on a new solar panel farm that is supposed to supply over 40% of the energy required for campus.

The University of Connecticut is in the top twenty greenest university campuses, according to the interesting engineering website for the year 2018. The university has many effective activities since 2002 in recycling projects, raising public awareness and the most important is the conservation of drinkable water and improvement of water quality projects.

After explaining the components and benefits of a smart building, it is important to apply this to real life examples. It is clear now that internet, technology and programming are facilitating tasks of student and staff life on campus; yet, the question here is what the role of artificial intelligence on smart campuses is?

The emergence of artificial intelligence solutions has provoked better standards of "smart" solutions. As stated previously, being smart is not just about being remotely connected. Furthermore, it is about analyzing problems, thinking faster stimulating and optimizing human quality of choices and decisions. AI helps campuses run themselves to a certain extent, by using wireless connections between devices and efficient computers. "Creating these 'smart campuses' is no longer an aspirational goal," the Cisco white paper notes. "It's a reality."

6 Models of Smart Arab Universities

Hamdan Bin Mohammed Smart University (HBMSU) is Dubai's first model of a smart university. It launched its first movement toward smart education in 2008. It is considered a smart environment community equipped with AI at the highest levels. As a smart hub, the university partnered with leading smart systems companies such as Signify, formally known as Philips Lightning, Siemens, Smart Citti and Trane.

Smart lighting and temperature controlling systems are working, interestingly, by using advanced technologies to help the building become a sustainable building.

Moreover, AI is considered to be a driving force for potentials of HBMSU by providing learners with a new GPS technology where they can be guided to their destinations on this smart campus.

Knowledge itself is being delivered to improve individual's intellectual and social behaviors adapting new educational methods. The university is providing distinctive learning experiences for students and working experiences for instructors and researchers. The university, walking in the steps and directions of His Highness the Sheikh Hamdan Bin Mohammed, to encourage innovation and technology by providing a variety of majors that can be accessed by local and distant students. This mobility of education is one of HBMSU's best features as a smart university. Openness and information accessibility are another feature of this smart university as it provides researchers with unique online databases along with many online videos for lectures that have been upload to share information.

The university is enhancing the quality of education and has an effective leadership role through the different smart leading programs such as the school of education, school of health and smart studies and the school of business and quality managements. Also, the university provides masters and Ph.D. programs taught by a diverse teaching staff that come from different backgrounds and degrees to enrich the learner's experiences.

The cloud campus is another domain where AI is reliable on advanced technologies to reshape education perceptions. The cloud campus includes the smart undergraduate programs and postgraduate programs. These function through high tech data portfolio including stored videos, lectures and conferences along with vast learning materials.

The cloud campus idea depends on high connection technologies, smart audio and video applications, virtual classrooms and instant chatting options all operated by smart innovative and intelligent applications.

7 Coronavirus (COVID-19) and Smart Universities

This paper is being written in conjunction with Covid19 circumstances early 2020, suitable conditions to highlight repercussions of epidemic situation in education at the global level. Timing was helpful as this chapter extract through real-world experience during the global epidemic period. Several universities and scientific institutions got affected during the embargo and the absence of physical presence in educational facilities all over the world. The Impact of Coronavirus (COVID-19) forced campuses shifting to remote learning; however there were many prepared universities with proficient strategies to pursue not only current semester moreover summer semester. Furthermore, many universities had the portability and electronic readiness to offer extra courses and degrees in AI. The circumstances of this epidemic reflected higher education preparations into three compilations, which are divided into:

1. Fully prepared for the transition to e-learning.
2. Semi-prepared for the transition to e-learning.
3. Not fully prepared for e-learning in all ways.

First, Covid 19 situation is not a high scale challenge for highly prepared institutes as they were equipped with high standards learning environments. Most of these universities already provide online courses and distanced programs in normal circumstances. Few weeks after campuses closure, well-equipped universities resumed teaching e-learning programs. Digital platforms existed already in these universities, and that was the first indicator of Smart Education. Courses and taught materials already had an online designed version and could be delivered through the practical simulation of normal classroom and smart software. Another indicator is the accessible online cloud-data storage; many researchers took the advantage of this feature and overcome interruptions. The University of Texas in Austin is a pioneer model of personalized mentored learning by offering seven months post graduate program in Artificial Intelligence and Machine Learning online. The university adapted the giving situation and took advantage for those interested as a live experience. This took place during the initial periods of the epidemic and during the closure of almost all countries. A third indicators for a Smart University in such circumstance is its encouragement the desire for education, seizing opportunities and not wasting time by offering free educational methods, free accessible uploaded lectures and scientific materials

Second, the semi-prepared universities are the universities took a while preparing their online platforms, cut out some programs for deficiencies in electronic systems or low attendance. Many of these universities had full time on campus programs which required under epidemic circumstances to customize courses where best content delivered for the majority of students. Financially, these universities suspended all activities due to courses withdrawals and unpredicted future enrollment is not clear yet.

Last, many universities had to go through the worst case scenario. Many of these facilities had to shut down for many reasons. Such universities have no preparation at all in providing e-leaning systems and have no interaction platforms from any kind. Or, students enrolled in such universities can't have a strong access for high speed internet and computers. For example, this crucial situation has affected learners on all levels in different states of India and China where the mandatory closure took please. Continuously internet or electricity interrupts, public transportation has stopped and the students 'family situations requires under these poor financial conditions a focus on saving money. Thus, studies shows decline demand of higher education globally. An online article by ET Government.com is illustrating the deteriorating situation in some states within India this by saying "Their low-income private and government school counterparts, on the other hand, have completely shut down for not having access to e-learning solutions. The students, in addition to the missed opportunities for learning, no longer have access to healthy meals during this time and are subject to economic and social stress." [8] Of course all practical majors and programs require face to face meeting between students and instructors has to be postponed. Medicine, architecture and most engineering majors could not be pursued as it depends on continues supervision and instant interaction on campuses.

International associations provided solutions and aids for distance learning defaulters and different groups of society like parents, students, instructors and researchers such as the United Nations Educational, Scientific and Cultural Organization (UNESCO). On their webpage, the UNESCO, provided numerous numbers of different platforms and applications for many different uses and mostly free of charge for users. Digital learning management systems, mobile phone reading and learning applications, offline functional training systems, massive open online courses, Self-directed learning resources, distance learning solutions and tools are only few examples for the type of assistant they provide worldwide. Furthermore, there are many psychosocial support links and guidance for those unfamiliar with e-learning. Many smart universities actually were using these recommend platforms and sites in normal situation and expanded their educational domains.

Through the scary global situation through the epidemic phase, e-learning considered to be an unequal opportunity for higher education's learners around the world. Learners at smart universities were lucky to pursue their education where others preferred to postpone it whenever situations turn optimistic.

8 Conclusion

AI, IoT and e-learning systems came as an educational breakthrough for many of smart learning organizations. They provide modern and effective solutions for most of this age learning problems. It became essential to develop digital platforms that use AI, those smart computer systems that use logic and reason simulating human actions could be the substantial motive for an outstanding academic performance.

There are many evidences supports urgency in applications of AI in higher education institutes such as increasing students retentions, increasing programs enrollments by outreaching students globally, promotes university's reputation and improving value. AI has created new generation of advanced learners using e-learning methodology despite its few setbacks it had served many fruitful advantages. The most important goal has to be noble to provide learning chances to all societies' segments.

Smart Universities require high level of technological recognitions and maintenance. "Smart" concept reveals savvier in computer technology not just remotely operating systems as it is mostly understood. Being "Smart" is a simulation of human behavior and the necessity of immediate intervention in various situations, as this concept is based on several elements.

IoT facilitated the implementation of e-learning model along with all digital tools such as the smart virtual platforms, cloud data storage and advanced sustainable materials.

In the future the expansion of Smart Universities and AI will allow machines to reduce time and effort and reflects exceptional progress in pedagogy systems and enhanced potentials than before.

References

1. Hoel, T., Mason, J.: Standards for smart education—towards a development framework. Smart Learn. Environ. **5**, 3 (2018). https://doi.org/10.1186/s40561-018-0052-3
2. McCarthy, J. (n.d.).: What is artificial intelligence? http://jmc.stanford.edu/artificial-intelligence/what-is-ai/index.html
3. Kiprop, J.: How many trees does it take to make 1 ton of paper? (2018, November 20). https://www.worldatlas.com/articles/how-many-trees-does-it-take-to-make-1-ton-of-paper.html
4. Kwok, L.F.: A vision for the development of i-campus, Smart Learning Environments—a Springer Open J. **2**, 2 Springer (2015)
5. Online Oxford Dictionary: Information on http://www.oxforddictionaries.com/definition/english/smart-home?q=smart+home
6. Rigano, C.: Using artificial intelligence to address criminal justice needs. NIJ J. **280** (2019). https:// www.nij.gov/journals/280/Pages/using-artificialintelligence-to-address-criminal-justice-needs.aspx
7. Zimmerman, E.: Higher Education Invests in Wi–Fi Technology for Smart Campus Projects. EDTechmagazine (29AD). https://edtechmagazine.com/higher/article/2019/03/higher-education-invests-wi-fi-technology-smart-campus-projects-perfcon
8. Aayog, N.I.T.I.: COVID-19 Pandemic: Impact and strategies for education sector in India. ETGovernment (16AD). https://government.economictimes.indiatimes.com/news/education/covid-19-pandemic-impact-and-strategies-for-education-sector-in-india/75173099

Artificial Intelligence Literature in Accounting: A Panel Systematic Approach

Ghassan H. Mardini and Amneh Alkurdi

Abstract The main objective of the current study is to extensively revise the literature review of AI and its impact on accounting. Moreover, this investigation leads to critically identify the research problems of AI in accounting that support researchers in investigating such research gaps in the near future. The methodology employed is the panel systematic dimensions approach that aims to address research problems by critically evaluating and integrating the findings of all of the relevant prior studies. Moreover, it contributes to our knowledge through achieving a well-established and systematic review, it also identifies relations and gaps and inconsistencies in the literature on AI and accounting in order to offer new research gaps.

Keywords Artificial intelligence · Accounting · Literature systematic approach

1 Introduction

In the recent decade, the use of Artificial Intelligence (AI) has increased [1]. AI enhances the capabilities of the computing world, bringing it to the next level; it enhances the predictions and risks and demonstrates how we should react to them [2]. This growth represents a major awareness of AI in accounting, with both regulators and clients demanding that auditing firms move to this new system in order to help them to finish their work within an acceptable period, in an accurate manner and at a reduced cost [3, 4] argued that accountants who are dealing with rote tasks in the accounting profession will perhaps replace their human capacities with AI in order to reach a better future. Nowadays, many auditing, as well as business firms, have

---

G. H. Mardini
Associate Professor of Accounting, College of Business and Economics, Qatar University, Doha, Qatar

A. Alkurdi (✉)
Assistant Professor of Accounting, Faculty of Administrative and Financial Sciences, Aqaba University of Technology, Aqaba, Jordan
e-mail: Akurdi@aut.edu.jo

A. Hamdan et al. (eds.), *The Fourth Industrial Revolution: Implementation of Artificial Intelligence for Growing Business Success*, Studies in Computational Intelligence 935,
https://doi.org/10.1007/978-3-030-62796-6_18

applied AI in order to analyse a large amount of data in a short time, while, without this kind of new of technology such a volume of work would not be easily acceptable by humans [5, 6]. For instance, the AI used for administrative and accounting tasks in general, leads to structural changes within firms [7]. Moreover, AI is a completely automated process through which to generate data; it is used to prepare taxation reports with high level of accuracy and in a timely way [8]. Furthermore, AI can be used to recognise and categorise external (financial reports) and internal (managerial reports) from different resources within the firm automatically and with accuracy so that they can be submitted to the head of each department [9]. AI is thus invading every industry on the globe, and it is making a significant impact on the accounting world, since it save time, money and provides accurate data to both the regulators and customers [10]. However, prior studies have found that AI leads to an emphasis on further competitive disadvantages between the auditing firms and business.

Moreover, prior studies tackle AI from different perspectives within the accounting fields (i.e., auditing, financial, and managerial). The main objective of the current study is to extensively revise the literature review on AI and its impact on accounting. Moreover, this investigation leads us to critically identify the research problems of AI in accounting that support the researchers in investigating gaps in the research in the near future. These two main objectives conclude that there are two research questions. First, what is the impact of AI on accounting? What are the directions for future research about AI and accounting? In order to answer these research questions, the current study has applied a panel systematic dimensions approach. [11] argued that systematic reviews aim to address research problems by critically evaluating and integrating the findings of all of the relevant prior studies. Moreover, the current study contributes to our knowledge through achieving a well-established and systematic review, our study contributes to identifying relations, gaps and inconsistencies in the literature on AI and accounting in order to suggest new research gaps [12]. Moreover, our paper provides practical implications for decision and policy makers. The systematic approach aims to find all of the available published work addressing the phenomena by using research terms [13]. Moreover, [12] argued that the best way to discover the majority of published work is to intensively search at least two different electronic databases. In this sense, the current study employed databases and indices such as Ebsco, Google Scholar, ProQuest, Scopus, and searched some key publishers' databases through Emerald, Inderscience, Springer, Wiley, Elsevier, and Taylor and Francis, in order to cover most of the published research about the impact of AI on accounting.

The remainder of the paper includes the main body of the literature in Sect. 2, while Sect. 3 provides the discussion, conclusions and recommendations.

2 Literature Review: A Panel Systematic Approach

A systematic review follows a qualitative or a quantitative research synthesis, the latter is performed through a meta-analysis that focuses on estimations [14]. Moreover, it is applicable, whether the research questions concerns central tendency research, group contrasts and an association between variables [15]. On the other hand, the qualitative approach of systematic review is appropriate in finding whether there are conceptual changes overtime, in case the researcher aims to develop a new theory and a detailed discussion of prior studies through panels (dimensions) in a narrative manner [16]. A systematic narrative review requires a thorough coverage of prior studies; it also requires descriptions of existing findings that provide something new that can be highlighted the discussion and conclusions sections of a systematic review paper [11]. Moreover, narrative review requires balancing group and individual illustrations of studies per panel, rather than showing each individual study in detail [14].

The current study adopted a qualitative approach with detailed discussion of prior studies through panels and narrative review, each panel lists most of the critical prior studies that discuss the uses of AI in the accounting fields. Specifically, prior studies can be divided into five major panels to have an overall knowledge of AI with (i) financial accounting, (ii) auditing, (iii) cost and managerial accounting (C&M), (iv) taxation and (v) public accounting. Table 1 lists the prior studies on AI and accounting using a systematic panels approach. The rest of the current section discusses prior studies in depth per panel.

Panel A of Table 1 summarises prior studies that mainly explore AI through analysing the financial accounting field by utilising a variety of research tools. For instance, some of the prior studies have investigated AI and its effect on financial accounting together with its educational applications in various countries (i.e., Europe, Australia, Japan, the USA and the UK). However, the subjects of financial accounting varied across the prior studies. For instance, in terms of financial accounting education, [21] investigated the impact of accounting tutoring based on AI tools in relation to the accounting cycle. They found that tutoring using an AI-based system improves the textual performance by 27%, if compared to the traditional study method that is based on textbooks and instructors' notes. This finding agreed with [31], whose work investigated the accounting cycle from a practical perspective. They investigated the usefulness of the cloud platform in building an integrated system that related to the steps of the accounting cycle. They built a practical model that covered all stakeholders' decision-making needs. They found that this model accelerates the accounting cycle period while offering a high level of accuracy and efficiency. They suggested that this model helps to reduce storage costs, to avoid high audit risks and fees, and aids the reconciliations' step and the comparability between financial statements. On the other hand, [23] investigated the financial accounting indicators that measure the financial health of firms adopting an AI financial solution that predicts financial crashes before their occurrence. They hypothesised that the good financial health of a firm's micro financial system contributed to the economy

Table 1 Review of AI and its accounting dimensions

Panel A: AI and financial accounting	Panel B: on AI and auditing	Panel C: AI and cost and managerial accounting	Panel D: AI and tax accounting	Panel E: AI and public accounting
[17] Elam and Konsynski (1987)	[33] Bailey et al. (1987)	[48] Duchessi et al. (1993)	[61] Kuzniacki (2018)	[63] Bailey (1986)
[18] Connell (1991)	[34] Hansen et al. (1992)	[9] Liebowitz (2001)	[62] Kuzniacki (2019)	[64] Hollander and Icerman (1991)
[19] Kim (2005)	[35] Wongpinunwatana et al. (2000)	[49] Miko and Szantai (2002)	[8] Biryukov and Antonova (2019)	[65] Sharafi et al. (2016)
[20] Binner et al. (2009)	[36] Koskivaara (2004)	[50] Metaxiotis et al. (2004)		
[7] Fethi and Pasiouras (2009)	[5] Baldwin et al. (2006)	[51] Lopez-Ortega et al. (2006)		
[21] Johnson et al. (2009)	[37] Bogolea et al. (2007)	[52] Lawrynowicz (2007)		
[22] Gruning (2011)	[38] Kirkos et al. (2009)	[53] Bouhouras et al. (2010)		
[23] Benhayoun et al. (2013)	[39] O'Leary (2010)	[54] Chou et al. (2010)		
[24] Minhas et al. (2013)	[40] Omoteso (2012)	[55] Więcek (2013)		
[25] Senoguchi and Kurahashi (2013)	[41] Syed (2014)	[56] Zahin et al. (2013)		
[26] Mathawala and Walter (1989)	[42] Issa et al. (2016)	[57] Neipp G. (1987)		
[27] Luchsinger and Van Blois (1989)	[4] Greenman (2017)	[58] Chou et al. (2015)		
[10] Chukwudi et al. (2018)	[6] Kokina and Davenport (2017)	[59] Kłosowski and Gola (2016)		
[28] Wall (2018)	[43] Sutton et al. (2017)	[60] Ivanov and Webster (2017)		
[29] Thiel and Raaij (2019)	[44] Amin (2019)			
[30] Wimmer and Rada (2019)	[45] Zhang (2019)			
[31] Faccia et al. (2019)	[46] Gotthardt et al. (2019)			
[2] Petkov (2019)	[47] Ukpong (2019)			
[32] Mohd et al. (2020)	[3] Munoko et al. (2020)			

in the context of the recent financial crisis. They used a Support Vector Machine as an AI tool with which to make these predictions. They found that this tool was useful in predicting the firms' risks in an accurate manner during the sample period 2009–2011.

The remainder of the prior studies in Panel A analysed AI's impact, as well as building ind-depth models for financial accounting using various aspects (i.e., in relation to accounting concepts, forecasting experiments, bankruptcy, voluntary disclosure and financial accounting scenarios). They found that AI is useful and makes a positive change in the financial accounting aspects. For instance, [22] developed an AI measurement of financial voluntary disclosures. He used the annual reports relating to ten different forms of financial information that were disclosed, based on this computerised technique and without human interaction. He found that the model was positively associated with financial information asymmetry. He suggests that this disclosure model may aid a firm through running it as an additional measurement prior to the preparations of the actual reports. More recently, [2] investigated the functionality of financial accounting through the AI tools in business firms. They have provided interesting findings that add further practical examples of the impact of AI in relation to the aspects of financial accounting. They ran many accounting tasks and journal entries using AI tools and models, and suggested that their findings are a starting point for the implementation of AI into firms and in assisting them to re-structure their financial accounting system to adapt to an AI implementation.

Panel B of Table 1 demonstrates the role and impact of AI in the auditing functions and processes, some specific factors that relate to external and internal auditing functions, which have been influenced through the use of AI. Prior studies have identified AI's role in these specific factors relating to auditing firms, such as expertise, service quality and processes. For instance, [35] investigated the impact of AI systems on auditing tasks. They adopted two problem-solving programs in a laboratory experiment utilising undergraduate auditing students. They found that the adopted AI systems enhanced the accuracy found in solving problems with structured tasks, while it showed a minor increase in the accuracy of the solution of unstructured tasks. Koskivaara [36] conducted an AI tool, called the artificial neural network, in the auditing process for the sampled firms. They found that this tool's architectures and learning parameters might serve as an analytical review instrument for the auditing process. They suggested that this tool needs further attention from academics in their teaching, and from practitioners in relation to their auditing processes and functions. More recently, [3] investigated AI in the auditing process in a more in-depth way. They found that AI is a technology whose use in increasing nowadays, but which might replace the human auditor in the next 10 years. This finding is interesting, since AI aims to mimic the skills and judgments of humans, and this will lead to a competitive disadvantage for the adopter in the next ten years. Briefly, the findings of prior studies, shown in Panel B, are unified in terms of their views and suggestions in relation to AI. Specifically, they found that AI in the auditing area is beneficial to its planning of risk assessments, concerning transactions with less of an audit paper base, to efficient financial periods and faster data analysis. They suggest that AI is continuing to replace the traditional human auditing functions and processes that may soon be fully awoken, which may lead to consequences for the traditional auditing profession.

Duchessi et al. [48] argued that AI would affect management accounting, the business organization and cost classification in the future. Panel C shows papers that have examined the role of AI in cost and managerial accounting. Specifically, prior studies of AI have investigated its impact on the main procedures, concepts and cost allocation of C&M. Moreover, it discusses the budgeting cycle and process. However, the findings vary between results that shown that there is a significant impact and those demonstrating that AI has no significant role in C&M. For instance, [55] investigated the impact of AI on the manufacturing process parameters that evaluate costs and their classification. He found that an AI solution needed some particular conditions in order to be effective, one of the main conditions that they proposed is effective for small-batch production lines only. He thus suggested that AI puts some constraints on cost classification and large manufacturing organisations. [54] tackled the production costs and processes in relation to different factors. Specifically, they assessed the prediction of production costs' accuracy and the time required to ascertain this when using AI tools, such as artificial neural networks and hybrid intelligence. They found that AI tools used for cost forecasting could benefit the managers and allow them to conclude on the costs through non-linear estimation relationships. In terms of a job order costing system, [59] examined the risks of cost estimation with AI for manufacturing job orders. They applied their model, with AI hybrid tools and Monte Carlo Simulation, to historical data. They found that the AI was a useful tool for making predictions of costs within the job order costing system. In alignment with [54], they suggested that their model is beneficial for predicting job order costs through non-linear estimation relationships.

On the contrary, some prior studies, shown in Panel C of Table 1, found that AI has drawbacks for the cost and management accounting systems. For instance, [9] found that AI has a negative impact on the knowledge management of cost classifications. Their AI model constrains the knowledge management of the sampled management practitioner against AI soft computing and virtual reality tools with high-speed responses, but with an enormous level of inaccuracy. Moreover, [51] documented that the product lifecycle system has difficulties with its customization and enhancement for the long term of product lifecycle that are exacerbated by the AI tools.

Panel D of Table 1 illustrates the few prior studies that have been conducted to date that have investigated the impact of AI on the tax accounting area, and the benefits of AI to tax regulators of formalizing a database that helps an accountant to be more efficient with the different services that they can provide to clients. However, the AI benefits to taxation were minimal, and require human engagement. For instance, [61] concluded that AI, such as neural networks, could work as a useful tool with which to reduce tax avoidance. Specifically, he suggested that AI could help tax advisers, tax managers, and tax systems to operate as a network that enhances and improves tax work to minimize illegal taxation behaviour. Yet, he argued that AI is beneficial to both the tax attorney and the professional accountant, but it cannot capture some tax behaviours, such as tax evasion, since it is classified as a non-routine task that requires planning and experience to be performed in the long term. [8] asked whether AI tools, such as neural network mathematical modelling usage, enhance the analysis of tax accounting. They agreed with [61], that the AI technological revolution could not control tax efficiently, because effective neural network modelling needs to construct fundamental concepts and the use of preliminary data processing, especially with non-routine taxation tasks.

Finally, Panel E discusses the few literary works on the impact of AI on productivity in governmental financial operation, the effectiveness of AI on public accounting examined on early stages, if compared to the remainder of the accounting fields. Prior studies have concluded that AI is very useful for public accounting. For instance, [63] discovered empirically that AI offered a valuable enhancement to governmental decisions and support systems. More recently, [67] investigated the impact of AI on government capital budgeting. They found that governments more efficiently operated their works related to capital planning by enhancing their practical decision models using AI systems. Another, more recent, study in the current decade, [65] found that advanced AI tools, such as local linear neuro networks and multi-layer perceptron models positively influenced the estimation of future cash flows in public accounting.

3 Discussion

A systematic qualitative literature review discusses the focal points of the phenomena that are being addressed [12]. Moreover, it leads to conclusions and recommendations and identifies the research gaps for future research papers. The current study has revised the literature review of AI and its impact on accounting in order to identify research problems relating to AI in accounting, and this will support researchers in conducting future research. From the intensive review adopted in the current study, there emerged great insights that need discussion. First, the findings of Panels A and B show a clear vision that AI will be beneficial to financial accounting and auditing in the future, it will dominate these two accounting fields. In particular, prior studies shown in Panel A found that AI is able to handle and process most of the financial accounting tasks that are carried out on an automated level without human interference. We can thus conclude that one of the key benefits of AI to financial accounting is the compliance of data processing and generation. Yet, AI cannot replace the fundamental accounting concepts that require human interaction. In terms of auditing (Panel B), the prior studies have agreed that AI's machine learning provides augmented analyses to the auditing processes, which may lead to consequences for the traditional auditing profession. We believe that AI can be considered a support tool to auditors, but it will not replace those auditors. This conclusion is correct, since we already have some AI tools, like Interactive Data Extraction, Analysis, and Audit Command Language, which are computer-aided solutions that assist the auditors to complete their tasks. In this sense, AI's machine learning for auditors can provide neither positive nor negative reports on the performance of the firm and its transparency and fair representation of its financial information. Moreover, advanced auditing applications identify the pattern of the transactions either to classify them as a normal transaction or to cause them to be red flagged as a transaction (a fraudulent transaction) for review. This important step in the auditing process may lead to data bias risk in cases where they are generated by AI tools without human interaction [6]. Even though AI tools generate the auditing data and process, a human auditor has to confirm its accuracy and its truthfulness in order to reflect the reality, and must thus then run a complete revision process. In summary, AI's machine learning invades the auditing process, and the human auditor role will be updated from being a performer in the auditing process to becoming a designer of the process, which will cover monitoring its effectiveness, accuracy and the freedom from biased data, with the interpretation of the results [45].

The findings of the prior studies that are listed in Panel C vary from dealing with the benefits and ethics of AI, to cost effectiveness and management accounting. These prior studies have agreed that AI improves the cost and managerial accounting by companies, since it helps to recognize, classify and assign the costs from different sources within the firm, to the right cost classification (i.e., direct material, manufacturing overheads) within different managerial systems and methods (i.e., job ordering system, product life cycle). Specifically, AI's machine learning has many implications for cost and managerial accounting. For instance, AI may be used to help to estimate the production costs, based on the re-classification of historical transactions. However, prior studies have argued that the natural nature of the costs may lead to bias in the AI's cost classifications, which requires human interaction [59]. In other words, AI can be used to analyse and classify the costs in order to run forecasting models; yet, the quality of the data set that is generated by the AI tools has the risk of being biased, which harms the quality of the predictions. We thus conclude that human validation of the classification of the cost is essential in order to revise the potential errors before the prediction process in order to enhance its quality.

There is a scant number of prior studies that have investigated the impact of AI on taxation (Panel D). However, these studies are helpful in beginning a discussion around the notion that AI tools are useful to tax accounting, but the implications of the different tools of AI need expertise from those with a related expertise in taxation. In other words, data generated by AI to prepare a tax report is assuring accuracy and will be generated quickly, since a tax report is considered a tax routine-task [8]. In other words, it is important for both tax authorities and firms to employ AI in their taxation system. For tax regulators, it can do numerous tasks with its analytical capability; understanding the client effectively, and can be alert to any future circumstance. For firms, AI can maintain their competitive advantages for their reputation. This means firms need to be more compliant with regulatory requirements, prepared for future risk, and their monitoring capabilities enhanced. Yet, human interaction, effectiveness and regulations are required for non-routine tax behaviours, such as tax avoidance and tax evasion. Tax regulators may design the tax report process to be run by the AI tools, but unusual tax behaviours required a full control process that is carried out by humans. In terms of AI with public accounting, Panel E has listed only a few papers about this relation. At the early stages of the AI revolution, prior studies found that it was useful to governance procedures and decisions. Nowadays, we believe that most of the governance services and payments are based on applications and AI advanced tools with an automated well-established control over their cash flows. More importantly, public accounting is widely interested in using AI's capabilities to generate efficient annual budget predictions within high-unpredictability circumstances. However, we believe that internal governance control is required to evaluate the effectiveness of governance procedures over the AI's machine learning applications and tools. Governmental control is required to choose the best method for governance procedures in order to reduce the risk of negative impacts on cash flows, and to weave in the equality of citizens' rights.

4 Conclusions and Future Research

In general, AI dominates all of the sciences. The accounting profession is not isolated from the impact of AI. Accounting researchers have explored the benefits provided by AI in performing huge tasks by using techniques such as fuzzy logic, neutral networks, hybrid systems, and other tools of AI that have never before been applied in an accounting field. As a conclusion, it is not surprising to see an increasing concern about the impact of AI on the accounting field, since AI has replaced humans in many tasks, which usually increases the company's productivity and efficiency and causes some cost reduction. Accounting, in general, and auditing practices especially, have benefited from the continuous improvements in AI technology. To answer the first research question of the current study, we thus conclude that the impact of AI on the five accounting dimensions that are listed in the current study is very beneficial and positive. Yet, human interaction, monitoring and revision are required. In accordance with our main conclusion, we believe that AI will definitely release accountants from routine accounting tasks and processes, but this does not mean that human accountants will be jobless in the future; instead, it will enhance their role. In other words, even though AI may perform all of the accounting calculations, processes, cost classifications and tax reports, a human accountant should be controlling, analysing and revising the process in order to draw up meaningful results and recommendations.

Accordingly, the role and impact of AI on accounting is still vague. In order to answer our second research question, we do recommend that future research be conducted. Through our intensive review of prior studies and its discussion of this topic, we have identified research problems and gaps for potential future research. First, we provide a call to innumerable researchers within this area to not only explore the benefits and impact of AI, but also to investigate its ethical, social and economic implications for accounting. Second, we found that human accountants cannot be replaced, thus, future studies may investigate the extent of human involvement in the AI era. Third, we call upon researchers to conduct future research on the impact of AI on accounting aspects that have not yet been investigated. For instance, the impact of AI on (i) earnings management and quality, (ii) the implementation of International Financial Reporting Standards, (iii) environmental (carbon) accounting, (iv) internal control systems, and (v) the internal auditing process. A limitation of the current study is that the search and coverage criteria of prior studies has an element of subjectivity; we may have missed a few studies that have not been covered in the current study. However, we have reduced this element through validating our search by re-searching the databases by all the authors. We have also enhanced the reliability of our coverage through searching in most of the databases and publishers that are recognized around the globe.

References

1. Smith, S.: Blockchain, Artificial Intelligence and Financial Services: Implications and Applications for Finance and Accounting Professionals, pp. 83–99. Springer, Cham (2020)
2. Petkov, R.: Artificial Intelligence (AI) and the accounting function-a revisit and a new perspective for developing framework. J. Emerg. Technol. Acc. **16**(2) (2019)
3. Munoko, I., Brown-Liburd, L., Vasarhelyi, M.: The ethical implications of using artificial intelligence in auditing. J. Bus. Ethics **163**(1), 1–26 (2020)
4. Greenman, C.: Exploring the impact of artificial intelligence on the accounting profession. J. Res. Bus. Econ. Manag. **8**(3), 1451–1454 (2017)
5. Baldwin, A., Brown, C., Trinkle, B.: Opportunities for artificial intelligence development in the accounting domain: the case for auditing. Intell. Syst. Acc. Finan. Manag.: Int. J. **14**(3), 77–86 (2006)
6. Kokina, J., Davenport, T.: The emergence of artificial intelligence: how automation is changing auditing. J. Emerg. Technol. Acc. **14**(1), 115–122 (2017)
7. Fethi, M., Pasiouras, F.: Assessing bank efficiency and performance with operational research and artificial intelligence techniques: A survey. Eur. J. Oper. Res. **204**(2), 189–198 (2010)
8. Biryukov, A., Antonova, N.: Expert Systems of Real Time as Key Tendency of Artificial Intelligence in Tax Administration, pp. 111–118. Springer, Cham (2018)
9. Liebowitz, J.: Knowledge management and its link to artificial intelligence. Expert Syst. Appl. **20**(1), 1–6 (2001)
10. Chukwudi, L., Echefu, S., Boniface, U., Victoria, C.: Effect of artificial intelligence on the performance of accounting operations among accounting firms in South East Nigeria. Asian J. Econ. Bus. Acc. **7**(2), 1–11 (2018)
11. Baumeister, R., Leary, M.: Writing narrative literature reviews. Rev. Gen. Psychol. **1**(3), 311–320 (1997)
12. Cooper, H.: Editorial. Psychol. Bull. **129**, 3–9 (2003)
13. Bem, D.: Writing a review article for psychological Bulletin. Psychol. Bull. **118**, 172–177 (1995)
14. Sternberg, R.: Editorial. Psychol. Bull. **109**, 3–4 (1991)
15. Lipsey, W., Wilson, D.: Practical Meta-Analysis. Sage Publications, London (2001)
16. Baumeister, F.: Writing a literature review. In: Prinstein, M.J., Patterson, M.D. (eds.) The Portable Mentor: Expert Guide to a Successful Career in Psychology, 2nd edn., pp. 119–132. Springer Science + Business Media, New York (2013)
17. Elam, J., Konsynski, B.: Using artificial intelligence techniques to enhance the capabilities of model management systems. Decis. Sci. **18**(3), 487–502 (1987)
18. Connell, N.: Expert systems in accountancy: a review of some recent applications. Acc. Bus. Res. **17**(67), 221–233 (1987)
19. Kim, D., Shin, H., Lim, M., Lee, J., Kim, H.: Designed strength identification of concrete by ultrasonic signal processing based on artificial intelligence techniques. Trans. Ultrasonics Ferroelectr. Freq. Control **52**(7), 1145–1151 (2005)
20. Binner, J., Gazely, A., Kendall, G.: An evaluation of UK risky money: an artificial intelligence approach. Glob. Bus. Econ. Rev. **11**(1), 1–18 (2009)
21. Johnson, B., Phillips, F., Chase, L.: An intelligent tutoring system for the accounting cycle: enhancing textbook homework with artificial intelligence. J. Acc. Educ. **27**(1), 30–39 (2009)

22. Gruning, M.: Artificial intelligence measurement of disclosure (AIMD). Eur. Acc. Rev. **20**(3), 485–519 (2011)
23. Benhayoun, N., Chairi, I., Gonnouni, A., Lyhyaoui, A.: Financial intelligence in prediction of firm's creditworthiness risk: evidence from support vector machine approach. Proc. Econ. Finan. **5**(13), 103–112 (2013)
24. Minhas, S., Poria, S., Hussain, A., Hussainey, K.: A review of artificial intelligence and biologically inspired computational approaches to solving issues in narrative financial disclosure. In: International Conference on Brain Inspired Cognitive Systems, pp. 317–327. Springer, Berlin, Heidelberg (2013)
25. Senoguchi, J., Kurahashi, S.: Prediction of financial crises using statistic model and intelligent technologies in ubiquitous environments. Int. J. Comput. Appl. Technol. **48**(2), 173–183 (2013)
26. Mathawala, Y., Walter, P.: Artificial intelligence: Implications for banking. Int. J. Comput. Appl. Technol. **2**(3), 191–197 (1989)
27. Luchsinger, P., Blois, V.: Developments in Artificial Intelligence in manufacturing. Int. J. Comput. Appl. Technol. **2**(1), 62–65 (1989)
28. Wall, L.: Some financial regulatory implications of artificial intelligence. J. Econ. Bus. **100**, 55–63 (2018)
29. Van Thiel, D., Van Raaij, F.: Artificial intelligence credit risk prediction: an empirical study of analytical artificial intelligence tools for credit risk prediction in a digital era. J. Risk Manag. Finan. Inst. **12**(3), 268–286 (2019)
30. Wimmer, H., Rada, R.: Applying artificial intelligence to financial investing. In: Advanced Methodologies and Technologies in Business Operations and Management. IGI Global – USA, pp. 1–16 (2019)
31. Faccia, A., Al Naqbi, K., Lootah, A.: Integrated cloud financial accounting cycle: how artificial intelligence, Blockchain, and XBRL will change the accounting, fiscal and auditing practices. In: 3rd International Conference on Cloud and Big Data Computing, pp. 31–37 Oxford, UK (2019)
32. Rahim, S., Mohamad, Z., Bakar, J., Mohsin, H., Isa, N.M.: Artificial intelligence, smart contract and Islamic finance. Asian Soc. Sci. **14**(2), 145–154 (2018)
33. Bailey, D., Meservy, D., Turner, H.: Decision support systems, expert systems, and artificial intelligence: realities and possibilities in public accounting. Ohio CPA J., 11–15 (1986)
34. Hansen, J., McDonald, J., Stice, J.: Artificial intelligence and generalized qualitative response models: an empirical test on two audit decision-making domains. Decis. Sci. **23**(3), 708–723 (1992)
35. Wongpinunwatana, N., Ferguson, C., Bowen, P.: An experimental investigation of the effects of artificial intelligence systems on the training of novice auditors. Manag. Audit. J. **15**(6), 306–318 (2000)
36. Koskivaara, E.: Artificial neural networks in analytical review procedures. Manag. Audit. J. **19**(2), 191–223 (2004)
37. Bogolea, D., Boyle, J., Shindyapin, V.: Artificial-intelligence-based energy auditing, monitoring and control. U.S. Patent Application, pp. 1–6 (2007)
38. Kirkos, E., Spathis, C., Manolopoulos, Y.: Audit-firm group appointment: an artificial intelligence approach. Intell. Syst. Acc. Finan. Manag. **17**(1), 1–17 (2010)
39. O'Leary, E.: The impact of gartner's maturity curve, adoption curve, strategic technologies on information systems research, with applications to artificial intelligence, ERP, BPM and RFID. Marshall School of Business Working Paper No. ACC, pp. 3–11 (2010)
40. Omoteso, K.: The application of artificial intelligence in auditing: looking back to the future. Expert Syst. Appl. **39**(9), 8490–8495 (2012)
41. Syed, M.: The role of artificial intelligence in the development of accounting systems: a review. IUP J. Acc. Res. Audit Pract. **13**(2), 7–19 (2014)
42. Issa, H., Sun, T., Vasarhelyi, A.: Research ideas for artificial intelligence in auditing: the formalization of audit and workforce supplementation. J. Emerg. Technol. Acc. **13**(2), 1–20 (2016)

43. Sutton, S.G., Holt, M., Arnold, V.: The reports of my death are greatly exaggerated"—Artificial intelligence research in accounting. Int. J. Acc. Inf. Syst. Elsevier **22**(C), 60–73 (2016)
44. Amin, E.: Financial fraud detection of the Egyptian companies annual reports using artificial bee colony algorithm. Int. J. Bus. Data Anal. **1**(2), 184–201 (2019)
45. Zhang, B., Dafoe, A.: Artificial intelligence: American attitudes and trends, pp. 1–110 (2019)
46. Gotthardt, M., Koivulaakso, D., Paksoy, O., Saramo, C., Martikainen, M., Lehner, O.M.: Current state and challenges in the implementation of robotic process automation and artificial intelligence in accounting and auditing. ACRN Oxford J. Finan. Risk Perspect. **8**, 31–46 (2019)
47. Ukpong, E.G., Udoh, I.I., Essien, I.T.: Artificial intelligence: opportunities, issues and applications in banking, accounting, and auditing in Nigeria. Asian J. Econ. Bus. Acc. **10**(1), 1–6 (2019)
48. Duchessi, P., O'Keefe, R., O'Leary, D.: A research perspective: artificial intelligence, management and organizations. Intell. Syst. Acc. Finan. Manag. **2**(3), 151–159 (1993)
49. Mikó, B., Szantai, M.: Artificial intelligence methods in early manufacturing time estimation. In: Proceedings of Gepeszet, pp. 535–539 (2002)
50. Metaxiotis, K., Ergazakis, K., Samouilidis, E., Psarras, J.: Decision support through knowledge management: the role of the artificial intelligence. Int. J. Comput. Appl. Technol. **19**(2), 101–106 (2004)
51. Lopez-Ortega, O., Sapidis, N., Wallace, D.: Challenges for developing intelligent, interactive and cooperative PLM systems: introductory article on applications of artificial intelligence and virtual reality to product lifecycle management. Int. J. Product Lifecycle Manag. **1**(3), 195–210 (2006)
52. Lawrynowicz, A.: Production planning and control with outsourcing using artificial intelligence. Int. J. Serv. Oper. Manag. **3**(2), 193–209 (2007)
53. Bouhouras, S., Labridis, P., Bakirtzis, G.: Cost/worth assessment of reliability improvement in distribution networks by means of artificial intelligence. Int. J. Electr. Power Energy Syst. **32**(5), 530–538 (2010)
54. Chou, S., Tai, Y., Chang, J.: Predicting the development cost of TFT-LCD manufacturing equipment with artificial intelligence models. Int. J. Prod. Econ. **128**(1), 339–350 (2010)
55. Więcek, D.: Implementation of artificial intelligence in estimating prime costs of producing machine elements. Adv. Manuf. Sci. Technol. **37**(1), 44–53 (2013)
56. Zahin, S., Latif, H., Paul, K., Azeem, A.: A comparative analysis of power demand forecasting with artificial intelligence and traditional approach. Int. J. Bus. Inf. Syst. **13**(3), 359–380 (2013)
57. Neipp, G.: Artificial intelligence and its impact on industry: A new dimension within the framework of computer–integrated manufacturing (CIM). Int. J. Technol. Manag. **2**(5–6), 743–760 (1987)
58. Chou, S., Lin, W., Pham, D., Shao, Y.: Optimized artificial intelligence models for predicting project award price. Autom. Constr. **54**, 106–115 (2015)
59. Kłosowski, G., Gola, A.: Risk-based estimation of manufacturing order costs with artificial intelligence. In: 2016 Federated Conference on Computer Science and Information Systems, pp. 729–732. FedCSIS, Gdansk (2016)
60. Ivanov, S., & Webster, C.: Adoption of robots, artificial intelligence and service automation by travel, tourism and hospitality companies – a cost-benefit analysis. In: International Scientific Conference, Contemporary Tourism – Traditions and Innovations, Sofia, Bulgaria, pp. 1–9 (2017)
61. Kuzniacki, B.: The artificial intelligence tax treaty assistant: decoding the principal purpose test. Bull. Int. Tax. **72**(9) (2018)
62. Kuzniacki, B.: The marriage of artificial intelligence and tax law: past, present, and future. Present Fut., 1–9 (2019)
63. Bailey, A.D., Meservy, R.D., Duke, G.L., Johnson, P.E., Thompson, W.: Auditing, artificial intelligence and expert systems. In: Decision Support Systems: Theory and Application, pp. 265–319. Springer, Berlin, Heidelberg (1987)

64. Hollander, S., Icerman, C.: Capital budgeting in governments: the feasibility of artificial intelligence technology. Expert Syst. Appl. **3**(1), 109–116 (1991)
65. Sharafi, A., Iranmanesh, H., Amalnick, M.S., Abdollahzade, M.: Financial management of Public Private Partnership projects using artificial intelligence and fuzzy model. Int. J. Energy Stat. **4**(2) (2016)

The Ethical, Professional Practices and Social Implications of Artificial Intelligence

Ethics of Artificial Intelligence and the Spirit of Humanity

Ismail Noori Mseer

Abstract The present world is experiencing the manifestations of the Fourth Industrial Revolution and the emergence of artificial intelligence products. Those that a person interacts with directly after the smartphone created an overall invasion of social, political, economic, and cultural details. The reality is that the achievements of artificial intelligence are not considered something new, but rather have their roots and important efforts. This was the subject of many projects affecting the directions of companies, giant institutions, and laboratories, while the attention and direct interest and care for it come from the nature of the relationship that has been growing between the ordinary person and the overall smartphone programs, which have been testing the trust in this relationship daily between a smart machine and a human. The study tries to focus on tracking the outcome of this relationship, especially on the moral level, so that the orientation is towards launching the most present question, which is related to who affects whom? The current reality indicates that the main influencer is based on the decisive human role in directing, managing, and programming the smart machine. Rather, the intelligence of the machine up to this point is based on the metaphor, given that the smartphone reflects the intelligence that has been reached by man, through the course of his technological development through history. But the nature of the interaction involved in the software matrix, which is the backbone of the computer's operation, makes it important to think later about the interactions that could result in more consequences.

Keywords The turing test · Internet of things · Artificial intelligence · Cypher · The three laws of Asimov

I. N. Mseer (✉)
Ahlia University, Manama, Bahrain
e-mail: imseer60@gmail.com

A. Hamdan et al. (eds.), *The Fourth Industrial Revolution: Implementation of Artificial Intelligence for Growing Business Success*, Studies in Computational Intelligence 935,
https://doi.org/10.1007/978-3-030-62796-6_19

1 Introduction

Why the ethics of artificial intelligence is an important topic? And is there a threat that could be observed around the products of this immense knowledge field, promising further achievement? In fact, the command does not stop at the observations of specialists but has become circulated in general. Reality began to produce more results that can be observed on the level of the daily life, as more daily habits and behaviors emanating from the human being were prevalent due to the trust in the established relationship between man and the intelligent machine to the extent that it became impossible to remove the man from the hegemony and control of this Arrogant device. Nowadays people can watch media and entertainment for long hours on their mobile devices, to the extent that threatens the habit of early waking that humanity is used to, making it harder to spend their normal daily life. Thus, psychologists have spoken out about the widespread computer addiction phenomenon. While the road accidents monitoring institutions began to indicate, clearly, an increase in the number of car accidents on public roads, due to the preoccupation of vehicle drivers with mobile devices. However, the bitter did not stop at this point, as the conditions of the cultural and social clash were aggravated by the spread of social media until it became a field for spreading rumors, a wide exhibition space for spreading talks full of hatred and racism, ostracism and exclusion, and aggravated conditions of privacy violation.

Scientists and specialists are heading towards a determined effort to confront these effects and work to develop appropriate solutions and treatments. However, the most important thing in all of this is based on the importance of looking carefully at the imminent moral prospect, especially since artificial intelligence works with energy, power, and capabilities that exceed the ability of humans to keep up with it. From this, thinking about the basic rights that humanity has long sought, is extremely important in light of the complexity of the websites and the kinds of multimedia platforms that have been publishing many details on reality. Hence the importance of reviewing the ethical rules and standards governing the level of relationship between the intelligent machine and the human element; rules that must review social values, rights, behaviors, customs, and practices. It is critical to monitor the conflicts of interest that can arise between the ordinary citizen and the giant corporations which their existence depends on the achievements of artificial intelligence.

The moral regulator is of exceptional importance in the way the public deals with the achievements of artificial intelligence. This is evident in how the audience responds to companies that rely on artificial intelligence in the way they work, such as Microsoft, Google, etc. Especially in terms of focusing on respecting consumer privacy and working hard to respect the principles and values that they believe in. In this, it is worth noting the crisis experienced by Facebook after being accused of exploiting the users' data and working to achieve commercial benefits from their data. From this reality emerges the moral side, considering the importance of focusing on values such as integrity, privacy, security, safety, transparency, and trust. The study aims to reinforce the ethical aspect of the activating method and the achievements of artificial intelligence; as they endeavor to establish the field for joint responsibility in facing the challenges the "Fourth Industrial Revolution" poses. It is a responsibility that does not stop at specialized institutions and companies, rather it requires a concerted effort, both private and public, and the importance of all disciplines to read the reality of the interaction resulting from artificial intelligence, given its direct relationship to the future of humanity.

2 Taking Control of Nature

The relationship between the human race and the planet Earth took the path it did due to the conditions of the human race's will to adapt. The history of humanity indicates the incentive to respond to the totality of material and emotional transformations. Thus, from the communal stage, the man began to understand the experience of living, with the first challenge, which is the instinct for survival, and the subsequent confrontations that nature continued to produce. While interacting was in his exotic conditions an expression of emotions, feelings, and psychological stimulation. Until the human established himself a system of living interactions, which were represented in culture, civilization, humanity, rationality, religion, and wars. While his idea of reassurance remained, so did his obsession with the presence on the planet Earth. And that will be expressed in the search for happiness at times, and the idea of utopia in other times. Those that have led the arguments of some philosophers. While the origin of all this was based on the activity of "the adaptation of nature", that human interacting has been practiced since the stone ages, where the aspiration of the primitive man towards the adaptation of the hard stone in an attempt to confront the details that abound in his environment and surroundings that he lives on. Up to the fourth industrial revolution, and its resulting gains and achievements, it became called the cyber era. And there was a need for a science of cybernetics, which the human race has reached in a never-ending series of enduring attempts of innovating, creating, thinking, brainstorming, bitterness, disappointments, losses, and successes.

Here is the person of the cyber age, his brightness surges and has reached the ability to summarize the totality of human innovations and discoveries over thousands of years through a smartphone, high-speed computer, high-tech scientific laboratories, optical fibers, nanotechnology, and the development he has achieved in the fields of predicting accurate weather forecasts, wind movement, hurricanes, and earthquakes. however, he is still unable to tackle present issues, like the Coronavirus. The World Health Organization even answered a question: (Do vaccines against pneumonia protect you from the new Coronavirus? The direct answer was issued on February 29, 2020, and via the WHO website; "No. Vaccines against pneumonia, Pulmonary, and Hepatitis Type B, do not offer protection against the new Coronavirus". The virus is so different that it needs a new vaccine. Researchers are trying to develop a vaccine against COVID-19, and WHO is supporting their efforts. (emergencies/ diseases/ new-coronavirus-2019)

This description does not mean that the human race has failed in dealing with the Coronavirus, as it has faced throughout history more severe and more catastrophic and disastrous fates. But observing this crisis is based on a method that represents the concept of the fourth industrial revolution and artificial intelligence. Until the near past, some people were convinced that the time for scientific miracles has come and that there is no obstacle in the face of this overwhelming industrial development.

The matter here does not fall into the trick of searching for strengths and weaknesses or questioning the imbalance of the development system that the human mind has reached in the field of the fourth industrial revolution, and what was achieved by the efforts of artificial intelligence. However, the study is concerned here with the moral topic, which relates to the way people view themselves first and view their surroundings second. The Coronavirus crisis will not be the first nor it be the last, life remains full of surprises and emergency events and catastrophes. Rather, reality continues to indicate conditions of human disability in the face of more incurable diseases such as cancer. While the psychological and material obstacles remain, and they have expanded in their heavy shadow on the overall details of contemporary life, despite the big steps that science and technology have been achieving. The matter is not an expression of disappointment in light of the terror sweeping the world, as the discovery of a vaccine may be announced within hours of editing this paper. As far as it is a direct reference to the reality of the details that abound in life by asking questions like what is good and what is bad, and how to find the will to live as described in the novel by Erich Maria Remarque (A Time to Love and a Time to Die) [1].

3 Finding the Key to Happiness

The Encyclopedia Britannica defines Artificial Intelligence (AI) as " the ability of a digital computer or computer-controlled robot to perform tasks commonly associated with intelligent beings. The term is frequently applied to the project of developing systems endowed with the intellectual processes characteristic of humans, such as the ability to reason, discover meaning, generalize, or learn from earlier experience". (britannica.com) Here we come face to face with the problem of the study as we look towards focusing on the effectiveness of performing tasks on behalf of humans. The moral problem here is based on the interpretation of the term artificial intelligence, without creating any confusion. The most obvious statement is based on (enabling the computer to perform related tasks) usually [with the help of the human being].

Is there another reliable definition associated with the term artificial intelligence that tries to distance itself from this goal? Researchers in many search institutions, scientific periodicals, and academic books almost unanimously adopt this definition. Whereas the current endeavor focused on enhancing the Cyber path where computer culture, information technology, and virtual reality as the implications unfold in the Internet of Things. The Internet of Things is the concept of connecting any device (as long as it has an on/ off switch) to the Internet. It is a giant network of things and people connected, all of whom collect and share data about the environment around them [2].

We can say in principle, at the very least, that artificial intelligence is based on man's command computers to take charge of all the activities usually associated with him, from thinking, discovering meaning, generalizing, and benefiting from experiences. And the development did not stop at this point, but the Internet of Things came to take the man out of the equation after the computer was assigned to open the doors of the house and clean it, to prepare food, and operate and stop all the buttons of the house; actually, there is nothing left but to carry the person to his bed and to tell him a bedtime story. The Internet of Things can make a massage for a man's body based on the accuracy generated by organized accounts, based on the study of the genetic map of the human being. Is there something more powerful than this organized, systematic, accurate, and based on a computer program that can hardly be mistaken?! The promising thing here is that the computer, which is just a deaf machine, can be equipped with the information that a person desires, and thus a professional will provide the computer, robot, or internet of things with the totality of habits, behaviors, desires, the genetic map, and the psychological structure of the customer who wants this service... And since the commissioning of all works was entrusted to the computer machine, the question of the future is based on the following: Who is responsible (on the level of future generations) for supplying smart machines, after the movement of people has been disrupted, and the computer has become a proxy for most business? The answer does not is one of the following (the assistant, the companion, the owner, the deputy) call it whatever you want, so that we remain in a circle that cannot be escaped, those that are based on the computer, and the multiples that arise from the approach and method of mechanical thinking that is mounted on it.

Science fiction has predicted that the human race will need robot facilities, which will facilitate the handling of living reality. And if these predictions had drawn some kind of a sarcastic smile on our faces, when we were watching science fiction films produced by Hollywood at the beginning of the second millennium; what happens in the year 2020 makes us almost convinced about the importance of these facilities to humans, which should be entrusted with the task of explaining, interpreting, facilitating, and simplifying. After the waves of software began to invade contemporary life with great abundance.

4 Being Cyber

If the world is experiencing moments of bliss, pleasure, and celebration due to the achievements of artificial intelligence in the present times, can it deal with the phenomenon based on positive reception? Given that humanity has reached one of the effective means that achieves its dream of obtaining perfection based on applications that cannot go wrong. There is no doubt that artificial intelligence carries more shipments of promise to advance, burn stages, and develop science and knowledge.

But the more present question remains stuck around; can all of this pass without consequences?! And that is the issue that the study attempts to ponder.

If mankind has grasped the achievements of artificial intelligence through the urgency of intense competition between the smartphone giants iPhone and Samsung, where the famous series of smartphone models. The two companies have doubled the payloads of profit, influence, and presence on the planet's population. They even left crumbs for the rest of the competitors in this field. This is on the level of common knowledge to the general public. Smartphones make up only the tip of the hidden iceberg of artificial intelligence applications, and it is the most complex and intricate field. The applications in space science, medicine, genetics, neural networks, human brain stimulation, movement, and thinking. Indeed, the human being lives in conditions that suggest reassurance.

We can assume the following scenario about an emergency that a single man was subjected to, and under urgent circumstances, he was obliged to travel to a remote country. The reference to history is very important because accelerating developments dictate specificity here. And let us choose the date of March 2020, for example, as this paper is being edited on this date. And let's go ahead with the single man who left everything behind and rushed to prepare his suitcase carrying his smartphone loaded with smart software. The hypothesis says that he received a phone call via WhatsApp telling him it is important to travel right away. The man immediately responded, hurrying to download one of the travel programs that provided him with an amazing price after making comparisons with the prices of competitors. He immediately rushes to the airport after calling an Uber ride. He did not hesitate to display the smartphone screen in front of the airline employee, who refers him to the plane gate number in the airport building, in an attempt to occupy his seat safely and comfortably in the departure plane. On arrival at the hotel, he completed the online check-in procedure. Remember that during his busyness, he forgot to turn off appliances, or inform the building guard of the importance of monitoring the apartment for fear of burglary. But he laughed when this thought comes to his mind.

These are the only remnants of the era before the smartphone. The whole thing is under control. The programs uploaded on the phone are everything, starting from monitoring the home and sending necessary alerts to contact the police in the event of an emergency, until the remote control that controls every button in the house. The home security mission has been secured, so what about work? This matter is more accessible and organized, he hurries towards his e-mail to schedule appointments and prepares a meeting through Skype program, so that the meeting and the conference with the coworkers can be done through video and sound. The interactions of artificial intelligence have reached a wide range and affect the overall range of behaviors and situations that humans express in the present day; were growing cyber-dependency has been dominating the system of ideas, values, and standards. It became very natural for your privacy to be compromised on Facebook, and you no longer have the right to express your dissatisfaction with whoever ignores you as he is busy with another conversation that is being held on his smartphone, it is a confusing condition, the one that forces you to accept new ethics which were previously unknown. Rather, the current disregard and in the shadow of the cyberethics, it makes you a stranger if you demanded a little respect that you were raised to at some point, before the development of artificial intelligence into human reality. It is the conditions of new values, behaviors, and customs, which are imposed by the cyber context.

5 Science Fiction Scenarios

The issue of artificial intelligence remains dependent on science fiction, and who can you imagine is the most ingenious in this field other than Hollywood and the American entertainment industry and its sumptuous and influential imagination; which was destined to recreate human taste over a hundred and more years. Here we stop with some of the films that dealt with this topic, for example. The popular "Matrix" films, a Hollywood production starts in 1999, starring Keanu Reeves. The first film talks about the matrix that was created by the computer after giving it full potential, until it managed to develop itself, to make the man its energy source [3]. From this reality, the aspiration is to invest in the superpowers of a computer hacker to face this situation.

Another pessimistic film is a one that warns against the consequences of the machinery controlling humans, as "I, Robot" starring Will Smith another Hollywood production from 2004. It tells about the robot's domination of human life in a way that threatens the existence of the human race until it is addressed by the accomplished investigator, heading to save the human race from the imminent danger of a robot. The film critic Peter Brad Shaw in the British newspaper "The Guardian" does not hesitate to use the term (Automaton-American) to describe the state of control the robots have reached in the film's scenario [4].

The film "Transcendence" was produced in 2014, starring Johnny Depp. Which narrates about the unbridled desire generated by a world of unique intelligence, and how it was required for the protagonist to achieve a system through which he can create a pairing between (knowledge and authority) to the extent that he constituted a true terror among the specialists who were unable to keep pace with him even after his death [5]. Commenting on the Transcendence film, Washington Post film critic Michael Sullivan points to how much evil can emerge, drawing on the applications of artificial intelligence. Sullivan admits that this field, in reality, has not reached the level the film offers. However, the question remains present regarding possible and thinkable consequences in the future (warnerbros.com).

The treatment of science fiction's consequences in cinema is also discussed in comedy films, in which these films interacted with a highly influential cultural treatment, for example, the film "Idiocracy" [6], a Hollywood production from 2006, directed by Mike Judge and starring Luke Wilson. The film tells the story of a young military man who wakes up five hundred years after a freezing experiment by the US military, to find the world is being controlled by smart machines. And how the protagonist discovers how the society became so incredibly stupid and lazy and excess have been exacerbated by the dependence of the machine by the man that he's easily the most intelligent person alive. The satirical treatment was not devoid of highly indicative monitoring of the political and economic system, and the hegemony and dominance exercised by power centers in the way of decision making.

6 Digital Ethics

The sudden and rapid transformations that foretold the processes associated with artificial intelligence have motivated scientists and researchers towards preparing for fronting this huge momentum of practices, which calls for stopping at the limits of the path and field of the smart machine in the reality of living. This is about ethical decision-making; this is what the Moral Machine (moralmachine.mit.edu) an electronic platform aimed at exploring the possible moral stance of autonomous vehicles. Experts have directed to develop the website in a variety of languages, where the audience can be targeted from all over the world, including most of the countries and nations of the world. What is remarkable is that the platform is based on transparency in the way it deals with the public, as it does not hesitate to present results, the ones that are drawn from the direct questionnaire to the public about ethical biases in the world. It works to monitor individual differences in the way you decide; and charting the path of differences between cultures, while making sure to come up with generalizations that are consistent with the norms of moral judgments globally. New production, insofar as it involves a creative idea based on the importance of establishing the (safety engineering science of artificial intelligence machines), where the main goal and the most important criterion is based on safe production and safety for the consumer. (Roman V. Yampolskiy).

The successive and rapid developments that artificial intelligence has been achieving are real, making the orientation toward deep and direct thinking by activating the American Medical Association (AMA) for example, and making the ethical factor inevitable [7]. This is in the design of the robots' new generations of productions that have made the issue of thinking directly related to the issue of (harmful and beneficial) caused by these very intelligent and highly accurate products. It is precisely on this that the scientific community, including specialists, scholars, thinkers, and philosophers, has gone to great lengths when studying the moral factor. Until the "Turing test" has become the methodological guide for programmers, computer scientists, and psychologists, as it focuses on the issue of the future relationship between humans and artificial intelligence machines [8]. What is especially concerning in the automation issues, which have become a concern that stimulates the overall human tendencies, is that they lie under various pretexts and justifications. At the forefront lies accuracy and regularity until the preference has become and has developed apparent about the direct comparison, between the work done by man, if compared to what the computer accomplishes in one field or another [9].

From an early period, the major manufacturing companies headed towards abandoning the human cadre and delegating robots with the task of accomplishing tasks, for considerations such as accuracy, speed, cost, accomplishment, quantity, and type, without having feelings, or sensations. Until this phenomenon invaded giant car production factories, and the robot became at the head of the whole production lines, and this scene began to circulate in various sectors, scientific, industrial, medical fields, laboratories, and major research centers. From this, the attendance became condensed in favor of the robot at the expense of the human individual. To the extent that the cries of the greatest human minds such as Bill Gates, Stephen Hawking and Elon Musk, have come up through the focused reference that this reality must be its destiny and that it represented a direct threat to the future of the human race, and the way he exists on the planet Earth may pave the way for the annihilation of mankind [10].

These cries of caution were not caused by personal concerns, as they were an expression of the reality of the steady development of applications of artificial intelligence, which went beyond the human mind in more fields and arenas. The matter here is based on the idea of leakage of applications of artificial intelligence in the field of control and systems, in which the smart machine went, forming the main pillar against the continuous feeding of superior programs. Through the availability of the ability to make decisions automatically, to the extent that the relationship was complicated and the human control over these machines began to decline slowly [11]. The issue here is not based on a miscalculation by those working in the field of artificial intelligence, or their lack of expectation. As far as this situation arises from the conditions of the higher administrations controlling the major companies of artificial intelligence, and the predominance of the obsession with profit on them, the matter will appear in many conditions and situations that contribute to the disruption of the scene. Artificial intelligence, in its search for perfection and excellence in achievement, does not stop long at the issue of budgeting in employment opportunities, for example. Its main objective is material and, therefore, the policy of dismissing and reducing excess labor and directing towards the use of robots is not negotiable. To show the conditions of social conflict and its consequences on the moral and ethical levels in human society.

The details that abound in reality are indicating the decline of human mental and physical skills in light of the steady reliance on the smart machine. And to monitor the vow of this situation has become clear and evident in the current reality, especially among the younger generation. This is revealed by the level of educational performance at the secondary school and university, where the exaggerated departure of young people from learning and the appalling compliance with the smartphone, which now strikingly controls the youth category.

While the conditions of indecent laziness emerged in the pursuit of more details that abound in life, to the extent that it became difficult for the teenager to make an effort to pay attention to the surroundings, or it became very easy for his answers to be ready, and he transmitted them in a provocative way to the smartphone. The crawl of software continues to multiply in a double manner, and conditions continue to indicate a state of surrender by the human community to the leakage of the clout of smart machines, which has been working to facilitate the requirements and needs of the human community. However, all this remains subject to internet service and networking conditions.

Though the threat remains and is present by internet hackers, those who do not hesitate to attack websites and work to disrupt life. Rather, the nature of the interaction in this vast cloud content, makes the individual interacting in his details and has fallen victim to the temptations that keep chasing him, those that come in the form of gifts, even for the matter seems and has become an expression of the story of the "Trojan Horse", the one that fortifies the enemy, then the enemy finds an opportunity to attack him [12]. The matter does not stop at the technical side but goes beyond it to the cognitive side where the shift in meanings. The direct experience indicated the role played by social media networks in supporting popular revolutions and working to overthrow governments, and this is what the Arab Spring experiment revealed. The issue of Julian Paul Assange on WikiLeaks [13] and the clashes between the official establishment and the attitudes of the opinion-holders on the open-source, and that the human community has the right to obtain information no matter how important it is, do not deviate from this context. It is the results of the information revolution, which has been working to transform the entirety of the established relations, and seeks to demarcate the new and existing relations based on the production of content by the masses [14].

7 When the Beast Roars

Amid the excessive celebration of the achievements of artificial intelligence, and the relentless aspiration of institutions and major manufacturing companies to develop, modernize and innovate, towards the production of more new generations of smart devices and complex programs, humanity will be facing the most important challenge. This is usually the unexpected outcome of the experiment. An order that scientists encounter in their high-level laboratories for so long. How many viruses leaked from the most prestigious laboratories in the most prestigious universities, and he posed a direct threat to human life. And how many experiments spiraled out of control and posed a terrifying threat. This is on the level of direct experiences, which are related to a limited scientific field, so how about us, and we are facing a broad path or stream represented in (artificial intelligence), which strives for most of the giant companies and institutions to engage in its manifestations and its details. The most important thing relates to the issue of growing development that reaches the level of fierce competition between giant companies, which have incredible capabilities and proficiencies that enable them to achieve advanced results in the field of artificial intelligence. Despite the positive conditions that include this trend, it is not without surprises and incalculable consequences; as in the case of Monster in Frankenstein [15].

The issue Stephen Hawking raises is concerned with a direct comparison of smart machines and human capabilities. Of course, it goes in favor of the machine, considering that smart machines are only an expression of the union of human minds to create these machines. While the most important anchor is stopping at the machine's ability to run itself and benefit from a redesign of the matrix that is directly growing in mathematical relationships, equations, algorithms, and highly complex neural networks. To be the case, the human race was shocked by its scientific elite, and it was powerless to stop the successive, unrecognized crawling of sophisticated artificial intelligence programs that can renew themselves [16].

8 Imagination and Reality

Now in the year 2020, when you start talking about machine ethics, many people who have high-profile and high-level degrees do not care about the topic and think it is irrelevant, and you can quickly notice the specter of a satirical smile on the face of the listener, to repeat the sentence denouncing and belittling (machine ethics?!). The matter can be generalized to the whole society so that your speech cannot be taken seriously, and at best they consider the conversation, as part of the science fiction. And if it is excessive exposure to the subject of legislation and laws governing cybercrime, or measures taken by governments to limit sites with harm, such as pornographic sites, or those sites that promote hatred, intolerance, racism, and crime. The direct response is usually and has been directed towards human efforts in developing, designing, and implementing them. that is what you refer to as nothing but talk about a deaf machine in which you can put what you want, and direct it in the way you wish. It is only a human-made machine and its control remains the human task. Meanwhile, hacking, distortion, and abusing by internet hackers, for example, are being overlooked. It remains subject to the efforts made by specialists in the field of artificial intelligence. Those who start the situation and look at the overall details, and are fully able to monitor the interactions resulting from the nature of the relationship between man and the intelligent machine [17].

In fact, despite their generous efforts to delineate and define the parameters of the machine and human ethics the difficulty remains facing specialists around the method of distinction, as it remains a dilemma tainted by a lot of complexity, and overlap. The issue is not a product of the present day, as far as its roots go back to the year 1950 when the Three Laws draft law proposed by the famous science fiction writer Isaac Asimov [18]. Which is summarized in: (The first law: It is not permissible for a robot to infect a person, or through inaction, allowing a person to harm. The second law: a robot must obey any orders that human's issue to it, except in cases where these orders may be attached to the first law. Third Law: The robot must protect its existence as long as this protection does not conflict with the first or second law [19]. In the end, the topic is about a machine that can be damaged, broken, and has a shelf life. How can a law be applied to a machine whose first law is based on a pledge not to harm the person who made it? The issue remains in the science fiction astronomy. And the conditions of the rule and the exception, how many customers paid huge amounts of money to buy a luxury car with fictional prices, they did not escape the consequences of failure and were not immune to catastrophic damage either. How can we believe in a law that regulates the relationship between man and the machine, and to what extent can the machine be fare and ensure that it continues to operate with the reliability and preparedness designed for it, and is it possible to avoid the side effects of heat, cold, or emergency conditions not originally thought?!. The matter here depends on the conditions of the relationship between imagination or reality [20].

Asimov is ultimately a novelist and writer based on his fertile imagination in distributing relationships. This fertility reached the limits of (creative imagination), which made the experts of artificial intelligence go towards adopting these laws and considering them as basic principles in the way of dealing with the question of the relationship between the smart machine and man [21]. Asimov's three laws remain subject to criticism and examination by specialists [22], where studies and readings abound as Eliezer S. Yudkowsky and his colleague Bostrom point to the issue of non-racial discrimination between the machine and the human being, in the calculation that the genes that make up the organism do not have a fundamental influence on the moral status, the most important thing depends on the importance of the distinction between (the moral duty) and (the moral status). And if the question is seen about the possible way of dealing with the artificial mind, the direct answer is based on the importance of the statement "we must deal with the artificial mind in the same way that we must deal with the human mind if the situation coincides" [23].

9 Interconnected Ethics Network

The talk about machine ethics and human morality is apparent on its face as an expression of the conditions of separation between the two ends of the relationship (machine and human). However, the reality of the relationship remains reliably connected. With the calculation that the initiative remains issued by the human side, considering its location based on invention, design, innovation, and development. It is the starting point and source of guidance, starting from its position as a maker and guide of the smart machine. And if there is talk about the ethics of artificial intelligence, then the matter remains for the person first. Where a direct reference to the three laws of Asimov that he planned, considering logical relationships based on effective science fiction. But in the end, it remains based on determining the behavioral framework of the automated product. While the attitudes of workers and specialists in the field of artificial intelligence focus on the importance of caring for the demarcation of guidelines and principles of ethics, on those directly involved in the effectiveness of artificial intelligence.

The matter here is based on an attempt to define the ethical directives of workers and specialists in this field, given that the ethics programmed in the machine is ultimately a representation of the ethical principles carried by the designer, implementer, and manufacturer of the smart machine [24]. The most present question remains in astronomy, how can legal principles be established at the time when a fictional imagination reincarnated the effect of setting a narrative plan for the path of the relationship between the robot and the human? Yes, the relationship posed by the laws of Asimov may seem logical, and carry realistic indications through which the scope of the relationship between the machine and the human can be restricted. But what are the guarantees that can be obtained, about stabilizing the peaceful relationship between the machine equipped with high intelligence and superhuman capabilities? Is it possible to be satisfied with adequate planning and programming, to ensure the safety and protection of handling robots?

10 Conclusion

Scientists and specialists are aware of the Fourth Industrial Revolution's connections, the promising manifestations of artificial intelligence, and the extreme and direct effects on shaping the future of humankind. This highlights the importance of stopping at the moral side. Morality here appears as a guide and a controller, seeking to work to avoid what is surprising and invisible. Thus, international institutions and organizations began to urge efforts towards creating an appropriate climate of fruitful and constructive cooperation, to establish a model of guidelines and strategies, especially in terms of empowering and strengthening human capacity for decision-making, especially as the digital world and its interactions have been attended on levels such as; prediction, machine learning, algorithms, and data. The aspiration here is based on the importance of distinguishing the range of opportunities and risks surrounding AI trends; especially on the design and usage levels. And to the extent that the importance of artificial intelligence expands, given its impact on the development and expansion of more fields and arenas, the moral concern remains present and important in guidance. Based on the dedication of the principles (support, awareness, and confidence). It is not based on a theoretical abstraction as much as it posed a deep-seated dilemma whose effects were in the automotive industry. The main pillar that distinguishes humans from the machine is originally based on the ability of humans to make judgments and self-discrimination; while the machine remains only a future tool whose existence is originally based on what information programmers provide.

Is it up to the specialists, programmers, computer scientists, and companies with a direct concern? Of course not, the issue is more complicated because it directly affects the future of mankind, and hence the importance of shared responsibility, that must be assumed by everyone, individuals, institutions, and governments, to create an ethical approach to which everyone rests. This is in the calculation that artificial intelligence represents a promising field that is distinguished by more potential, advantages, and impressive results that will have an extreme role in the heart of the equation of social, economic, and political relations. Therefore, the moral mission must focus its orientations on the implications of the labor market, unemployment and opportunities for the younger generation, and the way to deal with the experiences and competencies carried by the preartificial generation.

References

1. Wagener, H.: Understanding Erich Maria Remarque, p. 75. The University of South Carolina Press (1991)
2. Jen, C.: What is the Internet of Things? (2016). https://www.ibm.com/blogs/internet-of-things/what-is-the-iot/
3. Peter, B.: https://www.theguardian.com/film/2004/aug/06/willsmith. Accessed 6 Aug 2004
4. Peter, B: The Matrix trilogy review – pop-culture juggernaut still lands a thrilling punch. https://www.theguardian.com/film/2019/jul/12/the-matrix-trilogy-review. Accessed 12 July 2019
5. Michael, O.: (2014).https://www.washingtonpost.com/goingoutguide/movies/transcendence-movie
6. Matt, N.: Idiocracy Is a Cruel Movie and You Should Be Ashamed For Liking It. https://paleofuture.gizmodo.com/idiocracy-is-a-cruel-movie-and-you-should-be-ashamed-fo-1553344189. Accessed 29 July 2014
7. De Miranda, L.: 30-Second AI and Robotics: 50 Key Notions, Fields, and Events in the Rise Intelligent Machine Each Explained in Half Minute, p. 144. IVY Press, UK (2019)
8. Epstein, R., Grace, G.R. (eds.): Parsing the Turing Test: Philosophical and Methodological Issues in the Quest for the Thinking Computer, p. 5. Springer, USA (2009)
9. Colin, A., Varner, G., Zinser, J.: Prolegomena to any future artificial moral agent. J. Exp. Theoret. Artif. Intell. **12**(3), 25 (2000)
10. Mangal, S.K., Mangal, S.: Psychology of Learning and Development, p. 443. Phi Learning, Delhi (2019)
11. Bi, Z., Wang, X.: Computer Aided Design and Manufacturing, p. 372. Wiley, UK (2020)
12. Singer, P.W., Friedman, A.: Cybersecurity and Cyberwar: What Everyone Needs to Know, p. 62. Oxford University Press (2014)
13. Higgins, M.: Julian Assange: WikiLeaks Founder: Wiki Leaks Founder, p. 106. ABDO Publishing, Minnesota, USA (2012)
14. Fuchs, C.: Social Media: A Critical Introduction, p. 132. Sage, London (2017)
15. Stoker, B., Shelley, M., Stevenso, R.L.: Dracula, Frankenstein, Dr. Jekyll and Mr. Hyde: The Gothic Trilogy in Only one Volume. BoD Books on Demand, Paris, p 688 (2019)
16. Victor, L.: 5 Very Smart People Who Think Artificial Intelligence Could Bring the Apocalypse. https://time.com/3614349/artificial-intelligence-singularity-stephen-hawking-elon-musk/. Accessed 2 Dec 2014
17. Jobin, A., Ienca, M., Vayena, E.: The global landscape of AI ethics guidelines. Nat. Mach. Intell. **1**, 390 (2019)
18. Noessel, C.: Designing Agentive Technology: AI That Works for People, p. 160. Rosenfeld, New York (2017)
19. Sabater-Mir, J., Torra, V., Aguil, I. (eds.): Artificial intelligence research and development. In: Proceedings of the 22nd International Conference of the Catalan Association for Artificial Intelligence. IOS Press, Amsterdam, p. 20 (2019)
20. Christoph, S.: The Conversation US, Asimov's Laws Won't Stop Robots from Harming Humans, So We've Developed a Better Solution Instead of laws to restrict robot behavior, robots should be empowered to pick the best solution for any given scenario. https://www.scientificamerican.com/article/asimovs-laws-wont-stop-robots-from-harming-humans-so-weve-developed-a-better-solution/. Accessed 11 July 2017
21. Hamilton, J.: Robots and Androids. ABDO Publishing Company, USA (2007)
22. Agah, A., Cabibihan, J.J., Howard, A.M., Salichs, M.A., He, H. (eds.): Social Robotics: 8th International Conference, ICSR 2016, Kansas City, MO, USA, November, 1–3, 2016, Proceedings, Springer, Switzerland, P 167 (2016)
23. Bostrom, N., Yudkowsky, E.: The Ethics of Artificial Intelligence, p. 8 (2011)
24. van Rysewyk, S.P., Peter, M.P. (eds.): Machine Medical Ethics, p. 13. Springer, Switzerland (2014)

The Use of Artificial Intelligence in E-Accounting Audit

Hesham Zakaria

Abstract The chapter dealt with following the theoretical analytical approach in clarifying what is artificial intelligence in terms of its definition, characteristics and different types, and explaining the relationship between Artificial Intelligence and Accounting science with reliance on electronic and digital transactions, which in terms of both return and benefit, cost and burden, degree of risk, flexibility, and information characteristics Accounting, and the quality of Financial reports, with an explanation of the evolution of the Accounting Audit profession for electronic and digital workers, especially in the wake of what the current business environment witnessed from the pandemic of COVID-19 through the drivers of the need to develop Accounting Audit, which led to the emergence of the so-called E- Accounting Audit, which in turn depends on following types of intelligence The different synthetics in the field of Accounting Audit and the challenges and difficulties facing its application, and also determining the effective role of artificial intelligence in Accounting Audit through the method of work and programs used in its performance and showing the impact on the Internal Audit as a result of E- Accounting Audit.

Keywords Artificial intelligence · Accounting · Auditing · Types and images of AI · Business intelligence · Cloud computing · Neural networks · Digital interactions

1 Introduction

Accounting audit is a science of renewed science constantly under the shadow, and that in light of the science's endeavor to keep pace with the various variables and factors that the business environment is experiencing, As electronic and digital transactions come in as one of the most important of these problems in the absence of the tangible physical side For Accounting documents, and as a result, many academics and researchers have sought to recommend the use of Artificial Intelligence of its

H. Zakaria (✉)
Helwan University, Helwan, Egypt
e-mail: shark_h@hotmail.com

A. Hamdan et al. (eds.), *The Fourth Industrial Revolution: Implementation of Artificial Intelligence for Growing Business Success*, Studies in Computational Intelligence 935,
https://doi.org/10.1007/978-3-030-62796-6_20

various types and methods to help Accounting Audit in those electronic and digital transactions, which led to the emergence of the so-called E-Accounting Audit, which is the most modern aspect of keeping up with computerized or electronic Accounting Information Systems in The widespread use of computers to complete the Accounting work such as recording, tabulating, posting, summarizing and reporting, and therefore the performance of E- Audit will differ greatly from the usual Traditional Accounting Audit, especially after what the world witnesses of the spread of the COVID-19 pandemic, to identify the nature of the use of AI in accounting audit, the following will be addressed.

2 What Is Artificial Intelligence?

This aspect deals with defining the general framework of artificial intelligence in general and what it contains of characteristics, images and types that in turn work with many different scientific fields such as the field of Accounting and Auditing with increasing electronic, virtual and digital Financial transactions, which in turn affect the science of Accounting and the science of Accounting Auditing, Artificial intelligence can be identified through the following.

2.1 The Definition of Artificial Intelligence

It is defined as one of the forms of technological science that depends on simulating the ability to think and arrange data according to a specific scientific field based on what is available from previous experiences of similar situations and decisions [17], As it is a method or information system works to help in taking the appropriate decision according to behaviors similar to what Humans do in what is available to them in terms of learning, understanding, instructions and previous directions [9], Thus it is considered as the main pillar that includes electronic rules and methods based on the computer that It works according to the expectations and experiences previously drawn to reach the best decisions in the future with an objective degree and independent from the traditional non-synthetic influences [1]. Also it is seen as a system that is able to interpret external data correctly and learn from it and use those experiences to achieve specific goals and tasks in a manner flexible [4], as it is defined as the general framework for achieving a link between the hypothetical aspects of the planning process and the rationalization of decisions in a timely manner and the correct decision-making process And intact less error than expected and determined depending on the Human factor.

2.2 The Characteristics of AI

AI It contains many features that may agree with many of its types and images in each of:

(a) It is seen to him as a means, method, system, framework or technology through which reliable traditional data are converted into electronic data [2], It is self-configured and arranged according to the nature of the decision to be taken or applied and the decision is modified according to what is emerging Of influencing variables and factors.
(b) The public domain is considered to eliminate the human and traditional behavior witnessed by making incorrect and incorrect decisions based on the limitations of previous experiences [6], AI depends on feeding the computer with all the experiences and behavioral experiences of the human component on the computer Automated, that the AI takes over the arrangement of those behavior according to the nature of the decision and the variables of events and reach better results.
(c) It works to provide information electronically in a way that achieves many characteristics of information quality for any scientific field such as accessibility, comparison with similar information, timely availability, objectivity of information, and others [22].

2.3 Types of AI

AI has many types that all work through relying on computers and the technical and technological form in work, manufacturing, planning and decision-making, and perhaps for of the tendency of many countries and governments to digital transformation in providing their services, it is classified into two aspects.

(a) **First: The traditional types of AI**, Through it, artificial intelligence depends on replacing the way of thinking and arrangement from the human form to the automated form, and perhaps this aspect includes the following types:

- **Interactive Machines**, They are the oldest species, they have limited capacity, although they mimic the human mind, but they cannot benefit from their previous experiences in planning the next steps, because they lack the ability to learn [16], and work stops, This type is based on what is specified for its inputs according to a limited work plan.
- **Memory-Limited Machines**, They are more advanced and developed than interactive machines, they work by employing historical data with limited memory in the decision-making process [5], as it depends on the data stored in limited memory reflecting previous experiences to deal with future problems.
- **Mind Theory**, It is the most widespread in various fields on a large scale, as it depends on understanding and clarifying situations, entities and events and

interacting with them better by distinguishing between the needs, emotions, beliefs and intellectual processes of those events and situations [10], and perhaps Following this type of business intelligence can be relied upon in the field of Accounting, Auditing and taxation for what works in parallel, like human thinking, and this will be dealt with other types when dealing with the relationship with both Accounting and Auditing in the third section.

- **Self-Awareness**, it is considered the most advanced stage with regard to the first aspect, which requires the replacement of thinking and arrangement from the human form, as it in turn depends on exceeding the limits of assumptions to be very close to the way the human mind thinks and what it includes to realize and develop self-awareness so that the artificial intelligence has its emotions and needs And his beliefs and perhaps his own desires as well [12], although this type is seen as achieving sometimes superiority to what the human mind does in the decision-making process in complex situations and there is no previous experience or experiences to deal with such as The international level witnessing the spread of the COVID-19 pandemic.

(b) **Second: The Advanced Types of AI**, This aspect classifies AI into several images in terms of each of the more advanced types and in terms of the limited use of AI. Below, each of them will be addressed as follows.

- **In terms of limitations and level**, It is represented in classifying them according to the specific level of the mission or its goal, and those pictures can be identified as follows.

- **Limited AI**, the limited level that achieves AI is part of the task or a limited aspect of the task i.e. it is a partial level of dependence on artificial intelligence [23], and therefore this type does not exclude the human component from performance Other parts, Perhaps an example of this is what the Financial and banking entities and banks do.
- **General AI represents the general level**, which fully achieves AI, its ability to simulate capabilities and human reactions with a high degree and work to communicate with many links and relationships in many areas, which achieves a decrease in time depending on those levels [13].
- **Super AI**, represents the advanced and endless level of uses of AI, which is expected to include all scientific and practical fields, which is the increasing dependence on virtual reality and digital transactions [19], and perhaps this level will be more prevalent because of its superiority and progress from In terms of memory, processing speed, data analysis and the evolution of decision-making.

- **In terms of progress and development**, they are represented in the images related to the scientific fields, including the science of Accounting and Auditing, which decision-making depends on the achieved techniques for analyzing data and adding confidence and credibility to the information obtained by the decision-makers, and these images can be identified as follows:

- **Neural Networks**, are considered one of the forms of AI that work to simulate Human behavior in relation to information processing, as it works through three aspects, the first is a technical side, the second is statistical and the third is engineering so that each works on the similarity of Human performance and behavior in the interpretation of phenomena and events, forecasting and forecasting trends [21], and perhaps it has increased dependence on neural network technologies in forecasting the Financial hardships of Accounting units according to the statistical, algorithm and technical models it contains, which give a future vision of the institutional Financial situation for them.
- **Cloud Computing**, It is considered one of the forms of AI that is the technology or means on which Accounting units depend in the province and the transfer and delivery of information and data to the website (the cloud) is prepared by cloud computing services providers according to terms and arrangements that define how to retrieve and deal with that information Accounting among all parties authorized to do so [7], and thus it represents a platform for the protection and preservation of information and the retrieval and amendment immediately and to have a picture that makes it support the decision-making process and direct it in a timely manner.
- **Business Intelligence**, is considered one of the forms of AI as it works to employ and follow information technology in obtaining correct, sound and accurate information in an immediate and instantaneous manner, in turn achieving quality in terms of both honest representation and relevance and increasing its value among decision makers in all different fields that depend on data analysis Preparing information [8], and by following this type in the field of Accounting work, bias problems in Accounting Measurement with personal or subjective estimates will be eliminated, perhaps the most important thing related to following both the fair value and the added economic value, etc. Another basis of Accounting Measurement.
- **Expert Systems**, represent one of the most important forms of AI, As they are techniques and programs that redirect the production of Human behavior to achieve intellectual tasks in certain fields because they are specifically directed to dealing with assumptions at a high speed and at an ideal time compared to Human behavior, They are very similar to neural networks, But they are differ in that they are prepared to deal with more complex and diversified fields with the ability to amend and change laws and rules stored in them for of the variables and factors [3], and by following these images in the field of Accounting Auditing, They will contribute to Achieving one of the most important External Auditing standards associated with the Auditor, Which is both doing the required professional care as well as scientific and practical qualification, each of which is related to the availability of previous experiences achieved by relying on expert systems.

3 Assessing the Relationship of AI to Accounting

This aspect explains the relationship of the uses of AI and specifically the advanced level of it in Accounting science in general, as a start to move to the third section of the research to explain the role of AI in the field of Accounting Audit and determine the correlation between each of them, and perhaps it can be agreed that all the more sophisticated types of intelligence Synthetic agree in that relationship through the following:

- **Return and Benefit**, The relationship here is that if a comparison is made between the traditional methods used in Accounting science such as transferring, communicating, storing and protecting information and others and between the more advanced methods using AI, it becomes clear that the latter achieves a great return and benefits for all parties concerned with dealing with AI whether the information preparer or its Users because of the benefits it reflects, For example, the economic entities that provide Financial and non-Financial information on their websites help current and prospective investors in assessing Financial and non-Financial performance faster than traditional methods that are not based on AI.
- **Cost and Burdens**, Represented in I following the methods and images of AI compared to the benefits achieved with costs and burdens, it achieves the required economic feasibility, especially if the low degree of errors and the increase in objectivity are taken into account by the authors of the information and the speed of access to that information at the appropriate time.
- **Flexibility**, Represented by the fact that the business technology on which AI depends depends to a large extent on achieving flexibility through the fact that Financial and non-Financial information reflects the results immediately after using one of the methods or forms of AI, an example of which is the speed of Financial data published in Financial market sites to the changes in The prices of shares traded as soon as the variables are received.
- **Control the Degree of Risk**, Which is that relying on AI techniques will lead to a decrease in the degree of exposure to the risks associated with information and data that were faced by traditional methods of dealing with them in terms of retrieval, preservation, illegal entry, loss and other such images, as the AI will work To determine the authorized and parties to deal with that Financial and non-Financial information and data.
- **The Characteristics of Accounting Information**, That is a correlation between the introduction of AI methods and the quality of Accounting information, specifically across a set of characteristics defined by the Conceptual Framework for Financial Accounting issued in 2015 as the appropriate property and associated prediction, and the feature of honest representation of information prepared electronically What it includes is the absence of errors in the preparation of information. And its ability to verify for of what depends on AI from laws and procedures that are not subject to bias by the information provider, and the appropriate timing

of what is available in these methods of presenting information in an instantaneous manner.

- **The Quality of Financial Reports**, the relationship is that AI helps achieve the quality of financial reports. This is evident in the ease and simplicity of the presentation and disclosure of financial reports published electronically on the websites of the Accounting entities. Perhaps that relationship in turn is consistent with previous aspects such as return and benefit, cost and burden, and the degree of the risk and the characteristics of the Accounting information, each working to achieve the quality of financial reports.

The previous presentation to determine the nature of AI covered both the definition, characteristics, types and images included in AI, and perhaps the researcher will seek to demonstrate the use of AI in a more advanced way such as Neural Networks, Cloud Computing, Expert Systems, Business Intelligence, And The reflected benefits of The E- Accountant Audit for of the recent increase in business and digital transactions in the wake of the Corona COVID-19 pandemic, which will work to reduce dealing and work through the human component as much as possible in implementation of the recommended precautionary measures.

The following deals with the second section that explains the development of the profession of Accounting Auditing, so that E- Auditing is able to deal with digital and electronic transactions, to show the correlation between AI and Accounting Auditing, and this will be explained by the third section of the role of AI in Accounting Auditing.

4 Evolution of the Accounting Auditing Profession on the Impact of the Covid-19 Pandemic

This section deals with a presentation of the necessity and importance of developing the profession of traditional Accounting Audit to become electronic oriented to fit with electronic and digital transactions that are expected to increase dependence and resort to it in the wake of the spread of the COVID-19 pandemic, which requires reducing and reducing traditional dealings between Humans.

4.1 Motives Develop an Audit

It is considered one of the most important main reasons that contributed to the development of the traditional Audit is the progress that the business environment witnessed from the rapid developments and the necessity of adapting the economic factors surrounding the business environment from the developments of the Accounting profession in general and the Audit profession in particular, and perhaps the most important other reasons can be identified after the development and technological progress that The owner of the Accounting work environment in each of the following:

- The deficiency of the traditional Accounting Audit in relation to the activities and factors of electronic commerce led to the need to search for an alternative, which is the E-Accounting Audit as an alternative resulting from the development and technology of operating and electronic processing of Financial and Accounting transactions.
- The traditional Accounting Audit requires a longer time than usual and is due to many considerations such as the Auditor's experience and the extent of familiarity with the work environment and the provision of evidence and documents and follow-up of the work of the work team, while E-Accounting Audit works to achieve and complete the Audit process in the shortest time of the traditional Accounting Audit in addition to easy access to all Evidence of evidence and responsibilities.
- The traditional Accounting Audit depends on the evaluation of the Internal Control System and the extent of the ability and experience of the Auditor to determine an appropriate sample to judge the internal control system, while E-Accounting Audit depends on the evaluation of how the computerized data is processed and the information is rearranged in a way that reflects the improvement of the expression process in an impartial and objective manner.
- The conventional Audit does not achieve the required quality, due to the fact that the Auditor judges the validity and reliability of the Financial Statements based on what he does from his Audit in the light of the sample, which results in the fact that in the event of a sample size error, this is reflected in the quality of the Audit process while the E-Accounting Audit under The availability of electronic programs determines the sample accurately and appropriately, which leads to achieving a high quality process.

Consequently, E-Accounting Audit works through the use of many technical and technological methods, which in turn is one of the forms of AI with regard to providing experiences, forecasts, expectations and previous experiences that in turn support the performance of the Auditor in his report confirming the validity and integrity of Financial Statements in various traditional and electronic types In the statement of Financial position and business results of the Accounting unit subject to Audit.

4.2 Definition of E-Accounting Audit

Many scientific opinions dealt with the definition of E- Accounting Audit as a structured and objective process that aims to provide various evidence, traditional, computerized or electronic, through which an Internal Control System is evaluated for the statement and information and what is achieved from its protection and validity to reach the complete confirmation or a full percentage to the information required by users to make sure From the authenticity and ability to rely on financial statements, whether traditional or electronic [11], E-Accounting Audit depends on two objectives, The first is to review the computerized work environment, evaluate the information system and the deficiencies, It is included in the information system, while the other goal is to analyze the work system of the accounting unit in question to assess the difference between each of the computerized information work environment and traditional information and compare them to analyze the difference and work to explain it and reach agreement to confirm the validity of the information through reliance on methods of AI [14] in performing conscious tasks The Auditor's money and access to express a neutral professional and technical opinion.

The E-Accounting Audit is considered a sophisticated and continuous process through which the Auditor provides many evidences, sources and means to reflect and support the validity of the information presented in the Electronic Financial Report according to what is used from electronic methods based in its content on the methods of AI, and what is required by Users of these Reports in their various categories, given that the Electronic Financial Report spreads more easily and quickly than traditional reports, and therefore the Auditor's report is part of the Electronic Financial Report if it is based on E-Accounting Audit.

4.3 The Characteristics of Accounting Audit Under AI

Accounting Audit is generally characterized by many features and characteristics for of it being an organized process to collect, evaluate and examine evidence in a neutral and objective way in order to verify the truthfulness and correctness of the information provided by the Accounting unit and report on it with neutral professional opinions of the parties concerned with this information, as Auditing is based on methods of AI such as networks Nervousness will achieve the improvement of the performance of three pillars such as limited examination, verification and report [20]. Applying this to E-Accounting Audit, it will depend on the same pillars and pillars, in addition to that the references have many experiences and due diligence to deal with Electronic information systems in which the Accounting unit deals and eliminates the expectations gap, The Auditor in the absence of sufficient experiences to deal with these cases [15], then these characteristics can be explained in the following:

- The E-Accounting Audit works to achieve the economics of information through the Auditor's goal is to check the computerized or electronic Accounting information system and then it is possible to review more than one item or item by reviewing the electronic relationship with the program and evaluating the results and outputs.
- The E-Accounting Audit works to achieve effectiveness through that the Auditor seeks to evaluate the internal control system and the extent of this system's ability to detect errors and address them at all administrative levels.

- The E-Accounting Audit works to achieve efficiency through it helps to achieve all the goals of the organization without exaggeration or ignore one of those goals in achieving the rest, including E-Accounting Audit to judge those goals by linking them to results and outputs and what confirms the evidence and control means and evaluation of the control system the interior.
- The E-Accounting Audit works to achieve protection and safety through that it does not depend on Auditing Accounting or Financial transactions only, but rather extends to the investigation of protection and safety from keeping those documents and records and not being exposed to the risks of theft, loss or loss, and perhaps this highlights the important role that the Auditor plays to assess Internal control system.

4.4 Challenges of E-Accounting Audit

E-Accounting Audit is a recent and continuous development as previously mentioned, but despite the advantages it has achieved compared to the traditional Audit and identification in various aspects such as timing, economy, efficiency, effectiveness, and quality, it faces many limitations that can be summarized in the following:

- The absence of the tangible physical form of paper documents through the electronic Accounting processing stages, which is that the E-Accounting Audit is directed directly to the performance of the Audit work through computerized systems, and therefore the nature of the document takes the digital electronic form, which can be modified in any Time and impact on the final results of Accounting, Financial and non-Financial information.
- The nature of the work of electronic systems and computerized Accounting information systems requires upon completion of data and information for the Financial period not to be kept within the operating databases for more than one Financial period provided that it is preserved in the so-called electronic archive, which may be exposed in the event of a failure of security and protection to loss and loss And damage [18], and thus this represents a major challenge to the field of electronic Auditing work The Auditor in charge of Auditing must seek to identify and clarify this before performing the planning of the electronic Audit process.
- The failure of the internal control system represents a major challenge to the field of E-Accounting Audit work, given the adoption of the Audit work on its most important pillars, which is the evaluation of the internal control system, and that under the management of the Accounting Audit work, it is necessary to follow up how and easily access data, information and documents, whether paper or electronic and the role of unit management Accounting is the subject of scrutiny in achieving security and protection for that data and information and the ability of the internal control system to detect errors and deviations.

5 The Role of Artificial Intelligence in Accounting Auditing

E-Accounting Audit differs in its nature in terms of examination, verification and report from the traditional Accounting Audit, the first depends on the need to follow the Accounting and documentary course with through the traditional and technical side, while the second is only limited to the paper and traditional side and in most cases it consumes many efforts, time and care Due diligence and familiarity with an acceptable amount of experience to deal with.

Therefor considering the E-Accounting Audit an development of the traditional form of Accounting Audit, It is possible to address the methods and programs of AI that can be relied upon by the E-Accounting Audit in the performance of its tasks and actions.

5.1 The Way E-Accounting Auditing Works Under Artificial Intelligence

E-Accounting Audit as a relatively recent field includes several types and methods that the Auditor relies on in performing the Accounting Audit process and achieving its goals. These types and methods are as follows:

- **E-Accounting Audit with the same information system**, This type depends on the Auditor making use of the same Accounting or administrative information system in obtaining evidence and documents and evaluating the internal control system as the main and main procedure that the Auditor relies on in performing the Accounting Audit process. This requires The method is for the Auditor s to be familiar with and have sufficient knowledge of how the information system in question is working.
- **E-Accounting Audit of the environment surrounding the information system**, This type depends on the Auditor evaluating the business environment surrounding the information system, drawing on that in several considerations, the purpose of which is to evaluate the internal control system manually without using the information system except by comparing what has been obtained from the evidence and evidence of what the information system outputs provide, and perhaps this method takes more time than the previous method, due to the ability and experience of the Auditor to deal with each of the two methods.
- **E-Accounting Audit with the help of other subsidiary systems**, This type depends on the Auditor making use of sub-information systems that simulate the information system subject to Accounting Audit in parallel, and depends on feeding it with data that is part of that data that feeds the information system subject to Accounting Audit and comparison between each of The two results, and perhaps this type requires the Auditor to have complete confidence in the integrity and health of the information system subject to the Accounting Audit in order to simulate a similar sub-system that circulates a banner on its.

- **The Electronic Auditing Empirical**, This type depends on the Auditor choosing a set of data to work on testing them correctly and reaching the same results to achieve confidence in the outputs of the information system and rely on it in the performance of the Accounting Audit process, and perhaps this type differs from the E-Accounting Audit With the help of sub-systems, it is difficult to achieve the second without following the first, that is, as two stages.
- **Parallel E-Accounting Audit**, This type depends on the Auditor performing the Accounting Audit by comparing it with the same outputs with another system such as a bank account statement received from it with a report that explains the bank's movement between deposit or withdrawal and the extent of that movement's sequence together. This gives the Auditor confidence and a degree of dependence on the information system subject to the Accounting Audit in performing the Auditing process.
- **E-Accounting Auditing consecutively and observing**, This type depends on the Auditor achieving the completion of the electronic Auditing Auditing by selecting specific data that data is tested and follows the way it is handled and judged on the integrity and correctness of dealing with its.

5.2 Programs Used in E-Accounting Audit

The Auditor depends when performing the E-Accounting Audit on a set of programs that he designs himself to achieve the goal of the Audit process. These programs are as follows:

- **Auditing Programs, Transactions Tests, and Account Balances**. This type of program requires designing mathematical equations according to which the transactions are properly and accessed amounts. Perhaps this type of program is suitable for calculating taxes, delay penalties and calculated bank interest.
- **Auditing Programs and Testing the Internal Control System**, This type provides for achieving control, control, protection and integrity of the information and data system, procedures for entering databases and information and classification of responsibilities and specializations, and perhaps this type is considered as the core of the Audit work and it is the judgment and assesses the strength of the Internal Control System in reducing deviations and embezzlement.
- **Auditing Programs, Testing and Selecting the Sample of the Audit Process**, this type decides to choose the appropriate sample that represents the level at which the Auditor increases the evidence of verification and verification in relation to it, since the conversation in the field of E-Accounting Audit differs from the traditional Audit that determining the Audit sample is not determined randomly or diligently as We find it in the traditional Audit, but rather it is determined by a program that divides the balances and accounts into levels according to the amounts and the nature of the item, and then the sample is calculated for these considerations.

- **Auditing Programs and Analytical Tests**, this type requires that for of the increasing demand for verification services and Financial analysis, the E-Accounting Audit derives from one of the most important procedures for Auditing standards, which are analytical procedures that take place for of the use of mathematical and statistical approaches to clarify the vision of the Auditor, and then the observer Calculations depend on a set of statistical programs that help evaluate an item or item for short Financial periods and in the medium term.

5.3 The Effect of E-Accounting Audit on the Internal Audit Process

There are many factors and variables that affect the evaluation of the efficiency of E-Accounting Auditing and what is available from the degree of trust and dependence on verification and Audit services, where many scientific, practical, technical and professional opinions have identified several points that revolve around the evaluation of the efficiency of internal Auditing when following the approach and method of E-Accounting Audit which are:

- The specialized training work environment and the extent of familiarity with the Auditor with the environment and field of work of the technical system or computer and Accounting information systems pursued by the Accounting unit subject to Audit, and perhaps there is a direct relationship between the availability of experience with the Auditor and the quality and efficiency of the Internal Audit process in that he understands the nature Items and accounts that require obtaining evidence are extremely important.
- Improving and developing internal Audit procedures, in turn, is reflected in the better performance of the E-Accounting Audit process in that the E-Accounting Audit can represent a complement to internal Audit procedures and methods for of many studies seeking to address a more recent topic which is the extent of the External Auditor's reliance on work the Internal Auditor, and this may be evident in what the E-Accounting Audit does to evaluate the internal control system in all aspects and at a rate of 100% for computer accreditation and computerized information systems.
- E-Accounting Audit is considered an effective tool in evaluating the overall financial position of the Accounting unit in a suitable timing. Perhaps this shows that the use of Financial and Accounting treatments contributes to achieving the completion of the tasks and performing the E-Accounting Audit with less time and effort, which is reflected in the costs and burden of the External Audit and Internal Audit process on Limit both and increase the degree of confidence in the results of each of them when the Auditor showed a neutral professional opinion.
- E-Accounting Audit achieves accuracy and achieving correct results that reduce inequality and inconsistency if they are relied upon in the current, targeted and future decision-making process for their reliance on continuous and continuous

data and information analysis, which achieves the availability of financial reports and lists immediately following the introduction, amendment, or improvement of the information base. Both Internal Audit and External Audit.

6 Conclusion

E-Accounting Audit is the most recent trend in the development of the profession and the science of Auditing, whether it is directed like internal Auditing and is subject to the directives of management, the authority and the competent authority from within the Accounting unit in a way that affects the difficulty of achieving independence, but if it is directed to External Auditing, it makes the link between Internal Audit and External Auditing reflected In turn, that the E-Accounting Audit be achieved as a mediator between them to add confidence, accuracy and health to the tasks, work and efficiency of the verification and verification services, and perhaps this can be summarized through the following recommendations:

- Qualifying the internal Auditor and External Auditor to hold and establish training and qualifying courses that enable him to be familiar with the methods and aspects of E-Accounting Auditing of the use of methods, types and images of AI, for the ability to deal with various devices and databases of information to determine what he is doing of evidence to assess the strength and durability of the internal monitoring system in discovering Analysis and reduction of errors and deviations.
- Endeavoring to improve and develop the methods and procedures upon which the verification and Audit services depend on discovering and evaluating the Accounting and Financial transactions that take place through following automated and electronic systems in a manner that focuses on the essential points that agree with the requirements of relative importance from the viewpoint of both the Internal Auditor or the External Auditor If independence is achieved equally between them.
- Agreement between the authors, designers and programmers of Financial programs and Accounting information databases to achieve a degree of oversight, security and protection for the programs and technology of Accounting information systems so that the program is divided into a set of stages, levels and responsibilities that in turn eliminate understanding of the nature, terms of reference and responsibilities of each level in a way that helps the Auditor, whether internal or External to deal with that information.
- Activate the requirements of Auditing standards that require setting a scientific, professional and technical framework for E-Accounting Auditing, as all foreign countries and some Arab countries have set Accounting and Auditing Standards that define the scope and requirements for the application of those standards that the Auditor can deal with and follow.
- Electronic dealing and Accounting treatment of data and information also represents a major challenge for E-Accounting Audit, as it depends on being a designer, programmer and program developer with a great degree of familiarity with

Accounting Rules and Principles that are overlooked to the program or information base dealing with Accounting data and that for of an error With these rules and procedures, they affect the nature of the information that the Auditor expresses opinion on a neutral professional and technician.

References

1. Chowdhary, K.R.: Fundamentals of Artificial Intelligence. Springer India, ISBN: 978-81-322-3970-3 (2020). https://www.researchgate.net/publication/339527305_Fundamentals_of_Artificial_Intelligence. Accessed 2 May 2020
2. European Commission: On Artificial Intelligence - A European approach to excellence and trust. Brussels, 19.2.2020 COM, 2020, 65 final. https://ec.europa.eu/info/sites/info/files/commission-white-paper-artificial-intelligence-feb2020_en.pdf. Accessed 2 May 2020
3. Sasikumar, G., Babu, N., Chandra, G., Sandhya, G.: Artificial Intelligence Systems and Expert Systems: An Overview of Recent Trends and Roles in Information Science and Data Analytics. ICDSMLA 2019, pp. 1804–1812. https://link.springer.com/chapter/10.1007/978-981-15-1420-3_184https://link.springer.com/chapter/10.1007/978-981-15-1420-3_184
4. Haenlein, M., Kaplan, A.: A brief history of artificial intelligence: on the past, present, and future of artificial intelligence. California Manag. Rev. **61**(4), 5–14 (2019). https://journals.sagepub.com/doi/pdf/10.1177/0008125619864925. Accessed 2 May 2020
5. Heer, J.: Agency plus automation: designing artificial intelligence into interactive systems. pnas **116**(6) (2019). https://www.pnas.org/content/pnas/116/6/1844.full.pdf. Accessed 3 May 2020
6. Hristov, K.: Artificial intelligence and the Copyright survey. JSPG **16**(1) (2020). https://www.researchgate.net/publication/337387886. Accessed 2 May 2020
7. Ionescu, L.: Big Data, blockchain, and artificial intelligence in cloud-based accounting information systems. Anal. Metaphys. (18) (2019). https://www.ceeol.com/search/article-detail?id=816594. Accessed 5 May 2020
8. Jain, A., Shah, D., Churi, P.: A review on business intelligence systems using artificial intelligence. In: Advances in Intelligent Systems and Computing, International Conference On Computational Vision and Bio Inspired Computing, vol. 1108 ICCVBIC 2019, pp. 1023–1030. https://link.springer.com/content/pdf/10.1007%2F978-3-030-37218-7.pdf. Accessed 5 May 2020
9. Jordan, M.: Artificial Intelligence—The Revolution Hasn't Happened Yet (2018). https://hdsr.mitpress.mit.edu/pub/wot7mkc1/release/8. Accessed 1 May 2020
10. Kelly, T.M.: Ontology and method for creating trust-worthy, super-rational, theory-of-mind-capable artificial general intelligence. Worldwide applications, Publication of US20200097849A1 (2020). https://patentimages.storage.googleapis.com/5e/92/03/495a14d5c1c30a/US20200097849A1.pdf. Accessed 5 May 2020
11. Kaylee, M.G.: How artificial intelligence and machine learning will change the future of financial auditing: an analysis of The University of Tennessee's Accounting Graduate Curriculum. Chancellor's Honors Program Projects (2019). https://trace.tennessee.edu/cgi/viewcontent.cgi?article=3291&context=utk_chanhonoproj. Accessed 5 May 2020
12. Mansour, N., Rafeh, W.A., Zaatari, G.: The role of self-awareness, augmented artificial intelligence and enhanced leadership competencies in developing future academic physicians. Middle East J. Family Med. **17**(10), 27–35 (2019). https://web.a.ebscohost.com/abstract?direct=true&profile=ehost&scope=site&authtype=crawler&jrnl. Accessed 3 May 2020

13. Masengo, W.U., Abbu-Mahfouz, S., Adnan, M.I., Ramotsoela, T.D., Hancke, G.P.: A review of artificial intelligence based intrusion detection for software-defined wireless sensor network. In: 2019 IEEE 28th International Symposium on Industrial Electronics (ISIE), Vancouver, BC, Canada, 12–14 June 2019. http://researchspace.csir.co.za/dspace/bitstream/handle/10204/11178. Accessed 4 May 2020
14. Munoko, I., Brown Liburd, H.L., Vasarhelyi, M.: The ethical implications of using artifcial intelligence in auditing. J. Bus. Ethics (2019). https://link.springer.com/content/pdf/https://doi.org/10.1007/s10551-019-04407-1.pdf. Accessed 4 May 2020
15. Omoteso, K.: The application of artificial intelligence in Auditing: Looking back to the future. Expert Syst. Appl. **39**(9), 8490–8495 (2012). https://www.sciencedirect.com/science/article/abs/pii/S095741741200111X. Accessed 5 May 2020
16. O'Sullivan, S., Heinsen, H., Tenenholz, L., Chimelli, L., Amaro, E., Hilario, P., Jeanquartier, F., Jean, C., da Graça, M., Imran, M., Holzinger, A.: The role of artifcial intelligence and machine learning in harmonization of high-resolution post-mortem MRI (virtopsy) with respect to brain microstructure. Springer Brain Inf. **6**, 3 (2019). https://link.springer.com/content/pdf/https://doi.org/10.1186/s40708-019-0096-3.pdf. Accessed 5 May 2020
17. Yash Raj, S., Menhem, B., Shiko, M., von Georg, K.: Organizational decision-making structures in the age of artificial intelligence. California Manag. Rev. **61**(4) (2019). https://journals.sagepub.com/doi/pdf/https://doi.org/10.1177/0008125619862257. Accessed 1 May 2020
18. Stancheva, E.: How artificial intelligence is challenging accounting profession. J. Int. Sci. Publ. Econ. Bus. **12** (2018). ISSN 1314-7242. https://www.researchgate.net/publication/333728223. Accessed 5 May 2020
19. Ullman, S.: Using neuroscience to develop artificial intelligence. Science **363**(6428), 692–693 (2019). https://science.sciencemag.org/content/363/6428/692.summary. Accessed 4 May 2020
20. Vijayan, V., Rubasundram, G.A.: Artificial neural network (ANN): an artificial intelligent (AI) tool to predict fraudulent financial reporting and financial distress. Int. J. Psychosoc. Rehab. **24**(2), 1215–1226. 12p (2020). 1 Diagram, 3 Charts, 2 Graphs. https://web.b.ebscohost.com/abstract?direct=true&profile=ehost&scope=site&authtype=crawler&jrnl. Accessed 4 May 2020
21. Walczak, S., Narciso, V.: "Artificial Neural Networks", Advanced Methodologies and Technologies in Artificial Intelligence, Computer Simulation, and Human-Computer Interaction (2019). https://s3.amazonaws.com/academia.edu.documents/39320721. Accessed 4 May 2020
22. WIPO: Technology Trends 2019 - Artificial Intelligence (2019). https://www.pinterest.com/pin/736127501580344780/. Accessed 5 May 2020
23. Zhenan, S., Zhang, Z., Wei, W., Liu, F., Tan, T.: Artificial Intelligence: developments and advances in 2019. Front. Data Comput. **1**(2), 1–16 (2019). http://www.jfdc.cnic.cn/EN/abstract/abstract15.shtml. Accessed 4 May 2020

Web-Based Financial Disclosures by Using Machine Learning Analysis: Evidence from Bahrain

Adel M. Sarea, Suresh Subramanian, and Bahaaeddin Alareeni

Abstract This paper examines Electronic Financial Disclosure (EFD) by using machine learning analysis in listed firms in Bahrain bourse. Using machine learning techniques in Python Programming analysis is adopted to measure the effect of: Age, Liquidity, Leverage, Size, Industry and Profitability on the Electronic Financial Disclosure (EFD) through the Website of each firm listed in Bahrain Bourse (BB). The advantages of EFD is to predict better relation in firm characteristics and level of disclosure in banking sector in Bahrain. Which lead us to investigate the EFD by using machine learning analysis techniques. Further, in this research the sample size consists of all listed firms in Bahrain Bourse (BB) during 2017. The main finding is that profitability factor is having highest impact on the level of Electronic Financial Disclosure (EFD) which has been tested and predicted using machine learning. The implication of this paper helps firms in Bahrain to increase the level of (EFD) to reach full Web-Based Financial Disclosures to satisfy the stakeholders.

Keywords Web-based financial disclosures (EFD) · Machine learning · Regression analysis · Python programming · Bahrain Bourse (BB)

1 Introduction and Literature Review

This study seeks to contribute significantly to all aspects of firms listed in Bahrain Bourse (BB) through the Electronic Financial Disclosure (EFD) for all sectors by determining the level of disclosure of EFD through machine learning technique

A. M. Sarea (✉) · S. Subramanian
Ahlia University, Manama, Kingdom of Bahrain
e-mail: Adelsarea@yahoo.com

S. Subramanian
e-mail: Sureshsresearch@gmail.com

B. Alareeni
Middle East Technical University - Northern Cyprus Campus, Kalkanlı, Güzelyurt, KKTC via Mersin 10, Turkey

A. Hamdan et al. (eds.), *The Fourth Industrial Revolution: Implementation of Artificial Intelligence for Growing Business Success*, Studies in Computational Intelligence 935,
https://doi.org/10.1007/978-3-030-62796-6_21

implemented using Python Programming. The need of disclosing more information through EFD are increasingly raised in the previous studies conducted by different scholars from different countries [24].

In this regard, this research fills the gap in terms of analyzing the different factors such as Age, Liquidity, Leverage, Size, Industry and Profitability on the Electronic Financial Disclosure in banking sectors in Kingdom of Bahrain. Previous studies highlighted the importance of machine learning technique to analyze the financial sector [6, 20]. Therefore, in financial sector in Bahrain, this study will contribute to motivate researchers to conduct the research by applying machine learning techniques as a better alternative approach which would predict the impact of selected features towards the level of EFD.

This study attempts to evaluate the extent of the levels for all sectors listed in Bahrain Bourse (BB) by determining the level of disclosure of EFD. Furthermore, this study contributes to the financial literature, software programing and firms' characteristics in terms of the impact Age, Liquidity, Leverage, Size, Industry and Profitability on the EFD through the Website of each firm listed in Bahrain bourse.

Online based financial reporting disclosure is widespread amongst all firms of all sizes in most Gulf Cooperation Council (GCC) including Bahrain. The development of online financial reporting practice has been rapid, largely mirroring, and motivated by development of the World Wide Web since 1994, being the primary Internet medium for online reporting (Allam and Lymer 2003).

A number of studies of these developments have occurred over this time seeking to plot how firms are exploiting the media of online and how they are developing their financial reporting practices in response to this ubiquitous route to current and potential investors, and other stakeholders. In modern business environment the objective of online financial reporting is to assemble financial information useful for investors, information simplifying decisions related to investment and granting loans. In the last fifteen years, the Internet and applications of it have been increasingly widely employed in modern business operations. In developed countries, Internet is used with increasing frequency for financial reporting. In our days, numerous internet applications are successfully employed in business, such as e-commerce, internet banking and advertising (Kelton and Yang 2008). Therefore, our intention in this research paper is to investigate the level of online financial reporting disclosure in listed firms in Bahrain Bourse. On the other hands, the impact on the level of EFD for Bahraini firms in terms of Age, Liquidity, Leverage, Size, Industry and Profitability.

Sarea et al. [24] conducted the study to compare level of of electronic financial disclosure between Islamic banking vs conventional banking in Gulf cooperation council (GCC). Sarea et al. [24] indicated the level of electronic financial disclosure was 78.6%, in conventional banks compare to 73% by Islamic banks in GCC. However, EFD increase the efficiency and timeliness on obtaining financial information that will helps the investors and users in the process of decision making, provides information for Islamic banks and provides accessibility to investors and users.

In Kuwait, another study Alanezi [7] collected data form 179 firms listed on Kuwaiti joint-stock (KSX). The main findings, only 56% of the listed firms disclose online financial information whereas the rest did not disclose online financial information.

In related study, Joshi and Al-Modhahki [18] collected data of 75 firms in (Kuwait 42 and Bahrain 33). The main results, 47.6% of Kuwait firms owned website and disclose online financial information, while 48.5% of Bahrain firms did.

This objective can be summarized as research question of: What is the effect of (Age, Liquidity, Leverage, Size, Industry and Profitability) on the Electronic Financial Disclosure (EFD) through the Website of each firms listed in Bahrain bourse? The findings and recommendation of this research paper are therefore expected to fill the gap in literature review and contribute towards completely EFD for all firms listed in Bahrain Bourse.

2 Research Hypotheses and Variables

Several studies have addressed the issue of electronic disclosure in differnt countries such as Sarea et al. [24], Alrwahi and Sarea [12], Aljwader and Sarea [11] and Al-Sartawi [9]. These studies have investigated different aspects of online disclosure. The checklist was developed using 90 items to express the EFD by using machine learning technique.

2.1 Analysis Variables

The following section explains the dependent and independent variables used in this research, which consists of the following:

2.1.1 Firm Size

Debreceny et al. [14], Omran and Ramdhony [22], and Ferguson et al. [15] suggest that firm size as independent variable can explain and predict the relationship with the level of online disclosure. According to Agboola and Salawu [2], larger firms are more observable to disclose more information through the website. Moreover, Agboola and Salawu [2]" confirms the arguments in the literature give support to higher disclosure by larger firms. Adebimpe and Ikenna [1]" confirms that large sizefirms operate over wide geographical areas have disclosed more information to help for strategic purposes.

Therefore, based on the previous studies, the following hypothesis has been formulated:

H1: There is a relationship between the firm size of each firm and the level of online disclosure for the same firm.

2.1.2 Age

Age was one the elements to be tested with the level of online disclosure. In this regards, previous studies indicate that there is a "positive relationship between age of firms and the level of OFRD Akhtaruddin [5], Al-Shammari [10]. In addition to that, Older firms may disclose more information [1]. Therefore, older firms might have better established reporting systems than newly established firms.

Therefore, based on the previous studies, the following hypothesis has been formulated:

H2: There is a "relationship between the firm Age of each firm and the level of online disclosure for the same firm".

2.1.3 Profitability (Return on Assets-ROA)

Singhvi and Desai (1971) firms disclose more information when its profitability is above industry average in order to signal to the owners about its strong position to survive. Many studies have examined the relationship between profitability and the level of Electronic Financial Reporting Disclosure, such as Oyelere et al. [23], Agboola and Salawu [2], Agyei-Mensah [4], Adebimpe and Ikenna [1].

Therefore, based on the previous studies, the following hypothesis has been formulated:

H3: There is a relationship between the profitability of each firm and the level of online disclosure for the same firm.

2.1.4 Leverage

Alsaeed (2006) report that higher debt will cause high agency cost. Managers are encouraged to lower any agency cost by disclose information to the public. In another study conducted by Jumanji (2012), report indicates that leverage was insignificant to compliance levels by firms listed in Bahrain Bourse.

Therefore, based on the previous studies, the following hypothesis has been formulated:

H4: There is a relationship between the leverage of each firm and the level of online disclosure for the same firm.

2.1.5 Industry

Street and Gray (2001), Gallery et al. (2008), and Aljifri et al. (2014) state a positive correlation between adoption levels and industry type due to as some standards are more commonplace within certain industries, firms in those industries will comply more fully with IFRS (Alfaraih 2009). Therefore, industry category was added to this study and based on the previous studies, the following hypothesis has been formulated:

H5: There is a relationship between the industry of each firm and the level of online disclosure for the same firm.

2.1.6 Liquidity

Al Mutawa [8] studies state that there exist a relationship between liquidity and disclosure of listed firms in Kuwait financial market. Differently, Al Shammari [10] reported a negative association between liquidity and disclosure.

Therefore, based on the previous studies, the following hypothesis has been formulated:

H6: There is a relationship between the liquidity of each firm and the level of online disclosure for the same firm.

2.1.7 Electronic Financial Disclosure (EFD)

Sarea et al. [24] addressed the issues of current electronic disclosure in GCC and used check list consists of 90 items to express the EFD. In this research, machine learning technique is applied to find the co-relation between EFD and other variables mentioned in Table 2.

3 Research Methodology

3.1 Why Machine Learning Technique (MLT)?

Basic forms of machine leaning algorithms are commonly used by many companies including banking and financial sectors [19]. Machine learning and Artificial Intelligence techniques have been successfully applied in financial services especially in banking sectors to develop a more accurate and generally applicable prediction model [26].

To achieve the response levels in financial applications, deep learning techniques were implemented in many researches due to the advent efficacious hardware [13].

3.2 Benefits in the Current Field of Study

Machine Learning Technique has been applied to improve the feature prediction which are essential to identify the correlation between dependent and independent

Table 1 Sample size

Listed firms in Bahrain Bourse 2017	Excluded (due to unavailable of data)	Selected sample
48	12	36

variables. In addition, machine learning statsmodels and Ordinary Least Square (OLS) was used to fit the regression line, which helped us to predict that profitability has more impact on EFD as in the case of Bahrain financial sectors.

3.3 Study Sample

The study sample consists of all listed firms in Bahrain Bourse (BB) during 2017. Data obtained through secondary method that is from the website of individual firm listed in Bahrain Bourse Table 1.

3.4 Analyze the Data Using MLT

Step by step procedure was followed in this research to achieve our hypothesis, stating that Profitability has more impact on EFD.

The methodology flow is explained below:

As mentioned in Sect. 3.1, dataset was collected and to improve the dataset, it was initially verified for the null values and outliers. In the next step, features selection and importance was depicted using exploratory data analysis which was useful in this research to predict that Profitability has more importance than other independent variables. Research formula was framed and linear regression technique was applied to measure the correlation between dependent and independent variables stated in Sect. 2.1. In the final stage of this research, variables and its significance were measured using standard techniques and research findings were measured using the stats models in Python Programming language and was tested in the computer has Intel core i3 processor with 8 GB RAM.

3.5 Research Formula

The data obtained through secondary method from the website of individual firms listed in Bahrain Bourse.

1. Visited the website of each firm.
2. Checked both firm profile and Financial reporting.
3. Checked Age, Liquidity, Leverage, Size andIndustry.

4. Basing upon individual firm information under above steps assigned "1" for EFD and assigned "0" for negative findings.

The formula as follows:

$$EFD = \sum_{i=1} \frac{di}{n} \tag{1}$$

where

di: (1) if X firm discloses EFD otherwise (0).

n: Maximum score X firm.

3.6 Model

The following regression model was used to test the hypothesis. The regression model was developed using the EFD as dependent variable, type of firm characteristics as independent variables.

Simple linear regression model is applied in this research to find the co-relation

$$EFD_i = \beta_0 + \beta_1 Age_i + \beta_2 Liquidity + \beta_2 Leverage_i + \beta_4 Size_i + \beta_4 Industry_i + \beta_4 Profitability_i + \varepsilon_i \tag{2}$$

where the operationalization of all variables mentioned in Table 2.

It should be borne in mind that we are assuming there is information to disclose, and we are interested in how much of it is disclosed, and shown that importance of independent variables towards dependent variable and why.

Table 2 Summary of variables

Variable	Measurement
Firm size	Total assets
Leverage	Ratio of total debts to total assets
Firm age	Date of Establishment
Profitability	Return on Assets (ROA)
Industry	Dummy values
Liquidity	Current ratio
EFD	Dummy

4 Research Findings Through Machine Learning Prediction

Predictive analysis learning plays vital role in many phases of the financial systems, especially provides great support financial decision-making process using financial disclosure [21]. Traditional methods are usually able to correctly interpret the content, however, predicting the same is rarely true of computerized decision support systems, which struggle with the complexity and nature of the variables. Culkin and Das [13] stated that machines may be trained a lot faster than humans. Accordingly, financial disclosure can be trained and predicted by training the models.

4.1 Data Analysis

Bahrain is the financial hub in the region and attracts Foreign Direct Investment (FDI) from different continents. The data were collected from secondary source through the website of each firm in different sectors such as Commercial banks, Investment, Insurance, Service Sector, Industrial, Hotel and Tourism. However, mainly data collected from the financial reporting published in 2017. The sample size consists of 36 firms listed in Bahrain financial market.

Using the dataset available in Bahrain firms, with independent variables such as Age, Liquidity, Leverage, Size, Industry and Profitability and a dependent variable Electronic Financial Disclosure. This research analyzed the relationship between independent variables towards EFD.

4.2 Data Cleaning

As mentioned by Guyon et al. [16], data cleaning could get large information gain if systematically eliminating the suspicious pattern. However, this is dangerous since valuable informative patterns may also be eliminated.

Fig. 1 Data clean

```
Age              0.0
Liquidity        0.0
Leverage         0.0
Size             0.0
Industry         0.0
Profitability    0.0
EFD              0.0
dtype: float64
```

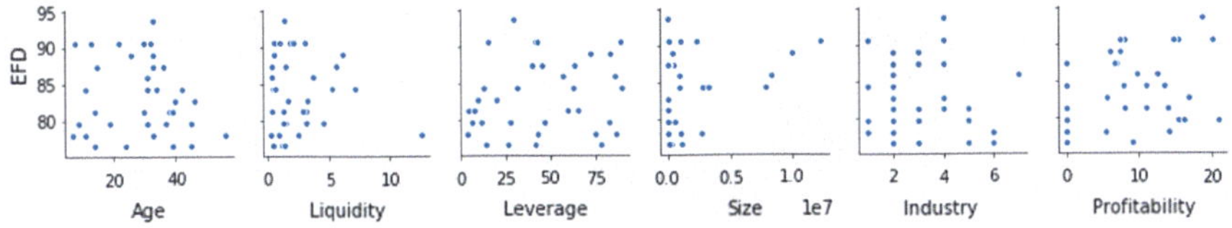

Fig. 2 Exploratory data analysis

Figure 1 shows that our dataset is clean which has no NULL values in the dataset, hence the data is clean and ready to use.

4.3 Outlier

An outlier is an observation which deviates so much from other observations as to arouse suspicions that it was generated by a different mechanism [3]. In this research, using MLT technique Outlier has been verified on the dataset and no considerable outlier has been found.

4.4 Feature Importance

The importance of independent variables towards the target variable has been shown using exploratory data analysis and pair plot method in the following sections.

4.4.1 Exploratory Data Analysis

To identify the features which might be theoretically important to generate the hypothesis and towards the target variable, exploratory data analysis (EDA) is used in many researches [17]. Accordingly EDA has been done and shown in Fig. 2.

Figure 2 depicts exploratory data analysis of independent variables towards the target variable.

4.4.2 Pair Plot

In Python, pair plot method helps to identify the trends for follow up analysis and displays the distribution of single variables and relationship between two variables. Accordingly, relationship between the EFD and independent variables have been analyzed and shown in Fig. 3:

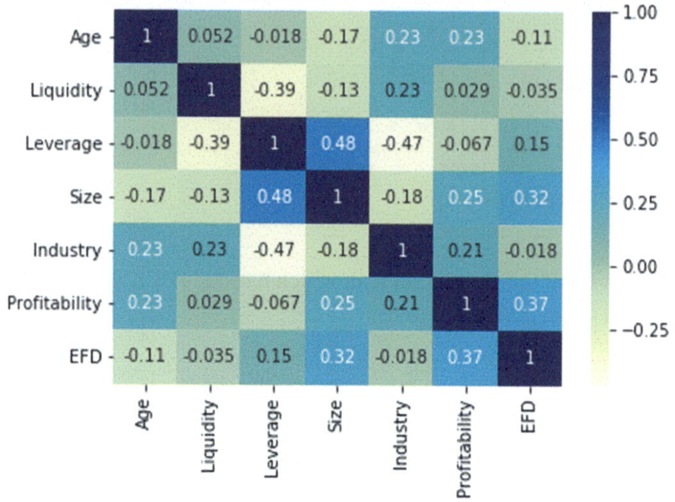

Fig. 3 PairPlot for Target and Independent variables relation

From Fig. 3 it seems that profitability is more related to target variable-EFD. Hence, simple linear regression will be performed in the following step using Profitability as our feature variable.

4.5 Regression

Simple linear regression model is applied in this research to find the co-relation, which was discussed in Eq. 1. Based on Fig. 3 and other points discussed on Sect. 4.3 and 4.4, it has been shown that profitability is the more likely to be related with EFD compare to other independent variables. Thus, in our the following equation has been formed

$$EFD_i = \beta_0 + \beta_1 Profitability \tag{3}$$

where

β_0 is the intercept and β_1 is the model co-efficient feature variable is profitability and EFD is the response variable.

In this research we explain various methods and how to interpret them substantively using visualizations from our Python software package. Model was built by statsmodel technique of python programming, accordingly consider

$$X = Bahrain_{firm}[Profitability] \tag{4}$$

$$Y = Bahrain_{firm}[EFD] \tag{5}$$

4.6 Regression Data Analysis Using Ordinary Least Square

In this section, dataset was split into training and test sets. Many researches split the ratio of 70% of data in train set and rest 30% of data in test set. By default, statsmodels library fits a line on the dataset which passes through the origin. But in order to have an intercept, constant value was added in the X_train dataset and Ordinary Least Square (OLS) was used to fit the regression line. The following Fig. 4 shows the summary of OLS regression results of our model applied in the training dataset.

The following section analyze the results produced through OLS model. To prove that regression equation mentioned in Eq. 3, the following parameters were concerned in this research; which were extracted from Fig. 4.

4.6.1 The Coefficients and Significance (p-Values)

The coefficient for Profitability is 0.3001, which means that the correlation between EFD and profitability is significant.

```
                            OLS Regression Results
==============================================================================
Dep. Variable:                    EFD   R-squared:                       0.147
Model:                            OLS   Adj. R-squared:                  0.109
Method:                 Least Squares   F-statistic:                     3.948
Date:                Fri, 18 Oct 2019   Prob (F-statistic):             0.0590
Time:                        12:09:50   Log-Likelihood:                -72.972
No. Observations:                  25   AIC:                             149.9
Df Residuals:                      23   BIC:                             152.4
Df Model:                           1
Covariance Type:            nonrobust
=================================================================================
                    coef    std err          t      P>|t|      [0.025      0.975]
---------------------------------------------------------------------------------
const            81.9394      1.516     54.035      0.000      78.802      85.076
Profitability     0.3001      0.151      1.987      0.059      -0.012       0.613
==============================================================================
Omnibus:                        9.218   Durbin-Watson:                   1.858
Prob(Omnibus):                  0.010   Jarque-Bera (JB):                2.133
Skew:                           0.016   Prob(JB):                        0.344
Kurtosis:                       1.569   Cond. No.                         16.4
==============================================================================
```

Fig. 4 OLS regression result for Train dataset

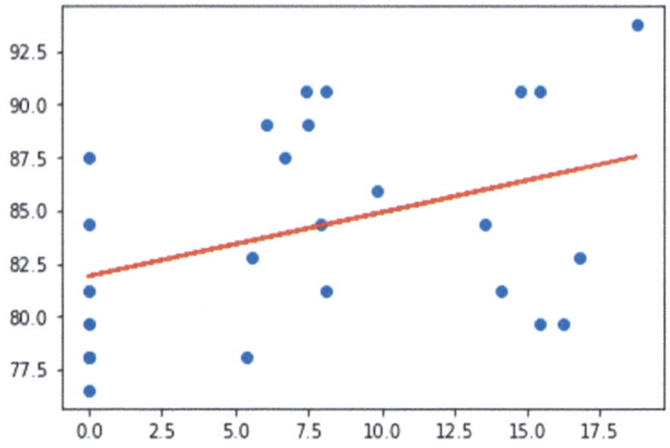

Fig. 5 Linear regression model

4.6.2 R-Squared

R-Squared value is 14.7%, though it is not high, Profitability has impact on EFD in minimal amount compared to other control variables.

4.6.3 F Statistic and its Significance

F-statistic is 3.948, which is relatively low value and it means that the model fit is statistically significant, and the explained variance isn't purely by chance and we have framed the new linear regression as

Substitute the values in Eq. 3 to visualize the linear regression as shown in Fig. 5.

$$EFD = 81.9394 + 0.3001 \times Profitability \tag{6}$$

4.7 *Model Evaluation*

One of the major assumptions of linear regression is to check if the error terms are also normally distributed. Histogram chart depicted in Fig. 6 shows that the errors terms are not having peak mean value as 0, however the curve distributed closer to normal distribution. Thus, this research model has normal distribution.

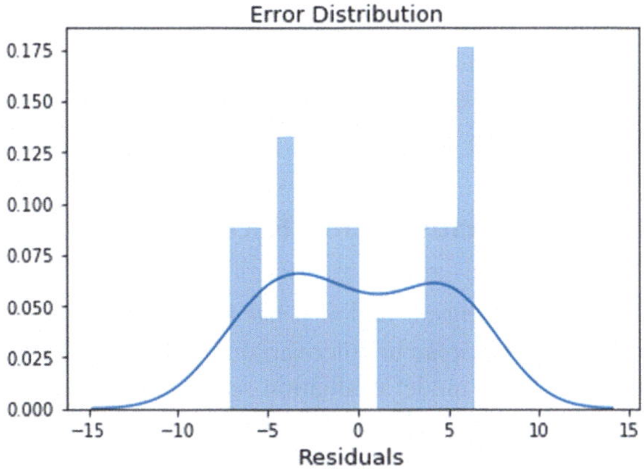

Fig. 6 Error terms distribution

4.8 Test Set Prediction

The test set prediction has been done in this research as similar steps followed in the training dataset. OLS regression results has been obtained and it has been shown in the Fig. 7.

As shown in Fig. 7, R-Squared, F-Statistics and Co-efficients values evidences that Profitability is having significant impact on EFD.

Our chosen objective function is the mean squared error (MSE), an objective function often chosen for regression problems. Accordingly Mean squared error value was found as 4.07896. In addition, R2-Score has been calculated to show goodness-of-fit measure for our linear regression model and it was found that 0.2322. As per

```
                            OLS Regression Results
==============================================================================
Dep. Variable:                    EFD   R-squared:                       0.232
Model:                            OLS   Adj. R-squared:                  0.147
Method:                 Least Squares   F-statistic:                     2.722
Date:                Fri, 18 Oct 2019   Prob (F-statistic):              0.133
Time:                        12:21:55   Log-Likelihood:                -31.073
No. Observations:                  11   AIC:                             66.15
Df Residuals:                       9   BIC:                             66.94
Df Model:                           1
Covariance Type:            nonrobust
=================================================================================
                    coef    std err          t      P>|t|      [0.025      0.975]
---------------------------------------------------------------------------------
const            78.4925      2.291     34.266      0.000      73.311      83.674
Profitability     0.3154      0.191      1.650      0.133      -0.117       0.748
==============================================================================
Omnibus:                        1.294   Durbin-Watson:                   2.248
Prob(Omnibus):                  0.524   Jarque-Bera (JB):                0.774
Skew:                           0.231   Prob(JB):                        0.679
Kurtosis:                       1.785   Cond. No.                         20.3
==============================================================================
```

Fig. 7 OLS regression test for test dataset

the definition of R-square, small R-squared values are not always a problem, and high R-squared values are not necessarily good. In this model, R2-score was 23.22 percentage which shows the significance importance of Profitability towards EFD.

5 Interpretation, Conclusion and Recommendations

As per the results shown in Figs. 3 and 4, it has been proved that Profitability has significant impact on EFD compare to other variables. Model has been evaluated and results were discussed in the model evaluation section and it was found that error terms are not reaching the mean value of 0.0, whereas the mean error value distribution is closer to normal distribution. R2 score and MSE proves that our Profitability is having more co-relation with the dependent variable in comparison with other independent variables.

This study contributes theoretically to "fill the gap in the literature". However, a few studies have dealt with the "level of EFD" in different sectors by using different techniques. This is the first study to use Machine learning through Python Programming analysis in order to measure the effect of: Age, Liquidity, Leverage, Size, Industry and Profitability on the Electronic Financial Disclosure (EFD) through the Website of each firm listed in Bahrain Bourse (BB).

The study creates awareness regarding "EFD" in listed firms in Bahrain Bourse as well as the effect of: Age, Liquidity, Leverage, Size, Industry and Profitability on the Electronic Financial Disclosure (EFD). The technique used in this research was based on a checklist developed and enhanced by "Sarea et al. [24], Al-Sartawi [9], Alrawahi and Sarea [12], and Al jawder and Sarea [11]" to calculate the level of EFD as well as Machine learning techniques. The outcome reports showed the effect of selected firm's factors in Bahrain Bourse on the EFD. The main findings shows that profitability factor is the highest impact on the level of EFD. This study recommends all firms listed in Bahrain Bourse to disclose more financial information through website to improve their exposure, credibility and transparency when compared to other countries in the region. This research as many of previous studies has such limitation, this research focuses on the impact of Age, Liquidity, Leverage, Size, Industry and Profitability on the Electronic Financial Disclosure (EFD) through the Website of each firm listed in Bahrain Bourse (BB) only. The current research could be expanded to cover more non-financial data by using different analysis techniques.

References

1. Adebimpe, O.U., Ikenna, E.A.: Internet financial reporting and company characteristics: a case of quoted companies in Nigeria. Res. J. Finan. Acc. **4**(12), 72–80 (2013)
2. Agboola, A.A., Salawu, M.K.: The determinants of internet financial disclosure: empirical evidence from Nigeria. Res. J. Finan. Acc. **3**(11), 2012 (2012)
3. Aggarwal, C.: Outlier analysis. In: Data Mining. Springer, Cham (2015)

4. Agyei-Mensah, B.K.: Corporate financial reporting: firm characteristics and the use of internet as a medium of communication by listed firms in Ghana. Afr. J. Bus. Manag. **6**(6), 2299–2309 (2012)
5. Akhtaruddin, M.: Corporate mandatory disclosure practices in Bangladesh. Int. J. Acc. **40**, 399–422 (2005)
6. Aktan, S.: Application of machine learning algorithms for business failure prediction. Invest. Manag. Finan. Innov. **8**(2), 51–65 (2011)
7. Al-Anezi, F.S.: Factors influencing Kuwaiti companies' internet financial disclosure. J. Econ. Admin. Sci. **25**(2), 1–23 (2009). http://www.emeraldinsight.com.libweb.uob.edu.bh/journals.htm?issn=1026-4116&volume=25&issue=2&articleid=17014762&show=html
8. Al-Mutawa, A.: Disclosure level and compliance with ifrss: an empirical investigation of Kuwaiti companies. Int. Bus. Econ. Res. J. **9**(5), 33–50 (2010)
9. Al-Sartawi, A.: Measuring the level of online financial disclosure in the Gulf Cooperation Council Countries. Corp. Ownership Control **14**(1), 547–558 (2016)
10. Al-Shammari, B.: Determinants of Internet financial disclosure by listed companies on the Kuwait Stock Exchange. J. Int. Bus. Econ. **7**(1), 162–178 (2007)
11. Al jawder, N., Sarea, A.: Determinations of internet financial reporting: evidence form Bahrain Bourse. Jordan J. Bus. Adm. **12**(4), 935–960 (2016)
12. Alrawahi, F., Sarea, A.: An investigation of the level of compliance with international accounting standards (IAS 1) by listed firms in Bahrain Bourse. Int. J. Islamic Middle Eastern Finan. Manag. **9**(2), 254–276 (2016)
13. Culkin, R., Das, S.R.: Machine learning in finance: the case of deep learning for option pricing. J. Invest. Manag. **15**(4), 92–100 (2017)
14. Debreceny, R., Gray, G.L., Rahman, A.: The determinants of internet financial reporting. J. Acc. Public Policy **21**(5), 371–95 (2002)
15. Ferguson, M.J., Lam, K.C.K., Lee, G.M.: Voluntary disclosure by state owned enterprises listed on the stock exchange of Hong Kong. J. Int. Finan. Manag. Acc. **13**(2), 125152 (2002)
16. Guyon, I., Matic, N., Vapnik, V.: Discovering Informative Patterns and Data Cleaning (1996)
17. Jones, Z., Linder, F.: Exploratory data analysis using random forests. In: Prepared for the 73rd Annual MPSA Conference (2015, April)
18. Joshi, P.L., Al-Modhahki, J.: Financial disclosure on the Internet: empirical evidence from Bahrain and Kuwait. Asia Rev. Acc. **11**(1), 88–101 (2003)
19. Lightbourne, J.: Algorithms & fiduciaries: existing and proposed regulatory approaches to artificially intelligent financial planners. Duke LJ **67**, 651 (2017)
20. Mathur, P.: Overview of Machine Learning in Finance. In Machine Learning Applications Using Python, pp. 259–270. Apress, Berkeley, CA (2019)
21. Nassirtoussi, A.K., Aghabozorgi, S., Wah, T.Y., Ngo, D.C.L.: Text mining for market prediction: a systematic review. Expert Syst. Appl. **41**(16), 7653–7670 (2014)
22. Omran, M., Ramdhony, D.: Determinants of internet financial reporting in African markets: the case of Mauritius. J. Dev. Areas Tennessee State Univ. Coll. Bus. **50**(4), 1–18 (2016)
23. Oyelere, P., Laswad, F., Fisher, R.: Determinants of internet financial disclosure by New Zealand companies. J. Int. Finan. Manag. Acc. **14**(1), 26–63 (2003)
24. Sarea, A.M., Al-Sartawi, A.M.M., Khalid, A.A.: Electronic financial disclosure: Islamic banking vs conventional banking in GCC. In: Conference on e-Business, e-Services and e-Society, pp. 279–291. Springer, Cham (2018, October)
25. Singhvi, S.S., Desai, H.B.: An empirical analysis of the quality of corporate financial disclosure. Acc. Rev. **46**(1), 129–138 (1971)
26. Tsai, C.F., Lu, Y.H., Hsu, Y.F.: Bankruptcy prediction by supervised machine learning techniques: a comparative study. In: Machine Learning: Concepts, Methodologies, Tools and Applications, pp. 668–683. IGI Global (2012)

Artificial Intelligence and Economic Development

Artificial Intelligence in Africa: Challenges and Opportunities

Emmanuel Ogiemwonyi Arakpogun, Ziad Elsahn, Femi Olan, and Farid Elsahn

Abstract The developments in artificial intelligence (AI) have the potential to disrupt and transform socio-economic activities across industries. While evidence is emerging that businesses and governments across the world are positioning themselves to maximise these potentials, evidence also indicates that countries in the Global North are better prepared to reap the benefits of AI even though a significant number of jobs that could be displaced in the process are in the Global South. Therefore, we posit that countries in the Global South such as those in Africa need to tackle governance issues and lack of institutional capacity to establish the building blocks to allow AI to flourish. It is important to also examine the roles of international communities' in bridging the technological gaps in Africa by adopting a problem-driven approach where local needs and problems are contextualised into AI policy formulation rather than a blanket 'copy-and-paste' practice that has limited the advancement of development policies in Africa. A problem-driven approach would help African countries to formulate robust AI policies that are relevant to their unique circumstances.

Keywords Africa · Artificial intelligence · Challenges · Digital divide · Opportunities

1 Introduction

Developments in artificial intelligence (AI) have attracted both academic and public attention to analyse and understand its potential to transform our societies and its disruptive impact on a range of industries. While the definition of AI varies across

---

E. O. Arakpogun (✉) · Z. Elsahn · F. Olan
Newcastle Business School, Northumbria University, City Campus East 1, Newcastle upon Tyne NE1 8ST, UK
e-mail: e.arakpogun@northumbria.ac.uk

F. Elsahn
Doha Institute for Graduate Studies, Doha, Qatar

A. Hamdan et al. (eds.), *The Fourth Industrial Revolution: Implementation of Artificial Intelligence for Growing Business Success*, Studies in Computational Intelligence 935,
https://doi.org/10.1007/978-3-030-62796-6_22

literature, the definitions are underpinned by the idea of AI as a "non-human intelligence programmed to perform specific tasks" [1]. We draw on existing literature [2–4] to posit that AI is a collection of information and communication technologies (ICTs) that mimic human intelligence to enable machines to facilitate our jobs better, create greater efficiencies and drive economic growth. Thus, various definitions of AI point to the capacity of machine to "perform specific roles and tasks currently performed by humans within the workplace and society in general" [1]. Although AI was first coined in 1956 by John McCarthy, the concept predates the 1950s [5–8]. Accordingly, while AI has existed for centuries and morphed from the first[1] to second[2] and third[3] industrial revolutions with limited attention, the emergence of the fourth[4] industrial revolution (4IR) has heightened attention due to the critical role of ICTs in accelerating socio-economic development. AI is accelerating the speed and scope of the 4IR at an unprecedented and exponential pace relative to the linear process of previous industrial revolutions [9]. From agriculture to banking, education to healthcare, the reach and usefulness of AI is still unfolding. It is estimated that by 2030, AI could contribute over $15 trillion to the global economy with more than 20% increase in the GDP of local economies as humans and robots 'work' in harmony to engineer solutions to challenging world problems [3, 10].

While governments and businesses across the world are beginning to position themselves to reap the opportunities of AI [4, 11], the development and deployment of AI technologies are uneven globally with a considerable gap between developed and developing countries [12]. Furthermore, it is argued that the increasing use of AI can have a negative impact on developing countries that host a significant number of jobs that could be displaced through automation. Therefore, it is vital that we investigate the readiness of developing countries by exploring the opportunities and challenges that AI can bring to these economies. To attain this feat, we focus on African countries. The overarching aim here is to provide insights into the African context—a region of the world that has not maximised the benefit of the previous industrial revolutions and often under-researched when it comes to the development of ICTs relative to the Global North [12, 13]. This chapter is organised as follows: we first take stock of recent developments in AI in Africa, followed by an examination of the potential economic, governance and social opportunities that it can provide to African nations. We then point out some of the challenges that policymakers need to be aware of as a result of AI deployment. We conclude by underscoring the need for policy initiatives to support the development of AI capabilities in the African context.

---

[1] First industrial revolution adopted water and steam engine to mechanise and advance production process.

[2] Second industrial revolution adopted the power of electricity to advance mass production.

[3] Third industrial revolution (3IR) adopted electronics and information technology to automate production.

[4] Fourth industrial revolution adopts a fusion of digital technologies to further advance production automation.

2 Artificial Intelligence in Africa: Distilling the Story so Far

As the development of AI continues to expand across the world, an ecosystem of AI is also emerging across Africa. According to Leila Janah, the founder and CEO of Samasource,[5] "*if you use a mobile phone or laptop's facial recognition features, drive a car, or shop online, there's a good chance that a person in East Africa helped train the algorithm that makes your technology work.*" [14]. Similarly, several local and international AI research hubs have been established. For example, the University of Cape Town (South Africa) established the Robotics and Agent Lab (RAL) in 2007 to focus the development of robotics and computational intelligence as well as building the capacity of children from low-income households by supporting their annual participation in RoboCup Junior—an annual international robotics competition [15, 16]. Data Science Africa (DSA), which has been operating since 2013, is a non-governmental organisation based in Kenya with a focus on promoting affordability, wider deployability and the suitability of AI solutions in Africa [17]. DSA provides a platform for AI practitioners and researchers across Africa to discuss and share knowledge of the development and usefulness of AI via summer schools and workshops, which have been held in Ethiopia, Ghana, Kenya, Nigeria, Tanzania and Uganda. Part of DSA achievements includes the training of over 200 people in machine learning techniques and data science applications using the internet of things (IoT) and social media data analytics. The outcomes of these trainings have resulted in the development of applications that facilitate agriculture, disaster management and healthcare [17].

Women in Machine Learning & Data Science (WiMLDS) also have chapters in Algeria, Botswana, Kenya, Morocco, Nigeria and Uganda to engage with and promote women participation in AI. WiMLDS is critical in amplifying the voice and input of women in the advancement of AI given the concerns raised on the lack of diversity and gender bias in AI [18]. Furthermore, IBM Research opened AI labs in Nairobi (Kenya) and Johannesburg (South Africa) in 2013 and 2016 respectively [19]. Google followed in 2019 and opened its AI lab in Accra (Ghana) [19]. Both Google and IBM labs have engineered AI solutions to improve food production and healthcare across Africa as outlined in Sect. 3. Given that majority of the academic and industry research hubs that focus on AI are predominantly located in advanced and wealthy locations such as Silicon Valley in California and Zhogguancun in Beijing, a recent shift to African countries is encouraging and has led to the emergence of around 100 AI start-ups across various regions in Africa, examples of which are presented in Table 1.

About 100 AI start-ups have emerged across Africa, raising over $140 million of seed-funding with the majority going into fintech in Nigeria. For example, Cellulant[6]

[5]Samasource is an AI data training company headquartered in San Francisco with operations in East Africa.

[6]Cellulant is a fintech start-up located in Nigeria. It uses AI to facilitate digital payments and transfers.

Table 1 Examples of AI start-ups across Africa

Country	Government AI readiness index ranking (out of 194 countries)	Number of AI start-ups	Start-up example	Year founded	Sector application	Description
Algeria	141	1	Monadim	2017	Marketing	Monadim combines data analytics and machine learning algorithms to provide a 'one-stop' digital platform for firms to manage business activities such as accounting, customer relationship management, purchasing, payroll, project management and inventory
Cameroon	119	1	Agrix Tech	2018	Agriculture	Agrix Tech uses data analytics and AI imaging techniques to detect crop diseases at early stages and propose viable treatments to farmers in local languages
Cote d'Ivoire	104	1	WeFlyAgri	2016	Agriculture	WeFlyAgri provides drone technologies and virtual reality services that are user-friendly for farmers to remotely monitor crop production from the sky
Egypt	111	5	WideBot	2016	Business service	WideBot provides a segmentation platform that focuses on over 400 million Arabic speakers using a customised chatbot to understand different Arabic dialects for targeted broadcast
			Niotek	2017	Manufacturing	Niotek combines hardware and software to power an IoT platform that relays real-time visibility and actionable insights to facilitates efficient and fast decision making for manufacturing operations

(continued)

Table 1 (continued)

Country	Government AI readiness index ranking (out of 194 countries)	Number of AI start-ups	Start-up example	Year founded	Sector application	Description
Ghana	75	5	Curacel	2016	Fintech	Curacel provides an automated platform that helps insurers to facilitate claims and fraud detection using algorithms that can detect fraud and errors in customer data
Kenya	52	9	PesaKit	2018	Fintech	PesaKit combines a chatbot with other AI techniques to provide a last-mile agent network to accelerate financial inclusion for the unbanked
Nigeria	107	20	Gradely	2017	Education	Gradely uses AI to study the learning pattern and output of students to engineer a personalised learning toolkit that helps schools and parents to intervene in real-time and plug the gaps in students learning
			Aajoh	2015	Health	Aajoh provides a platform that uses predictive analytics to improve healthcare access via mobile phones – where people can remotely receive medical diagnosis and treatment, and mitigate the lack of healthcare in various parts of Africa
South Africa	68	26	Accrad	2020	Health	Accrad developed an AI deep learning algorithm application called CheXRad to concurrently detect coronavirus (Covid-19) and up to 14 clinical diseases in chest radiography

Source created by the authors using data from a variety of sources including Alliance4AI.org and Oxford Insights have raised $47.5 million followed by Mines.IO[7] with $17.2 million [20]. While Nigeria has attracted the largest amount of seed-funding, South Africa has the highest number of AI start-ups (26) in Africa followed by Nigeria (20) and Kenya (9). Further, Tunisia and Zimbabwe have 6 AI start-ups apiece while Egypt and Ghana have 5 each. It is somewhat surprising that while Nigeria has attracted the largest AI seed-funding, Kenya, Tunisia, Mauritius, South Africa and Ghana rank higher (respectively) when it comes to government AI readiness in Africa [see 12 for details]. It is also interesting to note that while the fintech sector has attracted the largest amount of seed-funding, the healthcare sector has the highest number of AI start-ups (21) followed by agriculture (14) and manufacturing (5) sectors. What is more, most of the AI start-ups were founded about a decade after the advent of the 4IR, which began in 2000 [21]. An indication that African countries are already lagging behind the rest of the world. This is consistent with the 2019 government AI readiness index, which indicates that there is no African country in the top 50 from the 194 countries analysed [12]. This is reflective of [10] projection that Asia (especially China) and North America would have the largest economic gains from AI—nearly 70% of the projected $15.7 trillion estimated for AI contribution by 2030.

In order to avoid a repeat of the missed opportunities from the previous industrial revolutions that have left a negative legacy for African countries, governments must create an enabling environment for these AI start-ups to flourish and accelerate the socio-economic development of Africa. These AI start-ups have taken the first step to overcome the infrastructure and resource constraints prevalent across Africa to engineer local solutions [20, 22]. Strong political will and government leadership are now needed to complement and sustain the gains made so far. Specifically, AI funding is needed to help local start-ups expand capacity and reach because if one compares the $140 million that has been raised in Africa thus far to the projected $79.2 billion global AI spending for 2022 [23], one could see a huge AI funding gap between Africa and the rest of the world.

3 Opportunities

AI-powered 4IR has the capacity to improve various aspects of socio-economic development across the world. AI has the potential to improve employment, the advancement of medicine and the quality of lives, improve productivity and efficiency of global supply chains as well as raise global incomes [9]. Many of these opportunities are already emerging across Africa. Take **employment**, for example. Samasource has employed youth across Kenya and Uganda to 'train' data and transmit human intelligence to AI for big tech companies including Google, Microsoft and Yahoo [14, 24]. Over 11,00 youth are working on various projects across Kenya and Uganda with incomes that support, for example, the education of their siblings and overall living

[7]Mines.IO is a fintech start-up located in Nigeria. Mines provides a platform that uses financial analytics to develop credit rating and fraud detection.

condition of their families so much so that a network of over 50,000 people are now benefitting from this process [25]. The dependable income generated in the process also increases the purchasing power of people, which then helps them to gradually break the endemic cycle of poverty. For a country like Kenya with disproportionately high levels of youth unemployment of over 30% despite a growing literacy rate [25, 26], the jobs created by Samasource are critical to improving the quality of lives and maintaining social cohesion in Kenya.

In terms of promoting **food security**, Google AI lab has collaborated with farmers in rural Tanzania to create a machine-learning application called 'Nuru' (meaning light in Swahili) to diagnose early stages of cassava plant diseases for the advancement of the production of a common staple crop that provides food for over 500 million people [27]. Nuru works directly on farmers' cellphone even without internet connectivity and warns them to take early intervention by quickly identifying and managing cassava plant diseases, which, in turn, helps to maintain consistent food production. From the Sahel region to the Horn and Southern Africa, food insecurity is a major concern across the continent [28]. In 2018 alone, over 230 million people in sub-Saharan Africa (SSA) suffered from food insecurity [29]. While about 65% of global arable uncultivated land is situated in Africa, its governments collectively spend nearly $65 billion importing food in 2017 [30]. Furthermore, the locust outbreak in the Horn of Africa in 2020 is estimated to destroy over $8 billion worth of food and livestock [55]. The impact of food insecurity in Africa could be mitigated with the use of AI applications like Nuru and others in Table 1, crop diseases and disasters can be predicted, and farmers forewarned for better preparation. It is also useful to bear in mind that over 60% of Africa's employment comes from the agricultural sector [30]. Therefore, agriculture is a strategic sector that needs improvement across Africa and AI should be a critical part of the solution going forward.

Healthcare improvement is also not exempted. Following the delay in the reporting process of cancer diagnosis in South Africa as a result of the manual and unstructured pathology process, IBM Research pioneered a machine-learning system to automate the process and cut the reporting time from four to two years [19, 31]. The reduction of the reporting process offers invaluable information for the government to formulate national health policy and take timely decisions that would save lives. While South Africa is among the two countries in Africa (the other being Mauritius) that meets the minimum World Health Organisation's recommendation of 23 healthcare workers per 10,000 people [32], the country is still struggling to provide adequate healthcare to its citizens. Overall, access to healthcare in many African countries is acute. AI solutions like those engineered by IBM Research and other AI start-ups in Africa could be a 'game-changer' as the use of machine-learning systems and mobile phones can facilitate remote diagnosis and treatment for millions across Africa, particularly those living in remote and rural areas. Accordingly, from agriculture to banking, education to employment, healthcare to manufacturing, Table 1 is indicative of the opportunities and transformation offered by AI in Africa. But while there are several opportunities to be reaped from the development of AI in Africa, scholars have also pointed out several challenges and implications concerning, for example, structural inequalities and ethical concerns that policymakers need to take seriously [1].

Our country analysis in Sect. 2 indicates that the Fintech sector has attracted the highest investment in AI start-ups in Africa. **Financial inclusion** has thus been boosted by the advancement of technology in general and AI in particular. For example, it has been established that two-thirds of the over 1 billion people in Africa do not have access to traditional financial services [33]. The implication of this is that many people living across Africa are unbanked and excluded from the benefits of financial services, including the improvement of income earning potential for millions of people affected by poverty. For example, it has been found that an increase in financial inclusion by mobile money in Kenya has lifted over 180,000 women from poverty by enabling them to move from farming to developing small businesses [34]. AI is now enabling Fintech platforms such as Curacel and PesaKit (see Table 1) to accelerate digital financial inclusion for millions of unbanked across Africa.

4 Challenges

AI scholars [11, 35] have expressed concerns that the evolution of AI and other technological advancements could result in a series of unintended consequences. Broadly speaking, these concerns include the amplification of existing *structural inequalities, governance and regulation, and business and work disruptions* [35–37]. While these concerns could endanger any country in the world, we offer a nuanced discourse that focuses on African countries.

Structural inequalities are the disproportionate levels of access to socio-economic and political resources such as education, employment, income, ICTs and healthcare. When it comes to access to such socio-economic resources, African countries are among the least developed in the world [38]. Since AI is powered by a fusion of ICTs, the lack thereof is not the only disadvantage to prospective AI users but also AI developers with the implication that those suffering from a structural inequality like digital divides[8] are more likely to miss out on the critical opportunities of AI. Similar to the disproportionate levels of structural inequalities, African countries also lag behind other parts of the world when it comes to digital divides albeit with varying degrees across regions. For example, while Northern African countries average 68% average mobile phone penetration, SSA have 45% [39]. Digital divides in Africa are linked to issues such as inadequate telecoms network, lack of supporting infrastructure like electricity, unaffordability of smartphones and lack of digital skills [40]. Therefore, the Digital divides in Africa is symptomatic of deeper issues so much so that existing inequalities are then transferred into the digital space with the implication of poor AI readiness and many Africans falling through the net as indicated in the 2019 AI government readiness index [12]. With high levels

[8]Digital divides here refer to the lack of access and skills to and affordability of ICTs.

of structural inequalities and digital divides, [41] also highlights a high degree a lack of government preparedness for AI among African countries with no country in the continent among the top-ranked 50 positions. Therefore, there is a correlation between structural inequalities, digital divides and AI readiness so much so that the most disadvantaged would also miss out on AI opportunities such as employment, improvement in education and healthcare, and access to e-government services.

Challenges also abound for AI **governance and regulation**. The fusion of multiple technologies underpins the architecture of AI and its ability to drive the 4IR. This fusion is accelerating technology convergence in a manner that poses a new challenge to governments, particularly in the realms of regulation. It is argued that the institutional capacity (i.e., skills and financial resources) of governments to regulate previous industrial revolutions, was less challenging given the propinquity between the speed of public policy decision-making and the linear-mechanistic process that existed [9]. Accordingly, governments had more time to formulate relevant regulatory framework to set the rules of engagement using limited capacity to govern the industry [36]. However, with the fast-moving pace of modern technology coupled with constant convergence, regulation becomes increasingly challenging, complex and expensive for governments [42]. Therefore, relevant legal and regulatory frameworks to guide the rules of the game are difficult to formulate and could quickly become obsolete as technologies are becoming a moving target. The complexity that comes with the 4IR also requires a higher degree of institutional capacity, which is disproportionately lacking in African countries along with a fragile legal and regulatory framework [40, 43]. For example, only 19 of the 55 countries in Africa had enacted data security and privacy laws as of 2018 [44]. The headline figure is far meagre when it comes to AI—of all the 46 SSA countries, "only Kenya has an AI task force that is working towards a national strategy" [45].

The general lack of institutional capacity and AI regulatory framework across Africa could result in unintended consequences, including the inability of governments to guarantee the data security and privacy of citizens as well as mitigating the impact of cyber and national security. While the threat of cybersecurity has been around since the 3IR, the advancement that comes with AI and IoT further complexify personal and national security as critical and life-saving devices are wirelessly tethered together [46]. One implication of this is that state or individual actors could cripple critical infrastructure in a manner that threatens the existence of a nation. As technological advancement like AI develops, it heightens the threat levels of cybersecurity—this should also shift how governments respond to safeguard societies. Therefore, the paucity of comprehensive AI policy frameworks that are relevant to the African context is deeply concerning. While some have suggested that African countries should adopt the Convention on Cyber Security and Personal Data Protection of the African Union [45], we argue that such blanket 'copy-and-paste' practice has not helped the advancement of policies in the continent. From mimicking the tenets of wider economic reforms like the 'Washington consensus' [47] to telecoms liberalisation [48], copying-and-pasting policies have left much to be desired in developing economies such as those Africa [40, 49]. Instead of a copy-and-paste approach, African governments should adopt a problem-driven approach where local needs and problems are contextualised into policy formulation [50]. This approach would move AI policy in Africa beyond *institutional form* (how policy should look like) to *institutional function* (what policy should actually do) [51]. This, in turn, would help African countries to formulated robust AI policies that are relevant to their unique circumstances. However, African governments need stakeholder engagement to help reflect the inputs of relevant and a much broader community to build an agile institutional capacity capable of governing AI.

What about **Business and work disruptions?** According to Klaus Schwab, who introduced the 4IR concept, a common concern among business executives is that "*the acceleration of innovation and the velocity of disruption* [in the 4IR] *are hard to comprehend... even for the best connected and most well informed*" [9]. Similarly, there is evidence across Africa that the advancement of AI is beginning to challenge and underpin business models and processes. For example, over 70% of CEOs in South Africa have acknowledged the need to increase AI investment to boost competitiveness and productivity [3]. However, "*only about a third of these organisations are planning significant AI investments...*" [3]. This further underlines the huge AI funding gap between Africa and the rest of the world previously highlighted in Sect. 2. Low levels of AI investment would result in low adoption for businesses operating across Africa, making it very challenging to leverage on the opportunities offered by AI. Further, the evolution of AI could lead to the displacement of workers as automation could become substitutes for human labour [3, 36]. This would lead to wage stagnation and/or increase unemployment as the importance and negotiating power of labour becomes weak. This could increase the income gaps between the high-skilled and low-skilled workers, which could exacerbate structural inequalities. For example, [9] found that as the 4IR continues to emerge, the demand for high-skilled workers has increased while that of low-skilled workers, including those with less education, has decreased. Scholars [52] have highlighted the dangers of automation to the erosion of labour market and industrial policies in SSA and called for stronger institutions to enforce and protect workers' rights.

Another way to mitigate the displacement of workers in the face of automation is by reskilling the existing workforce on the one hand and repositioning the educational systems to cater for the needs of emerging digital opportunities on the other hand. The study of AI could be taken to the grassroots level and embedded in the curricula of primary and secondary schools so that people could learn about the usefulness of AI while also equipping young people with skills that are relevant to the 4IR. If such practice becomes widespread across Africa, this would help to mitigate the concerns around AI entrenching inequalities as more people would become knowledgeable of AI opportunities, increase its adoption, and create a pipeline of skills to compete in the emerging future of work.

5 Conclusion and the Way Forward

AI technologies have now become a central concern of academic and public discourse as they continue to disrupt industries and several aspects of our societies and daily life. Due to the forecasted contribution and impact of AI on future economies, academics, practitioners, and governments around the world are attempting to understand its implications and the ways through which its development can be nurtured. There remains however a huge gap between developed and developing countries when it comes to AI development and deployment which if not closed can lead to further inequalities between the Global North and South. Accordingly, several African countries such as Algeria, Egypt, Ghana, Kenya, Nigeria, and South Africa have established governmental ministries and agencies that are tasked with building national AI strategies to ensure that AI implementations are driving technological changes in businesses, economy, education, agriculture, and infrastructure [53]. However, most African countries are still faced with governance issues and lack of institutional capacity which limits their ability for building supporting institutional and technological infrastructure for AI development and deployment. This leadership challenge raises the question of what institutional development policies and models should be adopted to support AI start-ups in Africa. It is important to also examine the roles of international communities' in bridging the technological gap in Africa which is increasing the social divide [1].

The role of governments in nurturing a conducive environment for start-ups is well established in the innovation eco-system literature. As the innovation literature points out, "innovation, like regional competitiveness, will not be achieved by fiat but rather through a combination of public and private initiatives" [54]. A national innovation ecosystem is composed of complex linkages between a variety of actors including entrepreneurs, large corporations, universities, governments, and users. The system must provide the right incentives for actors to encourage them to engage in innovative activities. These incentives can take the form of funds and seed capital for entrepreneurs and universities research centres engaged in developing AI technologies; encouragement of collaboration between universities and the private sector; and provision of tax policies that rewards innovation in AI. It is important, however, that such policies are not "copy and pasted" by African countries from the west, or imposed by international organisations, rather they should be developed contextually to fit with the needs of their local contexts [50].

While there are considerable opportunities to be gained from AI in Africa, there are also several challenges and unintended consequences that might arise that policymakers need to take seriously. These challenges range from structural inequalities due to digital divides and the lack of digital skills among a large proportion of the African population, to the dangers of automation and the displacement of jobs that might affect many industries. Therefore, due diligence is needed to account for these challenges—for example, by focusing on AI technologies that can empower rather than displace workers as well as developing schemes that focus on bridging the digital divides in African economies. Furthermore, underpinning AI developments is the need for a digitally skilled workforce, as such, bridging the digital divides is vital for African countries to be better placed to benefit from advances in AI.

References

1. Dwivedi, Y.K., Hughes, L., Ismagilova, E., Aarts, G., Coombs, C., Crick, T., Yanqing, D., Rohita, D., John, E., Alex, E., Vassilis, G., Vigneshwara, I., Marijin, J., Paul, J., Arpan, K., Hatice, K., Bianca, K., Williams, M.: Artificial Intelligence (AI): Multidisciplinary perspectives on emerging challenges, opportunities, and agenda for research, practice and policy. Int. J. Inf. Manag. (2019)
2. Kok, J.N., Boers, E.J., Kosters, W.A., Van der Putten, P., Poel, M.: Artificial Intelligence: Definitions, Trends, Techniques, and Cases. UNESCO & Encyclopedia of Life Support Systems, Paris (2009)
3. Schoeman, W., Moore, R., Seedat, Y., Chen, J.Y.-J.: Artificial Intelligence: Is South Africa Ready?. Accenture and the University of Pretoria's Gordon Institute of Business Science, Pretoria (2017)
4. Singh, J.A.: Artificial Intelligence and global health: opportunities and challenges. Emerg. Topics Life Sci. **3**, 741–746 (2019)
5. Anyoha, R.: The History of Artificial Intelligence. Science in the News - Harvard University, pp. 1–19 (2017)
6. Plant, R.: An Introduction to Artificial Intelligence. American Institute of Aeronautics and Astronautics (2012)
7. Smith, C.: The history of artificial intelligence. Hist. Comput. - CSEP **590A**, 1–27 (2006)
8. Spector, L.: Evolution of artificial intelligence. Artif. Intell. **170**, 1251–1253 (2006)
9. Schwab, K.: The Fourth Industrial Revolution: what it means, how to respond. World Economic Forum: https://www.weforum.org/agenda/2016/01/the-fourth-industrial-revolution-what-it-means-and-how-to-respond/ (2016, January 14)
10. PWC: Sizing the Prize: What's the Real Value of AI for Your Business and How Can You Capitalise? PricewaterhouseCoopers (PWC) London (2017)
11. Xu, M., David, J.M., Kim, S.H.: The fourth industrial revolution: opportunities and challenges. Int. J. Finan. Res. **9**(22), 90–95 (2018)
12. Oxford Insights: Government Artificial Intelligence Readiness Index 2019. Oxford: Oxford Insights and the International Development Research Centre (IDRC). Oxford Insights and the International Development Research Centre (2019)
13. Amankwah-Amoah, J.: Technological revolution, sustainability, and development in Africa: Overview, emerging issues, and challenges. Sustain. Dev. **27**(5), 910–922 (2019)
14. Janah, L.: How East Africa Trains AI (2018, December 05). Linkedin: https://www.linkedin.com/pulse/how-east-africa-trains-ai-leila-janah/
15. Ferrein, A., Meyer, T.: A Brief Overview of Artificial Intelligence in South Africa. Association for the Advancement of Artificial Intelligence, pp. 99–103 (2012)
16. RoboCupJunior: What is RoboCupJunior? (2019). RoboCupJunior: https://junior.robocup.org/about/
17. DSA: About Data Science Africa (2020). Data Science Africa: http://www.datascienceafrica.org/
18. Feast, J.: 4 Ways to Address Gender Bias in AI. Harvard Bus. Rev. (2019)
19. Hao, K.: The future of AI research is in Africa. MIT Technol. Rev. (2019). https://www.technologyreview.com/2019/06/21/134820/ai-africa-machine-learning-ibm-google/
20. Tsado, A., Litombe, N., Leta, E., Giuliani, D.: AI Generation: Learnings from Alliance4AI's First 100 Startups in Africa (2019, September 13). Medium: https://medium.com/alliance4ai/ai-generation-learnings-from-alliance4ais-first-100-startups-in-africa-acfba0f753d1

21. Prisecaru, P.: Challenges of the fourth industrial revolution. Knowl. Horizons – Econ. **8**(1), 57–62 (2016)
22. Travaly, Y., Muvunyi, K.: The future is intelligent: harnessing the potential of artificial intelligence in Africa. Foresight Africa 2020: Top priorities for the continent 2020-2030, pp. 69–71. The Brookings Institution, Washington, D.C. (2020)
23. International Data Corporation: Worldwide Spending on Artificial Intelligence Systems Will Grow to Nearly $35.8 Billion in 2019, According to New IDC Spending Guide (2019, March 11). International Data Corporation: https://www.idc.com/getdoc.jsp?containerId=prUS44911419
24. Lee, D.: Why Big Tech Pays Poor Kenyans to Teach Self-driving Cars (2018, November 03). BBC: https://www.bbc.co.uk/news/technology-46055595
25. CNBC Africa: Artificial Intelligence: What Opportunities and Challenges Does it Bring to East Africa? (2020, January 16) CNBC Africa: https://www.cnbcafrica.com/east-africa/2020/01/15/artificial-intelligence-what-opportunities-and-challenges-does-it-bring-to-east-africa/
26. Statista: Kenya: Youth Unemployment Rate from 1999 to 2019 (2020). Statista: https://www.statista.com/statistics/812147/youth-unemployment-rate-in-kenya/
27. Alcober, F.: AI Takes Root, Helping Farmers Identify Diseased Plants (2018, June 20). Google: https://www.blog.google/technology/ai/ai-takes-root-helping-farmers-identity-diseased-plants/
28. Ehui, S.: Protecting Food Security in Africa During COVID-19 (2020, May 14). The Brookings Institution: https://www.brookings.edu/blog/africa-in-focus/2020/05/14/protecting-food-security-in-africa-during-covid-19/
29. Food and Agriculture Organization: The State of Food Security and Nutrition in the World: Safeguarding Against Economic Slowdowns and Downturns. Food and Agriculture Organization of the United Nations, Rome (2019)
30. Broom, D.: Millennials are Transforming African Farming (2019, July 12). World Economic Forum: https://www.weforum.org/agenda/2019/06/the-millennials-giving-african-farming-an-image-boost/
31. IBM Research-Africa: Could You Make a Critical Health Policy Decision Using Four-Year-Old Data? (2018, August 22). IBM Research-Africa: https://www.ibm.com/blogs/research/2018/08/cancer-machine-learning/
32. Jayaram, K., Leke, A., Ooko-Ombaka, A., Sun, Y.S.: Finding Africa's Path: Shaping BOLD Solutions to Save Lives and Livelihoods in the COVID-19 Crisis. McKinsey & Company, Nairobi (2020)
33. Demirgüç-Kunt, A., Klapper, L., Singer, D., Ansar, S., Hess, J.: The Global Findex Database 2017: Measuring Financial Inclusion and the Fintech Revolution. International Bank for Reconstruction and Development/The World Bank. The World Bank, Washington, DC (2017, April 19)
34. Suri, T.W.J.: The long-run poverty and gender impacts of mobile money. Science **354**(6317), 1288–1292 (2016)
35. Brynjolfsson, E., MacAfee, A.: The Second Machine Age: Work, Progress, and Prosperity in a Time of Brilliant Technologies. W. W. Norton & Company, London, New York (2014)
36. Schwab, K.: The Fourth Industrial Revolution. Crown Business, New York (2017)
37. World Economic Forum: The Fourth Industrial Revolution, by Klaus Schwab (2020). World Economic Forum: https://www.weforum.org/about/the-fourth-industrial-revolution-by-klaus-schwab
38. UNHDR: Human Development Indices and Indicators: 2018 Statistical Update. United Nations Human Rights Council, New York (2018)
39. GSMA Intelligence: The Mobile Economy. GSM Association, London (2020)
40. Arakpogun, E.O., Wanjiru, R., Whalley, J.: Impediments to the implementation of universal service funds in Africa—A cross-country comparative analysis. Telecommun. Policy **41**(7–8), 617–630 (2017)
41. Rutenberg, I.: Africa. In: Insights, O. (ed.) Government Artificial Intelligence Readiness Index 2019, pp. 9–11. Oxford Insights and Canada's International Development Research Centre (IDRC), Oxford (2019)

42. Janssen, M., Brous, P., Estevez, E., Barbosa, L. S., Janowski, T.: Data governance: Organizing data for trustworthy Artificial Intelligence. Gov. Inf. Q. (2020)
43. Fick, M., Akwagyiram, A.: In Africa, Scant Data Protection Leaves Internet Users Exposed (2018, April 04). Reuters: https://www.reuters.com/article/us-facebook-africa/in-africa-scant-data-protection-leaves-internet-users-exposed-idUSKCN1HB1SZ
44. Consumer International: The State of Data Protection Rules Around the World: A Briefing for Consumer Organisations. Consumer International, London (2018)
45. AI4D: A Roadmap for Artificial Intelligence for Development In Africa (2019, May 08). Artificial Intelligence for Development in Africa: https://ai4d.ai/blog-africa-roadmap/
46. Wilner, A.S.: Cybersecurity and its discontents: Artificial intelligence, the Internet of Things, and digital misinformation. Int. J. **73**(2), 308–316 (2018)
47. Williamson, J.: Democracy and the "Washington consensus". World Dev. **21**(8), 1329–1336 (1993)
48. Gillwald, A.: Good intentions, poor outcomes: Telecommunications reform in South Africa. Telecommun. Policy **29**(7), 469–491 (2005)
49. Thomas, J.W., Grindle, M.S.: After the decision: Implementing policy reforms in developing countries. World Dev. **18**(8), 1163–1181 (1990)
50. Andrews, M.: The Limits of Institutional Reform in Development: Changing Rules for Realistic Solutions. Cambridge University, Cambridge (2013)
51. Pritchett, L., Woolcock, M., Andrews, M.: Looking like a state: techniques of persistent failure in state capability for implementation. J. Dev. Stud. **49**(1), 1–18 (2013)
52. Gaus, A., Hoxtell, W.: Automation and the Future of Work in Sub-Saharan Africa. Konrad-Adenauer-Stiftung, Berlin (2019)
53. Rudman, R., Bruwer, R.: Defining Web 3.0: opportunities and challenges. Electron. Lib. **34**(1), 132–154 (2016)
54. Wessner, C.S.: Entrepreneurship and the innovation ecosystem policy lessons from the United States. Local Heroes in the Global Village, pp. 67–89. Springer, Boston, MA (2005)
55. Kray, H., Shetty, S.: The Locust Plague: Fighting a Crisis Within a Crisis (2020, April 14). The World Bank: https://blogs.worldbank.org/voices/locust-plague-fighting-crisis-within-crisis?cid=ECR_TT_worldbank_EN_EXT

Managing the Fourth Industrial Revolution: A Competence Framework for Smart Factory

Emanuele Gabriel Margherita and Alessio Maria Braccini

Abstract Several industrial initiatives around the globe usher in the Fourth Industrial Revolution that points at the deployment of a Smart Factory. A Smart Factory builds on several advanced digital technologies to integrate the production process within and across organizations. The literature asserts that Smart Factory increased efficiency and production outcome. While the technical aspects of the Smart Factory are quite extensively studied, few pieces of research paid attention to the competencies that organizations need to manage a Smart Factory. Therefore, we develop a competence framework for handling Smart Factory from a systematic literature review of Smart Factory empirical cases. We contribute to the literature proposing a framework with six competences areas.

Keywords Fourth industrial revolution · Smart factory · Competence · Industry 4.0 · Systematic literature review · Skill

1 Introduction

To preserve the competitiveness of the manufacturing industry, the German government and manufacturing associations of U.S. launched industrial initiatives ushering in the Fourth Industrial Revolution which aims at developing the Smart Factory [1, 2]. Within a Smart Factory, several leading-edge technologies—Big data analytics, Robotics, Cloud Manufacturing, Additive Manufacturing—are interconnected with human resources along the assembly line to increase efficiency, flexibility, and automation of the production process [1, 3]. The two industrial initiatives share the same principle of interconnection within the organization and along the supply chain [1, 4], while they differ on the leading technology. Industrial Internet, the U.S. initiative, builds a Smart Factory on Internet of Things infrastructure which integrates all the technology and people. The German initiative Industrie 4.0, on the

E. G. Margherita (✉) · A. M. Braccini
Department of Economics Engineering Society and Organization—DEIM, University of Tuscia, Via Del Paradiso, 47, 01100 Viterbo, Italy
e-mail: emargherita@unitus.com

A. Hamdan et al. (eds.), *The Fourth Industrial Revolution: Implementation of Artificial Intelligence for Growing Business Success*, Studies in Computational Intelligence 935,
https://doi.org/10.1007/978-3-030-62796-6_23

other hand, aims at developing a programmable cyber-physical system which aims at increasing the interconnection in the Smart Factory integrating the physical and virtual worlds [5]. Also, cyber-physical systems is that allows to address mechanical problems on the assembly line without human interaction but through autonomous machines that take on the job from the workforce [6].

The two initiatives are considered by most the most critical industrial initiatives for the Fourth Industrial Revolution and their blueprints have inspired other industrial plans around the world. As a matter of fact, several European countries launched their industrial initiative to develop the cyber-physical system with the label Industry 4.0 which also became a synonym of the Fourth Industrial Revolution.

Since the Fourth Industrial Revolution is mainly a technological wave, most of the studies so far concentrated on the development of Smart Factory technologies [3, 6] and the value creation of Smart Factory [7, 8]. Smart Factory adoption leads to enhanced organizational productivity and flexibility which places particular emphasis on sustainability practices [9].

While few pieces of evidence pay attention to how the organization manages the transition of Smart Factory. The technologies of Smart Factory creates a new level of socio-technical interaction between all the actors and resources involved in manufacturing [1]. Therefore, Smart Factory technologies transform work practices turning the organization into a more knowledge-intensive oriented and need the support of a high-specialized workforce [1, 8, 10].

To address the gap, we aim at identifying the specific requirement for organizations to take advantage of this transition through the study of competences [11, 12]. We consider competence as a set of knowledge, skill and personal traits that an organization should have for a significant performance [13]. We build over and systematize existing studies on competences for Smart Factory, which though surveys and non-systematic literature reviews, led to conflicting frameworks with competence different lists varying from study to study [11, 14–16].

We aim at defining in a systematic and rigorous way a comprehensive framework of competencies needed by organizations to handle Smart Factory. We perform a systematic literature review of Smart factory pioneering empirical case studies and seminal articles inspiring the Smart Factory. We identify the competencies for Smart Factory [17] through open coding using as a sensitive device the concept of competences and adopting validity principles of qualitative inquiry [18]. We answer the following research question: "*What competencies are needed for organizations to manage Smart Factory?*".

2 Theoretical Background

In this section, we present the Fourth Industrial Revolution, the concept of competences, and the existing competence framework for Smart Factory.

2.1 The Fourth Industrial Revolution

Similar to the previous industrial revolution which brought innovation in the way of production, the Fourth Industrial Revolution refers to the development of the Smart Factory (SF) [1, 3]. Due to the complexity of the innovation, several governments and manufacturing associations around the world launched industrial initiatives to promote the SF development.

The main industrial plans for SF are Industrial Internet by manufacturing associations in the U.S. and Industrie 4.0 by the German government. These initiatives are then followed by European countries that launched similar industrial initiatives with the label Industry 4.0 (I40) and Asian initiatives such as "Made in China 2025" in China and "Smart Manufacturing" in Korea [3]. Since most of the industrial plans employ the label Industry 4.0 (I40), this term is used as a synonym of the fourth industrial Initiative and SF.

The initiatives share the same principles which are the horizontal and vertical integration of production processes driven by share real-time data interchange and flexible manufacturing to enable SF [4, 19]. Vertical integration means that several organizational departments are integrated, while horizontal integration refers to the inclusion of several organizations composing the supply chain and the customer [1, 20].

The initiatives build on several leading technologies, including Big Data analytics, Additive manufacturing, Cloud manufacturing, and Robots to develop a SF [3, 8]. The primary technology of Industrial Internet initiative is the Internet of Things, which is an IT infrastructure enabling the collection and transmission of data between devices, resulting in identification, localization, tracking, and monitoring of objects [4]. While the primary technology paradigm of Industrie 4.0 is the cyber-physical system. Cyber-Physical system control pieces of machinery automatically in assembly lines. The way with which organizations use SF technologies allows them to address problems on the assembly line without human interaction but through autonomous machines that take on the job from the workforce [6].

The effective deployment of SF promises to deliver benefits in an economic, environmental and social contexts [20, 21]. SF allows to increase the efficiency and flexibility of the production process. SF allows to increase the production outcome and to produce small lots of customized products in order to fulfill different customer needs [1, 20]. The SF production process reduces natural resource usage with increasing data granularity and more sophisticated technologies [1, 8]. SF allows a more ergonomic workplace and oriented towards workforce needs [8, 22].

2.2 The Concept of Competences

The concept of competence is a traditional organizational topic in literature since the 90 s, and it is considered a critical organizational component to gain a competitive advantage [23]. In the literature, the concept competence appeared for the first time in 1973 [24], which refers to a set of knowledge, skills, traits, attitudes, self-concepts, values, or motives directly related to job performance leading to superior performance. Afterward, scholars from different disciplines sharpened the definition in different directions. In the IS context, competence studies dealt with the competence assessment to anticipate changes in the IS professions [25, 26].

There is not a collective agreement for the classification of competences. Scholars classify them based on the logic, theory, and purpose of the analysis [27]. However, there is a consensus that competence is composed of three main elements [23]:

- Knowledge: theory, concepts, and tacit knowledge which workers gain through experience.
- Skills: requirement for a given occupational area that the workforce affords to do.
- Attitude or personal traits: workforce characteristics for effective performance in a job.

Within the organization, the competencies are systematically managed to identify the current and future competencies required for the work and to assess available competences of the workforce [28]. According to Sandberg [17], competencies can be identified by the job analysis through three different methodological approaches [17]:

- Worker-Oriented Approach: It employs predefined categories to capture competencies from the workforce. This approach produces a too generic and abstract list of competencies, and therefore limited value as a basis for competence development.
- Work-Oriented Approach: It identifies activities that are central to accomplishing specific work and then transform those activities competences. This approach produces more concrete and specific competence. Here, the limitation regards the work activity list, which might sufficiently indicate the attributes required to accomplish those activities.
- Multimethod-Approach: It employs together the previous two approaches to identify more specific and comprehensive competences minimizing their weak points.

Finally, through competence (or skill) assessment, organizations can evaluate competence levels and identify competence gaps. Thus, organizations can fill the competence gap through vocational courses or hiring personnel [29]. Accordingly, this process allows the organization to maintain a high level of distinctive competencies which lead to competitive advantages [12].

2.3 Existing Competence Framework for Smart Factory

In literature, there are four attempts to develop a SF competence framework. Indeed, the two World Economic Forum competence frameworks have captured competences through surveys that have administered to Chief Human Resources and Chief Strategy Officers of leading global employers portraying their expectations regarding competencies for future jobs [11, 14]. Each report provided a framework of ten competencies, which include core competencies like complex problem solving, critical thinking, creativity and emotional intelligence. However, we find these frameworks challenging to follow for organizations since several competences varied from report to report, and this variance does not help organizations to create a path for acquiring all these competecnies. Also, these two frameworks are focus on competences related to the job of the future and therefore they are not center to SF. Table 1 depicts the last competence framework by World Economic Forum, which forecasts the most important competencies for 2020.

Hecklau et al. [15] proposed an SF competence framework presenting either the methodology for detecting competences by a literature review and how to manage the competencies gap by a radar chart. In this study, the authors employed a worker-oriented approach exploiting the PESTEL-Framework to detect competences based on a literature review of SF. They provided a competence framework including four primary categories, namely technical competences, methodological competences, social competences and personal competences. Each group contains a narrow list of competencies. The study emphasized on core competencies like problem-solving, coding skills, analytical skills as well as generic competencies like language skills and attitude to work under pressure. However, the focus is more on the methodological implications rather than on providing a competence framework since the SF literature review resulted non-systematic and limited to 7 studies.

Analogously, Erol et al. [16] present a learning approach to prepare future managers and workers for the fourth industrial revolution. The authors developed this approach for Austrian SF pilot projects. This approach highlights the importance of personal competencies, social competencies, action related competencies, and domain-related competences. Authors argued that these competencies are required for the workforce to deal with SF context characterized by an uncertain environment.

3 Research Method

Our investigation aims at developing a theoretical framework containing competences for SF. Figure 1 presents our data analysis. First of all, we performed a systematic literature review in February 2019, applying the protocol by Webster & Watson [30]. Table 2 shows the details of the literature search we performed over the SCOPUS database of indexed scientific publications to review literature discussing

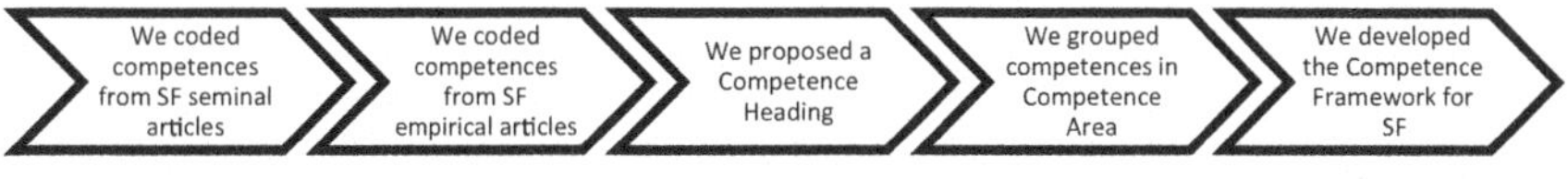

Fig. 1 Data analysis

the empirical SF case studies. The query contains "Smart Factory" and keywords that researchers often used as a synonym of SF like Industry 4.0, internet of things, and cyber-physical systems [9] as well as Industrial Internet [2]. Moreover, we used a set of secondary keywords, namely implementation, application, and adoption, since they point to empirical adoptions of the SF.

We included in the database only papers containing industrial empirical case studies that deployed SF. Instead, we excluded all theoretical and survey articles as well as 19 "false positive" papers that have the title in English but text in a different language. As can be noticed, there has been a large drop between the initial hits (386) to the first exclusion step (25). Still, we dropped several I40 applications in the first step since discussing applications of different sectors like smart building, agriculture and e-health. Indeed, the authors employed the I40 referring to the interconnection feature of the initiative making the query wider. Afterward, to include papers regarding the single technology of SF, we picked in the "title and abstract" step its literature review paper (e.g. a RFID literature review [31]) exploring the citations in the "backward and forward" search. The final query produced 18 entries that have used as a database to identify the SF competences. Lastly, to enlarge the number of potential competencies, we added in the database the two seminal papers of SF initiative [1, 2]. Secondly, adopting a work-oriented approach to define competencies, we identified SF competences from activities presented in the publications [17]. To maintain a qualitative rigor we follow the principles of qualitative inquiry [18] performing open coding on the literature sources, categorizing them in first and second level codes, and eventually identifying competences areas, labeling and describing competences.

We used the following definition of competence as a sensitive device: a set of knowledge, skills and attitude or personal traits that an organization should have for superior performance. Along the process, one researcher acted as a primary coder and performed open coding (first and second level coding) proposing competences. Whereas a second researcher questioned and verified the developed framework. We accomplished the final framework after three iteration rounds adopting a two-fold saturation criterion: (i) all the data sources could be coded with the set coding structure, and (ii) the research team agreed on the result of the coding.

We proposed a first-level coding from which we detected the competence and a second-level coding representing the competence heading. Thereafter we grouped competence into competence areas following aggregation criteria based on common sense and easily observable classes. Thirdly, we developed the competence framework following the criteria by Stuart [32]. Indeed, the competence framework should be *generalizable* across different organizations adopting SF. The competences and the competence area should be *simple* and *accurate* in order to be comprehensive and usable from various organizations. Accordingly, we present in Table 3 our competence framework for SF. From each competence, we propose a heading, a competence area and a description that is deduced by the literature review.

Table 1 Competence Framework 2020 for jobs of the future by World Economic Forum

Competence	Description
Complex problem solving	Identifying complex problems and reviewing related information to develop and evaluate options and implement solutions
Critical thinking	Employing reasoning to identify the strengths and weaknesses of alternative solutions, conclusions or approaches to problems
Creativity	Workers on this job try out their own ideas
People management	Motivating, developing, and directing people as they work, identifying the best people for the job
Coordination with others	Adjusting actions in relation to others' actions
Emotional intelligence	Ability to understand and manage your own emotions, and those of the people around you
Judgment and decision making	Considering the relative costs and benefits of potential actions to choose the most appropriate one
Service orientation	Actively looking for ways to help people
Negotiation	Bringing others together and trying to reconcile differences
Cognitive flexibility	Mental ability to switch between thinking about two different concepts

Table 2 Literature search for empirical smart factory case studies

Item	Description
Source	SCOPUS database
Query	TOPIC: "smart factor*" OR "industrial internet" OR "internet of thin*" OR "industry 4.0" OR "cyber physical system*" AND "implementation*" OR "application*" OR "adoption*" Refined by: LANGUAGES: (ENGLISH), Subject Area: Business, Management and Accounting, Source Type: Journals
Hits	386
Papers retained after: • Title and abstract selection • Full-text selection • Backward and forward search	25 10 18

4 Findings

We propose a competence framework for SF (in Table 3) based on a systematic literature review of empirical case studies.

The SF competence framework distinguishes six competence areas. The *inter-functional competence* area embraces technical and behavioral competencies at the individual level demanding a cross-functional approach to the management of increasing complexity of work practice and innovation practice. Competences such as

Table 3 Competence framework for the smart factory

Competence area	Competence heading	Description	Refs.
Inter functional	Risk-tolerance attitude	SF requires a workforce with risk tolerance attitude stretching firms beyond their comfort zone	[1, 2, 8, 33–37]
	Leadership and cross-discipline guidance	SF requires leadership roles that are explicitly cross-discipline for taking advantage of the initiative	
	Proficient IT use skills	SF requires the management of IT software and solutions to generate value from the initiative	
	Complex problem solving	SF requires workforce managing complex problem-solving owing to the increasing complexity of processes	
	Flexible work attitude	In SF, workforce with a flexible work attitude enables greater compatibility between work and personal needs	
	Self-development	SF requires a highly skilled and highly-motivated workforce to generate value	
	Interdisciplinary team orientation	SF requires workforce capable of working in an interdisciplinary team that draws on the expertise to generate value	
	Innovation attitude	SF requires workforce with an innovation attitude, capable of developing research projects	
Organizational management	Data management	SF requires the management of exploiting data sources of different types of machinery to enable self-decision making	[1, 8, 38–42]
	Business logistics skills	SF requires the workforce which manages novelty inventory management to handle the storage and the fruition of productions	

(continued)

Table 3 (continued)

Competence area	Competence heading	Description	Refs.
	Knowledge management	SF requires the management of the formalization of the knowledge foundations that data talent will need	
Business model innovation	Network creation	In a SF context, a key challenge will be to use the technology to create novel value networks	[1, 8, 38–42]
	Service and product orientation	SF generates value through the development and the realization of new services and products	
	Quality Focus	SF emphasizes quality assurance since the workforce performs routine tasks with more autonomy	
	Technology leadership	Within SF, managing investment plays a key role in leveraging hardware and software innovations	
	Sustainability	SF requires workforce capable of embracing sustainability practices within the organization's business model	
IS Design	Integration Skills	SF requires workforce capable of managing digital integration among advanced technologies	[37, 43–47]
	Interface design skills	SF requires workforce capable of managing new platforms that enable firms to build specific applications easy to use	
	Cybersecurity skills	SF requires workforce capable of managing the impact of new technologies on sensitive corporate data	
Organizational design	User-oriented Workplace Design	SF allows creating a safer and more attractive workplace oriented towards workforce interests	[1, 8, 41, 48]

(continued)

Table 3 (continued)

Competence area	Competence heading	Description	Refs.
	Workforce continuous improvement	SF operative issues enable companies to engage in constant learning to maintain a higher process quality	
Legal	Intellectual property management	SF requires workforce capable of facing intellectual property issues when innovations are developed in partnership	[1, 2]
	Legal liability	SF requires workforce capable of facing liability issues when the innovation is adopted in an organizational network	

problem-solving, self-development and risk tolerance attitude result essential in this initial stage of SF adoption. Likewise, proficient IT use skills are required to interact with advanced technologies. The *organization management* includes competences at the organizational level for managing recurring SF technology issues. Indeed, there is a constant need for data management capabilities to inform the choice of data for undertaking predictive analysis. The areas of *IS Design, Organizational Design* and *Legal Competences* classify competences to establish SF organizations. The IS foci is predominant since SF benefits derive from the interconnection and integration of advanced technologies as well as their effective employment through intuitive interface design.

Further commitment lies in the cybersecurity skills for protecting data which also implies a legal analysis regarding the connected intellectual property and the legal liability. Lastly, designers should pay particular attention to workplace design which is safer and engaging for workforce. The *business model innovation* area encompasses competences at the organizational level allowing to generate value through network creation and the development of new service & product [49]. In particular, these models have enriched whether organization leverages on advanced technology investment and sustainability mindset.

5 Discussion

Most of the literature regarding SF concentrates on the technical dimension of the initiative underpinning the idea that SF technologies deliver value mainly through the integration and automation of technologies [50, 51]. Our study debates this perspective showing how human competencies play a crucial role in a SF. Despite the competence frameworks show a lack of manual skills for the workforce, the primary skill for manufacturing organizations, we noted that there is an increasing request for competencies a higher level of knowledge. SF workforce operates in a more complex environment that requires problem-solving skills, a strong attitude towards innovation and attitude to work in an interdisciplinary team. This implies that the effective deployment of SF is challenging, and the role of the workforce is increasingly important to achieve it.

The competence framework of SF confirms the literature that SF should base its competitive advantage on technology leadership [1, 3]. Also, the framework shows that human resources with a sustainability mindset may develop an SF production process that contributes to the sustainable development of our society.

The framework also clarifies the fuzzy role of "data scientist" which is considered by most the most relevant job for the fourth industrial revolution [2]. Our competence framework for SF pinpoints proficient IT use and data management as his distinctive competences.

The framework also portrayed a socio-technical nature since there is a demand for competencies to improve work conditions through a safer workplace and workstations. Furthermore, SF technologies can also support a knowledge management system and undertaking a continuous learning approach for the workforce.

Moreover, according to the literature, SF integrates supply chain along with different organizations through the horizontal integration of advanced technologies [52]. This feature opens to several legal issues from intellectual property issues to legal liability. These concerns should be managed in advance for an effective implementation of SF within organizations.

Finally, we debate our work against the previous competence framework. Our work contributes to validating the World Economic Forum competence frameworks for the future of jobs, highlighting that problem-solving skills, people management, and integration skills are also proper skills for SF. Moreover, our study shows the importance of integration of technology skills and legal competencies, which are missed to the competence framework by Hecklau et al. [15]. While our study validates the competence categories of the framework by Erol et al. [16] since our competence areas, although with a different name, purports similar competencies.

6 Conclusion

The study aims at proposing a competence framework for handling SF based on a systematic literature review of SF empirical case studies. The competence framework for SF distinguishes six areas of competences that underlie a socio-technical perspective.

The study suggests implications both for practitioners and for research. Regarding the implications for practitioners, the SF competence framework provides guidance for organizations aiming at undertaking SF transition. Indeed, it is notably useful for the Human Resource department during the implementation stage, where it can be used as a required competence to seek in the labor market. It is also valuable for maintaining a high competence level in mature SF organizations preventing competence lack through the competence gap analysis [15]. Hence, the competence catalog can serve as a reference to develop tailor-made courses. The study is also useful for technical institutes and universities that may build a tailor-made study plan to develop SF workforce. The study highlights that organizations require several types of competencies to manage the transition towards SF. Therefore, we recommended that schools and universities create multidisciplinary paths for students to develop SF competences.

Regarding the implication for researchers, the SF competence framework sheds light on the nature of SF initiative. Several researchers claimed the determinist nature of the initiative conducting empirical studies where SF can emulate workforce tasks or decision making [3]. Nevertheless, these advanced technologies cannot automatically manage analysis on data. Therefore the SF requires a highly specialized workforce for handling them. Therefore, we confirm that human resources and these technologies are part of a sizeable socio-technical system within the SF where the conjoint optimization of the two systems leads to the effective implementation of the initiative.

Finally, we claim for further studies on SF initiative since SF allows to extend our IS focus on information encompassing those from the assembly line. Indeed, even though IS studies paid attention to the management of administrative information through ERP [53], our knowledge regarding information from the assembly line is limited. Further studies should cover this gap investigating how to manage that information as well as how to generate value employing both information sources.

To conclude, our study lays the foundation for our complete research project, which aims at investigating SF competences empirically and exploring pattern associations among them. Indeed, the next stage will concern administering a survey based on SF competence framework to organizations that adopted SF in order to validate it, proposing thus descriptive statistics regarding SF competences. Therefore, we encourage researchers to perform Fs/QCA [7, 54] otherwise, different quantitative methods to test our SF competence framework on a large sample.

References

1. Kagermann, H., Wolfgang, W., Helbig, J.: Recommendations for implementing the strategic initiative INDUSTRIE 4.0. Work Group Acatech, Frankfurt am Main, Ger (2013)
2. Evans, P.C., Annunziata, M.: Industrial Internet: Pushing the Boundaries of Minds and Machines (2012)
3. Kang, H.S., Lee, J.Y., Choi, S., et al.: Smart manufacturing: past research, present findings, and future directions. Int. J. Precis. Eng. Manuf. - Green Technol. **3**, 111–128 (2016)
4. de Sousa Jabbour, A.B.L., Jabbour, C.J.C., Foropon, C., Filho, M.G.: When titans meet—Can industry 4.0 revolutionise the environmentally-sustainable manufacturing wave? The role of critical success factors. Technol. Forecast. Soc. Change **132**, 18–25 (2018). https://doi.org/10.1016/j.techfore.2018.01.017

5. Waibel, M.W., Steenkamp, L.P., Moloko, N., Oosthuizen, G.A.: Investigating the effects of smart production systems on sustainability elements. Proc. Manuf. **8**, 731–737 (2017). https://doi.org/10.1016/j.promfg.2017.02.094
6. Lee, J., Bagheri, B., Kao, H.A.: A Cyber-Physical Systems architecture for Industry 4.0-based manufacturing systems. Manuf. Lett. **3** (2015)
7. Margherita, E.G., Braccini, A.M.: IS in the cloud and organizational benefits: an exploratory study. In: Lazazzara, A., Ricciardi, F., Za, S. (eds.) Exploring Digital Ecosystems. Lecture Notes in Information Systems and Organisation, vol. 33. Springer, Cham (2020)
8. Braccini, A.M., Margherita, E.G.: Exploring organizational sustainability of industry 4.0 under the triple bottom line: the case of a manufacturing company. Sustainability **11**, 36 (2019). https://doi.org/10.3390/su11010036
9. Piccarozzi, M., Aquilani, B., Gatti, C.: Industry 4.0 in management studies: a systematic literature review. Sustainability **10**, 3821 (2018). https://doi.org/10.3390/su10103821
10. Bonekamp, L., Sure, M.: Consequences of Industry 4.0 on human labour and work organisation. J. Bus. Media Psychol. **6**, 33–40 (2015)
11. World Economic Forum: The Future of Jobs: Employment, skills and workforce strategy for the fourth industrial revolution (2016)
12. Bergenhenegouwen, G.J., Ten Horn, H.F.K., Mooijman, E.A.M.: Competence development - a challenge for human resource professionals: core competences of organizations as guidelines for the development of employees. J. Eur. Ind. Train. **20**, 29–35 (1996)
13. Hartle, F.: How to Re-engineer your Performance Management Process. Kogan Page (1995)
14. World Economic Forum: The Future of Jobs Report 2018 (2018)
15. Hecklau, F., Galeitzke, M., Flachs, S., Kohl, H.: Holistic approach for human resource management in industry 4.0. Proc. CIRP **54** (2016)
16. Erol, S., Jäger, A., Hold, P., et al.: Tangible Industry 4.0: a scenario-based approach to learning for the future of production. Proc. CIRP **54**, 13–18 (2016)
17. Sandberg, J.: Understanding human competence at work: an interpretative approach. Acad. Manag. J. **43**, 9–25 (2000)
18. Corbin, J., Strauss, A.: Basics of Qualitative Research Techniques and Procedures for Developing Grounded Theory. SAGE Publications Inc., Thousand Oaks, California (2015)
19. Thoben, K.-D., Wiesner, S., Wuest, T.: "Industrie 4.0" and smart manufacturing—a review of research issues and application examples. Int. J. Autom. Technol. **11**, 4–19 (2017)
20. Fatorachian, H., Kazemi, H.: A critical investigation of Industry 4.0 in manufacturing: theoretical operationalisation framework. Prod. Plan. Control **7287**, 1–12 (2018)
21. Kiel, D., Muller, J.M., Arnold, C., et al.: Sustainable Industrial Value Creation: Benefits and Challenges of Industry 4.0 (2017)
22. Gabriel, M., Pessel, E.: Industry 4.0 and sustainability impacts: critical discussion of sustainability aspects with a special focus. Ann. Fac. Eng. Hunedoara: Int. J. Eng. **1** (2016)
23. Le Deist, F.D., Winterton, J.: What is competence? Hum. Resour. Dev. Int. **8**, 27–46 (2005)
24. McClelland, D.: Testing for competence rather than for intelligence. Am. Psychol. J. **20**, 321–333 (1973)
25. Bassellier, G., Benbasat, I.: Business competence of information technology professionals: conceptual development and influence on IT-business partnerships. MIS Q. **28** (2004)
26. Willcocks, L., David, F., Olson, N.: Implementing core IS capabilities: IT governance and management framework revisited. Eur. Manag. J. **24**, 28–37 (2006)
27. Markowitsch, J., Plaimauer, C.: Descriptors for competence: towards an international standard classification for skills and competences. J. Eur. Ind. Train. **33** (2009)
28. Sampson, D., Fytros, D.: Competence models in technology-enhanced competence-based learning. In: Adelsberger, H.H., Kinshuk, Pawlowski J.M., Sampson, D.G. (eds.) Handbook on Information Technologies for Education and Training. International Handbooks on Information Systems. Springer, Berlin, Heidelberg (2008)
29. Armstrong, M.: Armstrong' S Handbook of Human Resource Management Practice, 11th edn. Kogan Page (2010)

30. Webster, J., Watson, R.T.: Analyzing the past to prepare for the future : writing a literature review. MIS Q. **26** (2002)
31. Ben-Daya, M., Hassini, E., Bahroun, Z.: Internet of things and supply chain management: a literature review. Int. J. Prod. Res. (2017)
32. Stuart, R., Thompson, J.E., Harrison, J.: Translation. From generalizable to organization-specific competence frameworks. J. Manag. Dev. **14**, 67–80 (1995)
33. Sayar, D., Er, Ö.: The antecedents of successful IoT service and system design: cases from the manufacturing industry. Int. J. Des. **12**, 67–78 (2018)
34. Liang, Y.C., Lu, X., Li, W.D., Wang, S.: Cyber physical system and Big Data enabled energy efficient machining optimisation. J. Clean. Prod. **187**, 46–62 (2018)
35. Thiede, S.: Environmental sustainability of cyber physical production systems. Proc. CIRP **69**, 644–649 (2018)
36. Schulze, C., Thiede, S., Thiede, B., et al.: Cooling tower management in manufacturing companies: a cyber-physical system approach. J. Clean Prod. (2018)
37. Lee, J., Lapira, E., Bagheri, B., Kao, H.: Recent advances and trend in predictive manufacturing systems in big data environment. Manuf. Lett. **1** (2013)
38. Mourtzis, D., Fotia, S., Boli, N., Vlachou, E.: Modelling and quantification of industry 4.0 manufacturing complexity based on information theory: a robotics case study. Int. J. Prod. Res. (2019)
39. Pero, M., Rossi, T.: RFID technology for increasing visibility in ETO supply chains: A case study. Prod. Plan. Control **25**, 892–901 (2014)
40. Shin, S., Woo, J., Rachuri, S.: Predictive analytics model for power consumption in manufacturing predictive analytics model for power consumption in manufacturing. Proc. CIRP **15**, 153–158 (2014). https://doi.org/10.1016/j.procir.2014.06.036
41. Cui, L., Deng, J., Liu, F., et al.: Investigation of RFID investment in a single retailer two-supplier supply chain with random demand to decrease inventory inaccuracy. J. Clean. Prod. **142**, 2028–2044 (2017)
42. Goyal, S., Hardgrave, B.C., Aloysius, J.A., DeHoratius, N.: The effectiveness of RFID in backroom and sales floor inventory management. Int. J. Logist. Manag. **27**, 795–815 (2016). https://doi.org/10.1080/13675560600717763
43. Ardanza, A., Moreno, A., Segura, Á., et al.: Sustainable and flexible industrial human machine interfaces to support adaptable applications in the Industry 4.0 paradigm. Int. J. Prod. Res. (2019)
44. Kembro, J.H., Danielsson, V., Smajli, G.: Network video technology: exploring an innovative approach to improving warehouse operations. Int. J. Phys. Distrib. Logist. Manag. **47**, 623–645 (2017)
45. Shahbaz, M., Shaheen, M., Aslam, M., et al.: Data mining methodology in perspective of manufacturing databases. Life Sci. J. **9** (2012)
46. Yuan, Z., Qin, W., Zhao, J.: Smart manufacturing for the oil refining and petrochemical industry. Engineering **3**, 179–182 (2017)
47. Zhang, Y., Liu, S., Liu, Y., et al.: The 'Internet of Things' enabled real-time scheduling for remanufacturing of automobile engines. J. Clean. Prod. **185**, 562–575 (2018)
48. Reif, R. A.GW: An augmented reality supported picking system (2009). In: WSCG 2009 Full Pap proceedings, Plzen, Czech Republic, 2–5 Feb 2009
49. Braccini, A.M.: Value Generation in Organisations (2011)
50. Tangmunarunkit, H., Hsieh, C.K., Longstaff, B., et al.: Ohmage: a general and extensible end-to-end participatory sensing platform. ACM Trans. Intell. Syst. Technol. **6**, 1–21 (2015). https://doi.org/10.1145/2717318
51. Lee, C.K.M., Yeung, C.L, Cheng, M.N.: Research on IoT based cyber physical system for industrial big data analytics. IEEE Int. Conf. Ind. Eng. Eng. Manag., 1855–1859 (2015). https://doi.org/10.1109/IEEM.2015.7385969
52. Lasi, H., Fettke, P., Feld, T., Hoffman, M.: Industry 4.0. Bus. Inf. Syst. Eng. (2014)
53. Piccoli, G., Pigni, F.: Information Systems for Managers with Cases. Prospect Press (2016)
54. Ragin, C.C.: Fuzzy-Set Social Science. University of Chicago Press, Chicago/London (2000)

Analysis for the Knowledge Economy in GCC Countries

Araby Madbouly, Sameh Reyad, and Gopalakrishnan Chinnasamy

Abstract Gulf Cooperation Council (GCC) countries endowed by oil and gas. This towards their economies to be highly dependent on oil and gas extraction activities, which made its economy vulnerable to economic fluctuation as a result of uncertainty of oil and gas prices. This towards GCC governments to implement policies to support economic diversification to strengthen the business environment and increase the share of non-hydrocarbon sector. One of these sectors is the Knowledge economy which became one of the main engines of economic growth in recent decades. The purpose of this chapter is to analyse the knowledge economy in the GCC countries. The Knowledge Assessment Methodology (KAM) was developed to facilitate countries trying to make the transition to the knowledge economy. KAM consists of five main pillars: (i) Performance; (ii) Institutions; (iii) Education and HR index; (iv) Innovation system index and (v) Information Infrastructure index. The five pillars have been analysed for the GCC countries from 2010 to 2017. Although that KAM has been used widely, the last available data on the world bank databank was for 2012. This was a limitation of the study. To eliminate this limitation the researcher collected the data about the same five pillars from the global competitiveness report. The financings of this chapter that the macroeconomic environment and Institutions are the highest scores in the KEI for GCC, this reflects the stability of the macroeconomic environment and the being of strong the Institutional framework and facilitates the flow of new or existing knowledge. Innovation pillar is the latest contributor of KEI for GCC countries. There is the need for GCC countries to increase the capacity for innovation and the quality of scientific research institutions. Although the increase in the spending on R&D and attempts to increase the university-industry collaboration in R&D still more is required for these promising

A. Madbouly (✉) · G. Chinnasamy
Business and Accounting Department, Muscat College, Muscat, Sultanate of Oman
e-mail: araby@muscatcollege.edu.om

G. Chinnasamy
e-mail: gopalakrishnan@muscatcollege.edu.om

S. Reyad
College of Business and Finance, Ahlia University, Manama, Bahrain
e-mail: sreyad@ahlia.edu.bh

A. Hamdan et al. (eds.), *The Fourth Industrial Revolution: Implementation of Artificial Intelligence for Growing Business Success*, Studies in Computational Intelligence 935,
https://doi.org/10.1007/978-3-030-62796-6_24

countries. Many GCC countries have taken speedy steps in this direction, however more are still required specially in the area of availability of scientists and engineers, application of patents and intellectual property protection.

Keywords Knowledge economy · GCC · Knowledge economy index · Knowledge assessment methodology

1 Introduction

The knowledge became one of contributors in the productivity growth, and, in turn, to sustainable economic development. In other words, Knowledge economy is one of the main engines of economic growth, which can be one of the engines to the economic development in the GCC countries when they diversify their economies and increase the share of non-hydrocarbon sector in the GDP. In this direction, GCC countries started to build up human capital and move towards the knowledge economy. Some of these countries have "*articulated their visions on knowledge and innovation-based economy; others are implementing policies and programs to propel their economies toward that goal*" [8]. This has been translated into different initiatives by these countries which planned their initiatives considering all political and economic prespectives of their countries. Saudi Arabia has formed the "Saudi Vision 2030", Bahrain formed the "Bahrani Vision 2030", while Kuwait has its vision for 2035 called "Sate vision of Kuwait 2035". Other GCC countries have their respective initiatives for the same [1].

This chapter is an attempt to analyse the different pillars of the knowledge economy for the GCC countries to show the level of each pillar and enable the policy makers to assess the level of progress in this sector.

The analysis of the knowledge economy can be done via a benchmarking tool called the "Knowledge Assessment Methodology" (KAM) which was developed by the World Bank Institute through the Knowledge for Development Programme (K4D). The KAM is a user-friendly interactive Internet-based tool based on 148 structural and qualitative variables and a group of 146 countries. It enables to assess the readiness of a country or region for the knowledge economy. It is a allows to have a holistic view of the wide spectrum of all perspectives related to the knowledge economy [3, 6].

The knowledge-based economy can be assessed through five main pillars, they are:

- The overall macroeconomic performance that is reflected on the having stimulating environment to the knowledge creation and sharing;
- Institutional framework which enables the free flow of new or existing knowledge;
- The educated and skilled population in order to create, share and use knowledge;
- The level of innovation which is elaborated via a network of private enterprises, research centers, universities, etc. in order to adapt to local needs, and create new technologies; and

- The information infrastructure which facilitates the effective communication, dissemination and processing of knowledge.

Although the number of studies analysed the knowledge economy for many countries, there is scarcity of this type of studies for the GCC countries, which sheds the light to the need of having a study analyses the different pillars of knowledge economy in these countries indicating their strengths and weaknesses to enable the decision makers to design different policies foster this important part of the economy which is highly required within the direction of these countries towards economic diversification.

The remainder of this chapter is organized as follows. Section 2 presents the knowledge economy framework, followed by the methodology of assessing the Knowledge Economy in Sect. 3 while Sect. 4 is the analysis of KEI for GCC countries. Finally, the fifth chapter include the conclusion.

2 Knowledge Economy Framework

The first appear of the concept "knowledge economy" was beginning of 1960s and was credited to the pioneer economist Machlup [4]. The role of knowledge increased significantly and became more important as compared to other production factors. The knowledge society is not only limited to the concept of investment in R&D. It is extended to all aspects related to the contemporary economies where the value added is mainly generated by the knowledge. Huge number of studies trying to understand and explain this new phenomenon where the knowledge a critical determinant for the economic growth [2].

Despite the research for more than fifty years on the knowledge economy, still there is no widely accepted about it. It is still a buzzword (Smith, 2002). OECD defined the knowledge economy as *"Those which are directly based on the production, distribution and use of knowledge and information"* [7].

Powell, 2004 defined the knowledge economy as *"Production and services based on knowledge-intensive activities that contribute to an accelerated pace of technological and scientific advance as well as equally rapid obsolescence"*.

From holistic perspective, Brinkley [2] defined it saying *"Knowledge Economy is the utilization of knowledge as one of key engines of economic growth in a country or region. It is an economy where knowledge is acquired, created, disseminated and used effectively to enhance economic development"*.

Strengthen the Knowledge Economy in any country requires long-term investments in different areas related to sustainable development, such as education, human resource development, innovation and infrastructure of ICT. Further, having strong knowledge economy requires proper economic environment that facilitates the knowledge creation and knowledge sharing. *"The main drivers for a knowledge based economy include investments in: all levels of education; research and development (R&D), including capacity building and collaborative research; entrepreneurship;*

access to finance, including seed, angel, and venture capital; science parks and business incubators; and commercialization of proven technologies" [8].

The World Bank outlined the frame of the knowledge economy in the following main pillars:

- *Economic Performance:* This pillar refers to the overall macroeconomic policies which provide environment foster the knowledge creation and sharing.
- *Institutions:* This pillar is related to level of institutional system and its facilitation to the process of resources allocation and stimulation of innovation and creativity.
- *Education and Human Resource:* It's related to availability of skilled human resources who can efficiently participate in the knowledge creation and usage.
- I*nnovation system: This pillar is related to the availability of* higher education institutions, *r*esearch centers, consultants and different firms that can participate in the different processes related to knowledge.
- *Infrastructure for Information: This is related to the available infrastructure that* can facilitates the different processes related to the knowledge.

3 The Methodology of Assessing the Knowledge Economy

Although that measuring the performance of the knowledge economy may pose a greater challenge [5], it can be accessed via using the Knowledge Economic Index (KEI) which measures the level of knowledge as a source of income and its' role in the development of a country or region. KEI is the average of the performance score obtained by a country or region based on five pillars, which are: (i) perforce; (ii) institutions; (iii) Education and human resources; (iv) Innovation and (v) Information and Communications Technologies.

The KEI data was not available on the world bank database, this is the main limitation faced the researcher. To eliminate it, the researcher computed the index using the available data in the Global Competitiveness Report which is issued annually by the World Economic Forum. Five pillars were strived from the reports to assess the Knowledge economy in the GCC countries, as follows:

3.1 Performance

The overall macroeconomic performance which is stimulates the environment to the knowledge creation and sharing is assessed by the Macroeconomic environment pillar from the global competitiveness report. Business-friendly environment stimulates the process of technologies commercialization which lead to have innovation-based development [8]. This pillar assesses the stability of the macroeconomic environment in the countries. This index takes into consideration six indicators, they are: (i) Government budget balance; (ii) National savings rate; (iii) Inflation; (iv) Interest rate spread; (v) Government debt and (vi) Country credit rating.

3.2 Institutions

The institutions pillar from the global competitiveness report used to assess the Institutional framework and its' role in enabling the free flow of new or existing knowledge. This index is composite with weight age of 75% for public institutions and 25% for private institutions. The public institutions index includes (i) Property rights [Property rights and Intellectual property protection]; (ii) Ethics and corruption [Diversion of public funds, Public trust of politicians, Irregular payments and bribes, Undue influence, Judicial independence, Favoritism in decisions of government officials]; (iv) Government inefficiency [Wastefulness of government spending; Burden of government regulation; Efficiency of legal framework in settling disputes, Efficiency of legal framework in challenging regulations, Transparency of government policymaking] and (v) Security [Business costs of terrorism, Business costs of crime and violence, Organized crime, Reliability of police services]. Private institutions index consists of (i) Corporate ethics and (ii) Accountability [Strength of auditing and reporting standards, Efficacy of corporate boards, Protection of minority shareholders' interests, Strength of investor protection].

3.3 Education and Human Resource

Education and Human Resource pillar from the global competitiveness report used to assess how the education and other related HR perspectives can create educated and skilled population able to create, share and use knowledge. This index has three components with equal weight ages of each in the index. The components are: (i) Quantity of education [Secondary education enrollment rate, Tertiary education enrollment rate]; (ii) Quality of education [Quality of the educational system, Quality of math and science education, Quality of management schools, Internet access in schools] and (iii) On-the-job training [Local availability of specialized research and training services, Extent of staff

3.4 Innovation System

The innovation system pillar from the global competitiveness report used to assess the level of innovation which is elaborated via a network of private enterprises, research centers, universities, etc. in order to adapt to local needs, and create new technologies. The innovation has many benefits:

- Cost reduction for goods in widespread use
- The development of new goods and services [7]. This index consists of eight components with equal weight ages. They are: (i) Capacity for innovation; (ii) Quality of scientific research institutions; (iii) Company spending on R&D;

(iv) University-industry collaboration in R&D; (v) Government procurement of advanced technology products; (vi) Availability of scientists and engineers; (vii) Utility patents and (viii) Intellectual property protection.

3.5 Information and Communications Technologies

The technology readiness pillar from the global competitiveness report used to assess information infrastructure which facilitates the effective communication, dissemination and processing of knowledge. Two main components contribute in this index each has the same weightage, they are (i) Technological adoption [availability of latest technologies, firm-level technology absorption, FDI and technology transfer] and (ii) the ICT use [internet users, broadband Internet subscriptions, internet bandwidth, fixed telephone lines, mobile telephone subscriptions).

4 Analysis of KEI for GCC Countries

The researcher computed the KEI for the GCC countries from 2010 to 2017. The following sections show the analysis of the Knowledge Economy Index for the GCC countries followed by analysis of each pillar in this index.

4.1 KEI Index

Table 1 and Fig. 1 presents the Knowledge Economy Index (KEI) for the GCC countries from 2010 to 2017. Overall, the KEI for the GCC countries varies between 4.67 and 4.97. It was 4.67 in 2010 and increased to reach the maximum in 2014 where reached 4.97, then decreased gradually to 4.77 in 2017. Qatar got the highest KEI score for five years of the research period (2011, 2012, 2013, 2015 and 2016) while UAE got the highest KEI score for three years (2010, 2014, and 2017).

The index for 2017 shows that the highest knowledge GCC economy is UAE where it achieved 5.4, followed by Qatar in the second rank with score 5.33. Bahrain and Saudi Arabia are ranked in the third rank with score 4.68 followed by Oman in the fifth rank. Kuwait ranked in the sixth position. Figure 2 shows the GCC KEI for 2017. Overall, the KEI for all GCC countries is higher than the average of the world. The analysis of each pillar of KEI is shown in Tables 2, 3, 4, 5 and 6 and from Figs. 3, 4, 5, 6, and 7.

Table 1 KEI for GCC countries (2010–2017)

	2010		2011		2012		2013		2014		2015		2016		2017	
	Score	Rank	Score	Rank	Score	Rank	Score	Rank	Score	Rank	Score	Rank	Score	Rank	Score	Rank
Bahrain	4.68	3	4.62	5	4.68	5	4.66	5	4.58	5	4.61	4	4.52	4	4.68	3
Kuwait	4.25	6	4.29	6	4.28	6	4.31	6	4.29	6	4.43	5	4.34	5	4.16	6
Oman	4.59	5	4.71	4	4.78	4	4.83	4	4.66	4	4.37	6	4.23	6	4.37	5
Qatar	4.93	2	5.17	1	5.47	1	5.51	1	5.54	2	5.62	1	5.61	1	5.33	2
Saudi Arabia	4.64	4	4.97	3	5.13	3	5.00	3	5.12	3	4.99	3	4.65	3	4.68	3
UAE	4.96	1	5.01	2	5.21	2	5.27	2	5.62	1	5.41	2	5.31	2	5.40	1
Average	4.67		4.80		4.92		4.93		4.97		4.90		4.78		4.77	

Source Calculated by the author

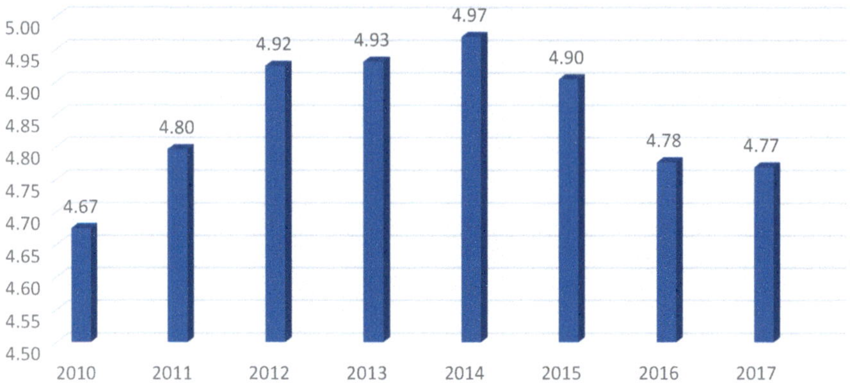

Fig. 1 Average KEI for GCC Countries (2010–2017). *Source* Calculated by the author

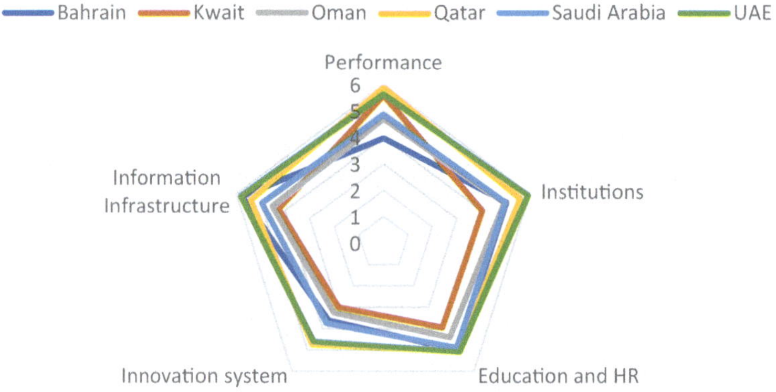

Fig. 2 Knowledge Economy Index (KEI) for GCC COuntries (2017). *Source* Table 1

Table 2 Macroeconomic environment index for GCC countries (2010–2017)

	2010	2011	2012	2013	2014	2015	2016	2017
Bahrain	5.65	5.15	5.50	5.90	5.19	4.60	3.88	3.98
Kuwait	6.42	6.59	6.58	6.70	6.73	6.72	6.31	5.60
Oman	6.11	6.48	6.56	6.64	6.56	5.99	4.46	4.70
Qatar	5.71	6.40	6.66	6.58	6.74	6.72	6.72	5.93
Saudi Arabia	5.35	6.09	6.55	6.69	6.67	6.63	4.71	4.87
UAE	5.65	6.14	6.41	6.42	6.63	6.53	5.28	5.63
Average	5.82	6.14	6.38	6.49	6.42	6.20	5.23	5.12

Source World Economic Forum, The Global Competitiveness Report, various issues

Table 3 Institutions index for GCC countries (2010–2017)

	2010	2011	2012	2013	2014	2015	2016	2017
Bahrain	5.02	5.29	5.13	4.77	4.70	4.92	5.04	5.04
Kwait	4.45	4.35	4.20	4.21	4.02	4.09	4.10	4.05
Oman	5.37	5.33	5.29	5.39	5.06	4.73	4.95	4.96
Qatar	5.55	5.39	5.77	5.95	5.90	5.86	5.70	5.60
Saudi Arabia	5.22	5.47	5.35	5.13	4.97	5.07	5.10	5.01
UAE	5.25	5.21	5.50	5.55	5.69	5.71	5.82	5.93
Average	5.14	5.17	5.21	5.17	5.06	5.06	5.12	5.10

Source World Economic Forum, The Global Competitiveness Report, various issues

Table 4 Education and HR index for GCC countries (2010–2017)

	2010	2011	2012	2013	2014	2015	2016	2017
Bahrain	4.64	5.00	4.93	4.52	4.66	4.82	4.90	4.99
Kwait	3.87	3.83	4.01	4.04	4.15	4.01	3.98	3.91
Oman	4.22	4.24	4.33	4.46	4.17	3.90	4.10	4.40
Qatar	4.84	4.62	4.94	5.11	4.99	5.12	5.41	5.01
Saudi Arabia	4.55	4.81	4.79	4.65	4.64	4.73	4.64	4.87
UAE	4.80	4.84	4.90	4.93	5.90	4.97	5.07	5.05
Average	4.49	4.56	4.65	4.62	4.75	4.59	4.68	4.71

Source World Economic Forum, The Global Competitiveness Report, various issues

Table 5 Innovation system index for GCC countries (2010–2017)

	2010	2011	2012	2013	2014	2015	2016	2017
Bahrain	3.21	3.2	3.13	3.17	3.32	3.41	3.61	3.6
Kwait	3.03	3	2.84	2.81	2.86	2.99	2.96	2.97
Oman	3.46	3.44	3.44	3.57	3.29	3.04	3.26	3.26
Qatar	4.11	4.69	4.71	4.80	4.88	4.98	4.87	4.68
Saudi Arabia	3.92	4.16	4.03	3.93	3.8	3.83	3.69	3.73
UAE	3.91	3.96	4.18	4.22	4.41	4.41	4.57	4.58
Average	3.61	3.74	3.72	3.75	3.76	3.78	3.83	3.80

Source World Economic Forum, The Global Competitiveness Report, various issues

4.2 Performance Index (Macroeconomic Environment)

The Average Macroeconomic environment index for GCC countries fluctuated between 5.12 and 5.82 during the research period. It increased from 5.82 in 2010 to 6.49 in 2013, then dropped from year after another to reach 5.12 in 2017. The overall

Table 6 Information Infrastructure index for GCC countries (2010–2017)

	2010	2011	2012	2013	2014	2015	2016	2017
Bahrain	4.88	4.48	4.72	4.95	5.01	5.29	5.15	5.77
Kwait	3.46	3.69	3.77	3.8	3.69	4.33	4.33	4.27
Oman	3.79	4.08	4.26	4.11	4.24	4.18	4.4	4.51
Qatar	4.44	4.74	5.29	5.1	5.17	5.41	5.35	5.41
Saudi Arabia	4.17	4.33	4.91	4.6	5.54	4.7	5.1	4.94
UAE	5.19	4.88	5.05	5.22	5.48	5.43	5.82	5.81
Average	4.32	4.37	4.67	4.63	4.86	4.89	5.03	5.12

Source World Economic Forum, The Global Competitiveness Report, various issues

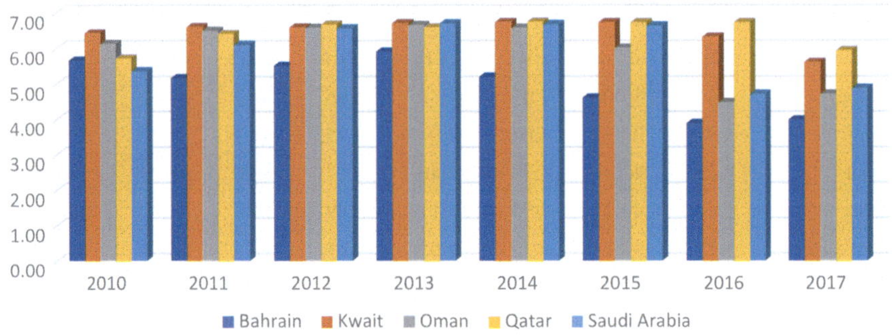

Fig. 3 Performance index for GCC countries (2010–2017)

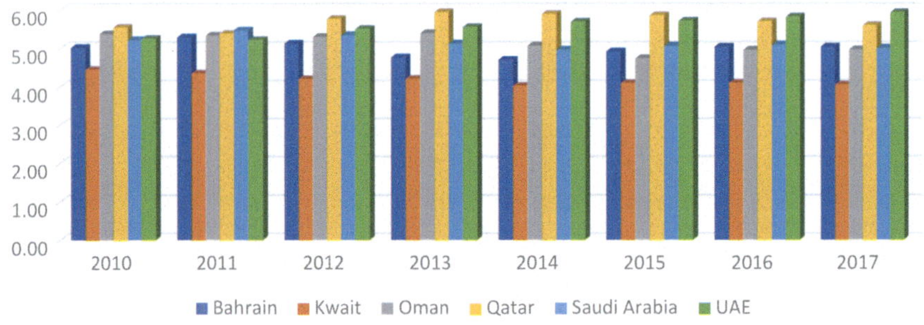

Fig. 4 Institutions index for GCC countries (2010–2017)

performance index dropped of all GCC countries after 2014 as a result of the fall in oil prices form $115 per barrel in June 2014 to reach $50pb at the start of 2015. Kuwait got the highest score in this pillar from 2010 to 0215, while Qatar is the highest in two years (2016 and 2017). In 2017, the highest GCC performance index shown in Qatar (5–93), followed by UAE (5.63). Kuwait ranked the third with performance index 5.60 while Saudi Arabia got the fourth rank with performance index 4.87 followed

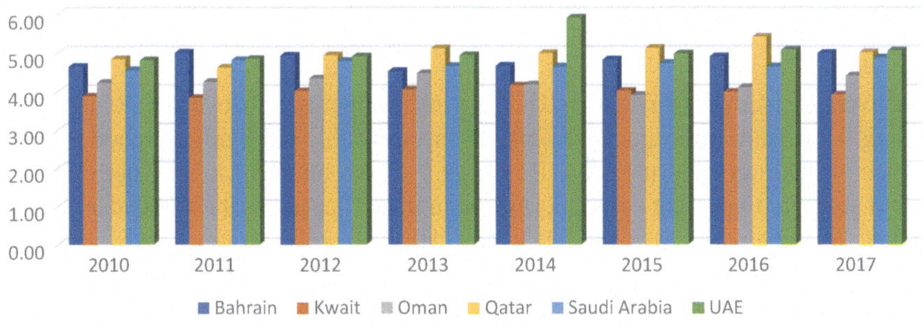

Fig. 5 Education and HR index for GCC countries (2010–2017)

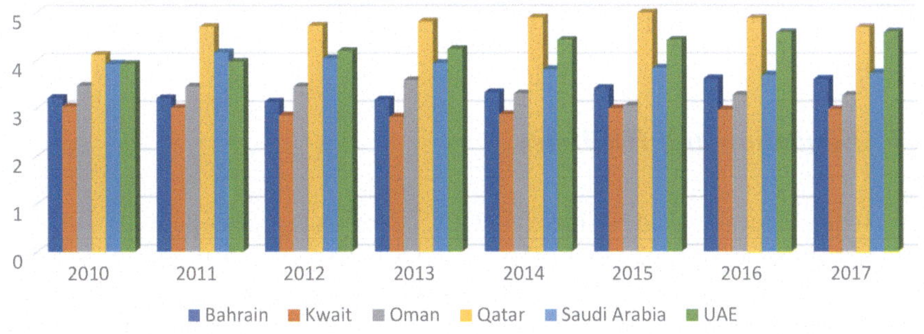

Fig. 6 Innovation system index for GCC countries (2010–2017)

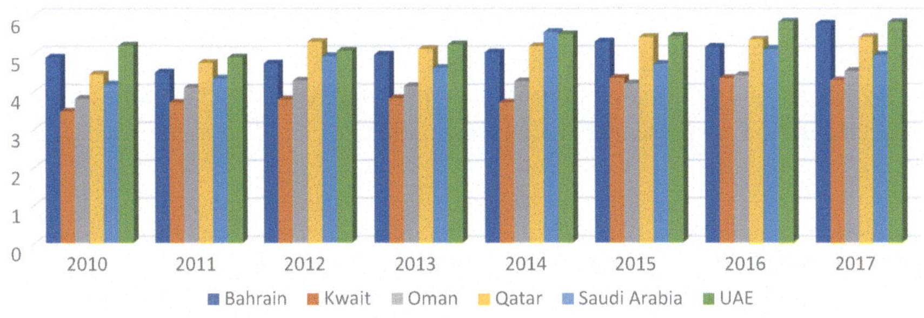

Fig. 7 Information Infrastructure index for GCC countries (2010–2017)

by Oman (4.70) and Bahrain in 6th rank with 3.98. Detailed performance index for GCC countries for the research duration are presented in Table 2 and Fig. 3.

4.3 Institutions Index

The average institutions index for GCC countries fluctuates between 5.06 and 5.21 during the period of the study. Overall, the score of this pillar reflects good performance of institutions as the score is at least 5.21 out of 7 which is higher than the world average. Oman got the highest score in this pillar in 2010 while Qatar get the highest score between 2011 and 2015 and UAE got the highest score of this index in 2016 and 2017.

In 2017, the highest value of the index shown in UAE (5.93) followed by Qatar (5.60) in the second rank. Bahrain came in the third position (5.04) followed by Saudi Arabia (5.01) and Oman in the fifth rank with index value 4.96 and Kuwait came in the 6th ranks with index value 4.5. Detailed institutions index for GCC countries for the research duration are presented in Table 3 and Fig. 4.

4.4 Education and HR Index

The average Education and HR index for GCC countries fluctuates between 4.49 and 4.71 during the period of the study. Qatar got the highest score in this index in 2010, 2012, 2013, 2015 and 2016 while Saudi Arabia got the highest score in 2011 and UAE got the highest score in 2014 and 2017. Overall, the score of this pillar reflects good Education and HR in GCC countries as the score is at least 4.71 out of 7 which is higher than the world average.

In 2017, the highest value of the index shown in UAE (5.05) followed by Qatar (5.01) in the second rank. Bahrain came in the third rank with index value (4.99) followed by Saudi Arabia (4.87), then Oman (4.40) in the fifth rank and Kwait in the sixth rank (3.91). Detailed Education and HR index for GCC countries for the research duration are presented in Table 4 and Fig. 5.

4.5 Innovation System Index

The average innovation system index for GCC countries fluctuates between 3.61 and 3.80 during the period of the study. Although the score of the index show gradual increase from 2010 to 2017, its' value does not cross 4 for the average GCC, which reflects medium level of performance of this pillar. Qatar got the highest score in the innovation index among all the GCC countries for the research period. In 2017, the highest value of the index shown in UAE (5.93) followed by Qatar (5.60) in the second rank. Bahrain came in the third position (5.04) followed by Saudi Arabia (5.01) and Oman in the fifth rank with index value 4.96 and Kuwait came in the 6th ranks with index value 4.5. Detailed Innovation system index for GCC countries for the research duration are presented in Table 5 and Fig. 6.

4.6 *Information Infrastructure (Technology Readiness)*

The average for Information Infrastructure index for GCC countries increase between 2010 and 2017 from 4.32 to 5.12. This reflects improvement in the index performance for the GCC countries. UAE got the highest score for the technology readiness for all the research period, except 2012 where Qatar got the highest score.

In 2017, the highest value of the index shown in UAE (5.81) followed by Bahrain in the second rank (5.77) then Qatar (5.41) in the third rank. Saudi Arabia ranked in the fourth rank where the index equals (4.94) followed by Oman (4.51) and Kuwait (4.27). Detailed Information Infrastructure index for GCC countries for the research duration are presented in Table 5 and Fig. 6.

5 Conclusion

Overall, the KEI of GCC countries is good, which reflects proper outcomes resulted from the economic diversification programmes deployed by GCC countries. The macroeconomic environment and Institutions are the highest scores in the KEI this resulted from a stability of the macroeconomic environment in the countries and the being of strong the Institutional framework and facilitates the flow of new or existing knowledge. This may be fostered by having more diversified economy. This matches the findings of Amin [1] which argued that Opening trade barriers between GCC members and also increase intra-trade between them in goods and services would make them a prominent player in the non-oil market. *"The initiative for diversifying the economy would play a crucial role in bringing them closer to one another"* [1].

Innovation pillar is the latest contributor of KEI for GCC countries. Although the score of the index show gradual increase from 2010 to 2017, its' value does not cross 4 (out of 7) for the GCC countries, which reflects medium level of performance of this pillar. This reflects the need of GCC countries to increase the capacity for innovation and the quality of scientific research institutions. Although that there is increase in the spending on R&D and attempts to increase the university-industry collaboration in R&D still more is required for these promising countries. Many GCC countries have taken speedy steps in this direction, however more are still required specially in the area of availability of scientists and engineers, application of patents and intellectual property protection. This result came in the same line with [5] results which found that "there is a potentiality of GCC members to allocate higher R&D expenditure by government and higher education".

References

1. Amin, T.: Prospects of knowledge economies in the Gulf. Prospects of Knowledge Economies in The Gulf. Observer Research Foundation. January (2020)
2. Brinkley, I.: Defining the Knowledge Economy. Knowledge Economy Programme Report, The Work Foundation (2006)
3. Chen, D.H.C., Dahlman, C.J.: The Knowledge Economy, the KAM Methodology and World Bank Operations. World Bank Institute, Washington, D.C. (2006)
4. Dang, D., Umemoto, K.: Modeling the development toward the knowledge economy: a national capability approach. J. Knowl. Manag. **13**(5), 359–372 (2009). https://doi.org/10.1108/13673270910988169
5. Hossain, A.: Evolution of mutual knowledge-based economy in regional integration: an experience from the cooperation council of Arab States of the Gulf. J. Knowl. Econ. **6**, 790–817 (2015). https://doi.org/10.1007/s13132-013-0148-5
6. Nancu, D.: The Knowledge Assessment Methodology of Countries. Ovidius University Annals, Economic Sciences Series XV **2**, 156–165 (2015)
7. OECD: The knowledge-based economy. OECD/ GD. (1996)
8. Schwab, K.: The Global Competitiveness Report. World Economic Forum. Various Issues

Machine Learning in Credit Risk Modeling: Empirical Application of Neural Network Approaches

Mohammad Shamsu Uddin

Abstract Motivated by massive development and the ongoing practice of machine learning (ML) approaches in credit risk modeling, this chapter addresses theoretical aspects of machine learning and credit default prediction. This chapter also discusses the properties of mostly used and robust machine learning approaches in credit default prediction. The objective of the chapter does not do empirical analysis, however, to show the practical application of ML approaches in credit default prediction, to the end, this chapter presents an empirical example of trendy classifier neural network approaches on real-world credit datasets.

Keywords Machine learning · Credit risk · Neural network · Multilayer perceptron

1 Introduction

The modeling of credit risk data has always been a key contemporary research item, a crucial issue for financial organizations, as well as a yardstick to the economic health of a country, at large. In that end, credit default prediction (CDP) is a systematic analysis conducted by issuers or lenders and related financial organizations to evaluate the financial soundness of the customers to repay the loan. The purpose of CDP is not only limited by whether credit application should be accepted, but it also related to credit extension, credit limit as well as behavioural modeling, like collection scoring. CDP is not only important for traditional credit activities, but it also significant for modern credit affairs, for example, selection of customers, measurement of risk, supervision of loan, assessment of overall loan performance, and portfolio risk management [1]. To that end, for an extended period, many efforts have been dedicated to establishing a suitable credit default prediction model by financial organizations, credit managers,

M. S. Uddin (✉)
School of Economics and Management, Dalian University of Technology, Dalian 116024, People's Republic of China
e-mail: msuddin16@yahoo.com

Department of Business Administration, School of Business and Economics, Metropolitan University, Bateshwar, Sylhet-3103, Bangladesh

A. Hamdan et al. (eds.), *The Fourth Industrial Revolution: Implementation of Artificial Intelligence for Growing Business Success*, Studies in Computational Intelligence 935,
https://doi.org/10.1007/978-3-030-62796-6_25

lenders, regulators, stakeholders of the financial markets, as well as academics and financial economists [2–5]. As a result, the extensive literature on credit risk has appeared [6, 7].

Prior studies defined credit scoring as a two-class binary problem, such as classification a loan into two categories; default and non-default (for a survey of earlier studies, see [8, 9]). Before the automated credit scoring system, the risk of potential customers was assessed via underwriters; generally, a lending decision was made via judgment process, and authority (underwriter) evaluated applicants based on 5Cs criteria. Later, few traditional statistical classifications approach, for instance, logistic regression (LR), discriminate analysis (DA), probit regression (PR), have been used in CDP [10–14]. Afterward, because of the intensive computational requirements, artificial intelligence approaches, such as support vector machine (SVM), neural network (NN), rough set (RS), decision trees (DT), random forest (RF) and gradient boosting (GB) have been introduced in the credit risk modeling process [13, 15–21].

This chapter is designed to present a short summary of machine learning, then the theoretical aspects of credit default prediction describe in different forms. Subsequently, mostly used statistical and machine learning approaches are discussed. To the end, the empirical applications of a robust machine learning approach present via NNs on real-world credit scoring datasets.

2 Machine Learning

Machine learning (ML) is a data analytics technique and one aspect of artificial intelligence (AI) that offers mechanisms to accomplish a particular objective without requiring explicit instructions. In the era of big data, ML has become the leading technique for acquiring the expected solution in many research fields, such as credit scoring or (more broadly) in computational finance, computational biology, image processing, energy production, and natural language processing. Due to the massive improvements of ML algorithms and the advent of enormous frequencies and dimensionality of customer data modeling, credit scoring via ML approaches has become increasingly crucial [22–24].

Generally, ML approaches can be separated into supervised and unsupervised algorithms. The key distinction between these two methods is that supervised learning is conducted using either known input and output data or prelabeled instances. In contrast, unsupervised learning learns from unknown data and unlabeled cases. The most common supervised algorithms are the support vector machine (SVM) [25, 26], artificial neural network (ANN) [27], decision tree (DT) [28], and random forest (RF) [29]. In contrast, some of the more popular unsupervised ML algorithms are K-means [30], Kohonen's self-organizing maps (SOM) [31], hierarchical clustering [32], and isolation forest [33]. Both supervised and unsupervised approaches have been applied widely to credit risk prediction. Supervised methods are usually applied to credit customer risk assessments in binary format. The existing credit literature has revealed the efficiency of supervised methods for credit scoring [34–36]. Unsupervised ML approaches generally refer to clustering techniques and are typically used as an additional tool with supervised methods [37, 38]. However, little evidence has been reported regarding the problem of predicting credit risk [39, 40].

3 Credit Default Prediction

3.1 Definition of Credit Default Prediction

Credit issuing to customers is considered as a principal function of the lending organization that makes a profit for the lender, for the all other stakeholders of the lending institutions like depositors, shareholders and as to contributing the nations as well. For this reason, proper credit default prediction or credit scoring is very significant for sustainable economic development. Credit default prediction is a comprehensive process, which assists in identifying the most suitable customer for potential lending [41]. Credit risk modeling is essential for multiple types of credit affairs, for example, selecting potential customers, evaluating existing customer risk, loan supervision, comprehensive customer reviews, and predicting financial risk [42]. It also aids banks in reducing the loss of non-performing loans by making a more systematic credit decision. Moreover, in some perspective, any little development for the credit evaluation models would generate enormous earnings for the banks [8]. Therefore, credit scoring continuously regarded as a significant trendy study area, and it turns into more and more meaningful for the financial sector [43, 44].

This chapter uses credit default prediction as a substitute for credit scoring. Basically, credit scoring is the route of evaluating creditworthiness by lending institutions [45]. According to the suggestion of Anderson [46], we can divide credit scoring into two parts, credit and scoring. The first component 'credit' represents "buy now, pay later" this word comes from the Latin' credo,' which means 'I believe' or 'I trust in.' The second part 'scoring' represents "the use of the numerical tool to rank order cases according to some real or perceived quality in order to discriminate between them, and ensure objective and consistent decisions." Therefore, we can define credit scoring is a comprehensive process of evaluating credit applicants or customers to make proper credit decisions. As such, [41] also mentioned credit scoring determined who will receive credit, the appropriate amount of credit, and potential operational strategy to ensure the profitability of the lenders, as well as borrowers. Some other previous research studies also presented a detailed discussion of credit scoring, such as books by [46–48]; and research papers by [49–52].

3.2 Background of Credit Default Prediction

The term credit or borrowing and lending is related to human civilization and has a long history like trade and commerce. Although the idea of credit starting from about 2000 BC or before, however, the credit scoring history is not as old as credit, it had been initiated about 60 years ago. Accordingly, banks or other lending institutions collected information and used to prepare credit scores for all individual applicants [41, 46, 53]. Afterward, the function of credit scoring has been extensively expanded in various fields. Particularly in the twenty-first century, credit scoring has been expanded more than before due to the massive development of technologies and to fulfill the demand of new age. Therefore the literature of credit risk modeling is limited, while few contemporary pieces of literature exist on credit default prediction via different statistical and machine learning approaches [16, 17, 20, 54–59].

Before the automated credit scoring system, the risk of potential customers was assessed via underwriters, typically on the experience of underwriters. The personal relationship between applicant and lender was the key source of information [46]. Generally, the lending decision was made via judgment process and authority (underwriter) evaluated applicants based on 5Cs criteria:

Character is the credit candidate or else any family member known to the lending institutions? Whether applicants have the willingness to pay potential credit or other monetary habits are sensible for granting the loan.

Capital-Capital represents about deposit and expected loan amount ratio of a potential credit customer.

Collateral-Collateral refers to security offerings by applicants for the expected loan amount.

Capacity-Capacity indicates the financial ability to repay the loan amounts.

Condition-Condition not only represents applicants; it is basically related to the current economic situation for business survival.

It is assumed that the credit-granting decisions via judgment process are not consistent and reliable. As such, [60] has mentioned some fundamental limitations, such as; decisions are severely affected by management attitude, discontinuity in the application of earlier decisions, risk of revenue loss due to limited human capacity in the application.

The demand for credit products was substantially increased in the public community; in this regard, to protect the public interest regulatory framework also developed. These regulatory changes also helped to developed suitable credit scoring systems, such as to ensure the interest of fairness, equality, and transparency; the United States Congress established new law, known as the Equal Credit Opportunity Act (ECOA), in 1974. This act mainly protects credit applicants from discriminations in the credit granting processes, such as race, religion, or colour. On the other hand, Basel Accords are that have been implemented more than 100 countries over the world, played a significant role in developing banking rules and standards.

Later on, statistically developed credit default prediction techniques were developed to ensure creditors to follow regulatory requirements. As such, [60] mentioned that electronic devices such as computers presented necessary means to use automated processes. This process substantially reduced credit evaluation cost and customer default losses compared to prior judgment process [47, 61]. With the development in huge computational capacity and progress in statistical methodologies, financial institutions began to apply comprehensive credit evaluation techniques, such as the default risk prediction [8], methods for fraud detection [62], customer response to the advertisement of a new product [63], for customer retention [64], customer attrition [64], product usage intensity [65], profit and product default scoring [48]. For data mining and operational research methods, credit scoring is the most popular application area compared to other filed of applications [66].

3.3 Advantages of Credit Default Risk Prediction

Comparative studies between credit default risk prediction and other traditional techniques, such as with judgment processes, are not adequate in the existing literature. Only a very few studies, for example, [68 and 69] compared modern credit scoring and traditional judgment techniques and point out advantages for both parties in the lending process borrower and lender.

In the modern credit default risk prediction or credit scoring, final models have been designed to include significantly correlated predictors. While in the judgment method, there is no such variable selection techniques are available [67]. Credit scoring approaches usually utilized accepted and rejected applications in the decision-making process, like bias correction. Conversely, judgment processes generally based on accepted applicants and who consequently defaulted [67]. In the big data era, credit scoring can initiate with high dimensional data sets; for example, more than thousands of variables. However, in the judgment techniques, an underwriter cannot consider several predictors, except for a few common indicators. In addition, different credit scoring models can use the same data set with the same weight for comparison, where it is improbable to be so for judgment techniques [67, 68]. Some other advantages of credit scoring have been presented by [68], these are efficient modeling and helpful for the decision making route, minimum modeling cost and efforts, low-level error rate, modeling via real data, adjustment of cut off score based on the business environment and so on.

3.4 Limitations of Credit Default Risk Prediction

In the credit default risk prediction, usually historical data are used to predict default customers. In this process, the predictors are considered to be stable over a specific point. This technique creates an error in the modeling, except if it is regularly updated. Most of the credit risk prediction models only use dichotomous outcomes: either one default or non-default. Definitely, in the real world, there is another possible status that may exist, such as delay in payment, late interest payment, or complete default.

The major limitation of credit risk modeling is considered a credit score of particular applicants simply via summarizing points received from application forms. However, this score basically depends on system design. For s sole cut-off technique, the candidate's full score is evaluated to a specific cut-off standard. When the credit score is more than the cut-off, credit is approved, otherwise not. In addition, advanced prediction modeling is used as a two-stage prediction method. For example, the applicant's credit score evaluated with two cut-off standards. The application automatically accepted if the score goes beyond the top cut-off value, while if it is less than the lower cut-off, the application is automatically rejected. In the case of a score among two cut-off values, re-evaluation is required based on other information, such as credit history. Therefore, if the new score is above standard, the credit application is accepted, otherwise not [69].

Besides, credit modeling uses some predetermined or specific predictors, while sometimes significant predictors which closely related to loan repayment may ignore. Furthermore, in customized credit risk modeling systems by chance, an applicant has more similar characteristics of default or non-default customer, which cannot provide a real sense for accurate decision making. In addition, credit risk models are different from market to market and are not standardized as well. Therefore, are costly to buy and consequently to train credit professionals; sometimes, it may reject an applicant for insignificant issues, such as profession, address for customized prediction process [70].

4 Major Statistical and Machine Learning Approaches in Credit Default Prediction

4.1 Logistic Regression

The logistic regression (LR) is an analytical and mostly used method which is appropriate for classification and regression problem in case of the dichotomous dependent variable. A lot of sophisticated machine learning approaches have been developed. However, until now, due to its unique properties, it has been considered an industry-standard in credit risk modeling [2]. The appropriate balance of accuracy in the prediction outcomes, efficiency in modeling, and interpretability of outcomes for real-world users are the main issues that make LR different from other approaches and ensure long time application in the practical field [71]. LR model classifies the probability of default event to occur is defined as [72]

$$\mathrm{Log}\big[\mathrm{p}(1-\mathrm{p}/\mathrm{p})\big] = \beta_{0+}\beta_1\mathrm{X}_1 + \beta_2\mathrm{X}_2 + \cdots + \beta_\mathrm{n}\mathrm{X}_\mathrm{n} \tag{1}$$

where p represents the probability of default, β_0 is intercepted, β_i is the coefficient of independent predictors X_i (i = 1 … 2 … n) in addition to log [p (1 − p/p)] stands for the dependent variable.

4.2 Discriminant Analysis

The discriminant analysis (DA) considered as another classical approach developed by Fisher [73]. There are several applications of DA in the different study areas, such as in statistics, pattern recognition, and machine learning, for verifying a linear relationship among the variables to differentiate or segregate two or more types of objects. DA applied to predict a categorical dependent variable by particular or several independent predictors, for instance, binary or continuous variables. The function of DA is known as discriminant function assessment, is useful in evaluating whether a set of features is proficient in predicting class relationships. The discriminant function of the DA model is described as follows:

$$Z = a_0 + a_1X_1 + a_2X_2 + \cdots + a_nX_n \tag{2}$$

where a_1, a_2....a_n is the discriminant coefficient and X_1, X_2.....X_n is the discriminating variables.

4.3 Decision Tree

In the field of data mining and credit risk modeling, the decision tree (DT) is a regularly applied technique. This method is also popular in other areas of application, such as operation research area. Mainly in case of decision analysis, this technique helps in selecting a method most likely to achieve an objective. For example, according to the contribution of different input predictors, this model predicts the objective predictors. A decision tree comprises three kinds of nodes, for example, decision nodes-generally stand for by squares, chance nodes-usually stand for by circles and end notes- usually hold for by triangles [74]. The DT is considered most suitable for the field of credit risk modeling, as in credit risk modeling, it needs to identify two groups instances, such as 'non-default customers' and 'default customers.' The calculation process of DT method goes around each part to identify best one and afterward chooses the successful sub-tree to provides the most appropriate "good" and 'bad' customers depends on its velocity arises from overall miscalculation and least misclassification cost [75, 76].

4.4 Support Vector Machines

Cortes and Vapnik [77] developed support vector machines (SVM), which considered popular and mostly used machine learning technology in the different real-world study fields. In the credit risk ground, it has been widely applied due to its advanced classification ability and comparatively easy construction than its close counterpart artificial neural network (ANN) and other classifiers [78, 79]. The purpose of the SVM is to reduce the generalization error of the upper bound, which depends on the structural risk minimization approach. In the SVM, initially it is necessary to utilize training instances to approximate a function for evaluation. The function presented in the following:

$\int : \mathrm{R}^{\mathrm{N}} \rightarrow \{1, -1\}$, which are k N-dimensional patterns X_i and class labels Y_i, where

$$(\mathrm{X}_1, \mathrm{Y}_1), \cdots (\mathrm{X}_\mathrm{k}, \mathrm{Y}_\mathrm{k}) \in \mathrm{R}^{\mathrm{N}} X\{1, -1\} \tag{3}$$

According to the Eq (3), the SVM classifier should satisfy the following formulation:

$$\mathrm{W}^{\mathrm{T}}\varphi\,(X_{\mathrm{i}}) \;+\mathrm{b} \geq 1 \mathrm{if}\, y_{\mathrm{i}} = +1 \tag{4}$$

$$\mathrm{W}^{\mathrm{T}}\varphi\,(X_{\mathrm{i}}) \;+\mathrm{b} \leq -1 \mathrm{if}\, y_{\mathrm{i}} = -1 \tag{5}$$

This is the equivalent to the next equation.

$$y_i\left[\mathrm{w}^{\mathrm{T}}\varphi\,(x_i) \;+\mathrm{b}\right] \geq 1, i = 1, 2, \ldots, k \tag{6}$$

The non-linear function φ will map the original space a high-dimensional feature space. The hyper plane will be assembled by the revealed inequalities, can be defined as

$$\mathrm{w}^{\mathrm{T}}\varphi\,(x_{\mathrm{i}}) \;+\mathrm{b} = 0 \tag{7}$$

The main purpose of this process is to find a hyper plan which backed by support vector with intension to identify two classes of predictors with maximum margin. Accordingly depends on the feature of the support vectors, the new input sample label can be estimated. Several kernels (function) of SVM, such as linear, polynomial, sigmoid and radial basis function (RBF) can be applied to map input instances to the high-dimensional feature space [80].

4.5 *Artificial Neural Network (ANN) Approaches*

4.5.1 Multilayer Perceptron

The structure of an Artificial Neural Network (ANN) is developed on the basis of the biological neural frameworks. ANN can effectively be applied in credit risk and several other study areas, for example, data classification and clustering, prediction of time series data, pattern recognition, signal processing, and others. ANN's training method basically related to altering relations among the neurons. Feed Forward Neural Network (FFNN) and Feed Back Neural Network (FBNN) are the two types of ANNs that depend on the system topology. In FFNN, the data stream in the network is unidirectional with no response loops, whereas, in FBNN, the stream is bi-directional with response loops. Multilayer Perceptron (MLP) belongs to FFNN, and it has three layers; input layer, hidden layer, and output layer. In the input layer, inputs are received, and the numbers of neurons are the same as the features of the datasets. The hidden layer is the essential part of MLP used for mapping and transferring function between input and output layer. Finally, in the output layer is used to provide the outcome of the network.

In MLP networks, neurons in the layer are completely interrelated by numeric weights; every neuron holds summation and activation functions. The summation function is the summary of the product of inputs, weights, and bias as exposed in Eq. (8). Where w_{ij} is the relationship weight linking I_i to neuron j, β_j is a bias term, and n is the entire quantity of neuron inputs. Activation functions will receive the output of the summation function as an input. Typically, the S-shapes curved sigmoid function is used as the non-linear activation function. The sigmoid function is shown in Eq. (9). Consequently, the outcome of the neuron j can be described as in Eq. (10).

$$S_j = \sum_{i=1}^{n} w_{ij} I_i + \beta_j \tag{8}$$

$$f(x) = \frac{1}{1 + e^{-x}} \tag{9}$$

$$y_j = f_j\left(\sum_{i=1}^{n} w_{ij} I_i + \beta_j\right) \tag{10}$$

While the formation of ANN is designed, the learning procedure of this approach is functional to facilitate the parameters of the network (set of weights). Alternatively, these weights are curved and modernized to estimate the outcomes and reduced little error standard. One of the key MLP training methods is supervised learning. The objective of supervised learning is to reduce the error among the expected and computed outcomes. Backpropagation is regarded as one of the general supervised learning algorithms based on Gradient technique, which is a process of discovering the derivative of ANNs objective function regarding the weights and bias that replace among layers. This technique is ineffective when the search space is large, and the technique is appropriate solely for differentiable objective functions.

4.5.2 Radial Basis Function Network

There is another type of artificial neural network based on basis function for the activation function during modeling of the mathematical problem. The linear arrangement of the radial basis function is used for generating outputs of the neuron parameters and inputs. Broomhead and lowe [81] developed RBF for various functions, such as classification, time series prediction, estimation, and system control. The RBF is somewhat similar to MLP, such as; MLP uses sigmoid function as a transfer function in every node. Conversely, in all hidden nodes, RBF uses different radial basis functions; for example, Gaussian function. The RBF network has a simple architecture with a single hidden layer like MLP. However, there is the little dissimilarity among MLP and RBF, such as in MLP every node (i.e., hidden note and output node) has the same transfer function as sigmoid function. On the other hand, in RBF all hidden node has its radial basis function like Gaussian function.

4.5.3 Adaptive Neuro-Fuzzy Inference System

For modeling of non-linear dynamic systems, Takagi-Sugeno-Kang (TSK) inference system is considered the most useful fuzzy inference system and powerful instrument. The significant improvement of TSK system modeling is that it is a 'multimodal' method which can merge linear sub-models to explain the stellar performance of the complex non-linear dynamic system [82]. Studies have employed one of the fashionable neuro-fuzzy methods, adaptive neuro-fuzzy inference systems (ANFIS), for regression, modeling, prediction, and control problem [83, 84]. In a five-layered network configuration, ANFIS utilized TSK kind fuzzy inference system. ANFIS describes twofold sets of parameters, i.e., premise parameters and consequent parameters. The if-then rules Fuzzy systems describe the affiliation among the parameters. To give fuzzy if-then rules example, if we judge the fuzzy inference structure with two inputs x and y and one output z, then the first order Sugeno model the two rules can be offered as follow:

$$\begin{aligned} &\text{Rule-1 : If } x = A_1 \text{ and } y = B_1 \text{ then } f_1 = p_1 x + q_1 y + r_1 \\ &\text{Rule-2 : If } x = A_2 \text{ and } y = B_2 \text{ then } f_2 = p_2 x + q_2 y + r_2 \end{aligned} \tag{11}$$

where x and y are independent predictors, A_i and B_i are fuzzy sets, p_i, q_i, r_i are the parameters of the dependent variable. The similar ANFIS design, which combination of five layers, is described in some earlier studies [83, 84].

4.5.4 Deep Neural Network

A deep neural network (DNN) is another version of an artificial neural network (ANN) with some improvement, such as more layers between the input and output layer. In comparison to other ANN models, for example, MLP and RBF, the DNN modeling approach is complicated. The modeling techniques of DNN are made-up to simulate the motion of the human brain. This chapter discusses Deep belief networks (DBN) as the DNN model by following [85]. Hinton et al. develop DBN. Basically, it is a graphical method with a multilayer. The restricted Boltzmann machine (RMB) is the key tool in the DBN training process. The training process of the DNN method discussed some previous studies, such as [86, 87]. In the beginning, every layer independently trained as an RMB. The RMB is developed by contrastive divergence (CD) algorithm, which is an unsupervised technique [88, 89]. Modeling time and overfitting problems are the main challenges for DNN in the training process. DNNs are riskier for overfitting because of more layers compare to other ANN models.

5 Empirical Examples from Neural Network Approaches

Recently, Chi et al. [17] performed a comprehensive study and compared sixteen feature selection based hybrid models, such as the combination of logistic regression (LR), discriminant analysis (DA), and decision trees (DT) with four techniques of neural network (NN): an adaptive neuro-fuzzy inference system (ANFIS), deep neural network (DNN), radial basis function network (RBF) and multilayer perceptron (MLP). That study, primarily used four conventional methods DA, two approaches of DT and LR for feature selection and credit default prediction. Significant selected variables by conventional methods have been utilized as the input predictors of four AI (NNs) classifiers. To end with, twenty classifiers outcomes of every dataset described in the perspective of ten different performance measures. Model 1, 2, 3, and 4 are developed from the following methods DA, DT: CHAID, DT. Ex.CHAID and LR, respectively. To provide a summary and empirical application of NNs approaches, this chapter graphically shows four measures, namely accuracy, AUC, type I error, and type II error. The examined average findings of five datasets are present in the following Figs. 1, 2, 3 and 4.

The empirical findings confirm that LR and MLP based hybrid classifier (MLP_4) outperformed other baseline and hybrid models. In addition to this, MLP based other hybrid classifiers also confirmed advanced performance in comparison to other models. The average accuracy of the best model MLP_4 reached about 98%, which provides supremacy of the classifiers on all possible outcomes. Except for DNN based models, all other classifiers also exposed better classification performance is about more than 80% rate. The area under the curve (AUC) is another global performance measure; basically, measure the discrimination capacity of the classifier develops based on receiver operating characteristics (ROC) curve. The AUC performance also authenticates MLP_4 has an excellent discriminant power compared to other

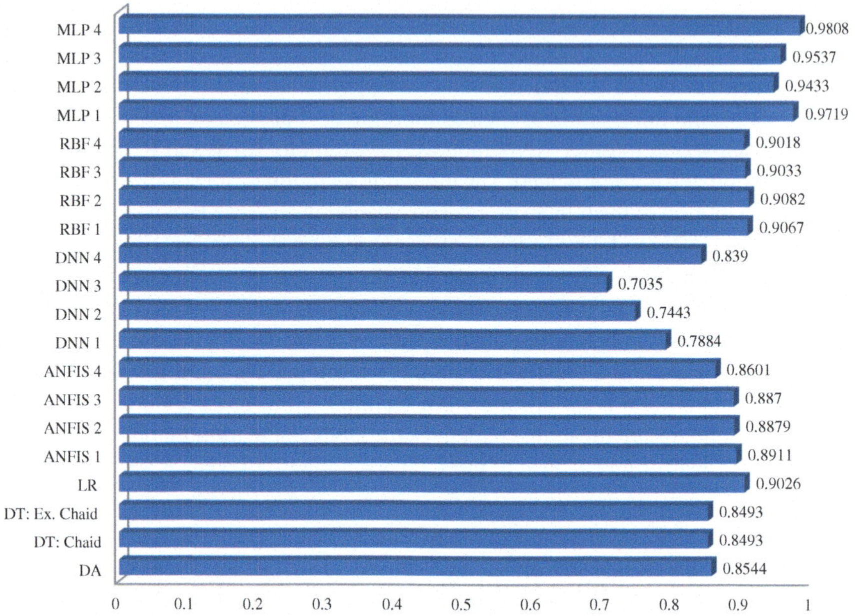

Fig. 1 Average predicted accuracy of all classifiers on five credit datasets

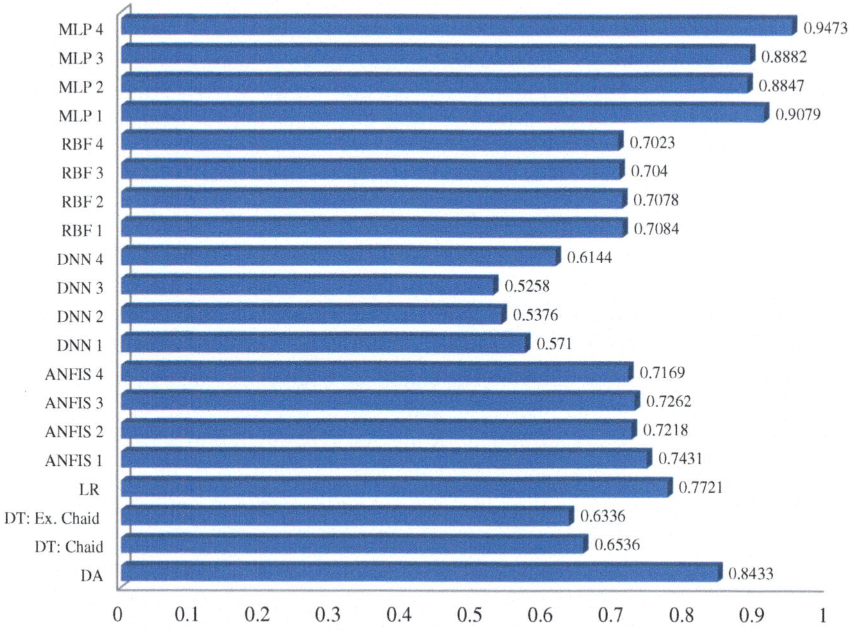

Fig. 2 Average predicted AUC of all classifiers on five credit datasets

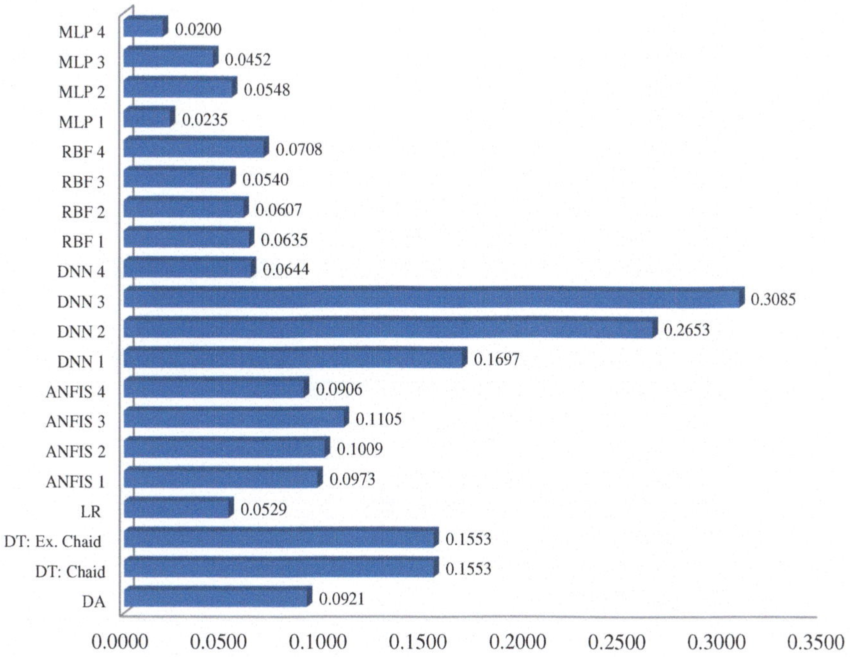

Fig. 3 Average predicted type I error of all classifiers on five credit datasets

classifiers. Some classification errors occur in the classification process. Type I error appears in the case when non-default customers are misclassified as defaults.

On the other hand, it is considered a type II error when default customers misclassify as non-default. It is assumed that the type II error is more costly than the type I error. In that study in both cases, MLP_4 presents better performance. There are four conventional models utilized for feature selection and classification. Among them in maximum cases, LR outperforms other conventional classifiers, and LR model is closely followed by DA method. Moreover, in some cases, LR outperforms some hybrid models also. For example, regarding average accuracy and AUC, LR provides better results than ANFIS and DNN based model. The ANFIS and DNN are robust approaches in other study areas; however, in credit risk modeling, their application is restricted.

According to overall findings, the effect of feature selection on the classifier's performances is summarized as follows.

- ANFIS related models: In accordance with average findings of five credit approval datasets, ANFIS related methods provide moderate results compared to its counterparts. This NN and fuzzy combined classifier demonstrated competitive results with RBF based methods. Nevertheless, about AUC, it confirmed a superior performance than RBF.

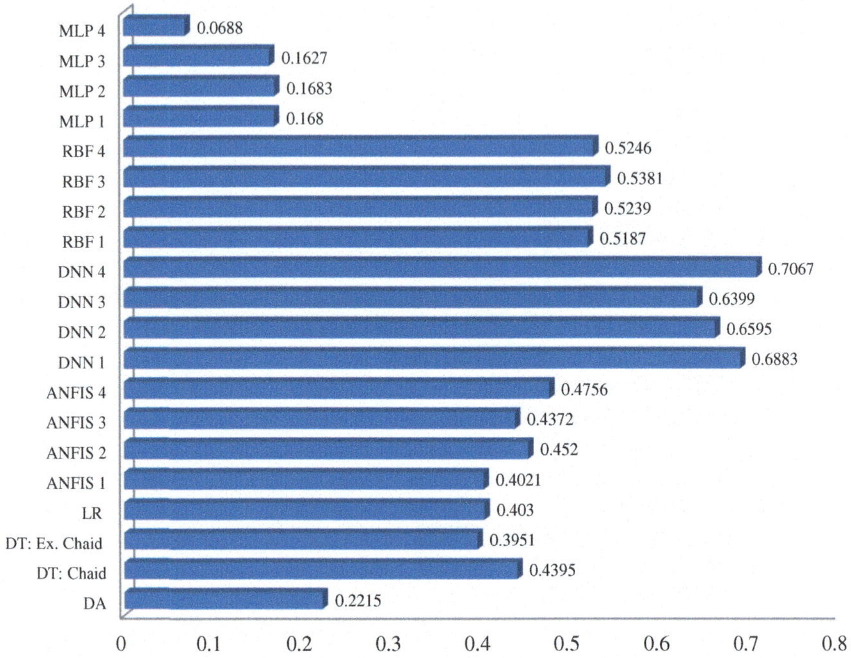

Fig. 4 Average predicted type II error of all classifiers on five credit datasets

- DNN based Models: A deep neural network (DNN) is a feed-forward, artificial neural network, usually consists of the multilayer. In recent times DNN has attached a significant concentration of researchers due to its prevailing effectiveness in modeling and forecasting. Through it is a very proficient technique for other study fields, but maybe due to the problem environment in this study, DNN provides low efficiency compared to its counterparts.
- RBF based classifiers: The average results demonstrated that RBF based classifier present modest efficiency in maximum cases. The performance of this model is somewhat similar to ANFIS based models. In a few points, it is superior to ANFIS and vice versa. There are significant differences between MLP based models with RBF based models.
- MLP based Classifiers: All MLP based models perform better than other conventional and feature selection based models. According to five credit approval datasets, the MLP_4 model significantly outperforms other models. The DA based other classifier MLP_1 also demonstrated stable performance all over the study. Other MLP related models showed better prediction performance as well.

6 Conclusion

This chapter attempt to provide a scenario of machine learning approaches in credit risk modeling, and an empirical example is presented to understand its real-world application. So far, this chapter is only gone over some most used machine learning algorithms in credit risk modeling. However, there are lots of alternatives that can be applied to different problem domains. There is no consensus for selecting an algorithm for all types of solutions. Some algorithms, such as ANFIS and DNN, are robust classifiers, but in the exampled study [17] confirms the minimum capacity of those models in credit risk modeling compared to other models. Other previous studies also supported these findings [90–92]. There is no simple solution to get the algorithm for a particular problem. In that end, the data-driven method can be a feasible option to decide algorithms. For that, we need to know the characteristics' of datasets, such as how noisy it is, how it can affect final outcomes and so on. Based on the above scenario, we can reach a feasible solution via applying a related algorithm with different probable parameters.

In addition, it should be mentioned that the approaches discussed in this chapter are well known, mostly used in credit risk modeling. Many studies have done on those approaches and recommended the robustness of those technologies in the prediction process. However, some limitations also mentionable, for instance, these 'black box' approaches have the massive computational capacity, but their interpretability of outcomes is very limited. As credit risk modeling is a pure real-world problem, interpretability is very significant along with high classification accuracy. Some research has been done to improve this; nevertheless, comprehensive research is required to solve such a crucial problem.

This chapter deals with data classification techniques. Now a day, it is a significant research agenda and assumed that data is very important, like other valuable assets. As such, data science and big data handling techniques significantly attached the concentration of different stakeholders of various industries to make a proper trade-off between risk and return and ultimately for effective decision making. Due to the massive computational capacity of artificial intelligence (AI) techniques, it is considered as the main tool for big data problems. Data science or big data can play a significant role for the Fourth Industrial Revolution. Therefore AI or machine learning is the principal part of the Fourth Industrial Revolution; it would be an integral part of human civilization. Due to the impressive progress of AI, it can be concluded that it will change the industrial production process, management philosophies, human life and workspace as well.

References

1. He, H., Zhang, W., Zhang, S.: A novel ensemble method for credit scoring: adaption of different imbalance ratios. Expert Syst. Appl. **98**, 105–117 (2018)
2. Lessmann, S., Baesens, B., Seow, H.-V., Thomas, L.C.: Benchmarking state-of-the-art classification algorithms for credit scoring: An update of research. Eur. J. Oper. Res. **247**, 124–136 (2015)
3. Djeundje, V.B., Crook, J.: Incorporating heterogeneity and macroeconomic variables into multistate delinquency models for credit cards. Eur. J. Oper. Res. **271**(2), 697–709 (2018)
4. Djeundje, V.B., Crook, J.: Dynamic survival models with varying coefficients for credit risks. Eur. J. Oper. Res. **275**(1), 319–333 (2019)
5. Bhattacharya, A., Wilson, S.P., Soyer, R.: A Bayesian approach to modeling mortgage default and prepayment. Eur. J. Oper. Res. **274**(3), 1112–1124 (2019)
6. Duffie, D., Singleton, K.J.: Credit Risk, Pricing, Measurements, and Management. Princeton University Press, Princeton, New Jersey (2003)
7. Jones, S., Hensher, D.A.: Advances in Credit Risk Modelling and Corporate Bankruptcy Prediction. Cambridge University Press, Cambridge, UK; New York (2008)
8. Hand, D.J., Henley, W.E.: Statistical classification methods in consumer credit. J. R. Stat. Soc.: Ser. A (Stat. Soc.) **160**(3), 523–541 (1997)
9. Rosenberg, E., Gleit, A.: Quantitative methods in credit management: a survey. Oper. Res. **42**(4), 589–613 (1994)
10. Nikolic, N., Zarkic-Joksimovic, N., Stojanovski, D., Joksimovic, I.: The application of brute force logistic regression to corporate credit scoring models: evidence from Serbian financial statements. Expert Syst. Appl. **40**(15), 5932–44 (2013)
11. Shi, B.F., Chi, G.: A model for recognizing key factors and applications thereof to engineering. Math. Probl. Eng. **1**, 368–81 (2014)
12. Shi, B.F., Wang, J., Qi, J.Y., Cheng, Y.Q.: A novel imbalanced data classification approach based on logistic regression and fisher discriminate. Math. Probl. Eng. **6**, 1–12 (2015)
13. Jones, S.: corporate bankruptcy prediction: a high dimensional analysis. Rev. Acc. Stud. **22**, 1366–1422 (2017)
14. Hung, K., Cheng, H.W., Chen, S., Huang, Y.: Factors that affect credit rating: an application of ordered Probit models. Romanian J. Econ. Forecast. **16**(4), 94–108 (2013)
15. Pai, P.F., Tan, Y.S., Hsu, M.F.: Credit rating analysis by the decision-tree support vector machine with ensemble strategies. Int. J.Fuzzy Syst. **17**(4), 521–30 (2015)
16. Uddin, M.S., Chi, G., Habib, T., Zhou, Y.: An alternative statistical framework for credit default prediction. J. Risk Model Valid. **14**(2), 1–36 (2020)
17. Chi, G., Uddin, M.S., Abedin M.Z., Yuan, K.: Hybrid model for credit risk prediction: an application of neural network approaches. Int. J. Artif. Intell. Tools **28**(5), 1950017, 1–33 (2019)
18. Bai, C.G., Shi, B.F., Liu, F., Sarkis, J.: Banking credit worthiness: Evaluating the complex relationships. Omega **83**, 26–33 (2018)
19. Nie, G.L., Rowe, W., Zhang, L.L., Tian, Y.J., Shi, Y.: Credit card churn forecasting by logistic regression & decision tree. Expert Syst. Appl. **38**(12), 73–85 (2011)
20. Jones, S., Johnstone, D., Wilson, R.: An empirical evaluation of the performance of binary classifiers in the prediction of credit ratings changes. J. Bank. Finan. **56**, 72–85 (2015)
21. Jones, S., Wang T.: Predicting private company failure: a multi-class analysis. J. Int. Finan. Mark. Inst. Money (2019)
22. Bequé, A., Lessmann, S.: Extreme learning machines for credit scoring: an empirical evaluation. Expert Syst. Appl. **86**, 42–53 (2017)
23. Dahiya, S.: A feature selection enabled hybrid-bagging algorithm for credit risk evaluation. Expert Syst. **34**(6), e12217 (2017)
24. Xia, Y., Liu, C., Da, B., Xie, F.: A novel heterogeneous ensemble credit scor-ing model based on bstacking approach. Expert Syst. Appl. **93**, 182–199 (2018)

25. Sun, J., Li, H.: Dynamic credit scoring using B & B with incremental-SVM-ensemble. Kybernetes **44**(4), 518–535 (2015)
26. Xu, X., Zhou, C., Wang, Z.: Credit scoring algorithm based on link analysis ranking with support vector machine. Expert Syst. Appl. **36**(2), 2625–2632 (2009)
27. West, D.: Neural network credit scoring models. Comput. Oper. Res. **27**(11), 1131–1152 (2000)
28. Zhou, L., Si, Y.W., Fujita, H.: Predicting the listing statuses of Chinese-listed companies using decision trees combined with an improved filter feature selection method. Knowl.-Based Syst. **128**, 93–101 (2017)
29. Malekipirbazari, M., Aksakalli, V.: Risk assessment in social lending via random forests. Expert Syst. Appl. **42**(10), 4621–4631 (2015)
30. Kodinariya, T.M., Makwana, P.R.: Review on determining number of cluster in k-means clustering. Int. J. Adv. Res. Comput. Sci. Manag. Stud. **1**(6), 90–95 (2013)
31. Yan, A., Nie, X., Wang, K., Wang, M.: Classification of Aurora kinase in- hibitors by self-organizing map (SOM) and support vector machine (SVM). Eur. J. Med. Chem. **61**, 73–83 (2013)
32. Ruppert, D.: The elements of statistical learning: data mining, inference, and prediction. J. Am. Stat. Assoc. **99**(466), 567 (2004)
33. Liu, F.T., Ting, K.M., Zhou, Z.H.: Isolation forest. In: Proceedings-IEEE International Conference on Data Mining. ICDM (2008)
34. Ben-david, A., Frank, E.: Accuracy of machine learning models versus "hand crafted" expert systems—A credit scoring case study. Expert Syst. Appl. **36**(3), 5264–5271 (2009)
35. Sohn, S.Y., Kim, D.H., Yoon, J.H.: Technology credit scoring model with fuzzy logistic regression. Appl. Soft Comput. J. **43**, 150–158 (2016)
36. Twala, B.: Multiple classifier application to credit risk assessment. Expert Syst. Appl. **37**(4), 3326–3336 (2010)
37. Luo, S., Cheng, B., Hsieh, C.: Prediction model building with cluster- ing-launched classification and support vector machines in credit scoring. Expert Syst. Appl. **36**(4), 7562–7566 (2009)
38. Yu, L., Yue, W., Wang, S., Lai, K.K.: Support vector machine based multi- agent ensemble learning for credit risk evaluation. Expert Syst. Appl. **37**(2), 1351–1360 (2010)
39. AghaeiRad, A., Chen, N., Ribeiro, B.: Improve credit scoring using transfer of learned knowledge from self-organizing map. Neural Comput. Appl. **28**(6), 1329–1342 (2017)
40. Huysmans, J., Baesens, B., Vanthienen, J., Van Gestel, T.: Failure prediction with self organizing maps. Expert Syst. Appl. **30**(3), 479–487 (2006)
41. Thomas, L.C., Crook, J.N., Edelman, D.B.: Credit Scoring and Its Application. Society for Industrial & Applied Mathematics (2002)
42. Hu, L., Gao, W., Zhao.K., Zhang, P., Wang, F.: Feature selection considering two types of feature relevancy and feature interdependency. Expert Syst. Appl. **93**, 423–434 (2018)
43. Sun, J., Lang, J., Fujita, H., Li, H.: Imbalanceenterprise credit evaluation with DTE-SBD: decision tree ensemble based on SMOTE and Bagging with differentiated sampling rates. Inf. Sci. **425**, 76–91 (2018)
44. Ying, Z., Habib, T., Chi, G., Uddin, M.S.: Real-world credit scoring: a comparative study of statistical and artificial intelligent methods. Int. J. Knowl. Eng. Data Mining **6**(1), 32–55 (2019)
45. Hand, D.J., Jacka, S.D.: Statistics in Finance. Arnold Applications of Statistics, London (1998)
46. Anderson, R.: The Credit Scoring Toolkit: Theory and Practice for Retail Credit Risk Management and Decision Automation. Oxford University Press, New York (2007). Thomas, L.C., Edelman, D. B., Crook, L. N. 2002. Credit Scoring and Its Applications Philadelphia: Society for Industrial and Applied Mathematics. (41)
47. Mays, E.: Credit Scoring for Risk Managers: The Handbook for Lenders. Thomson/South-Western, OH, USA (2004)
48. Thomas, L.C.: Consumer Credit Models: Pricing, Profit, and Portfolios. Oxford University Press, USA (2009)
49. Thomas, L., Oliver, R., Hand, D.: A survey of the issues in consumer credit modelling research. J. Oper. Res. Soc. **56**, 1006–1015 (2005)

50. Crook, J., Edelman, D., Thomas, L.: Recent developments in consumer credit risk assessment. Eur. J. Oper. Res. **183**, 1447–1465 (2007)
51. Thomas, L.: Operations research in consumer finance: Challenges for operational research. J. Oper. Res. Soc. **61**, 41–52 (2009)
52. Van Gool, J., Verbeke, W., Sercu, P., Baesens, B.: Credit scoring for microfinance: is it worth it? Int. J. Finan. Econ. **17**, 103–123 (2011)
53. Lewis, E.M.: An Introduction to Credit Scoring. Fair, Isaac & Co., Inc, California (1992)
54. Bastani, K., Asgari, E., Namavari, H.: Wide and deep learning for peer-to-peer lending. Expert Syst. Appl. **134**, 209–224 (2019)
55. Masmoudi, K., Abid, L., Masmoudi, A.: Credit risk modeling using Bayesian network with a latent variable. Expert Syst. Appl. **127**, 157–166 (2019)
56. Wang, Y., Xu, W.: Leveraging deep learning with LDA-based text analytics to detect automobile insurance fraud. Decis. Support Syst. **105**, 87–95 (2018)
57. Ala'raj, M., Abbod, F.M.: Classifiers consensus system approach for credit scoring. Knowl. Based Syst. **104**, 89–105 (2016)
58. Malhotra, R., Malhotra, D.K.: Evaluating consumer loans using neural networks. Omega **31**, 83–96 (2003)
59. Thomas, L.C.: A survey of credit and behavioral scoring: forecasting financial risks of lending to customers. Int. J. Forecast. **16**(2), 149–72 (2000)
60. Hand, D.: Modelling consumer credit risk. IMA J. Manag. Math. **12**, 139–155 (2001)
61. Greenspan, A.: Remarks by U.S. Federal Reserve Chairman Alan Greenspan. Speech at the American Bankers Association, Phoenix, Arizona (2002). http://www.federalreserve.gov/boarddocs/speeches//20021007/default.htm. Accessed 29 Jan 2013
62. Phua, C., Lee, V., Smith, K., Gayler, R.: A comprehensive survey of data mining-based fraud detection research. CoRR, abs/1009.6119 (2010)
63. Lee, H., Cho, S.: Focusing on non-respondents: Response modeling with novelty detectors. Expert Syst. Appl. **33**, 522–530 (2007)
64. Zhao, Y., Li, B., Li, X., Liu, W. Ren, S.: Customer churn prediction using improved one-class support vector machine. Adv. Data Mining Appl., 731–731(2005)
65. Haenlein, M., Kaplan, A., Beeser, A.: A model to determine customer lifetime value in a retail banking context. Eur. Manag. J. **25**, 221–234 (2007)
66. Baesens, B., Mues, C., Martens, D., Vanthienen, J.: 50 years of data mining and OR: upcoming trends and challenges. J. Oper. Res. Soc. **60**, 16–23 (2009)
67. Crook, J.N.: Credit scoring: An overview. Working paper series No. 96/13, British Association, Festival of Science. University of Birmingham, The University of Edinburgh (1996)
68. Chandler, G.G., Coffman, J.Y.: A comparative analysis of empirical vs. judgemental credit evaluation. J. Retail Bank. **1**(2), 15–26 (1979)
69. Capon, N.: Credit scoring systems: A critical analysis. J. Mark. **46**(2), 82–91 (1982)
70. Al Amari, A.: The credit evaluation process and the role of credit scoring: A case study of Qatar. Ph.D. Thesis, University College Dublin (2002)
71. Crone, S.F., Finlay, S.: Instance sampling in credit scoring: An empirical study of sample size and balancing. Int. J. Forecast. **28**, 224–238 (2012)
72. Atiya, A.F., Parlos, A.G.: New results on recurrent network training: Unifying the algorithms and accelerating convergence. IEEE Trans. Neural Netw. **11**, 697–709 (2000)
73. Fisher, R.A.: The use of multiple measurements in taxonomic problems, Ann. Hum. Genet. **7**(2), 179–188 (1936)
74. Green, S.B., Salkind, N.J., Akey, T.M.: Using SPSS for Windows and Macintosh: Analyzing and Understanding Data. Prentice Hall, New Jersey (2008)
75. Kamiński, B., Jakubczy, M., Szufel, P.: A framework for sensitivity analysis of decision trees. CEJOR **26**(1), 135–159 (2018)
76. Breiman, L. et al.: Classification and Regression Trees. Wadsworth, Belmont, CA (1984). Thomas, L.C.: Consumer Credit Models: Pricing, Profit, and Portfolios: Pricing, Profit and Portfolios, p. 48. Oxford University Press, Oxford (2009)
77. Cortes, C., Vapnik, V.: Support-vector networks. Mach. Learn. **20**, 273–297 (1995)

78. Bao, W., Lianju, N., Yue, K.: Integration of unsupervised and supervised machine learning algorithms for credit risk assessment. Expert Syst. Appl. **128**, 301–315 (2019)
79. Danenas, P., Garsva, G.: Selection of Support Vector Machines based classifiers for credit risk domain. Expert Syst. Appl. **42**, 3194–3204 (2015)
80. Zhou, L., Lai, K.K., Yu, L.: Least squares support vector machines ensemble models for credit scoring. Expert Syst. Appl. **37**(1), 127–133 (2010)
81. Broomhead, D.S., Lowe, D.: Multivariate functional interpolation and adaptive network. Complex Syst. **2**, 321–355 (1988)
82. Yen, J., Wang, L., Gillespie, C.W.: Improving the interpretability of TSK Fuzzy models by combining global learning and local learning. IEEE Trans. Fuzzy Syst. **6**(4), 530–537 (1998)
83. Jang, J.S.R.: ANFIS: adaptive-network-based fuzzy inference system. IEEE Trans. Syst. Man Cybern. **23**(3), 665–685 (1993)
84. Jang, J.S.R., Sun, C.T., Mizutani, E.: Neuro-Fuzzy and Soft Computing: A Computational Approach to Learning and Machine Intelligent. Prentice Hal (1996)
85. Tanaka, M., Okutomi, M.: A novel inference of a restricted Boltzmann Machine. In: International Conference on Pattern Recognition (ICPR2014), August, 2014
86. Hinton, G., et al.: Deep neural networks for acoustic modeling in speech recognition: The shared views of four research groups. IEEE Signal Process. Mag. **29**(6), 82–97 (2012)
87. Hinton, G.E., Salakhutdinov, R.R.: Reducing the dimensionality of data with neural networks. Science **313**, 504–507 (2006)
88. Hinton, G.E.: Training products of experts by minimizing contrastive divergence. Neural Comput. **14**(8), 1771–1800 (2002)
89. Tieleman, T., Hinton, G.: Using fast weights to improve persistent contrastive divergence. In: International Conference on Machine Learning (ICML). ACM, pp. 1033–1040 (2009)
90. Akkoç, S.: An empirical comparison of conventional techniques, neural networks and the three stage hybrid Adaptive Neuro Fuzzy Inference System (ANFIS) model for credit scoring analysis: The case of Turkish credit card data. Eur. J. Oper. Res. **222**, 168–178 (2012)
91. Shihabudheen, K.V., Pillai, G.N.: Recent advances in Neuro-fuzzy system: a survey. Knowl.-Based Syst. **152**, 136–162 (2018)
92. Goodfellow, I., Bengio, Y., Courville, A.: Deep Learning. MIT Press, Cambridge (2016)

Big Data, Deep Learning and Business Success

Sentiment Analysis of Arabic Sequential Data Using Traditional and Deep Learning: A Review

Thuraya. M. Omran, Baraa T. Sharef, and Crina Grosan

Abstract With the emergence of social media and review sites peoples express their opinions toward entities, generating a huge amount of data or what is called big data that comes in non structured form of sequential data such as tweets or reviews. The availability of big data leads to the excitement in Artificial Intelligence and many applications such as Sentiment Analysis (SA). Although many studies conducted in SA, however majority of them focused on English, while that consider the Arabic one are very limited due to many challenges like variation of dialects, morphological attributes, and the lack of Arabic sources and corpora, despite the spread of the Arabic language and its frequent use in social media. The objective of this review is to highlight different studies of Arabic sequential data that utilized traditional and deep learning techniques.

Keywords Big data · Sentiment analysis · Machine learning · Deep learning

1 Introduction

Social media have contributed in producing a huge amount of data through its contents like tweets, posts, comments, and reviews, either via providers of social media or via websites of reviews. These huge or a massive amount of data is called big data [1], which needs a processing application differs from that used for traditional data [2]. The availability of big data leads to the thrilling in the artificial intelligence (AI) field

T. M. Omran · C. Grosan
College of Engineering, Design and Physical Sciences, Department of Computer Science, Brunel University London, London, UK
e-mail: Thuraya.Omran@brunel.ac.uk

C. Grosan
e-mail: Crina.Grosan@brunel.ac.uk

B. T. Sharef (✉)
College of Information Technology, Department of Information Technology, Ahlia University, Manama, Kingdom of Bahrain
e-mail: baraa.nct@gmail.com

A. Hamdan et al. (eds.), *The Fourth Industrial Revolution: Implementation of Artificial Intelligence for Growing Business Success*, Studies in Computational Intelligence 935,
https://doi.org/10.1007/978-3-030-62796-6_26

and data analysis [1]. There are many applications of social media big data such as SA and trending discovery [1].

SA is a computational study of peoples' sentiments, opinions, attitudes, and appraisals toward entities such as services, products, events, topics, individuals, and organizations [3, 4], whether these opinions were favorable or not [5].

The general process of SA was presented by the researcher in four steps 1. data extraction 2. data preprocessing 3. data analysis 4. useful identification of knowledge (Bhuta et al. 2014; El-Masri et al. 2017; Zia et al. 2018) as cited in [6]. In most of the works of research, the term sentiment classification is used instead of SA to denote the step of data analysis of SA process (Singh et al. 2013; Liang and Dai 2013; Akaichi 2013; Antai 2014; Colace et al. 2013) as cited in [6].

There are many tools and techniques for treating the textual data of user's sentiments such as: Machine learning, Lexicon based technique [7] and hybrid technique [8].

The machine learning techniques were categorized by [8, 9] to: supervised, unsupervised, and semi supervised or hybrid.

Despite of different studies using machine learning methods, they are limited by the dataset size that affects the generalization process of machine learning approach, especially when applying the techniques of deep learning [10].

Deep learning is a powerful technique of machine learning by which the learning process occurs through multiple layers of data features representation to produce state of the art prediction results [3].

Although many studies conducted in SA, however, majority of them focused on English, while the one that consider the SA in Arabic contents are very limited, due to many challenges like variation of dialects, morphological attributes, and the lack of Arabic sources and corpora [10] despite the spread of the Arabic language and its frequent use in social media. The objective of this review is to highlight different studies of Arabic sequential data that used machine learning or deep neural networks, through gathering published studies according to Arabic SA challenges they addressed, which make the future researchers get benefit and direct their efforts to bridge available gaps in this field.

The rest of the paper is organized as follow: Sect. 2 gives theoretical background of SA. Section 3 describes some of the Arabic language characteristics and its affect on SA. Section 4 lists some of challenges of the Arabic SA. Section 5 presents the related works in the field of Arabic SA. Finally Sect. 6 concludes the chapter.

2 Theoretical Background

SA is a study field, of current researches related to social science and management, which interlocks with data mining, natural language processing (NLP) and text mining [2].

2.1 Basics of Sentiment Analysis

Previously, in order to make a decision regarding any product or service, it is done through groups' opinions polls or via surveys, but now with the presence of social media, and the other means of expressing opinions on the web like discussion forums or blogs, a need arises to automatically detect and summarize these opinions in intelligent way that can provide the polarity of textual data and can be done through SA system [5].

Thus when doing SA, it should be taken into consideration the meaning of terminologies and the differentiation between them such as; opinions and facts, opinion holder, subjectivity classification and sentiment classification [5], object, features and model.

2.2 Levels of Sentiment Analysis

According to [6, 9] SA can be studied at three levels: document level, sentence level, and aspect level.

- *Document level*: the piece of text is subjective and deal with one entity. This piece of text is assigned to a polarity class such as neutral, negative, or positive.
- *Sentence level*: where classification should occur for each sentence in two steps, in the first step, the sentence is classified as subjective or objective. In the second step, the subjective sentence is classified as negative, positive or neutral. This level of sentiment is very challenging because the sentiment depends on the context of sentence constituents.
- *Aspect level*: It is the fine grain level of sentiment, where two tasks take a place. The first task is the aspect extraction and the second task is the aspect sentiment classification [9] and entity extraction [6]. For example, the sentence "iphone camera resolution is great, but battery life is short", "the iphone" should be identified by the extractor as an entity, "the battery and camera resolution" should be identified as aspects, and the sentiment towards these aspects has to be classified by the classifier to positive and negative respectively.

2.3 Classifications of Sentiment Analysis Approaches

There are many tools and techniques that can be used for treating the textual data of user's sentiments such as: machine learning, lexicon based technique and hybrid technique [8].

1. Machine learning
   There are many definitions for Machine learning, differ in wording but ultimately lead to the same meaning; for example, [11] defined it as a field of developing two

types of algorithms supervised and unsupervised for the purpose of clustering, classifying or predicting, whereas [12] defined it as optimized performance using two types of data: example data and past experience data. The machine learning techniques were categorized by [8, 9] to: supervised, unsupervised, and semi supervised or hybrid.

- *Supervised*: set of covariates (xs) or features to predict the output(y), where there are observations with both x and y (training data), and the goal is to predict the (y) value in (test data) based on a given values of x. The observations are supposed to be independent [11]. Turban et al. [13] defined it as a learning type where the training data include both of class attribute (output variable) and the independent variables (features variables), supervised ML can be categorized into two methodologies: classification and regression. The classification is to distinguish the label of data via a model in order to predict the class of unknown object label [14], while it was defined by [15] as the most commonly technique used for predicting the target value from a given data that contain independent variables (features) and a dependent output attribute. Keeping in mind that the output result in classification process is always categorical value (label or class), while in the regression process the output is usually a numerical value. Supervised ML based on large labeled data that are used in training the classifier, for the purpose of creating a model that has the ability to predict the polarity of new texts. Examples of supervised machine learning are K-Nearest Neighbor (KNN), Naïve Bayes (NB), Artificial Neural Networks (ANN), and Support Vector Machine (SVM) [9]. This approach is characterized by the need of more efforts to be made for annotating collected data
- *Unsupervised*: machine learning with no explicit target value, which means that the label of classes is unknown. The learning occurs using clustering where objects that have same properties are gathered or collected in one group [15]. Machine learning input represented at unlabeled data, where the hidden structure is discovered using certain algorithms [8]. Unsupervised ML uses lexicons, where the polarity scores as negative, positive or neutral is assigned to each word. The lexicon could be created using corpus or from existing dictionaries. The unsupervised learning model needs a wide coverage of sentiment words instead of labeled data which was the case in supervised model [9].
- *Semi-supervised*: machine learning that deal with both of supervised and unsupervised techniques. According to [12] it is a learning method that learns knowledge from both labeled and unlabeled data.

2. Lexicon based technique of SA uses a predefined list of words, each word is accompanied with a specific sentiment. To clarify more, it can be said it is a way for determining the polarity via a matching between opinions words in a dictionary of sentiment and data set [16].

3. Hybrid technique of SA combines lexicon and machine learning approach in order to have the potential to improve the performance of sentiment classification [16]. Hybrid or semi supervised ML uses less annotated data and large amount of unlabelled data. The classification process based on using supervised model that is trained on labeled data in classifying the unlabeled data [9]. Lexicon based technique can be categorized into dictionary based and corpus based approaches. According to [17], the dictionary based approach involves a set of sentiment words that are collected in a manual way in order to be like a seed, and then the set of these words is expanded by adding the antonyms and synonyms of the seed words using well known thesaurus or dictionaries and robust resources. One disadvantage of the dictionary based approach is its inability to determine the context or domain of the opinion words which is overcome by corpus based approach which identifies the context of specific opinion words based on the pattern of words co-occurrence and syntactic in the text using linguistics constraints. Finally, the hybrid technique, which combines both of machine learning and lexicon based techniques, has the potential to improve the classification performance, while it has the limitation of assigning neutral score for noisy review because of the inability of sentiment detection [16].

2.4 Sentiment Analysis Process

2.4.1 Text Pre-processing

In text analysis, there are certain essential steps that should be followed in order to extract useful information from data especially the unstructured ones, whose content is full of noise like special characters, redundant and meaningless words, etc. In order to get cleaned useful data, the dataset should pass through a pre-processing process that contributes in reducing the overall time of processing process. According to [18] the type of pre-processing process depends on the source of collected data, for example if the collected data are tweets, extra pre-processing will be needed to remove some parts that compromise the tweet body like hash tag, mentions, links and retweets. Some of the common pre-processing techniques are:

- Tokenization: where the text is divided into chunks like symbols or words called tokens. These tokens will be fed to further process for the purpose of mining or parsing. As stated by [19]: for example the sentence (most of our high school teacher are working in a hard way) after the tokenization process will be ('most', 'of', 'our', 'high', 'school', 'teacher', 'are', 'working', 'in', 'a', 'hard', 'way').There are so many open source tools for tokenization that were compared by [20] such as textblob word tokenize, pattern word tokenize, MBSP word tokenize, mila tokenizer, Nlpdotnet tokenizer and word tokenization with python NLTK.
- Stop word removal: it is a procedure to remove all useless and meaningless words for the purpose of text classification [21] for example words like 'are','a' 'the'.

- Converting text to lower case in order to calculate occurrences of the similar word.
- Stemming: it is an information retrieval and linguistic morphology term used for reducing the inflected or derived words (adverb, verb, adjective, noun) to their root or stem [19] by removing the Affix and minimizing the number of words to match their stem in accurate way [22] for the purpose of reducing the processing time of final input of any query and increasing the number of retrieved documents in the systems of information retrieval (IR) [23]. There is a stemming language like "Snowball" and many stemming algorithms which are classified by [23, 24] into Truncating (Removal of affix), Statistical and Mixed, with subcategories.
- TF-IDF: it is an approach used for constructing a feature vector; it stands for Term Frequency-Inverse Term Frequency. In other words, it is a technique used to transform the information of textual shape to a sparse feature or a vector space model (VSM). The VSM is an algebraic method used to represent the textual information as a vector of numbers that reflect the features extracted from the text or document, but how can it be done? The first step to answer this question is to remove all stop words, and then create a dictionary (index vocabulary) of the terms that present in the document by converting them to a dimension in VSM [25].

2.4.2 Text Representation

In natural language processing, text representation plays a major and important role, according to [26], it maps the various length of documents, paragraphs and sentences to fixed length vectors, which in turns their quality affect the quality of model's performance. The most commonly used models of text representation are bag-of-words and n-grams which represent the baseline for recent researches due to their robustness and simplicity.

Bag of words represents the text as a fixed length vector, where the grammar and the order of words are ignored, while in n-grams model the word and the consecutive ones are considered. It is agreed by [27, 28]. They stated that N-gram model is a strategy or method for checking persistence of n words or sounds of a given text or speech used for predicting the following thing in grouping.

In order to analyze the text for SA, the n-gram model is used to achieve this purpose. Unigram represents n-gram of size 1, Bigram represents n-gram of size 2, and Trigram represents n-gram of size 3, and so on.

In 2018 Zhang et al. presented another method of text representation through word embedding. Word Embedding is a mechanism used for modeling of language and feature engineering. It transforms text vocabulary or words to vectors of continuous numbers. The idea of word embedding based on embedding from high dimensional high sparse space to lower dimensional dense space of vectors [3]. The learning of word embedding can be achieved using matrix factorization or neural network, one example of word embedding techniques is Word2Vect [3]. The Word2Vect representations are used for semantic word distribution, where every word in text is represented in a vector form. The vectors that are closed together are treated as similar words [29].

The Word2Vec has two architectures: Skip Gram (SG) and Continues Bag of Word (CBOW). The skip gram is used to predict the surrounding text from a given word, whereas the CBOW is used to predict the target word given a context within specified length of window as [29].

2.4.3 Feature Extraction

Feature extraction is defined by [30] as a process of transforming features with high dimensional space to a feature space with low dimensions, by applying mapping process. All of [30–32] listed some of the most popular techniques such as linear discriminant analysis (LDA) and latent semantic analysis (LSA).

2.4.4 Feature Selection

Feature selection is important to get a reduced feature subset in size that is characterized by relevant and irredundant features, which contributes in getting accurate data models with enhanced learning efficiency, increased predicting accuracy and reduced complex learned results. In another definition that is more precise by [33], it is a combination of search technique and evaluation scale, the search engine for suggesting new features subset and the evaluation scale to score the various feature subsets.

3 Arabic Language in Sentiment Analysis

Arabic language is ranked as the fourth predominantly language used in the web, it is one of six official foremost languages in the united state, while it is one of the official languages of 27 countries [9]. Arabic language is a Semitic language [10]. According to [9, 10], it has three forms 1. Classical: is the written language of Quran (the holy book of Islam) 2. Modern Standard Arabic (MSA) the most universal variety of used in formal communication, speech, and writing newspaper and books. 3. Dialectical is the form used daily in written or spoken communication between most of the individuals. Arabic dialects differ from one country to another and from region to another in the same country [9].

Dialects are more understood when it is annotated with diacritics that indicate the meaning clearly. For example the word that means felt (شَعَرَ) has the meaning of hair when it is written like (شَعْر) or the meaning of poetry when it is written like (شِعْر). According to [10], Arabic language is composed of 28 letters, used in writing from right to left, each letter has different written form depending on its position in the word, at the end, middle or at the beginning.

Arabic language is characterized by many aspects of morphology such as agglutination, inflection, and derivation which results in complex structure of Arabic words, which represents a main difficulty in Arabic SA for the researcher [9].

4 Challenges of Arabic Sentiment Analysis

Detecting of sentiment orient in a text depends on the presence of positive and negative words that express the sentiment. The identification of these words faces challenges. They are categorized by [34] to linguistic and contextual challenges. The Linguistic challenges such as:

- Diacritics marks absence.
- Agglutination to the complex morphology of Arabic language.

While the contextual challenges such as:

- Comparing between two things or persons.
- Mismatching between the opinion polarity analysis and sentiment words score.
- Changing of word polarity depending on the domain.
- Availability of negation words which alter the polarity of the words that come after it.

While [6] categorized other successive challenges to a processing of the languages as;

- Stemming in lexicon based approaches.
- Polysemy in corpus based approaches.

Training of word embedding, extracting features, creating lexicons and building model, are all challenges within the languages especially when the data resources are limited in a targeted language [35] as Arabic, where the annotated data are rare, when compared to the English language [36].

Baly et al. [37] had mentioned that exerted efforts to create a dataset, especially on tweets, are very limited to certain dialects like Jordanian and gulf. They also added, the diversity in a culture between Arab countries makes the trained model of sentiment to be applicable on dialects of one region while not applicable for the other region.

Soufan [36] and Baly et al. [37], added the nature of the dialects, used in social media specially twitter, is un standardized. Factors as using of arabizi, special characters like mentions, hash tag, URL, miss spelling of words which abide to the restricted length of tweets, may impact the sentiment in implicit way, particularly when modeling text semantics.

Further details are invoked by [36, 38, 39] regarding MSA and Arabic dialects' challenges as, one word has multi and various meaning which differs from one dialect to another. The morphology property, grammar rules, Arabic diacritics, affixing, suffixing, prefixing, and position of the word in the sentence, all of which result in different meaning for the word. The nature of the sentence as being verbal or nominal also represents a challenge. All these challenges and difficulties contribute in changing the opinion polarity.

Elnagar et al. [40] mentioned that names and idioms also represent a challenge in Arabic SA. For example the name 'سعيد' that means 'happy' in English may leads to a trigger of false positive sentiment.

More open issues and challenges are listed by [39] as follows:

- Translating mechanism of a figurative language while keeping its actual essence.
- Detecting sentiment in irony, sarcasm and figurative expression.
- Identifying euphemism, hyperbole, and metonymy.
- Setting a mechanism for understanding the rhetorical questions.
- Polysemy and synonymy.

In their survey, [6] inferred more challenges such as:

- The small size of manual Arabic resources and datasets give accurate and best results in SA, which is not adequate for machine deep learning because of its small size.
- Most of the works in reviewed approaches were in document level and sentence level, where aspect level seems as unexplored.

5 Related Work

In the literature, there are many studies arisen in the field of Arabic SA, although they differ in their goals, they ultimately aim at improving the classification of sentiment by finding solutions to the difficulties and challenges facing the process of SA, which vary in using approaches and learning methods. This section is devoted to present some of these studies, which tackle some of the challenges mentioned in the previous section; such as parsing of a language and morphological complexity, negation handling, and scarcity of Arabic resources, in addition to the classification approach, the dataset utilized with each approach, and the evaluation metrics.

5.1 Language Parsing and Morphological Complexity

Some of the conducted studies dealt with the parsing of a language and morphological complexity, like the ones carried out by [29, 41, 42].

In the study carried out by [41] four different architectures were explored, three of them based on Deep Auto Encoder (DAE), Deep Neural Network (DNN) and Deep Believe Networks (DBN), while the fourth one based on Recursive Auto Encoder (RAE) used to compensate the loss of text handling in the three models.

Each model fulfills certain tasks, DBN for applying a generation phase of pre training before feeding step, DAE for generating a reduced dimensionality of representations and RAE used for parsing words in best order that reduces the error of reconstruction the same word order, i.e. giving the best parsing tree. RAE and DAE are unsupervised technique of learning that used for providing compact presentation of input sentences. DAE parses the words of whole sentences at once, while RAE considers the order and context of sentence parsing.

It is worth to mention that [41] focused on sentence level of sentiment classification of Linguistic Data Consortium Arabic Tree Bank (LDC-ATB) dataset which was used to evaluate the proposed models. It is also worth noting that ArSenL lexicon sentiment scores were used for features vectors.

The results findings showed the better representation of the input sparse vector was obtained by DAE, while the best F1-score was obtained by RAE with improvement of 9% compared with the literature models.

In 2107 a recursive deep learning model was presented by [42]. It was called "AROMA", i.e. (A Recursive deep learning model for Opinion Mining in Arabic). The AROMA was proposed to tackle the limitations and challenges of the RAE model in performing the SA in Arabic such as parsing of language and morphological complexity. The AROMA proposed model addressed some challenges as implementing morphological tokenization, and modeling semantic composition at morpheme level.

In order to model the sentiment and semantic composition [42] used a parser of phrase structure to generate a tree of syntactic parsing. To evaluate the proposed model, different types of datasets were used such as (tweets, online comments extracted from Qatar Arabic Language Bank (QALB) and newswire extracted from Arabic Tree Bank (ATB) with different styles of Arabic writing dialects and standard.

The experiments results showed that the proposed model outperformed 1. RAE model in tweets, QALB and ATB regarding the accuracy metric improvement by 7.2%, 8.4% and 12.2% respectively. 2. The literature models on same dataset (tweets, online comments extracted from Qatar Arabic Language Bank (QALB) and newswire extracted from Arabic Tree Bank (ATB) by 7.6%, 1.7% and 7.3% respectively.

Other authors like [29] used ensemble model of LSTM integrated with CNN. [29] study objectives were: 1. Investigating the benefits of combining the mentioned models and reporting the obtained results using various Arabic datasets. 2. Considering the morphological variance of certain words utilizing different levels of sentiment classification.

In their study, [29] used four datasets where one dataset is a subset of another. These datasets are: 1. Arabic Health Service Dataset (AHS), unbalanced dataset contains 2026 tweets (628 positive and 1398 negative). 2. Sub dataset of (AHS) contains 502 positive tweets and 1230 negative tweets. 3. Twitter Dataset (Ar-Twitter) contains 1000 positive and 975 negative tweets covering different scopes and topics in Arabic such as arts, communities and politics. 4. Arabic Sentiment Tweets Dataset (ASTD) contains a total of 2479 tweets, 795 positive tweets and 1684 negative ones. For each dataset, there is three different levels of SA: character level for the purpose of increasing feature number per tweet, character N-gram level (ch 5 gram-level) and word level.

The proposed model of [29] is composed of multi layers: input layer, convolutional, Max-pooling, LSTM, a fully connected layer and the output layer. The input layer is represented by a matrix of fixed dimension with different vector embedding based on the level of SA. In order to produce a feature map, filters slide over the matrix in the convolutional layer, where different features could be obtained using various sizes of the filters. The max pooling computes the maximum value and assigns it as a feature to a certain filter. The Maximum pooling layer will feed the LSTM network, which in turns gives the final output as negative or positive.

The results obtained by [29] model showed that the performance accuracy has improved by achieving 0.9424 regarding (AHS) dataset and 0.9568 regarding sub AHS dataset compared with other previous models that achieved 0.92 for (AHS) and 0.95 for the sub dataset.

5.2 Negation Handling

Another issue of the SA is the negation handling, it was tackled in works done by [34, 43, 44].

In the work of [34], a supervised approach was developed for the purpose of considering the negation context. To achieve their objective and to cope with the scarcity of Arabic resources in lexicons, [34] constructed new lexicons composed of words of existing resources integrated with a translated one. In addition, [34] used a corpus, consisted of reviews covered four scopes (restaurants, movies, productions, and hotels).

A preprocessing process was applied to the corpus and the lexicon. Eliminating diacritics, elongation, numbers, and non Arabic words were the basic filtering elements for the corpus, followed by stemming and lemmatizing steps, which resulted in three different versions of the corpus: a version composed of words, another composed of stems, and a third composed of lemmas. While the preprocessing process of lexicons represented at: 1. removing numbers, words that are redundant, dialectal, and non Arabic. 2. excluding the redundancy between trigrams, bigrams and words while keeping the unigram. 3. excluding the negation words in the trigrams and the bigrams, changing their polarities and moving them to the positive lexicon. Similar to the corpus, a stemming and lemmatizing process was applied to the lexicon, which resulted in a stemmed and lemmatized versions of lexicons that are specific in the similar domains of the corpus, namely (restaurants, movies, productions, and hotels).

The supervised approach, developed by [34] gave the attention to the context of negation, hence a list of negation words formed with its stemmed and lemmatized versions.

In their supervised approach, [34] associated a feature vector with each review, and applied the SVM to build the model. Since the model performance is influenced by the feature selection process, four types of feature selection were applied namely: word based, stem based, lemma based, and class semantic based each one of them with certain characteristics.

Six vectors of characteristics were tested, one of which was the one that considered the negation context, which gave a value of 93.21, 84.14, 82.96 and 83.68 in accuracy metric per the hotels, the productions, the movies and the restaurant domains respectively. It is worth to mention that more experiments were conducted by [34] to study the influence of lexicon variations and model types.

For the purpose of improving SA in Arabic, [43] proposed a combined approaches to extract opinions from online reviews of movies. In order to achieve their objective, [43] 1. used different n-grams models including skip-ngram scheme. 2. Investigated the significance of Part-Of-Speech (POS) tagging in exploiting subjective words like verbs, nouns, and adjectives. 3. Extracted reviews summaries and opinion conclusion using lexicon. 4. Combined approach 2 and 3 utilizing a voting rule.

Brahimi et al. [43] built their data set by collecting 1000 reviews of movies from both of cairo and el cinema websites, in a period spanning from 24 April to 21 June 2016. These reviews are in MSA and dialects. The collected dataset is called Arabic movie review dataset (ARMD).

Before applying their proposed approaches, [43] performed preprocessing steps by removing unrelated text and noise, followed by a tokenization process. They applied their first approach of the n-gram word utilizing bigrams of words to tackle the negation and intensification problem. In addition a one skip-bigram is also used to compensate the information missed by the bigrams.

In the second approach [43] applied three parts of speech tags (verbs, adjectives, and nouns) added to the usage of bigram representation. To reduce the number of features, TF-IDF was used to remove the features that occur rarely in the dataset. The SVM classifier was used in both of the first and second approach.

One of the contributions tasks in the study of [43] is the generation of summary of sentiments or opinion conclusion by concentrating on the summarization of reviews which can be achieved in two methods namely 1. Producing opinions related to the aspects and features of target entity. 2. Extracting the useful text of the reviews. In their experiments [43] considered two states: the first state for identifying the opinion conclusion through words of evaluation like "bad movie" or bad. The second state is extracting the opinion conclusion by seeking the target words that explicitly available in the comment using a list of words such as (series and movies). The Nile University's Arabic lexicon (NiLUlex) was used to classify the sentiment summary and conclusion of opinion.

To improve the sentiment prediction, [43] hybridized the approach of sentiment summary and that one of the POS tagging at classification level and obtained the overall opinion using the rule of voting. To evaluate the classifier performance [43] employed 10 folds cross validation and f1-measure to measure the classifiers performance. Brahimi et al. [43] applied their approach on two datasets, the ARMD they collected and opinion corpus of Arabic (OCA) that is publically available. According to the results obtained, F1-measure achieved the highest value at 91.98% when the

POS model are adjective, noun and verbs with bigram using OCA, while the highest F1-measure value was 84.64% with identical previous model using the ARMD.

5.3 *Scarcity of Arabic Resources*

The scarcity of Arabic resources represents another challenge that faces the SA, which is somewhat treated by [10, 37, 38, 44].

Algburi et al. (2019) conducted a study to assess the performance of three machine learning algorithms, using a dataset they have created from different sources covering various domains, for the purpose of addressing the gap of the scarcity of Arabic sentiment dataset.

The research methodology followed by [44] represented at: data collection, data preprocessing, feature extraction, specifying the classification algorithm, and finally the evaluation of the classifier performance.

The collected dataset by [44] including 36,000 sentences in different domains such as movie reviews from elcinema.com, hotel reviews from TripAdvisor.com, restaurant reviews from qaym.com, whereas product attraction review and product reviews from amazon.ae, in addition to tweets collected using Twitter API. All the reviews and tweets are unannotated. (Algburi et al. 2019) did not mention the way of their dataset annotation.

The processing steps included 1. Tokenization process, 2. normalization by removing unnecessary symbols from the words, 3. removal of stopwords such as (هو , حتى, عن, لما, هذا, عليه,من), 4. stemming by returning the words to their roots.

The feature extraction methods used by [44] were the unigram and bigram, whereas the Ngram was used to handle the negation issues.TF-IDF was also used in [44] to measure the importance of words in the dataset. Three learning algorithms were used namely: Support Vector Machine (SVM), Naïve Bayes (NB), and Decision Tree (DT).The evaluation metrics in the experiments were the accuracy, precision, recall and F1-measure.

The results obtained indicated that SVM outperformed the other classifiers, 0.96% in accuracy while it was 0.99% in each of precision, recall and F1-measure.The results obtained achieved by NB was 0.96, 0.97, 1.00, and 0.98% in accuracy, precision, recall, and F1-measure respectively, where the lowest performance was by DT which achieved 0.82% in accuracy, 0.88% in precision, and 0.87 in both of recall and F1-meausre.

Algburi et al. (2019) concluded that in order to help the machine learning algorithm to do appropriate classification of words that have the same letters and different meaning (orthography property), the classifiers should be provided with a correct set of morphological features. It is worth to mention that the sentiment in [44] was at sentence level.

Mohammed and Kora (2019) constructed a corpus composed of 40 k tweets and compared the achievement of three deep neural networks namely convolutional neural network (CNN), long short term memory (LSTM), and recurrent convolutional neural network (RCNN). The 40 k tweets are mixed of Egyptian dialects and MSA. The 40 k tweets were collected using twitter API in a period spanning from 11 April to 12 December 2015, covering multi topics such as proverbs, poetry, sarcastic jokes, social topics, politics, health, sports, opinions about products. The collected tweets were filtered, processed, annotated manually, and validated by two experts.

The manual filtering and processing include removing of: spam tweets, duplicated tweets, gulf Arabian countries tweets, words' diacritics, and elongation. In addition to, replacing different forms of letters by one form, adding space to combined words, and correcting process for wrongly written words or words with missed letters.

For the purpose of classifying the corpus tweets to positive and negative, [10] proposed an approach of three deep learning models namely CNN, LSTM, and RCNN. The RCNN is a neural network that contains the LSTM as a layer with the layers of CNN. The CNN used for extracting features in a strong way, while the LSTM layer takes the role of memorizing and applying the architecture of recurrent neural network on extracted features.

In the proposed approach, [10] used Aravec which is a pretrained CBOW model for generating the matrix of word embedding. (Mohammed and Kora, 2019) used three different data splits (60%, 40%), (70%, 30%), and (80%, 20%) on each model for training and testing purpose. 10% was considered as a validation data for tuning the model hyper parameters. The evaluation metrics were f1-score, precision, recall, and the accuracy. Data augmentation with shuffling was used for the objective of changing words order randomly in a small window of text sequence. The obtained results showed that the LSTM outperformed the other two models by achieving an accuracy of 81.3% compared to 78.46% and 75.72% for accuracy measure achieved by RCNN and CNN respectively. The LSTM also achieved the highest accuracy of 88.05% when applying the technique of data augmentation.

Alahmary et al. [38] created Saudi dialect dataset and called it 'Saudi Dialect Corpus from Twitter' (SDCT), for the objective of enhancing the sentiment analysis at sentence level of Saudi dialect using an approach of deep learning. 60,000 tweets from different scopes were collected using twitter API, only 32063 tweets were extracted to create the dataset. In order to classify the tweets manually, an annotation process took place which results in 17707 positive tweets and 14356 negative one.

After the annotation process, a preprocessing process started as follow: a-Removing all special symbols like (#, &, %, $), diacritics, punctuations, single Arabic letters, and the letters which are non Arabic. b-Normalization by replacing multi variant of (آ,إ,أ) letter to be one letter (ا), (ة) by (ه), and (ئ, ي) by(ى). c- Repeated characters are replaced by one, for example شكرررررا by (شكرا).

To perform sentiment analysis of Saudi dialects [38] employed 1. CBOW which is a model of Word2Vec for the purpose of learning the words vector representation in unsupervised way. 2. LSTM and Bidirectional LSTM (Bi-LSTM) are deep learning models in supervised manner. 3. SVM for the purpose of evaluating and comparing its performance with LSTM and Bi-LSTM.

For the aim of exploring of how sentiment analysis of the Saudi dialects could be enhanced using deep learning, [38] presented their conducted experiment as follows: a-Training of the CBOW in an unsupervised way using gensim library of python, and feed it to the Bi-LSTM via TensorFlow.b-Selecting 70% of the dataset as training part, while 30% as testing. c-Utilizing SVM after applying vectorization using CountVectorizer and TF-IDF.

The obtained results showed that Bi-LSTM outperformed the other classifiers in terms of accuracy, which was 94% where as the accuracy of the LSTM was 92 and 86.4% for the SVM.

Baly et al. [37] created Multi Dialect Arabic Sentiment Twitter Dataset (MD_ArSenTD)

(MD_ArSenTD) is composed of tweets from 4 regions (Gulf, Levant, North Africa, and Egypt), Both of Gulf and Levant tweets covered 4 countries. Gulf tweets covered (Kuwait, KSA, Qatar, and UAE), where Levant tweets covered (Jordan, Lebanon, Palestine, and Syria).The covered countries of North Africa were (Algeria, Morocco, Tunisia), and Egypt, with no specification regarding the involved countries, resulting in a total of 12 covered countries.

Tweets in the (MD_ArSenTD) were assigned sentiment labels using a scale of (5) points, for the purpose of providing information of intensity beside the polarity.

Baly et al. [37] focused in describing the characteristics and specificities such as discrete feature and structure of Egyptian and UAE tweets, and highlighting the discussed topics in tweets of both countries.

The (MD_ArSenTD) was created by the following steps: 1. Retrieving of tweets: 470 k of tweets from 12 countries was collected using Twitter4J API, through a period spanning from first of March 2017 till the end of April 2017.To enforce the retrieving of the tweets from the 4 regions' countries, a specific-gio locations were used. 2. Selection of the tweets is as follows: a- A target size of 14400 for the (MD_ArSenTD) was set. b- the number of tweets per country was set to 1200. c- Removing all duplicated tweets and that ones with less than 30 characters. d- Applying a pre trained model which won SemEval-2017 task4 for the remaining of tweets. e-the top 1200 tweets with high confidence which were predicted as negative, positive, and neutral were selected, for the purpose of decreasing irrelevant tweets. 3. CrowedFlower was used by [37] in annotating the selected tweets for both of dialect and sentiment. Regarding the sentiment notation, the annotators were asked to use the polarity of 5 points scale represented at (very negative, negative, neutral, positive, and very positive). Regarding the dialect notation, the annotators were guided to identify the country and region for each country, otherwise choosing foreign language or MSA.

The performance of the annotators was monitored by using a test of gold set of 100 tweets per country, and calculating Kohen's Koppa which was 0.65 in sentimental annotation, and 0.8 in the annotation of region level dialects.

Two types of models were in the focuses of [37] that are: 1. Feature engineering by evaluating equivalent model to the one that won SemEval-2017 task4, who trained the SVM with a group of hand crafted features that covered semantic, syntactic and surface information. The evaluated model extracts features such as emoticons, presence, URL, mentions of user, ngram of lemma, counting of: POS, both of questions and exclamation marks, negated context, positive and negative emoticons. 2. Deep learning, throughout evaluating LSTM, using SKIP_GRAM model of word2vec to generate the input feature.

Baly et al. [37] used two types of embedding, specific and generic, for the objective of assessing how the sentiment analysis is influenced by the dialect.

An effective approach was proposed by [45] to show that classification ability based on automatically annotated dataset containing emojis outperforms the one annotated manually. So they collected 134,194 emotional Arabic tweets using twitter trending hash tag to scan the tweets for more frequent emojis, which categorized later on to only sadness, joy, disgust or anger. After categorizing emojis, the tweets of the dataset were labeled according to the emojis contained in it. For example if the tweet contains ❤❤😭 then the category of the tweet is ($2*3 = 6 - 4 = 2$) which is joy. (Hussien et al. 2016) Collected 122,000 tweets using crawler, and extracted 22,752 ones that have emojis to be annotated automatically. The extracted tweets fall into the mentioned categories as 7878 tweets of sadness, 2874 anger tweets, 1533 disgust and 10,467 joying ones.

2025 tweets, that are free from emojis, are categorized manually to 450 sadness tweets, 620 anger, 360 disgust and 630 joys. These manually labeled data are divided into 80% as a training part and 20% as a testing. It is worth to mention that the same testing part was used by the classifiers in both approaches the automatic and manual.

After collecting the dataset, it passes through a preprocessing steps and feature extraction method using bag of words (BOW), in addition to, term frequency-inverse document frequency (TF-IDF).

The machine learning algorithms, support vector machine (SVM) and multinomial naïve base (MNB) were trained for the purpose of producing two models: 1. trained classifier for manually labeled data (TCMLD) and 2. trained classifier for automatically labeled data (TCALD). The classifiers were implemented in WEKA data mining libraries. fivefold cross validation and average f1 measure, precision and recall were calculated to evaluate the classifiers performance. The findings of the experiments showed the automatically labeled dataset classified by the SVM and the MNB obtained higher values in all metrics, specifically 4% achieved by the SVM regarding f1 measure than the findings obtained with manual labeled one. The MNB achieved values of 0.757, 0.7256, and 0.7534 in precision, recall, and f1 measure respectively.

In a distinctive way, [2] carried out a study using in-house designed and constructed tool called Arabic Opinion Polarity Identification (AOPI) for the purpose of; 1. identifying the polarity of collected dataset 2. comparing the effectiveness of AOPI with already another two existed SA tools (SentiStrength and SocialMention). Using a crawler [2] collected 3015 opinions covering three domains (weather, food, and sport) from Facebook pages to create a balanced benchmark dataset.

After collecting the data, 10 lexicons were created by [2], two lexicons (negative, positive) for each domain, two lexicons for positive and negative emotions, in addition to, two general lexicons. The in-house tool (AOPI) was designed and developed by one author of [2] to classify the polarity of input data in three ways (positive, negative, or neutral), whether the input data was in dialect or Modern Standard Arabic (MSA) form.

After Tokenization process, the preprocessing steps followed by [2] were as follows: 1. Removal of Arabic stopwords. 2. Normalization by converting some forms of letters to other. For example (أ, إ, آ) are converted to bar alif (ا), (ى) is converted to (ي), TaaMarboota (ة) is converted to (ه). 3. Removal of consonants and all diacritical symbols such as Damma, Fatha, Kasra, and sukuun. A Term Frequency (TF) was in AOPI for searching similar words in the lexicons.

To evaluate the AOPI, [2] conducted two parts of experiments. In the first part, they presented the results of the other two existed online SA tools the SentiStrength and SocialMention using the original and a stemmed of their collected dataset, applying the accuracy metric. The results provided indicated that the SentiStrength tool was unable in identifying the polarities of Arabic dialects opinions, while the second tool the SocialMention was only able to assign polarity to just 50% of opinions, due to the availability of dialect percentage that exceeds 55%.

In the second part of experiments, two versions of a collected dataset were used, the first version is the original dataset and the second is a stemmed one by light stemmer. In addition to, running AOPI tool two times, one is with specifying the review domain while, the other is without specifying the review domain. The classification algorithms were SVM, K-Nearest Neighbor (KNN), and NB.

The prediction metrics are: True Positive (TP), True Negative (TN), False Positive (FP), and False Negative (FN) of Receiver Operating Characteristics (ROC), in addition to F1-Measure, recall, and precision. The results showed that supervised learning model SVM achieved the best results as follows: 1. when the domain is not specified, the SVM achieved 53.561% using the original dataset, and 53.9745% for the stemmed dataset. 2. When the domain is specified, the SVM yields 75.882, 52.1386, and 65.970% accuracy in the domains of food, sport, and weather respectively using the original dataset, whereas its achievements were 76.3314, 53.4483 and 70.8709% in food, sport, and weather respectively using the stemmed dataset.

By over viewing the current review, it can be discerned that most researchers agree that deep learning algorithms such as CNN, RNN, and LSTM, give better results especially when large dataset are available, while state-of- the-art results can be achieved with traditional algorithms when using SVM or NB.

Another thing to be mentioned that in most of the current review studies, there was no mentioning to the software applied during the implementations, and no validation set were applied, despite its significance role in the accuracy of prediction process of classification (Table 1).

6 Conclusion

Nowadays, the internet provides wide horizons for people, to express their opinions, through social media, blogs, and review sites. These opinions generate big sequential data, which require the presence of sentiment analysis systems. The sentiment analysis systems are represented in traditional or deep approaches, whether they were supervised, unsupervised or hybrid approaches. The presence of abundant

Table 1 Summary of Arabic sentiment classification works

Refs.	Tackled challenges	Datasets/lexicons	Algorithm	Performance metrics
[41]	Morphological complexity, Language parsing	LDC-ATB/ ArSent lexicon	DAE, DNN, DBN, RAE,	Acc: 0.74 F1: 0.73
[42]	Morphological complexity, Lexical sparsity	QALB, ATB	AROMA	Acc: 0.86 F1: 0.84
[29]	Morphological complexity	AHS, Ar-Twitter, ASTD/lexicons	LSTM, CNN	Acc: 0.95
[34]	Context negation	Reviews cover different domains/lexicons	SVM	Acc: 0.93
[43]	Context negation, Intensification problem	Reviews	SVM, Lexicon based	F1: 0.91
[10]	Scarcity of Arabic resources	Tweets		Acc: 81.3
[38]		Tweets	SVM, LSTM, Bi-LSTM	Acc: 0.94
[37]		Tweets	SVM, LSTM	Acc: 0.70 F1: 0.69
[34]		Reviews cover different domains/lexicons	SVM	Acc: 0.93
[44]	Context negation, Orthographic property, Scarcity of Arabic resources	Tweets, Reviews	SVM, NB, DT	Acc: 0.96 Pre: 0.99 Rec: 0.99 F1: 0.99
[45]	Aannotation	Tweets	SVM, MNB	Pre: 0.757 Rec: 0.7526 F1: 0.7534
[2]	Stemming	Facebook comments	SVM, KNN, NB, Lexicon based	Acc: 0.76

corpora and lexicons play a fundamental role in SA, particularly for a language full of challenges, complex, and morphological such as Arabic.

This review paper highlighted different Arabic studies, that in general aimed improving sentiment analysis, while tackling some challenges of sentiment analysis within mostly used forms of Arabic language namely dialect and modern standard Arabic (MSA). At the same time this review paper shows that Arabic SA attracted many researchers and became the focus of their attention, which can be noticed by good obtained results, using different learning approaches whether they are: 1. traditional techniques such as SVM, NB, and KNN, that can be supervised by using labeled datasets or unsupervised using semantic lexicons or combination of both.

2. Deep learning techniques like RNN, CNN, LSTM, and DBN which require an availability of ample large resources that needs perfect annotations, comprehensive preprocessing, and validated adjustment of training parameters to end with state of the art models. However more studies are needed to be conducted to tackle other challenges of Arabic SA.

References

1. Ghani, N.A., Hamid, S., Hashem, I.A.T., Ahmed, E.: Social media big data analytics: a survey. Comput. Hum. Behav. **101**, 417–428 (2019)
2. Al-Kabi, M., Alsmadi, I., Khasawneh, R.T., Wahsheh, H.: Evaluating social context in arabic opinion mining. Int. Arab J. Inf. Technol. **15**(6), 974–982 (2018)
3. Zhang, L., Wang, S. and Liu, B., 2018. Deep learning for sentiment analysis: A survey. Wiley Interdisciplinary Reviews: Data Mining and Knowledge Discovery, p. e1253
4. Munezero, M., Montero, C.S., Mozgovoy, M., Sutinen, E.: Exploiting sentiment analysis to track emotions in students' learning diaries. In: Proceedings of the 13th Koli Calling International Conference on Computing Education Research, pp. 145–152. ACM (2013, November)
5. Pawar, A.B., Jawale, M.A., Kyatanavar, D.N.: Fundamentals of sentiment analysis: concepts and methodology. In: Sentiment Analysis and Ontology Engineering, pp. 25–48. Springer International Publishing (2016)
6. Guellil, I., Azouaou, F., Mendoza, M.: Arabic sentiment analysis: studies, resources, and tools. Soc. Netw. Anal. Mining **9**(1), 56 (2019)
7. Ain, Q.T., Ali, M., Riaz, A., Noureen, A., Kamran, M., Hayat, B., Rehman, A.: Sentiment analysis using deep learning techniques: a review. Int. J. Adv. Comput. Sci. Appl. **8**(6), 424 (2017)
8. Ahmad, M., Aftab, S., Muhammad, S.S., Ahmad, S.: Machine learning techniques for sentiment analysis: a review. Int. J. Multidiscip. Sci. Eng **8**(3), 27–32 (2017)
9. Boudad, N., Faizi, R., Oulad Haj Thami, R., Chiheb, R.: Sentiment analysis in Arabic: a review of the literature. Ain Shams Eng. J. (2017)
10. Mohammed, A., Kora, R.: Deep learning approaches for Arabic sentiment analysis. Soc. Netw. Anal. Mining **9**(1), 52 (2019)
11. Athey, S.: The impact of machine learning on economics. In: Economics of Artificial Intelligence. University of Chicago Press, Chicago (2017)
12. Chao, W.L.: 2011. Machine Learning Tutorial
13. Turban, E., Sharda, R., Aronson, J.E., King, D.: Business Intelligence: A Managerial Approach, pp. 58–59. Pearson Prentice Hall, Upper Saddle River, NJ (2008)
14. Kaur, S., Deol, R.: Students feedback for mining their opinions using supervised learning algorithm. Int. J. Eng. Sci., 12845 (2017)
15. Kalarani, P., Selva Brunda, S.: An overview on research challenges in opinion mining and sentiment analysis. Int. J. Innov. Res. Comput. Commun. Eng. **3**(10) (2015)
16. Alessia, D., Ferri, F., Grifoni, P., Guzzo, T.: Approaches, tools and applications for sentiment analysis implementation. Int. J. Comput. Appl. **125**(3) (2015)
17. Rajput, Q., Haider, S., Ghani, S.: Lexicon-based sentiment analysis of teachers' evaluation. Appl. Comput. Intell. Soft Comput. **2016**, 1 (2016)
18. Altrabsheh, N., Cocea, M., Fallahkhair, S.: Learning sentiment from students' feedback for real-time interventions in classrooms. In: Adaptive and Intelligent Systems, pp. 40–49. Springer, Cham (2014)
19. Sailaja, D., Kishore, M.V., Jyothi, B., Prasad, N.R.G.K.: An overview of pre-processing text clustering methods. Int. J. Comput. Sci. Inf. Technol. **6**(3), 3119–3124 (2015)

20. Vijayarani, S., Janani, R.: Text mining: open source tokenization tools–an analysis. Adv. Comput. Intell. **3**(1), 37–47 (2016)
21. Krouska, A., Troussas, C., Virvou, M.: The effect of preprocessing techniques on Twitter Sentiment Analysis. In: 2016 7th International Conference on Information, Intelligence, Systems and Applications (IISA), pp. 1–5. IEEE (2016, July)
22. Nayak, A.S., Kanive, A.P.: Survey on pre-processing techniques for text mining. Int. J. Eng. Comput. Sci. **5**(6) (2016)
23. Jivani, A.G.: A comparative study of stemming algorithms. Int. J. Comp. Tech. Appl. **2**(6), 1930–1938 (2011)
24. Atharva, J., Nidhin, T., Megha, D.: Modified Porter Stemming algorithm. Int. J. Comput. Sci. Inf. Technol. **7**(1), 266–269 (2016)
25. http://blog.christianperone.com/2011/09/machine-learning-text-feature-extraction-tf-idf-part-i
26. Li, B., Liu, T., Zhao, Z., Wang, P., Du, X.: Neural Bag-of-Ngrams. In: AAAI, pp. 3067–3074 (2017, February)
27. Tiwari, P., Mishra, B.K., Kumar, S., Kumar, V.: Implementation of n-gram methodology for rotten tomatoes review dataset sentiment analysis. Int. J. Knowl. Discov. Bioinf. (IJKDB) **7**(1), 30–41 (2017)
28. Tripathy, A., Agrawal, A., Rath, S.K.: Classification of sentiment reviews using n-gram machine learning approach. Expert Syst. Appl. **57**, 117–126 (2016)
29. Alayba, A.M., Palade, V., England, M., Iqbal, R.: March. Improving sentiment analysis in Arabic using word representation. In: 2018 IEEE 2nd International Workshop on Arabic and Derived Script Analysis and Recognition (ASAR), pp. 13–18. IEEE (2018)
30. Cateni, S., Vannucci, M., Vannocci, M., Colla, V.: Variable selection and feature extraction through artificial intelligence techniques. Multivariate Analysis in Management, Engineering and the Sciences, pp. 103–118 (2012)
31. Biricik, G., Diri, B., Sönmez, A.C.: Abstract feature extraction for text classification. Turkish J. Electr. Eng. Comput. Sci. **20**(Sup. 1), 1137–1159 (2012)
32. Zareapoor, M., Seeja, K.R.: Feature extraction or feature selection for text classification: A case study on phishing email detection. Int. J. Inf. Eng. Electron. Bus. **7**(2), 60 (2015)
33. Anuradha, C., Velmurugan, T.: Feature selection techniques to analyse student acadamic performance using Naïve Bayes classifier. In: 3rd International Conference on Small and Medium Business, Vietnam, pp. 345–350 (2016, January)
34. Touahri, I., Mazroui, A.: Studying the effect of characteristic vector alteration on Arabic sentiment classification. J. King Saud Univ.-Comput. Inf. Sci. (2019)
35. Can, E.F., Ezen-Can, A., Can, F.: Multilingual Sentiment Analysis: An RNN-Based Framework for Limited Data (2018). arXiv preprint arXiv:1806.04511
36. Soufan, A.: Deep learning for sentiment analysis of Arabic text. In: Proceedings of the ArabWIC 6th Annual International Conference Research Track, p. 20. ACM (2019, March)
37. Baly, R., Badaro, G., El-Khoury, G., Moukalled, R., Aoun, R., Hajj, H., El-Hajj, W., Habash, N., Shaban, K.: A characterization study of arabic twitter data with a benchmarking for state-of-the-art opinion mining models. In: Proceedings of the Third Arabic Natural Language Processing Workshop, pp. 110–118 (2017)
38. Alahmary, R.M., Al-Dossari, H.Z., Emam, A.Z.: Sentiment Analysis of Saudi Dialect Using Deep Learning Techniques. In: 2019 International Conference on Electronics, Information, and Communication (ICEIC), pp. 1–6. IEEE (2019, January)
39. Alsayat, A., Elmitwally, N.: A comprehensive study for Arabic sentiment analysis (Challenges and Applications). Egyptian Inf. J.
40. Elnagar, A., Khalifa, Y.S., Einea, A.: Hotel Arabic-reviews dataset construction for sentiment analysis applications. In: Intelligent Natural Language Processing: Trends and Applications, pp. 35–52. Springer, Cham (2018)
41. Al Sallab, A., Hajj, H., Badaro, G., Baly, R., El-Hajj, W., Shaban, K.: Deep learning models for sentiment analysis in Arabic. In: Proceedings of the Second Workshop on Arabic Natural Language Processing, pp. 9–17 (2015, July)

42. Al-Sallab, A., Baly, R., Hajj, H., Shaban, K.B., El-Hajj, W., Badaro, G.: AROMA: a recursive deep learning model for opinion mining in Arabic as a low resource language. ACM Trans. Asian Low-Resour. Lang. Inf. Process. (TALLIP) **16**(4), 25 (2017)
43. Brahimi, B., Touahria, M., Tari, A.: Improving sentiment analysis in Arabic: A combined approach. J. King Saud Univ.-Comput. Inf. Sci. (2019)
44. Algburi, M.A., Mustapha, A., Mostafa, S.A., Saringatb, M.Z.: Comparative analysis for Arabic sentiment classification. In: International Conference on Applied Computing to Support Industry: Innovation and Technology, pp. 271–285. Springer, Cham (2019, September)
45. Hussien, W.A., Tashtoush, Y.M., Al-Ayyoub, M., Al-Kabi, M.N.: Are emoticons good enough to train emotion classifiers of arabic tweets?. In: 2016 7th International Conference on Computer Science and Information Technology (CSIT), pp. 1–6. IEEE (2016, July)

A Light Spot on the Role of Artificial Intelligence and Deep Learning in Social Networks

Tamam Alsarhan

Abstract Artificial Intelligence (AI) is the broad science that concerned with teaching machines to "think" or perform tasks like humans. Recently, AI is driving the progress in various up-to-date applications in natural language processing, computer vision, speech recognition and expert systems. The area of the smart applications in our daily life is predicted to expand in the future due to the remarkable evolution of AI. Deep learning, the most advanced form of AI, has developed hand in hand with the digital era, casting its shadow on various areas such as entertainment, finance, healthcare and social networks. With the explosion of data as well as the progressive progress in hardware field, deep learning can be considered as the future of AI. This chapter provides a brief history review of AI as well as a condensed explanation of deep learning. Furthermore, it sheds a light around the usage of deep learning algorithms in social networks.

Keywords Artificial intelligence · Deep learning · Social networks

1 Introduction

The goal of artificial intelligence is to let computers perform tasks that are traditionally done by humans. When AI was officially born, its main goal was to tackle problems that are difficult for humans to solve, but easy for computers to handle, such as the numerous calculations. Nowadays, the challenge is to let computers perform tasks that are easy for humans to perform but difficult to describe, such as recognizing objects in images or organizing spoken words. Generally, AI focuses on designing algorithms that are capable of analyzing and predicting from data. Basically, AI is composed of two subsets mainly, deep learning and machine learning. The primary aim of machine learning is to learn from raw data, find patterns using statistics and make predictions without being explicitly programmed. Hence, machines

T. Alsarhan (✉)
Computer Science Department, The School of Electronic Information and Electrical Engineering, Shanghai Jiao Tong University, 800 Dongchuan RD. Minhang District, Shanghai, China
e-mail: tamamhazza@sjtu.edu.cn

A. Hamdan et al. (eds.), *The Fourth Industrial Revolution: Implementation of Artificial Intelligence for Growing Business Success*, Studies in Computational Intelligence 935,
https://doi.org/10.1007/978-3-030-62796-6_27

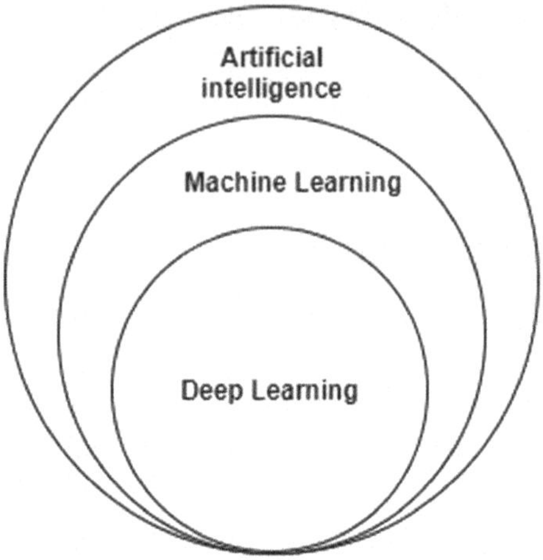

Fig. 1 Artificial Intelligence, machine learning and deep learning

are being automatically learned without human assistance. Many up-to-date services are powered by machine learning algorithms including search engines like Google, recommendation systems in YouTube and voice assistants in smartphones like Siri or Alexa. In contrast with machine learning which imposes the organized data to run through predefined equations, deep learning focuses on learning some parameters during training the deep model which consists of many layers. To make things clear, the relationship between artificial intelligence, machine learning and deep learning is summarized in Fig. 1.

More than half a century ago, the term AI used as a theoretical aspect of computer science. The question "Can computers think like humans?" was repeatedly asked by many scientists and mathematicians at that time. In fact, this decades-old question is under investigation until this moment. In 1950, Alan Turing [1] discussed in his paper the ability of designing intelligent machines as well as testing their intelligence. Thus, that paper was the first to introduce the AI concept to the general public. Six years later, researches started getting serious about AI as some algorithms had been developed to solve mathematical problems and geometrical theorems. In the late 1960s, the intention was directed to machine vision learning and robotics. The big hope of creating intelligent machines was obstructed by a serious challenge, represented by the limited abilities of computer to process enormous amount of data at that time. Although AI has been extremely improved, it is still not prefect. The most prominent stations in AI through history are summarized in Fig. 2.

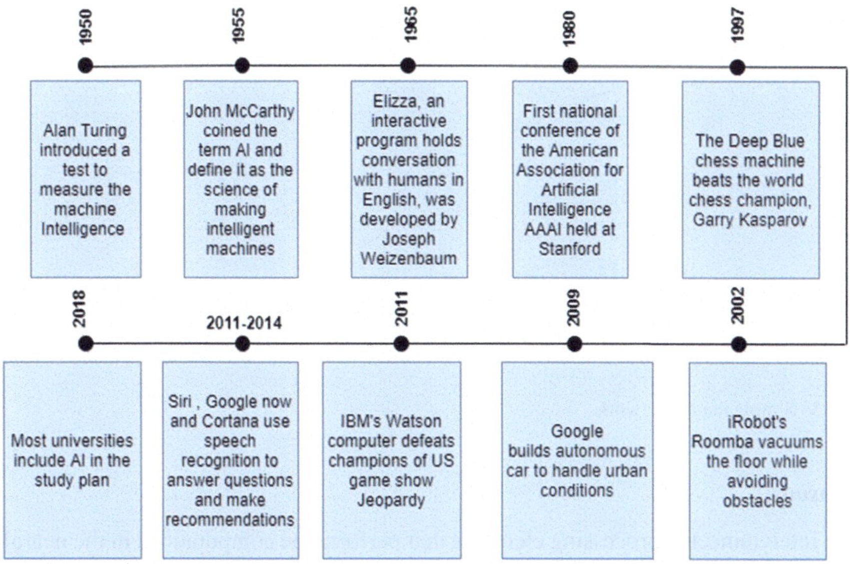

Fig. 2 Timeline of artificial intelligence

2 Deep Learning: Basics

> Incorporating general intelligence, bodily intelligence, emotional intelligence, spiritual intelligence, political intelligence and social intelligence in AI systems are part of the future deep learning research.
>
> —Amit Ray, Compassionate Artificial Intelligence.

Deep learning, a class of machine learning, refers to representational-learning models consist of non-linear modules designed to learn representations or extract high level features from raw data. The technical term "deep" refers to the number of modules (layers) in the model. With stacking multiple modules, very complex function is learned. Essentially, the backbone of deep learning is Artificial Neural Network (ANN), which is a computing system inspired by the brain structure of humans. In human nervous system, the flow of information occurs between the nervous cells while in ANN through neurons (see Fig. 3).

The architecture of ANN, the multi-layer networks, is made up of three components:

1. Neurons
2. Weights
3. Biases

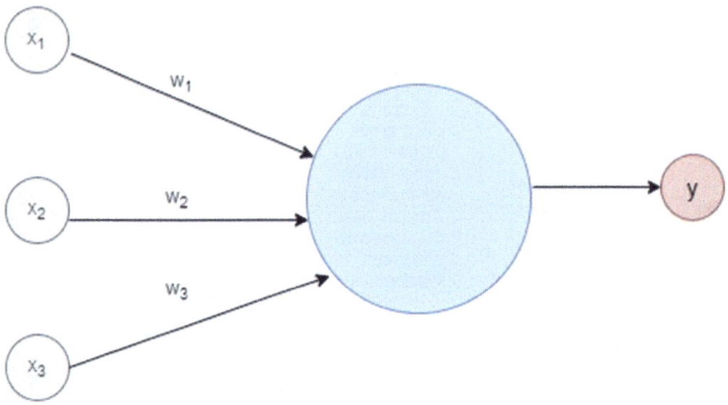

Fig. 3 Simple neural network

Neurons

The interconnected processing elements that perform the computations in the neural network are the neurons. Structurally, they are arranged in a series of layers. Each neuron is given an input (data) from the previous layer, produces an output (modified data) and forwards it to the next layer. The formula of computation is given in Eq. 1.

$$y = \sum_{i} wx + b \tag{1}$$

where x is the input, w is the weight and b is the bias.

Weights: When transferring data from one neuron to another, each connection is assigned a weight as shown in Fig. 3. Weights are the learnable parameters in ANN.

Biases: Biases are constants added to the result of multiplying the input with the weights, representing what the ANN "thinks".

As previously mentioned, the term "deep" in deep learning refers to the depth of ANN. The depth is measured by the number of layers. The difference between simple and deep ANN is illustrated in Fig. 4. On Left: simple ANN consists of one

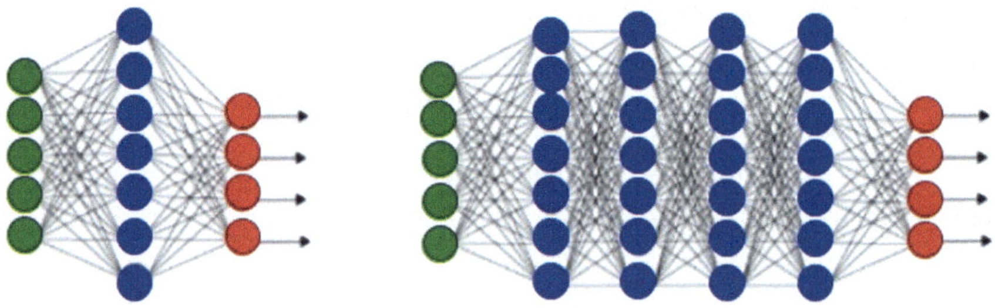

Fig. 4 Left: simple ANN. Right: deep ANN

input layer (the green circles), one hidden layer (the blue circles) and one output layer (the red circles). On right: deep ANN with one input layer, four hidden layers and one output layer.

2.1 Learning Process of Neural Network

Learning neural network-based model involves mapping inputs to labels. This is done by feeding the model with data (samples), which can be of any type as numbers, images, texts, audios or videos and their labels. Normally, data is divided into training set, which contains samples and their labels and testing set, which contains only the samples. In image classification task, the model is fed with images (such as animal's images) and their labels (such as "cat" or "dog"), after some time it will be fed with the images without labels and it is supposed to classify the images into their specific categories. In other words, it is supposed to predict the exact label of each image (whether the image represents a cat or a dog or anything else).

The first phase of learning any neural network is the forward pass. The learning is achieved by training the neural network model using the training data. During the training, the deep learning model learns data representations using the mathematical framework formulated in Eq. 1. Afterwards, the predicted output $y\tilde{}$ is being calculated. A loss function is used to estimate the error between $y\tilde{}$ and the actual y (the true label). The loss function reflects how good/bad the model is. Ideally, the less the error is, the best the model is. Essentially, the learning process is about finding the optimal values of the weights and the biases that make the difference between y and $y\tilde{}$ as close as possible to zero. Once the loss function is calculated the **backward pass** phase starts. In this phase, the loss information propagates backward through all the neurons in the hidden layers. Next, the model updates the training parameters (weights and biases) based on the propagated information with the help of the gradient descent technique. Basically, the gradient descent changes the weights with the help of the calculation of the derivative of the loss function.

To make things clear, an example of a classification problem processed by deep learning ANN is formulated in Fig. 5. First, the input is provided to the network, a new representation is being learned using multi-stage layers. The final prediction is achieved in the output layer.

3 Recurrent Neural Network

The architecture of ANN assumes that the input neurons are independent of each other which is not always the case. In our daily life, we have a kind of dependency in our speaking. Moreover, the translators have a context in which predicting the next word in a sentence depends on the previous context. Hence, Recurrent neural networks (RNNs) are the subset of ANN that specialized for data that have a context

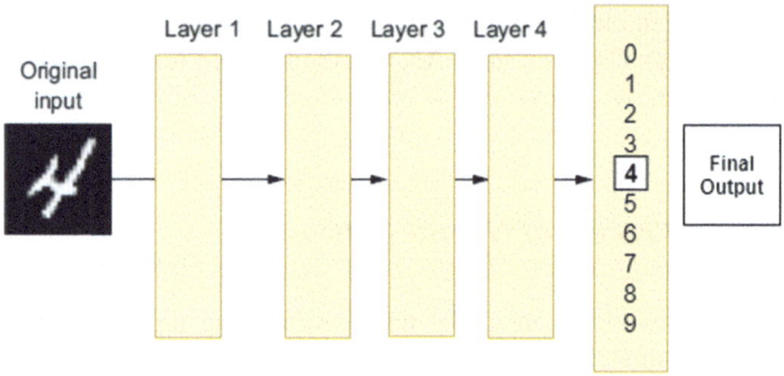

Fig. 5 Image classification of digits using ANN

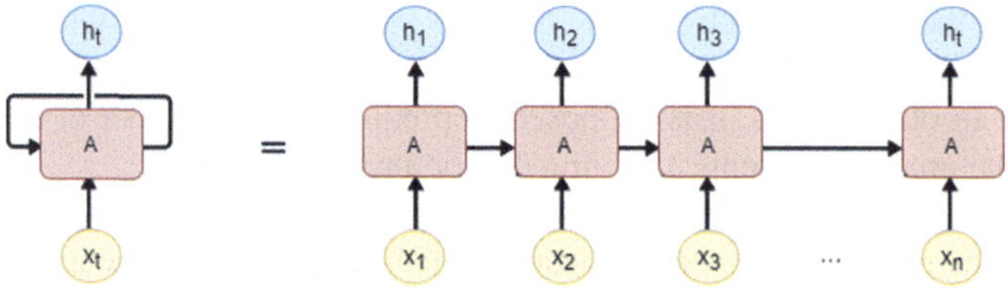

Fig. 6 An unrolled version of recurrent neural network

or dependency between its data points over the time, such as the time series or the sequenced data. Basically, RNN has the capability of remembering the previous input states from sequential data. The power of RNN is its capability of remembering the context during the training which makes it convenient to be used in a wide range of applications such as machine translation [2], speech recognition [3], music composition [4] and human activity recognition [5, 6]. Figure 6 illustrates the structure of RNN, where A represents a neural network, x represents the input at time step (t), h represents the hidden state. The input of the second stage consists of the current input x_2 and the previous hidden state h_1 and so on. By repeating this procedure over the time, the model ensures that the current time step contains information about the previous time step and so on which makes it intuitively "remembers" the past.

3.1 *Different Types of RNN*

RNN comes up with three different models, mainly Vanilla RNN, Long short-term memory (LSTM) and gated recurrent unit (GRU). Vanilla RNN suffers from Long-Term Dependencies problem which was explored in 1994 [7]. It has been noticed that when the sequence gets longer, Vanilla RNN become unable to remember previous hidden states. Thus, LSTM was proposed in 1997 [8] to tackle this issue using the gating scheme. Then, a dramatic variation on the LSTM, called GRU, was introduced in 2014 by Cho et al. [9].

4 Convolutional Neural Network

Another class of ANN is the Convolutional Neural Network (CNN or ConvNet), which is universally used for image recognition [10–13], face recognition [7, 14], object detection [15, 16] and many other applications, causing remarkable advancements in computer vision area.

CNN algorithm is the core algorithm used for image search in Google, image tagging in Facebook and product recommendations in Amazon.

4.1 Applications

4.1.1 Image Classification

CNN takes an input image, processes it and outputs a class label of the image. Basically, computers "see" images as a matrix of number (pixels) based on its resolution. Any image is represented as $h\ w\ c$, where h denotes the height, w denotes the width and c denotes the number of channels. If (c = 1), the image is a grey scale image and while if (c = 3), the image is a RGB image. Processing an image using the CNN algorithm is performed using a series of convolutional layer(s), activation layer(s), pooling layer and fully connected layer. The rectified linear unit (ReLU) and the softmax function are the two popular activation functions used along with the convolutional layers. Generally, convolutional layers are used to extract features from the input image by multiplying it with a filter contained weights to be learned during the training process. Pooling layers are used to reduce the number of parameters when the images are too large. Activation functions are used to add some kind of non-linear property to the neural network. Finally, fully connected layer used a flattened version of feature matrix to perform the prediction (classification) along with the softmax activation function which predicts weight for all the possible classes. The probabilities are summed to 1 and the predicted class has the highest value. See Fig. 7.

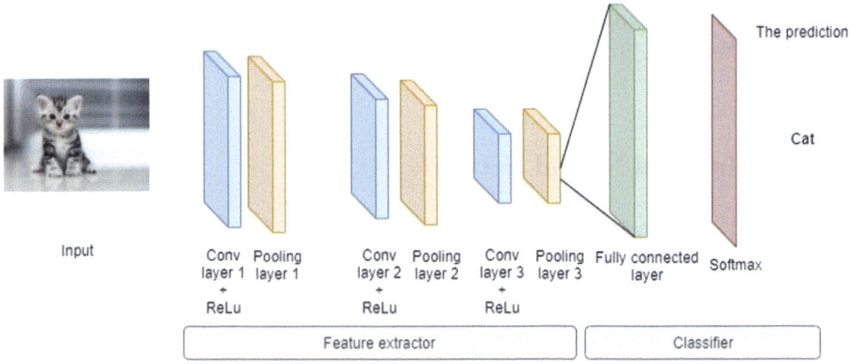

Fig. 7 CNN architecture for image classification task

Fig. 8 Object detection example

4.1.2 Object Detection

One of the most interesting and challenging computer vision tasks is object detection which involves identifying an object with a bounding box of a certain class in an input image. This task involves object localization and image classification at the same time. Localization any object requires finding the location of the object in a single image as shown in Fig. 8. Object detection is useful in many applications such as face detection, video surveillance and aerial image analysis.

4.1.3 Image Captioning

This task can be summarized by one question " What do you see in the picture?". The idea of image caption is to generates a textual description of an image (see Fig. 9). To generate the caption, this task involves a combination of computer vision and natural

Fig. 9 Image captioning example

language processing. Image captioning can be useful in many future applications such as self driving cars, where describing the scene around the car can boost the automatic driving system. Moreover, image captioning can help giving aid to the blinds by describing the road for them so they can travel without any need to help from people.

5 Deep Learning in Social Media

Nowadays, owning a social media account is as common as owning a mobile phone. Hence, it is not surprising that managing the massive amount of data on social media is a tough process. Traditional methods for analyzing these data are inefficient. Deep learning succeeded in giving structure to unstructured data which make it the key base of the popular social networks. Additionally, deep leaning has become a valuable tool for big companies like Facebook, Twitter, Google and Youtube. Following are some examples of deep learning- based applications in social media.

5.1 Image Tagging

This task involves object recognition as well as sentiment analysis of an image. Currently, Google, Facebook and Amazon are using CNN algorithm in visual search tagging. Facebook uses machine learning algorithms plus deep learning algorithms to detect faces, which is known as the facial recognition task. What Facebook has built is considered to be the largest name-to-face database in the world. The idea of facial recognition is to scan people's faces and identify them using deep learning neural networks. This feature in Facebook let any user be notified of photos they appears in, even without being tagged.

When a Facebook user is tagged in a photo, machine learning algorithms used in Facebook transform the tagged photo to a numerical representation of the user face, called "template". Hence, every uploaded photo will have a unique template. If a new photo is uploaded, Facebook will compare its representation with the existing representations and suggest a tag if any match is exist.

5.2 Visual Search

Earlier, searching for a visual content has relied on the text-based search, where the quality of the search is depend on the keywords of the query. As one picture is worth more than one word, in some cases a long detailed description is needed. In the world of artificial intelligent, visual search has been emerged and streamlined the way we search and hence, remarkably improved the search results. Visual search uses real-world pictures rather than keywords for online search, to recognize a target object and filter out irrelevant targets. As a result, the sense of visual discovery to the online world is achieved by visual search. The theory behind the modern visual search is to use AI to learn machines to identify objects in images, and then returns a list of related results. Its essential impact has been shown in a host of applications, especially eCommerce industry. Thus, many companies including Google, Amazon,Bing and Pinterest have developed impressive computer vision capabilities to serve the visual search. The stimuli of the visual search engines is the neural networks which utilize deep learning technology, so the systems are continuously learning and extending their field of experience.

5.3 Recommendation Engines

A recommendation engine is a filtering system that uses different algorithms to recommend the most relevant item from a host of products, services, information to users based on analysis of data [17]. It first captures the history of the user and performs a recommendation. Through the lens of user's past behavior, preferences and interests, recommendation engines became the essential way to expose to the digital world. Many companies use the recommendation engine to provides their customers with personalised information and solutions. The good point about recommendation engines is that they can save the user's time and delivers a better results.

5.4 *Bad Content Detection*

Social media is essential for each one of us at present. With the explosion of data posted every day, it's impossible to manually filter the content in. Social networks struggle to keep their platform off bad content, which includes nudity, hate speech, fake accounts, terrorism, graphic violence, and others. Recently, AI serves to proactively enforce community standards and detect bad contents. For example, Facebook's AI system uses computer vision algorithm (CNN) to detect and remove any content of nudity or graphic violence. Moreover, social networks use AI to detect any text can be considered as hate speech or terrorism contents. Furthermore, Facebook's AI system hunts for fake accounts by training the system to search for any signal that would indicate illegitimacy. These signals include any abnormal activity like trying to reach many more other accounts than usual.

5.5 *Semantic Analysis*

Text is the heart of the online communication in social networks. Comments in Face- book, Instagram and LinkedIn, tweets in Twitter and all kinds of communication on social media depends on texts. Hence, analysing the behavior of users on social media could be useful for the networks themselves to stay ahead by understanding people efficiently. Currently, almost all companies, factories, online supermarkets depends partially or totally on social media marketing. Understanding user's behavior on social media is valuable for any institution who wants to increase their Return on investment (ROI). The process of extracting information from a given text, emotions, appraisals, and attitudes towards entities is called semantic analysis. In short, it assigns a polarity (e.g., positive, negative, neutral) to a specific text. Earlier, companies relied on written surveys and questionnaires to understand consumer's feedback about a product. In the digital era, artificial intelligence have made it possible to understand consumers behavior using online-analysing text from social media accurately.

5.5.1 Semantic Analysis Techniques

Analysing a text involves understanding its contents and assigning an indicator to its context, which is simply a text classification. To achieve that, artificial intelligence-based methods need to convert words to vectors of continuous real numbers, called word embedding as an input to the model (e.g. "Pen"(0.20,. . ., 0.11, . . .)). Afterward, feed them into the deep model, which can be CNN or RNN.

1. Convolutional neural network-based model

For classification task, CNN is useful for feature extraction from the texts. It is possible to learn CNN-based model certain sequences of words, which can be indicators to a specific topic.

2. Recurrent neural network-based model

After obtaining a feature vector for each word in the text, a feature representation is obtained by encoding the feature sequence using RNN. Further, simple neural network is used to transform the encoded sequence information to output, which can be text classification.

5.6 Text Translation

In the era of digital communication and multi-mediality, online communication between people all around the world requires online translation to remove the language barriers. Hence, the communication between people from different language- backgrounds would be easily achieved. Social media uses AI for text translation from one language to another.

Facebook Artificial Intelligence Research (FAIR) announced that language translation is important to Facebook's mission of making the world more open and connected as half of Facebook users are non-English speakers. Machine translation is a challenging task as it requires producing translations that more closely resemble localized text. Previously, Facebook used RNNs algorithm, which was tested and worked well. RNN-based translation in Facebook uses LSTM network which translates the text information linearly and methodically from left-to-right or right-to-left order, depending on the original language of the text. Moreover, Google search through a database of texts and suggest the closest translation to the users using RNN network.

Right now, Facebook decided to use CNN algorithm as it showed promising results and more closer results to the localized text. As mentioned in Sect. 4, CNN is widely used for image recognition tasks. When it comes to machine translation, CNN process information (texts) in a different way from RNN. Hierarchically, CNN looks for non-linear relationships in texts, capturing the contextual meaning from the text and translating it accordingly. Basically, Facebook's CNN algorithm uses the multi-hop attention and gating mechanism, imitating the translation procedure done by humans. Generally, when someone translates a text, he attempts to read it multiple times to grasp the closest meaning. Similarly, multi-hop attention mechanism frequently checks the sentence and estimates choices about the translations. After per- forming the translation in multi-hop, the contextually or hierarchically relationship in the texts can be figured out by CNN algorithm. Although RNN served machine translation in Facebook, CNN-based approach achieved state-of-the-art accuracy at nine times faster than RNN.

6 Conclusion

Exploring the applications of the broad science of artificial intelligence and deep learning in social media is a journey, and reading this chapter is merely one step towards that. Keeping ourselves up-to-date with this broad science is a challenging task, as it is considered to be one of the most active research areas these days. In this chapter we briefly introduced a history review of artificial intelligence as well as a condensed explanation of deep learning and its algorithms. Moreover, we shed a light around the usage of deep learning algorithms in social networks, mainly in image tagging, recommendation engines, visual search, bad content detection, semantic analysis and text translations. In future, the area of social networks will definitely expand. Hence, the development in deep learning algorithms will leave its unique impacts on it.

References

1. Turing, A.M.: Mind **LIX**(236), 433 (1950). https://doi.org/10.1093/mind/lix.236.433
2. Li, X., Wu, X.: Constructing long short-term memory based deep recurrent neural networks for large vocabulary speech recognition. IEEE International Conference on Acoustics, Speech and Signal Processing (ICASSP), Brisbane, QLD, pp. 4520–4524 (2015). https://doi.org/10.1109ICASSP.2015.7178826
3. Cho, K., van Merrienboer, B., Gulcehre, C., Bahdanau, D., Bougares, F., Schwenk, H., Bengio, Y.: Learning phrase representations using rnn encoder-decoder for statistical machine translation. In: Proceedings of the 2014 Conference on Empirical Methods in Natural Language Processing, EMNLP 2014, October 25–29, 2014, Doha, Qatar, A meeting of SIGDAT, a Special Interest Group of the ACL, pages 1724–1734 (2014)
4. Liu, T., Ramakrishnan, B.: Bach in 2014: Music composition with recurrent neural network, CoRR, vol. abs/1412.3191 (2014)
5. Alsarhan, T., Alawneh, L., Al-Zinati, M., Al-Ayyoub, M.: 2019 IEEE Sensors, 1–4 (2019)
6. Wang, J., Chen, Y., Hao, S., Peng, X., Lisha, H.: Deep learning for sensor-based activity recognition: Survey, Pattern Recognit. Lett **119**, 3–11, ISSN 0167–8655 (2017). https://doi.org/10.1016/j.patrec.2018.02.010
7. Bengio, Y., Simard, P., Frasconi, P.: Learning long-term dependencies with gradient descent is difficult. In: IEEE Transactions on Neural Networks **5**(2), 157–166 (1994). https://doi.org/10.1109/72.279181
8. Hochreiter, S., Schmidhuber, J.: Neural Comput. **9**(8), 1735–1780 (1997). https://doi.org/10.1162/neco.1997.9.8.1735. URL https://doi.org/10.1162/neco.1997.9.8.1735
9. Cho, K., van Merrienboer, B., Bahdanau, D., Bengio, Y.: On the Properties of Neural Machine Translation: Encoder-Decoder Approaches. https://doi.org/10.3115/v1/W14-4012
10. Castillo Camacho, I., Wang, K.: Proceedings of the ACM Workshop on Information Hiding and Multimedia Security. IHMMSec'19, pp. 107–112. Association for Computing Machinery, New York, NY, USA (2019). https://doi.org/10.1145/3335203.3335723. https://doi.org/10.1145/3335203.3335723
11. Ng, Y.S., Xue, W., Wang, W., Qi, P.: Proceedings of the 5th International Workshop on Multimedia Assisted Dietary Management, MADiMa '19, pp. 33–41. Association for Computing Machinery, New York, NY, USA (2019). https://doi.org/10.1145/3347448.3357168. https://doi.org/10.1145/3347448.3357168

12. Hashimoto, N., Fukushima, D., Koga, R., Takagi, Y., Ko, K., Kohno, K., Nakaguro, M., Nakamura, S., Hontani, H., Takeuchi, I.: Multi-scale Domain-adversarial Multiple-instance CNN for Cancer Subtype Classification with Unannotated Histopathological Images. 3851–3860. https://doi.org/10.1109/CVPR42600.2020.00391
13. Li, Q., Shen, L., Guo, S., Lai, Z.: Wavelet integrated CNNs for noise-robust image classification. Proc. IEEE/CVF Conf. Comput. Vis. Pattern Recognit. (CVPR) 7245–7254 (2020)
14. Liu, W., Wen, Y., Yu, Z., Li, M., Raj, B., Song, L.: Sphereface: Deep hypersphere embedding for face recognition. In: CVPR (2017)
15. Ding, M., Huo, Y., Yi, H., Wang, Z., Shi, J., Lu, Z., Luo, P.: In: Proceedings of the IEEE/CVF Conference on Computer Vision and Pattern Recognition (CVPR) Workshops, 1000–1001 (2020)
16. Kim, Y., Kang, B.N., Kim, D.: Learning Relationship between Convolutional Features for Multi-Scale Object Detection The European Conference on Computer Vision (ECCV) (2018)
17. Zhang, S., Yao, L., Sun, A., Tay, Y.: Deep Learning Based Recommender System: A Survey and New Perspectives. ACM Comput. Surv. **52**(1), 38. https://doi.org/10.1145/3285029

Artificial Intelligence, Big Data, and Value Co-creation: A Conceptual Model

Muneer Abbad, Faten Jaber, Kholoud AlQeisi, and Shorouq Eletter

Abstract The development of Artificial Intelligence (AI) and big data technologies in the ecosystem has reshaped the concept of value co-creation. However, this concept needs to be renovated from a digital perspective. This study conceptualizes the dynamic relationships between AI, big data, and value co-creation. Its theoretical framework adopts S-D logic to connect the two players (customers as creators and firms as users) of big data. This framework shows that big data is a crucial driver in transferring the concept of value in marketing to value co-creation. It also suggests that value co-creation is composed of two dimensions: co-production and value in use. Finally, the framework posits AI technologies as a critical link between the two players of big data that facilitates value co-creation and, ultimately, leads to improvements in customer satisfaction and firm performance.

Keywords Value co-creation · Conceptual framework · Marketing · AI · Big data · S-D logic

1 Introduction

Artificial intelligence (AI) and big data technology have transformed the marketing landscape. Marketers can benefit from the knowledge obtained from an analysis of big data to create value for their businesses. The term 'perceived value' donates a trade-off between benefits received and costs [14]. Vargo and Lusch [37] describe

M. Abbad (✉)
Community College of Qatar, P.O. Box 7344, Doha, Qatar
e-mail: muneer.abbad@ccq.edu.qa

F. Jaber
Oxford Brookes University, Oxford OX3 0BP, UK

K. AlQeisi · S. Eletter
Al Ain University, P.O. Box 112612, Abu Dhabi, UAE

A. Hamdan et al. (eds.), *The Fourth Industrial Revolution: Implementation of Artificial Intelligence for Growing Business Success*, Studies in Computational Intelligence 935,
https://doi.org/10.1007/978-3-030-62796-6_28

value as "always intangible, heterogeneously experienced, co-created, and potentially perishable". However, there are multiple facets of value for different stakeholders such as service providers and users [36]. The value of products is shared and co-created among different stakeholders as part of the essential marketing process of value exchange [33]. In line with Service-Dominant (S-D) Logic, which focuses on the value exchange process and interactions among service providers and customers, the intended value of a product is extended into a co-created bundle of benefits as part of a long-range process [37]. McColl-Kennedy et al. [23] define value co-creation (VCC) as "benefit realized from the integration of resources through activities and interactions with collaborators in the customer's service network".

Notably, the concept of VCC has attracted substantial interest from scholars; however, this stream of research notwithstanding, an integrative framework for measuring VCC has yet to emerge [6, 13, 15, 20, 26, 40]. Previous research on VCC has given rise to several differing conceptualizations. Some researchers have built on S-D while others have moved in a different direction (see [30, 36]. Such authors have developed a framework for measuring VCC that comprises three phases: pre-delivery (information search and collation, cooperation, and cerebral activities); co-delivery (co-learning, co-production, changes in habits, and a combination of complementary activities), and post-delivery (connection). Tommasetti et al. [36] acknowledge that phases are individualized and do not imply a sequential order.

Despite the continuing interest in understanding VCC, research scrutinizing this concept in the current digital ecosystem is scarce. Hence, in this study, we analyze the drivers and outcomes of value co-creation in digital ecosystems. In this digital age, the primary driver that transforms value creation into VCC between companies and customers is AI technologies that feed on big data [16, 17]. For example, customers share their data, including their location, interests, and behavior, which is then analyzed by AI agents working in firms that seek to use big data in ways that improve their decision-making [5, 35]. In this study, we first describe the dynamic relationships between AI, big data, and VCC. The theoretical background to this research is then discussed in detail. The conceptual theoretical framework is then presented, following which conclusions are drawn.

2 Artificial Intelligence, Big Data, and Value Co-creation

Artificial intelligence (AI) is a broad field within computing science that focuses on the design of machines capable of mimicking human intelligence [25]. Such machines exhibit an increasing capability to perform specific roles and tasks akin to those performed by humans in the workplace and society in general [9]. AI technology has existed for decades and has experienced both AI winters and springs. However, the availability and power of big data along with advances in computing power and storage capabilities have contributed to a resurgence in AI in recent years [8]. During this time, AI has rapidly expanded into a position in which it has been able to undergo transformations for the augmentation and potential replacement of human industrial, intellectual, and social tasks and activities. Nowadays, most people adapt to AI and interact with technology on a daily basis. Moreover, AI technology has led to astonishing breakthroughs in algorithmic machine learning and autonomous decision-making, creating opportunities for ongoing innovation. AI has also exerted a significant impact on industries ranging from finance, healthcare, manufacturing, retail, supply chain, logistics, and utilities, all of which have been potentially transformed by the onset of AI technologies [9]. AI has been successfully employed in numerous applications from self-driving cars to providing customer services. Firms are now utilizing AI to market their services [28]. Thus, AI is making our work faster and easier, which enhances the efficiency and effectiveness of businesses.

Machine learning has now emerged as a subfield of AI. It refers to the ability of a machine to learn from data. Such learning occurs by training the machine to perform a specific task using an adaptive approach such as deep learning without explicit programming [25]. In addition, Big Data Analytics (BDA) has emerged as a technology-driven ecosystem. The revolution of BDA has been driven by underpinning waves of innovation, analytic services with intelligence, and remarkable advances in technology. Furthermore, BDA has captured the attention of academic researchers, industry practitioners, and government organizations. Many organizations now use BDA to extract knowledge from data in an interpretable and appropriate format to support better decision-making [32].

Jagadish et al. [16] define the process of creating value from big data as a "multi-step process: acquisition, information extraction and cleaning, data integration, modeling and analysis, and interpretation and deployment" (p. 86). The realization of the grand potential of data analytics relies on its ability to extract value from massive data; machine learning lies at the core of this process due to its ability to learn from data and provide insightful decisions and predictions [19]. AI and machine learning play a significant role in helping firms to create value through intelligent data analysis and by capturing interpretations of increasingly available raw data [27].

During this era of big data, organizations have been integrating enormous data resources, data analytics models, and machine learning to capture, visualize, and analyze the underlying patterns in data and improve the quality of business decisions [10]. They have also been utilizing big data to shift from intuitive- to data-based decision making [18]. Data-driven decisions or evidence-based decisions are more effective, which will have a positive influence on all aspects of organizations. Firms employ data analytic experts to reveal hidden patterns in data and transform them into valuable business insights [1]. Firms use machine learning to imperceptibly collect consumer data from multiple sources, pool those data, and then mine them to deliver consumer insights in real-time [18]. However, the value created through big data is dependent not only on data or technology but also on organizational contexts and managerial actions [24].

3 Theoretical Framework

3.1 Players in Value Co-creation

The logic of S-D provides a proper understanding of how value is created through what is fundamentally exchanged between firms and customers. Marketers create and deliver value and benefits for customers, and facilitate the exchange [3]. S-D logic elucidates how possibilities for value co-creation can be shaped through big data contexts and an open-ended understanding of the potential sources of strategic advantage and a boundless view of market possibilities [3]. In this study, we build a conceptual framework of VCC from data generated by customers and used by firms to construct VCC from an S-D perspective. The proposed model will revamp the concept of VCC to emphasize two players: customers as creators and firms as users of big data.

3.2 Dimensions of Value-Co-creation

Ranjan and Read [30] derived a conceptual framework based on extant research on VCC that postulates two core conceptual dimensions: co-production (covering the issue of exchange) and Value-in-Use (ViU) which captures the shared reality of experience and use. Their work identified three components measuring co-production: Knowledge (creativity and ideas), Equity (power-sharing, alignment, transparency, and access) and Interaction (dialogue and interaction); and three components measuring ViN: Experience (benefits, empathy, use-value, co-experience and value experience), Personalization (consumer orientation and uniqueness), and Relationship (collaboration, interdependence, engagement, network, involvement, and enduring exchange). The VCC framework was then tested in a field study as a higher-order structure impacting consumer satisfaction. Utilizing the same dyadic structure of Co-production and ViN, Dollinger et al. [7] specified the anticipated benefits of VCC in the education sector for students (graduate capabilities, satisfaction, and quality interaction) and institutions (student-university identification, university image, and student loyalty). Thus, AI is an important information technology that creates value and knowledge from big data [29].

3.3 Value Co-creation and Big Data

This study proposes a theoretical model based on the big data classifications and platforms developed by Xie et al. [41]. In our model, these big data resources and platforms deal with AI to create value from two players (firms and customers) and

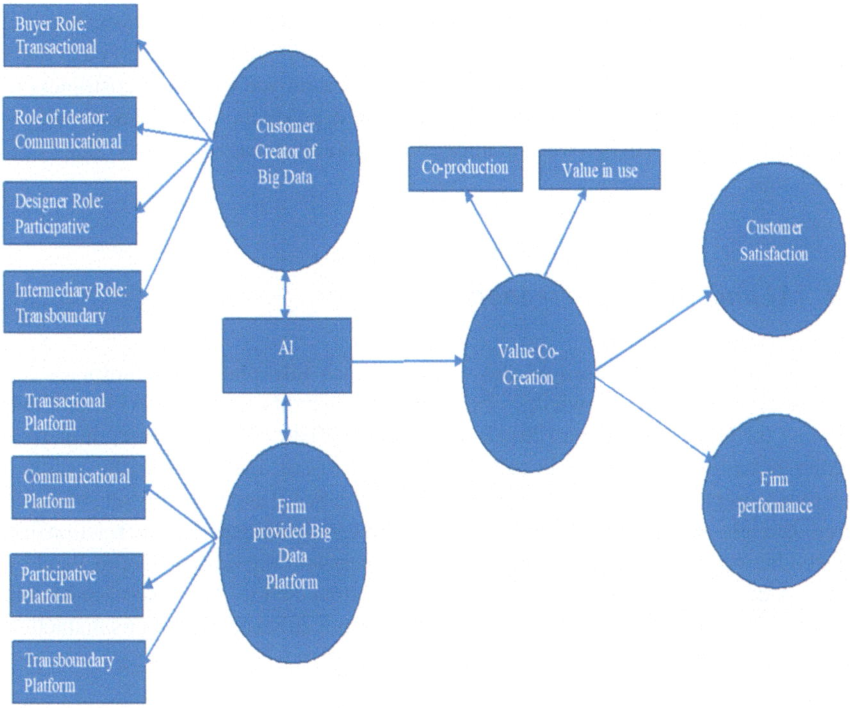

Fig. 1 Proposed model

achieve the following outcomes: customer satisfaction and firm performance, as shown in Fig. 1.

Xie et al. [41] proposed a theoretical framework on VCC to understand the interconnections between firms and customers based on big data cooperative assets by transforming big data resources in the context of S-D logic. They identified four big data resources that create big data from customers and four associated types of digital platforms provided by firms that contain big data.

Xie et al. [41] explained the four big data resources as follows. First, transactional big data derives from customer purchasing behavior by recording prices, products, locations, and other related data. Second, communication big data resources derive from interactions between customers and companies in different ways: message, telephone calls, websites, clubs, and so on. Third, participative big data is generated when customers actively participate in product or service development (e.g., customizing designs or voting for specific product locations). Finally, transboundary big data is generated from different service ecosystems shared by customers who then transfer knowledge across the boundaries of these ecosystems.

In association with these big data resources, Xie et al. [41] identify and explain four types of big data platforms provided by firms. First, a transactional platform enables the collection of transactional big data, supporting customer purchases, and sharing data with different departments to analyze customer satisfaction. Second, a communication platform is used to facilitate communications between customers and companies. Third, a participative platform is employed to attract customers to actively participate in developing or designing new products or services. Finally, a transboundary platform helps companies collect data from customers who share new knowledge.

3.4 AI and Value Co-creation

Several studies have focused on the ways in which information technologies can be employed to create knowledge and actionable insights, of which AI is one that requires further investigation [29]; Bergeret et al. 2020; [34]. However, numerous studies have highlighted the benefits of AI in terms of dealing with the volume, velocity, variety, and veracity of unstructured data [2, 8, 29] and transforming unstructured data into structured information [22]. Therefore, we argue that the relationship created between big data generated from customers and big data platforms used by firms, as described in Xie et al. [41], should be connected to AI to increase its role in VCC. This value will not be sufficient if it does not enhance customer satisfaction and firm performance (Fig. 1).

4 Theoretical Framework

Researchers have noted that research on the dynamic and interdependent construct of VCC has been carried out in isolation from the broader relational context and therefore does not fully capture the impact of other value processes [6, 40]. This has resulted in a failure to capture the value of process integration. Building on the work of Xie et al. [41], this study incorporates the VCC dyad structure, as depicted in Fig. 1.

Xie et al. [41] argue that big data should be classified based on value to understand the process of VCC. Their classifications were based on four customer roles: buyer, ideator, design, and intermediary. From these, they identified four types of big data: transactional, communication, participative, and transboundary. Firms can provide big data platforms in the context of VCC to enhance their performance by adding strategic value, finding new markets, and/or developing new products and services. Firms will also improve their internal business processes and work efficiently and effectively by utilizing AI tools. The data-analysis capabilities can then be employed through AI to create the required value that enables firms to increase their performance.

5 Discussion and Conclusions

Big data encompasses raw, structured, and unstructured data. Businesses collect and integrate big data from a variety of sources, including clouds, the internet of things, sensors, customers, suppliers, competitors, business processes, and internal databases. Numerous studies have focused on using different technologies in the management of big data to help firms respond to market changes [4, 11, 42] and thus attain a competitive advantage. Firms can increase their performance by utilizing these technologies to collect data, communicate with customers, and develop efficient ways to sense and respond to changes in customer preferences [11, 31, 39].

AI technologies provide a means to analyze the big data assets firms possess, which can be viewed as value potential or value enabler. Firms and marketers need to analyze such data to extract knowledge and develop better informed and actionable decisions that will activate the value of data for firms and customers.

Big data collected from customers is now playing a substantial role in powering strategic resources for companies through an analysis of customers' behavior and unforeseen patterns [21]. Firms are using AI to increase customer satisfaction through VCC by applying different analytical tools. Managing big data and using AI will also enhance the performance of firms. For example, Lee et al. [21] described different examples that demonstrated the outcomes of using big data and AI. They explained how AI could help in increasing patients' satisfaction by applying automated learning algorithms in hospital data and insurance claims to discover hidden patterns in big data. They concluded that the financial performance of companies would increase once big data is managed at the top executive level. They also explained how AI is used with data from online customer surveys to help credit card companies improve fraud detection and risk management. They further explained how to use unstructured feedback data from patients to detect weaknesses in healthcare institutions and manage the supply chain with partners and customers.

AI has evolved from a mere technological alignment to enhancing organizational performance and engendering core competences. New digital marketing tools employed in search engine optimization and social media marketing are replacing traditional marketing practices [38]. Organizations can now use AI marketing to engage their customers in strategic tasks ranging from nurturing and follow up to segmentation salesmanship and customer service and gratification [12]. The AI-Marketing fusion is becoming a promising paradigm in terms of both stature and functionality for successful business organizations [38]. This study thus supports Vishnoi et al. [38] claim that "the fusion of customer-centricity and data-centricity is the new future-centricity of successful business conglomerates". Rather than persuading a customer to buy a specific product, marketers must now extract the knowledge hidden in big data to open up new opportunities for innovation and growth.

References

1. Abbady, M., Akkaya, M., Sari, A.: Big data governance, dynamic capability and decision-making effectiveness: Fuzzy sets approach. Decis. Sci. Lett. **8**(4), 429–440 (2019)
2. Berger, J., Humphreys, A., Ludwig, S., Moe, W.W., Netzer, O., Schweidel, D.A.: Uniting the tribes: using text for marketing insight. J. Mark. **84**(1), 1–25 (2020)
3. Bettencourt, L.A., Lusch, R.F., Vargo, S.L.: A service lens on value creation: marketing's role in achieving strategic advantage. Calif. Manag. Rev. **57**(1), 44–66 (2014)
4. Chuang, S.H., Lin, H.N.: The roles of infrastructure capability and customer orientation in enhancing customer-information quality in CRM systems: empirical evidence from Taiwan. Int. J. Inf. Manage. **33**(2), 271–281 (2013)
5. Clarke, R.: Big data, big risks. Inf. Syst. J. **26**(1), 77–90 (2016)
6. Corsaro, D.: Capturing the broader picture of value co-cre-ation management. Eur. Manag. J. **37**(1), 99–116 (2019)
7. Dollinger, M., Lodge, J., Coates, H.: Co-creation in higher education: towards a conceptual model. J. Mark. High. Educ. **28**(2), 210–230 (2018)
8. Duan, Y., Edwards, J.S., Dwivedi, Y.K.: Artificial intelligence for decision making in the era of Big Data–evolution, challenges and research agenda. Int. J. Inf. Manage. **48**, 63–71 (2019)
9. Dwivedi, Y.K., Hughes, L., Ismagilova, E., Aarts, G., Coombs, C., Crick, T.,... Galanos, V.: Artificial Intelligence (AI): Multidisciplinary perspectives on emerging challenges, opportunities, and agenda for research, practice and policy. Int. J. Inf. Manag., 101994 (2019)
10. Eletter, S.F., Yasmin, T., El Refae, G.A.: Marketing intelligence in the era of Big Data. TEM J. **8**(3), 938 (2019)
11. Elliot, S.: Transdisciplinary perspectives on environmental sustainability: a resource base and framework for IT-enabled business transformation. MIS Q. **35**(1), 197–236 (2011)
12. Fowler, W.: Inside big data (2018, August 31). www.insidebigdata.com: https://insidebigdata.com/2018/08/31/ai-taking-lead-generation-nearest-future/
13. Galvagno, M., Dalli, D.: Theory of value co-creation: a sys-tematic literature review. Manag. Serv. Q. **24**(6), 643–683 (2014)
14. Itani, O.S., Kassar, A.-N., Loureiro, S.M.C.: Value get, value give: the relationships among perceived value, relationship quality, customer engagement, and value consciousness. Int. J. Hosp. Manag. **80**, 78–90 (2019)
15. Jaakkola, E., Helkkula, A., Aarikka-Stenroos, L.: Service experience co-creation: conceptualization, implications, and future research directions. J. Serv. Manag. **26**(2), 182–205 (2015)
16. Jagadish, H.V., Gehrke, J., Labrinidis, A., Papakonstantinou, Y., Patel, J.M., Ramakrishnan, R., Shahabi, C.: Big data and its technical challenges. Commun. ACM **57**(7), 86–94 (2014)
17. Jobs, C.G., Aukers, S.M., Gilfoil, D.M.: The impact of big data on your firms marketing communications: a framework for understanding the emerging marketing analytics industry. Acad. Mark. Stud. J. **19**(2), 81 (2015)
18. Kietzmann, J., Paschen, J., Treen, E.: Artificial intelligence in advertising: How marketers can leverage artificial intelligence along the consumer journey. J. Adv. Res. **58**(3), 263–267 (2018)
19. L'heureux, A., Grolinger, K., Elyamany, H.F., Capretz, M.A.: Machine learning with big data: challenges and approaches. IEEE Access **5**, 7776–7797 (2017)
20. Leclercq, T., Hammedi, W., Poncin, I.: Ten years of value co-creation: an integrative review. Recherche Appl. Mark. (English Edition) **31**(3), 26–60 (2016)
21. Lee, Y., Madnick, S.E., Wang, R.Y., Wang, F., Zhang, H.: A cubic framework for the chief data officer: Succeeding in a world of big data (2014)
22. Liu, X.: Analyzing the impact of user-generated content on B2B Firms' stock performance: Big data analysis with machine learning methods. Ind. Mark. Manag. (2019)
23. McColl-Kennedy, J.R., Vargo, S.L., Dagger, T.S., Sweeney, J.C., Van Kasteren, Y.: Health care customer value cocreation practice styles. J. Serv. Res **15**(4), 370–389 (2012)
24. Müller, S.D., Jensen, P.: Big data in the Danish industry: application and value creation. Bus. Process Manag. J. **23**(3), 645–670 (2017)

25. Nilsson, N.J.: Principles of Artificial Intelligence. Morgan Kaufmann (2014)
26. Oertzen, A.S., Odekerken-Schröder, G., Brax, S.A., Mager, B.: Co-creating services – conceptual clarification, forms and outcomes. J. Serv. Manag. **29**(4), 641–679 (2018)
27. O'Leary, D.E.: Artificial intelligence and big data. IEEE Intell. Syst. **28**(2), 96–99 (2013)
28. Overgoor, G., Chica, M., Rand, W., Weishampel, A.: Letting the computers take over: using AI to solve marketing problems. Calif. Manag. Rev. **61**(4), 156–185 (2019)
29. Paschen, J.: Creating market knowledge from big data: Artificial intelligence and human resources. Doctoral dissertation, KTH Royal Institute of Technology (2020)
30. Ranjan, K.R., Read, S.: Value co-creation: concept and measurement. J. Acad. Mark. Sci. **44**(3), 290–315 (2016)
31. Roberts, N., Grover, V.: Leveraging information technology infrastructure to facilitate a firm's customer agility and competitive activity: An empirical investigation. J. Manag. Inf. Syst. **28**(4), 231–270 (2012)
32. Saggi, M.K., Jain, S.: A survey towards an integration of big data analytics to big insights for value-creation. Inf. Process. Manage. **54**(5), 758–790 (2018)
33. Sarkar, S., Banerjee, S.: Brand co-creation through triadic stakeholder participation. Eur. Bus. Rev. **31**(5), 585–609 (2019)
34. Singh, J., Flaherty, K., Sohi, R.S., Deeter-Schmelz, D., Habel, J., Le Meunier-FitzHugh, K., Onyemah, V.: Sales profession and professionals in the age of digitization and artificial intelligence technologies: concepts, priorities, and questions. J. Pers. Sell. Sales Manag. **39**(1), 2–22 (2019)
35. Tirunillai, S., Tellis, G.J.: Mining marketing meaning from online chatter: strategic brand analysis of big data using latent dirichlet allocation. J. Mark. Res. **51**(4), 463–479 (2014)
36. Tommasetti, A., Troisi, O., Vesci, M.: Measuring customer value co-creation behavior: developing a conceptual model based on Service-Dominant Logic. J. Serv. Theory Pract. **27**(5), 930–950 (2017)
37. Vargo, S.L., Lusch, R.F.: Service-dominant logic: continuing the evolution. J. Acad. Mark. Sci. **36**(1), 1–10 (2008)
38. Vishnoi, S.K., Bagga, T., Sharma, A., Wani, S.N.: Artificial intelligence enabled marketing solutions: a review. Indian J. Econ. Bus. **17**(4), 167–177 (2018)
39. Wang, E.T., Hu, H.F., Hu, P.J.H.: Examining the role of information technology in cultivating firms' dynamic marketing capabilities. Inf. Manag. **50**(6), 336–343 (2013)
40. Wulandari, N., Nasution, R.A.: Integrated value co-creation and affective commitment in banking industry. Bus.: Theory Pract. **20**, 509–551 (2019)
41. Xie, K., Wu, Y., Xiao, J., Hu, Q.: Value co-creation between firms and customers: The role of big data-based cooperative assets. Inf. Manag. **53**(8), 1034–1048 (2016)
42. Zhao, X., Xue, L.: Competitive target advertising and consumer data sharing. J. Manag. Inf. Syst. **29**(3), 189–222 (2012)

The manufacturer's authorised representative in the EU is Springer Nature Customer Service Centre GmbH, Europaplatz 3, 69115 Heidelberg, Germany. If you have any concerns regarding our products, please contact ProductSafety@springernature.com

Printed and bound by CPI Group (UK) Ltd, Croydon, CR0 4YY
30/09/2026
02225687-0012